Lecture Notes in Computer Science

Lecture Notes in Computer Science

Vol. 23: Programming Methodology. 4th Informatik Symposium, IBM Germany Wildbad, September 25–27, 1974. Edited by C. E. Hackl. VI, 501 pages. 1975.

Vol. 24: Parallel Processing. Proceedings 1974. Edited by T. Feng. VI, 433 pages. 1975.

Vol. 25: Category Theory Applied to Computation and Control. Proceedings 1974. Edited by E. G. Manes. X, 245 pages. 1975.

Vol. 26: GI-4. Jahrestagung, Berlin, 9.–12. Oktober 1974. Herausgegeben im Auftrag der GI von D. Siefkes. IX, 748 Seiten. 1975.

Vol. 27: Optimization Techniques. IFIP Technical Conference. Novosibirsk, July 1–7, 1974. (Series: I.F.I.P. TC7 Optimization Conferences.) Edited by G. I. Marchuk. VIII, 507 pages. 1975.

Vol. 28: Mathematical Foundations of Computer Science. 3rd Symposium at Jadwisin near Warsaw, June 17–22, 1974. Edited by A. Blikle. VII, 484 pages. 1975.

Vol. 29: Interval Mathematics. Procedings 1975. Edited by K. Nickel. VI, 331 pages. 1975.

Vol. 30: Software Engineering. An Advanced Course. Edited by F. L. Bauer. (Formerly published 1973 as Lecture Notes in Economics and Mathematical Systems, Vol. 81) XII, 545 pages. 1975.

Vol. 31: S. H. Fuller, Analysis of Drum and Disk Storage Units. IX, 283 pages. 1975.

Vol. 32: Mathematical Foundations of Computer Science 1975. Proceedings 1975. Edited by J. Bečvář. X, 476 pages. 1975.

Vol. 33: Automata Theory and Formal Languages, Kaiserslautern, May 20–23, 1975. Edited by H. Brakhage on behalf of GI. VIII, 292 Seiten. 1975.

Vol. 34: GI – 5. Jahrestagung, Dortmund 8.–10. Oktober 1975. Herausgegeben im Auftrag der GI von J. Mühlbacher. X, 755 Seiten. 1975.

Vol. 35: W. Everling, Exercises in Computer Systems Analysis. (Formerly published 1972 as Lecture Notes in Economics and Mathematical Systems, Vol. 65) VIII, 184 pages. 1975.

Vol. 36: S. A. Greibach, Theory of Program Structures: Schemes, Semantics, Verification. XV, 364 pages. 1975.

Vol. 37: C. Böhm, λ-Calculus and Computer Science Theory. Proceedings 1975. XII, 370 pages. 1975.

Vol. 38: P. Branquart, J.-P. Cardinael, J. Lewi, J.-P. Delescaille, M. Vanbegin. An Optimized Translation Process and Its Application to ALGOL 68. IX, 334 pages. 1976.

Vol. 39: Data Base Systems. Proceedings 1975. Edited by H. Hasselmeier and W. Spruth. VI, 386 pages. 1976.

Vol. 40: Optimization Techniques. Modeling and Optimization in the Service of Man. Part 1. Proceedings 1975. Edited by J. Cea. XIV, 854 pages. 1976.

Vol. 41: Optimization Techniques. Modeling and Optimization in the Service of Man. Part 2. Proceedings 1975. Edited by J. Cea. XIII, 852 pages. 1976.

Vol. 42: James E. Donahue, Complementary Definitions of Programming Language Semantics. VII, 172 pages. 1976.

Vol. 43: E. Specker und V. Strassen, Komplexität von Entscheidungsproblemen. Ein Seminar. V, 217 Seiten. 1976.

Vol. 44: ECI Conference 1976. Proceedings 1976. Edited by K. Samelson. VIII, 322 pages. 1976.

Vol. 45: Mathematical Foundations of Computer Science 1976. Proceedings 1976. Edited by A. Mazurkiewicz. XI, 601 pages. 1976.

Vol. 46: Language Hierarchies and Interfaces. Edited by F. L. Bauer and K. Samelson. X, 428 pages. 1976.

Vol. 47: Methods of Algorithmic Language Implementation. Edited by A. Ershov and C. H. A. Koster. VIII, 351 pages. 1977.

Vol. 48: Theoretical Computer Science, Darmstadt, March 1977. Edited by H. Tzschach, H. Waldschmidt and H. K.-G. Walter on behalf of GI. VIII, 418 pages. 1977.

Vol. 49: Interactive Systems. Proceedings 1976. Edited by A. Blaser and C. Hackl. VI, 380 pages. 1976.

Vol. 50: A. C. Hartmann, A Concurrent Pascal Compiler for Minicomputers. VI, 119 pages. 1977.

Vol. 51: B. S. Garbow, Matrix Eigensystem Routines – Eispack Guide Extension. VIII, 343 pages. 1977.

Vol. 52: Automata, Languages and Programming. Fourth Colloquium, University of Turku, July 1977. Edited by A. Salomaa and M. Steinby. X, 569 pages. 1977.

Vol. 53: Mathematical Foundations of Computer Science. Proceedings 1977. Edited by J. Gruska. XII, 608 pages. 1977.

Vol. 54: Design and Implementation of Programming Languages. Proceedings 1976. Edited by J. H. Williams and D. A. Fisher. X, 496 pages. 1977.

Vol. 55: A. Gerbier, Mes premières constructions de programmes. XII, 256 pages. 1977.

Vol. 56: Fundamentals of Computation Theory. Proceedings 1977. Edited by M. Karpiński. XII, 542 pages. 1977.

Vol. 57: Portability of Numerical Software. Proceedings 1976. Edited by W. Cowell. VIII, 539 pages. 1977.

Vol. 58: M. J. O'Donnell, Computing in Systems Described by Equations. XIV, 111 pages. 1977.

Vol. 59: E. Hill, Jr., A Comparative Study of Very Large Data Bases. X, 140 pages. 1978.

Vol. 60: Operating Systems, An Advanced Course. Edited by R. Bayer, R. M. Graham, and G. Seegmüller. X, 593 pages. 1978.

Vol. 61: The Vienna Development Method: The Meta-Language. Edited by D. Bjørner and C. B. Jones. XVIII, 382 pages. 1978.

Vol. 62: Automata, Languages and Programming. Proceedings 1978. Edited by G. Ausiello and C. Böhm. VIII, 508 pages. 1978.

Vol. 63: Natural Language Communication with Computers. Edited by Leonard Bolc. VI, 292 pages. 1978.

Vol. 64: Mathematical Foundations of Computer Science. Proceedings 1978. Edited by J. Winkowski. X, 551 pages. 1978.

Vol. 65: Information Systems Methodology, Proceedings, 1978. Edited by G. Bracchi and P. C. Lockemann. XII, 696 pages. 1978.

Vol. 66: N. D. Jones and S. S. Muchnick, TEMPO: A Unified Treatment of Binding Time and Parameter Passing Concepts in Programming Languages. IX, 118 pages. 1978.

Vol. 67: Theoretical Computer Science, 4th GI Conference, Aachen, March 1979. Edited by K. Weihrauch. VII, 324 pages. 1979.

Vol. 68: D. Harel, First-Order Dynamic Logic. X, 133 pages. 1979.

Vol. 69: Program Construction. International Summer School. Edited by F. L. Bauer and M. Broy. VII, 651 pages. 1979.

Vol. 70: Semantics of Concurrent Computation. Proceedings 1979. Edited by G. Kahn. VI, 368 pages. 1979.

Vol. 71: Automata, Languages and Programming. Proceedings 1979. Edited by H. A. Maurer. IX, 684 pages. 1979.

Vol. 72: Symbolic and Algebraic Computation. Proceedings 1979. Edited by E. W. Ng. XV, 557 pages. 1979.

Vol. 73: Graph-Grammars and Their Application to Computer Science and Biology. Proceedings 1978. Edited by V. Claus, H. Ehrig and G. Rozenberg. VII, 477 pages. 1979.

Vol. 74: Mathematical Foundations of Computer Science. Proceedings 1979. Edited by J. Bečvář. IX, 580 pages. 1979.

Vol. 75: Mathematical Studies of Information Processing. Proceedings 1978. Edited by E. K. Blum, M. Paul and S. Takasu. VIII, 629 pages. 1979.

Vol. 76: Codes for Boundary-Value Problems in Ordinary Differential Equations. Proceedings 1978. Edited by B. Childs et al. VIII, 388 pages. 1979.

Lecture Notes in
Computer Science

Edited by G. Goos and J. Hartmanis

117

Fundamentals of Computation Theory

Proceedings of the 1981 International
FCT-Conference, Szeged, Hungary
August 24–28, 1981

Edited by Ferenc Gécseg

Springer-Verlag
Berlin Heidelberg New York 1981

Editor
Prof. Ferenc Gécseg
József Attila University, Bolyai Institute
Aradi vértanúk tere 1, H-6720 Szeged, Hungary

CR Subject Classifications (1981): 5.1, 5.2, 5.3, 4.2, 4.34

ISBN 3-540-10854-8 Springer-Verlag Berlin Heidelberg New York
ISBN 0-387-10854-8 Springer-Verlag New York Heidelberg Berlin

Printed in Germany

Printing and binding: Beltz Offsetdruck, Hemsbach/Bergstr.
2145/3140-543210

PREFACE

This volume constitutes the proceedings of the Conference on Fundamentals of Computation Theory (Algebraic, Arithmetic and Logical Methods in Computation Theory) held in Szeged, Hungary, August 24-28, 1981. The conference is the third in the series of the FCT-Conferences initiated by our Polish colleagues in Poznan, 1977.

The papers in this volume are the texts of invited addresses and shorter communications falling in one of the following three sections: A. Algebraic and Constructive Theory of Machines, Computations and Languages; B. Abstract Algebra, Combinatorics and Logic in Computation Theory, C. Computability, Decidability and Arithmetic Complexity. The shorter communications were selected by the Organizing and Program Committee out of more than 100 submitted papers.

This Third Conference was organized by the Institute of Mathematics of the József Attila University (Szeged) in co-operation with the Computer and Automation Institute of the Hungarian Academy of Sciences (Budapest). The Organizing and Program Committee consisted of J. Berstel, L. Budach, R.G. Buharajev, the late C.C. Elgot, Ju.L. Ershov, F. Gécseg (chairman), J. Hartmanis, G. Hotz, M. Karpinski, L. Lovász, O.B. Lupanov, I. Munro, M. Nivat, Z. Pawlak, A. Pultr, A. Salomaa and H. Thiele. J. Demetrovics held the post of organizing secretary.

Thanks are due to the members of the Organizing and Program Committee for their work in evaluating the submitted papers and for their kind co-operation in preparing the scientific program of the Conference. I would like to thank K. Dévényi, Z. Ésik and especially Gy. Horváth for their friendly assistance in all organizational matters.

-*-*-*-

Finally, I would like to commemorate most respectfully Professor Calvin C. Elgot who worked enthusiastically for the success of the FCT-Conferences. His decease is a heavy loss for computer scientists all over the world. I would like to devote this volume to his memory.

Szeged, May 1981

Ferenc Gécseg

CONTENTS

APPENDIX

OBSERVABILITY AND NERODE EQUIVALENCE IN CONCRETE CATEGORIES

J. Adámek
Faculty of Electrical Engineering
Technical University Prague,
Czechoslovakia

Abstract. Functorial automata are studied in a concrete category K with structured hom-sets. For each functor $F : K \longrightarrow K$, which respects this structure, the observability morphisms of F-automata are defined analogously to those of sequential automata. If each F-automaton has an observable reduction, the minimization problem is both much simplified (in fact, translated to the image factorization of the observability morphisms) and made global. We prove that this is the case iff each behavior has a Nerode equivalence. First, we present a survey of the related results on minimal realization and Nerode equivalence.

I. Minimal realization (a survey)

I.1 Let K be a category and $F : K \longrightarrow K$ a functor. An F-algebra is a pair (Q, δ) consisting of an object Q in K and a morphism $\delta : FQ \longrightarrow Q$. An F-automaton is an F-algebra with an object Y and an (output) morphism $y : Q \longrightarrow Y$. An initial F-automaton has, moreover, an object I and an (initialization) morphism $i : I \longrightarrow Q$. This is the concept introduced by M. A. Arbib and E. G. Manes [6]; all the notions of this section are from [6].

Examples. (i) Sequential Σ-automata are F-automata in the category of sets, K = Set, where $F = S_\Sigma$ is the following functor

$$S_\Sigma Q = Q \times \Sigma \qquad \text{and} \qquad S_\Sigma f = f \times id_\Sigma$$

(for each set Q and each map f). Here I is a singleton set mapped by i to the initial state.

(ii) Linear sequantial Σ-automata, where Σ is a module over a commutative ring R, are F-automata in the category of R-modules, K = R - Mod. Here, again, $F = S_\Sigma$ with $S_\Sigma Q = Q \times \Sigma$ and

$S_\Sigma f = f \times id_\Sigma$. The morphism $\delta : Q \times \Sigma \longrightarrow Q$ here decomposes as follows:

$$\delta(q,\sigma) = \delta_1(q) + \delta_2(\sigma) \quad (\text{for all } q \in Q; \sigma \in \Sigma)$$

for unique linear maps $\delta_1 : Q \longrightarrow Q$ and $\delta_2 : \Sigma \longrightarrow Q$ ..

(iii) Bilinear sequential Σ-automata are F-automata in R-Mod where $F = V_\Sigma$ is the tensor-product functor :

$$V_\Sigma Q = Q \otimes \Sigma \quad \text{and} \quad V_\Sigma f = f \otimes id_\Sigma \; .$$

I.2 A <u>homomorphism</u> from an F-algebra (Q,δ) into an F-algebra (Q',δ') is a morphism $f : Q \longrightarrow Q'$ in K with $f \cdot \delta = \delta' \cdot Ff$. A <u>morphism of automata</u> is a homomorphism commuting with the outputs $(y = y' . f)$ and, for initial automata, with the initializations $(i' = f \cdot i)$.

A <u>reduction</u> of an F-automaton M is an F-automaton M′ together with a morphism $f : M \longrightarrow M'$ which is a regular epi (a coequalizer) in K . A reduction $f_0 : M \longrightarrow M_0$ is <u>minimal</u> if each reduction $f : M \longrightarrow M'$ of M can be further reduced to M_0 , i.e., there exists $h : M' \longrightarrow M_0$ with $f_0 = h \cdot f$.

We shall assume that K has <u>regular factorizations</u>, i.e., each morphism f factorizes as $f = m \cdot e$ where e is a regular epi and m is a mono.

I.3 A <u>free F-algebra</u> generated by an object I in K is an F-algebra $(I^\#, \varphi)$ together with a universal morphism $\eta : I \longrightarrow I^\#$. The universality means that for each F-algebra (Q,δ) and each morphism $f : I \longrightarrow Q$ there exists a unique homomorphism

$$f^\# : (I^\#, \varphi) \longrightarrow (Q, \delta)$$

with $f = f^\# \cdot \eta$. The functor F is called a <u>varietor</u> if $I^\#$ exists for each I .

<u>Exmaples.</u> (i) The functor S_Σ : Set $\longrightarrow$ Set is a varietor.The free algebra on one generator (say, 0) is the string algebra Σ^* with

$$\varphi : \Sigma^* \times \Sigma \rightarrow \Sigma \quad ; (\sigma_1, \ldots, \sigma_n; \sigma) \longmapsto (\sigma_1 \ldots \sigma_n \sigma)$$

and with $\eta(0)$ the void string. Generally, $I^{\#} = I \times \Sigma^{*}$.

(ii) The functor S_{Σ} : R-Mod $\longrightarrow$ R-Mod is a varietor with

$$I^{\#} = I[z] \times \Sigma[z]$$

(where $I[z]$ is the module of all polynomials with coefficients in I ; analogously $\Sigma[z]$). Here we have

$$\varphi : I^{\#} \times \Sigma \longrightarrow I \quad ; \quad (a(z),\ b(z);\ \sigma) \rightarrow (z \cdot a(z),\ z \cdot b(z) + \sigma)$$

while $\eta(i) = (i, 0)$.

I.4 A construction of the free algebra has been exhibited in [1]. Assume K cocomplete. For each object I define a transfinite chain $W_0 \rightarrow W_1 \rightarrow \ldots$ in K by induction as follows :

$$W_0 = I \ ;$$

$$W_{n+1} = I + FW_n$$

(+ denotes the coproduct) and, for each limit ordinal k ,

$$W_k = \operatorname{colim}_{n<k} W_n \ .$$

The first morphism $w_{01} : I \longrightarrow I + FI$ is the first injection; given $w_{n,m} : W_n \longrightarrow W_m$ then $w_{n+1\ m+1} = 1_I + Fw_{nm}$ and, for each limit ordinal k , $w_{nk} : W_n \longrightarrow W_k$ are the colimit injections. We say that the free algebra <u>exists constructively</u> if there exists an ordinal n such that $w_{n\ n+1} : W_n \longrightarrow I + FW_n$ is an isomorphism. (If $n = \omega$, the free algebra is said to <u>exist recursively</u>.) Then $I^{\#} = W_n$ and φ ; η are the components of the morphism $w_{n\ n+1}^{-1} : I + FW_n \longrightarrow W_n$. See [11] for a discussion of the constructive varietors.

I.5 Let F be a varietor. For each initial automaton $M = (Q, \ , Y, y, I, i)$ we can extend i to a homomorphism $i^{\#} = \rho$: $(I^{\#}, \varphi) \longrightarrow (Q, \delta)$, called the <u>run morphism</u> of M . If ρ is a regular epi then the automaton is said to be <u>reachable</u>. E.g., in the case of sequential automata, $\rho : \Sigma^{*} \longrightarrow Q$ assings to each input sequence $\sigma_1 \ldots \sigma_n \in \Sigma^*$ the resulting state when these inputs are successively applied (in the initial state $i(0)$).

The behavior morphism of an initial F-automaton is the morphism

$$b = y \cdot \rho : I^\# \longrightarrow Y .$$

Conversely, for each behavior morphism $b : I^\# \longrightarrow Y$ we want to find its minimal realization which is a reachable automaton M_0 with behavior b such that each reachable automaton with this behavior can be reduced to M_0

Example : for sequential Σ-automata, the minimal realization of a behavior map $b : \Sigma^* \longrightarrow Y$ is obtained by a factorization of the algebra Σ^* under the Nerode equivalence E : given $v_1, v_2 \in \Sigma^*$ then $v_1 \mathrel{E} v_2$ iff $b(v_1 w) = b(v_2 w)$ for all $w \in \Sigma^*$. Here E is a congruence on Σ^* and the quotient algebra Σ^*/E defines the minimal realization.

I.6 The existence of minimal realizations is discussed in [12], [5] and [3] . Roughly speaking, if F is a varietor preserving regular epis then F has minimal realizations (for all behaviors) iff F preserves cointersections of regular epis. A more "descriptive" condition is the preservation of directed unions : under special additional hypothesis, fulfilled e.g. by Set and R-Mod (for each field R), F has minimal realizations iff F preserves directed unions.

I.7 Nerode equivalences have been generalized in several directions. The external equivalences, which we present now, are from [6] . They exist seldom; therefore, a more widely applicable model has been suggested in [4] : the inner equivalences. This concept will not be used now.

Let F be a varietor and let $b : I^\# \longrightarrow Y$ be a behavior . A pair of morphisms $e_1, e_2 : E \longrightarrow I^\#$ is said to be externally b-equivalent if (1) $b \cdot e_1 = b \cdot e_2$; (2) $(b \cdot \varphi) \cdot Fe_1 = (b \cdot \varphi) \cdot Fe_2$; (3) $(b \cdot \varphi \cdot F\varphi) \cdot F^2 e_1 = (b \cdot \varphi \cdot F\varphi) \cdot F^2 e_2$; etc.

$$\begin{array}{ccccccccc} & & F^2E & & FE & & E & & \\ & & {\scriptstyle F^2e_1}\Big\downarrow\Big\downarrow{\scriptstyle F^2e_2} & & {\scriptstyle Fe_1}\Big\downarrow\Big\downarrow{\scriptstyle Fe_2} & & {\scriptstyle e_1}\Big\downarrow\Big\downarrow{\scriptstyle e_2} & & \\ \cdots & \xrightarrow[F^2\varphi]{} & F^2 I^* & \xrightarrow[F\varphi]{} & F I^* & \xrightarrow[\varphi]{} & I^* & \xrightarrow{b} & Y \end{array}$$

The <u>external Nerode equivalence</u> of b is the largest externally b--equivalent pair $e_1, e_2 : E \longrightarrow I^*$ (i.e., for each externally b-equivalent pair $e_1', e_2' : E' \longrightarrow I^*$ there exists a unique $d : E' \longrightarrow E$ with $e_1' = e_1 \cdot d$ and $e_2' = e_2 \cdot d$).

In the case of sequential automata, the Nerode equivalence is an equivalence relation $E \subseteq \Sigma^* x \Sigma^*$. The projections $e_1, e_2 : E \rightarrow \Sigma^*$ define a b-equivalent pair : given $v_1 = e_1(x)$ and $v_2 = e_2(x)$ for some $x \in E$ then condition (1) says that $v_1 E v_2$ implies $b(v_1) = b(v_2)$; condition (2) that $v_1 E v_2$ implies $b(v_1 \sigma) = b(v_2 \sigma)$ for all $\sigma \in \Sigma$; etc.

I.8 We say that a functor F has <u>external Nerode realization</u> if each behavior $b : I^* \longrightarrow Y$ has an external Nerode equivalence as well as a minimal realization. It has been proved in [4] that, under weak additional hypotheses, this implies that the minimal realization is actually constructed through the Nerode equivalence. This means that the coequalizer $c : I^* \rightarrow Q$ of $e_1, e_2 : E \longrightarrow I^*$ is a congruence, i.e. there exists $\delta : FQ \rightarrow Q$ with $c : (I^*, \varphi) \longrightarrow (Q, \delta)$ a homomorphism. Note that $b \cdot e_1 = b \cdot e_2$ implies the existence of $y : Q \rightarrow Y$ with $b = y \cdot c$. Then the automaton

$$(Q, \delta, Y, y, I, c \cdot \eta)$$

is the minimal realization of b .

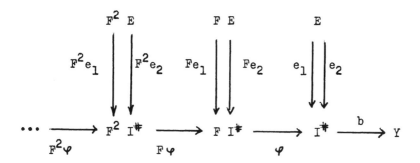

Unluckily, the external Nerode equivalence exists rather exceptionally. This follows from the observation that if it does exist, realization becomes universal (see [10]) and, e.g., in the categories Set and R-Mod , this implies that the functor F preserves unions.

II. Observability

II.1 Observability of sequential automata. Each state q of a sequential Σ-automaton $M = (Q, \delta, Y, y)$ defines a response map

$$b_M(q) : \Sigma^* \rightarrow Y$$

assigning to every input string $\sigma_1 \dots \sigma_n \in \Sigma^*$ the resulting output, when this string is applied in the initial state q . By varying $q \in Q$, we obtain the so-called observatility map

$$b_M : Q \rightarrow \hom(\Sigma^*, Y)$$

of the automaton M . To find the minimal reduction of M , consider the kernel equivalence of b_M $[q_1 \sim q_2$ iff $b_M(q_1) = b_M(q_2)]$. Clearly, $\sim$ is a congruence on M and the quotient automaton $M/\sim$ is the minimal reduction of M .

Note that $\Sigma^* = 1^*$ for the singleton set $1 = \{0\}$. Each state $q \in Q$ corresponds to a morphism $h_q : 1 \rightarrow Q$ with $h_q(0) = q$; this can be extended to a homomorphism

$$h_q^{\#} : (1^*, \varphi) \rightarrow (Q, \delta)$$

and, clearly,

$$b_M(q) = y \cdot h_q^{\#} .$$

To generalize this, we introduce "hom-objects" in a category.

II.2 We are going to work within a concrete category, i.e., a category K together with a faithful functor from K to Set. This functor assigns to each object A its (underlying) set $|A|$ and to each morphism $f : A \rightarrow B$ its (underlying) map $f : |A| \rightarrow |B|$ (denoted by the same symbol - this is correct due to the faithfulness).

A map $h : |A| \times |B| \longrightarrow |C|$ is called a <u>bimorphism</u> if (i) $h(a,-): B \longrightarrow C$ is a morphism for each $a \in |A|$ and (ii) $h(-, b) : A \longrightarrow C$ is a morphism for each $b \in |B|$. The set of all morphisms from A to B is denoted by hom (A,B). For each set X and each map $k : X \longrightarrow \mathrm{hom}(A,B)$ we denote by

$$\hat{k} : X \times |A| \longrightarrow |B|$$

the map defined by

$$\hat{k}(x,a) = (k(x))(a) \quad \text{for all} \quad x \in X \text{ and } a \in |A|.$$

The following concept is, essentially, that studied by B. Banaschewski and E. Nelson in [8].

<u>Definition.</u> A concrete category K is said to have <u>structured hom-sets</u> if for each pair of objects A,B there is an object Hom(A,B) the underlying set of which is hom(A,B), such that

(a) for each object C, a mapping $k : |C| \longrightarrow \mathrm{hom}(A,B)$ is a morphism $k : C \rightarrow \mathrm{Hom}(A,B)$ iff the mapping $\hat{k} : |C| \times |A| \rightarrow |B|$ is a bimorphism;

(b) there exists an object 1 such that the functor

$$\mathrm{Hom}(1, -) : K \longrightarrow K$$

(assigning to morphisms $f : A \longrightarrow B$ the morphism $\mathrm{Hom}(1,f): \mathrm{Hom}(1,A) \rightarrow \mathrm{Hom}(1,B)$ with $\mathrm{Hom}(1,f)(h) = f \cdot h$) is naturally equivalent to the identity functor.

<u>Examples.</u> (i) Set has structured hom-sets (trivially); 1 is any singleton set.

(ii) R-Mod has structured hom-sets : hom (A,B) is an R-module under the pointwise operations. Here 1 = R (as a module over itself) : points $b \in |B|$ correspond to linear maps $h_b : R \longrightarrow B$ where

$$h_b(r) = r \cdot b \qquad \text{for each} \quad r \in R .$$

(iii) The category of topological spaces has structured hom-sets with the topology of pointwise convergence on hom (A,B) and with 1 any singleton space.

(iv) The category of groups does not have structured hom-sets: we have a natural pointwise operation on homomorphisms but the result fails to be a homomorphism, in general. Analogously in other categories of algebras.

Remarks. (i) In (a) put $C = \text{Hom}(A,B)$ and $k = \text{id}$; then $\hat{k} = \text{ev} : |\text{Hom}(A,B)| \times |A| \to |B|$ is the evaluation map, $\text{ev}(f,a) = f(a)$. Hence, ev is a bimorphism, i.e., for each $a \in |A|$,

$$\text{ev}(-, a) : \text{Hom}(A,B) \to B$$

is a morphism.

(ii) Due to (i) we can define the contravariant hom-functor

$$\text{Hom}(-, B) : K \to K \qquad (B \in K^{obj})$$

which assigns to a morphism $f : A_1 \to A_2$ the morphism $\text{Hom}(f,B)$: $\text{Hom}(A_2,B) \to \text{Hom}(A_1,B)$ defined by $h \mapsto h \cdot f$.

These functors turn colimits to limits. I.e., given a diagram $D : \mathcal{D} \to K$ and its colimit $X = \text{colim}\, D$ (with injections ε_d : $: Dd \to X$, $d \in \mathcal{D}^{obj}$) then the diagram $\text{Hom}(-, B) \cdot D : \mathcal{D} \to K$ has a limit $\text{Hom}(X,B)$ (with projections $\text{Hom}(\varepsilon_d, B)$).

(iii) The map

$$e : A \to H(H(A,B), B); \quad e(a): f \mapsto f(a) \quad (a \in |A|;\ f:A \to B)$$

is a morphism, since $\hat{e} = \text{ev}$. Hence, the above condition (i) is a simplified formulation of the functionality, introduced in [8].

(iv) The natural isomorphisms of Q and $\text{Hom}(1,Q)$ will be denoted by $q \mapsto h_q : 1 \to Q$ $(q \in |Q|)$. Note that for each morphism $f : Q \to Q'$ and each $q \in |Q|$ we have

$$f \cdot h_q = h_{f(q)} .$$

II.3 Let K have structured hom-sets and let F be a functor with a free algebra 1^{*}. Then the observability maps of F-automata are defined as in II.1. Given $M = (Q, \delta, Y, y)$, for each state $q \in |Q|$ we extend h_q to $h_q^{*} : (1^{*}, \varphi) \longrightarrow (Q, \delta)$ and put

$$b_M(q) = y \cdot h_q^{*} : 1^{*} \longrightarrow Y .$$

This defines the <u>observability map</u>

$$b_M : |Q| \longrightarrow \text{hom}\,(1^{*}, Y) ;$$

in case b_M is one-to-one, the automaton M is said to be <u>observable</u> .

We are interested to know whether b_M is a morphism

$$b_M : Q \to \text{Hom}\,(1^{*}, Y) .$$

The answer is negative even in the basic example of linear sequential automata on a single input :

<u>Example.</u> Consider S_R : R-Mod $\to$ R-Mod ; here $I^{*} = R[z] \times R[z]$. Let $M = (Q, \delta, Y, y)$ be an automaton with $\delta_1 = 0$ (see I.1 Example (ii)). For each $f : R \to Q$ the free extension $f^{*} : R[z] \times R[z] \longrightarrow Q$ is defined by $f^{*}(a, b) = f(a_0) + \delta_2(b_0)$ .In particular, with each state $q \in |Q|$ we associate the morphism $h_q : R \to Q$, defined by $h_q(a_0) = a_0 q$ $(a_0 \in R)$; then $h_q^{*}(a,b) = a_0 \cdot q + \delta_2(b_0)$. Thus,

$$b_M : |Q| \longrightarrow \text{hom}\,(R[z] \times R[z], Y)$$

assing to each state $q \in |Q|$ the morphism

$$b_M(q) : (a,b) \mapsto (a_0 \cdot q + \delta_2(b_0)).$$

Clearly, b_M fails to be linear in general. (Coinsider $Q = Y = R$ with $\delta_2 = y = id_R$, then $b_M(1) : (a,b) \mapsto a_0 + b_0$ and $b_M(2) : (a,b) \mapsto 2a_0 + b_0$, thus, $b_M(2) \neq 2b_M(1)$.)

The reason for this failure is that S_{Σ} is not linear on hom-objects. Therefore, we introduce the corresponding condition.

II.4 <u>Definition.</u> Let K be a concrete category with structured hom-sets. A functor $F : K \longrightarrow K$ is said to be <u>inner</u> if for each pair of objects A,B the map $f \mapsto Ff$ defines a morphism

$$F_{A,B} : \mathrm{Hom}\,(A,B) \rightarrow \mathrm{Hom}\,(FA,FB) .$$

<u>Theorem.</u> Let K be a concrete category with structured hom-sets. Let $F : K \longrightarrow K$ be an inner functor with a constructive free algebra $1^{\#}$. Then for each F-automaton $M = (Q, \delta, Y, y)$ the observability map is a morphism

$$b_M : Q \rightarrow \mathrm{Hom}\,(1^{\#}, Y) .$$

Sketch of proof. By hypothesis, $1^{\#} = W_{n_0}$ for some ordinal n_0. Given an algebra (Q, δ) and a morphism $h_q : 1 \longrightarrow Q$ $(q \in |Q|)$ its extension $h_q^{\#} : (1^{\#}, \varphi) \rightarrow (Q, \delta)$ is defined by transfinite induction : $h_q^{\#} = h_q^{(n_0)}$ where

$$h_q^{(0)} = h_q : 1 \longrightarrow Q ;$$

given $h_q^{(n)}$ then

$$h_q^{(n+1)} : 1 + FW_n \longrightarrow Q$$

has components $h_q : 1 \rightarrow Q$ and $\delta . Fh_q^{(n)} : FW_n \rightarrow Q$;

$$h_q^{(k)} = \operatorname*{colim}_{n < k} h_q^{(n)} \quad \text{if } k \text{ is a limit ordinal.}$$

We shall prove by induction that for each n, the map $q \mapsto h_q^{(n)}$ defines a morphism $v_n : Q \longrightarrow \mathrm{Hom}\,(W_n, Q)$; then also b_M is a morphism since $b_M(q) = y . h_q^{\#} = y \circ h_q^{(n_0)}$ implies

$$b_M = \mathrm{Hom}\,(1, y) \circ v_{n_0} : Q \rightarrow \mathrm{Hom}\,(1^{\#}, Y) .$$

First, $v_0 : Q \longrightarrow \mathrm{Hom}(1,Q)$ is the (canonical) isomorphism. Assume v_n is a morphism. By Remark II.2(ii), we have

$$\mathrm{Hom}(W_{n+1}, Q) = \mathrm{Hom}(1 + FW_n, Q) = \mathrm{Hom}(1, Q) \times \mathrm{Hom}(FW_n, Q) .$$

It suffices to prove that the two components of v_{n+1} are morphisms; since $v_{n+1}(q) = h_q^{(n+1)}$ has components h_q and $\delta . Fh_q^{(n)}$, the components of v_{n+1} are $v_0 : Q \longrightarrow \mathrm{Hom}(1, Q)$ and

$$\mathrm{Hom}(FW_n, \delta) . F_{W_n,Q} . v_n : Q \to \mathrm{Hom}(FW_n , Q) .$$

Finally, if k is a limit ordinal then

$$\mathrm{Hom}(W_k ,Q) = \mathrm{Hom}(\operatorname*{colim}_{n<k} W_n ,Q) = \lim_{n<k} \mathrm{Hom}(W_n ,Q) .$$

Since the components of v_k are the morphisms $v_n : Q \longrightarrow \mathrm{Hom}(W_n ,Q)$, also v_k is a morphism. This concludes the proof.

II.5 <u>Remark.</u> The observability morphisms are compatible, i.e., for each morphism of automata $f : M \longrightarrow M'$ the following triangle

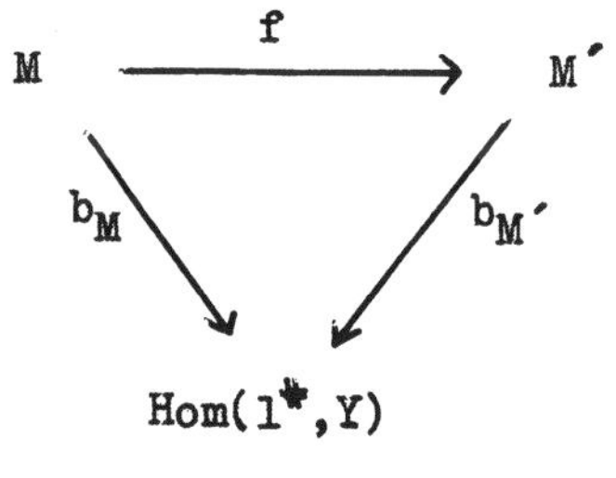

commutes. Indeed, for each $q \in |Q|$ we have $h_{q'} = f . h_q : 1 \longrightarrow Q'$, where $q' = f(q)$. Then $h_{q'}^{\#} = f . h_q^{\#}$, since $f . h_q^{\#}$ is a homomorphism extending h_q. Thus, $b_M(q) = y . h_q^{\#} = y' . f . h_q^{\#} = y' . h_{q'}^{\#} = b_{M'}(q')$.

Another model of observability has been suggested by V.Trnková [9]: the colimit of the category of all F-automata, as a diagram in K ; the colimit injections are the observability morphisms. It is an open problem whether, for each inner functor F which a constructive $1^{\#}$, the object $\mathrm{Hom}(1^{\#}, Y)$ is this colimits. For the categories

Set and R-Mod the affirmative answer to this problem is due to V. Trnková. Her paper [9] initiated the present approach.

II.6 An important question is whether the observability morphisms can be used to find the minimal reduction as the image factorization (the way we have seen it in case of sequential automata). Surprisingly, this is true iff the functor F has external Nerode realizations.

Definition. A functor F is said to have global minimization if each F-automaton has an observable reduction.

Remark. If F has global minimization then for each automaton M we just factorize the observability morphism b_M as $b_M = m \cdot e$, with e a regular epi and m a mono. Then, necessarily, m is the observability map of the observable (hence minimal) reduction of M. By the Remark II.5 the minimization becomes global in the sense that each morphism of automata $f : M \to M'$ defines the corresponding morphism of their minimal reductions, $f_0 : M_0 \to M_0'$

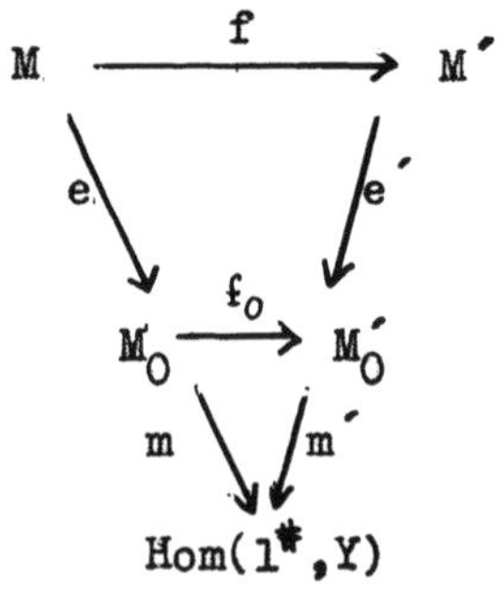

II.7 Theorem. Let K be a concrete category with regular factorizations and structured hom-sets. Let K have kernel pairs and their coequalizers which are, as maps, onto.

Let F be an inner varietor with a recursive free algebra 1^*. Then F has global minimization iff F has external Nerode realizations.

Sketch of proof. I. Let F have global minimization. For each behavior $b : I^{\#} \longrightarrow Y$ we form the "free" automaton $M_b = (I^{\#}, \varphi, Y, b)$. Let $c : M_b \longrightarrow M$ be the observable reduction of M_b. Its kernel pair $e_1, e_2 : E \longrightarrow I^{\#}$ is easily seen to be b-equivalent. We prove that e_1, e_2 is the Nerode equivalence. If $e_1', e_2' : E' \longrightarrow I^{\#}$ is another b-equivalent pair then, to prove that it factorizes through the kernel pair of c, it suffices to prove that $c \cdot e_1' = c \circ e_2'$. The observability morphism b_M of M is one-to-one and so we only have to prove that $(b_M \circ c) \cdot e_1' = (b_M \cdot c) \cdot e_2'$. By Remark II.5, $b_M \circ c$ is the behavior morphism of M_b. Hence, we are to prove that, for any $x \in |E'|$, the states $q_1 = e_1'(x)$ and $q_2 = e_2'(x)$ have the same behavior, i.e.,

$$b \cdot h_{q_1}^{\#} = b \cdot h_{q_2}^{\#} : 1^{\#} \longrightarrow Y .$$

We use the fact that $1^{\#} = W_{\omega}$: it suffices to prove $b \cdot h_{q_1}^{(n)} = b \cdot h_{q_2}^{(n)}$ for each $n < \omega$. We proceed by induction. Since $b \cdot e_1' = b \cdot e_2'$, clearly

$$b \circ h_{q_1}^{(0)} = b \cdot h_{q_2}^{(0)} : 1 \rightarrow Y .$$

Since $b \cdot \varphi \cdot Fe_1' = b.\varphi.Fe'$ and since (clearly) $h_{q_1} = e_1' \cdot h_x$ and $h_{q_2} = e_2' \cdot h_x$, we get

$$\begin{aligned} b \cdot \varphi \cdot Fh_{q_1} &= b \cdot \varphi \cdot Fe_1' \cdot Fh_x \\ &= b \cdot \varphi \cdot Fe_2' \cdot Fh_x \\ &= b \circ \varphi \circ Fh_{q_2} \end{aligned}$$

and this implies $b \cdot h_{q_1}^{(1)} = b \cdot h_{q_2}^{(1)}$. Etc.

II. Let F have Nerode realizations. Each F-automaton $M = (Q, \delta, Y, y)$ can be turned into a reachable initial automaton $M^i = (Q, \delta, Y, y, Q, id_Q)$. Since M^i has a reduction which is minimal (the minimal realization of its behavior), it suffices to prove that every minimal automaton is observable.

Let M be the minimal realization of a behavior $b_0 : I^{\#} \to Y$. Let $e_1, e_2 : E \to I^{\#}$ be the Nerode equivalence of b_0 and let $c : I^{\#} \to Q$ be its coequalizer. To prove that b_M is ono-to-one, we shall show that given states q_1, q_2 in the free automaton M_{b_0} with the same behavior, then $c(q_1) = c(q_2)$. (Then b_M is one-to-one, since $b_M \cdot c$ is the observability map of M_{b_0}, and $b_M(c(q_1)) = b_M(c(q_2))$ implies $c(q_1) = c(q_2)$; by hypothesis, c is onto). Consider the pair $h_{q_1}, h_{q_2} : 1 \to I^{\#}$. Since $b_0 \cdot h_{q_1} = b_0 \cdot h_{q_2}$, this pair is b_0-equivalent : $h_{q_i} = h^{\#}_{q_i} \cdot \eta$ implies $b_0 \cdot h_{q_1} = b_0 \cdot h_{q_2}$; $\varphi \cdot Fh_{q_i} = \varphi \cdot Fh^{\#}_{q_i} \cdot F\eta = h^{\#}_{q_i} \cdot \varphi \cdot F\eta$ implies

$$\begin{aligned}(b_0 \cdot \varphi) \cdot Fh_{q_1} &= b_0 \cdot h^{\#}_{q_1} \cdot \varphi \cdot F\eta \\ &= b_0 \cdot h^{\#}_{q_2} \cdot \varphi \cdot F\eta \\ &= (b_0 \cdot \varphi) \cdot Fh_{q_2} \ ;\end{aligned}$$

etc. Thus, the pair h_{q_1}, h_{q_2} factorizes through e_1, e_2; then $c \cdot e_1 = c \cdot e_2$ implies $c \cdot h_{q_1} = c \cdot h_{q_2}$, equivalently, $c(q_1) = c(q_2)$. This concludes the proof.

References

[1] J. Adámek: Free algebras and automata realizations in the language of categories. Comment. Math. Univ. Carolinae 15(1974), 589-602.

[2] J. Adámek: Cogeneration of algebras in regular categories. Bull. Austral. Math. Soc. 15 (1976), 55-64.

[3] J. Adámek: Realization theory for automata in categories. J.Pure Appl. Algebra 9 (1977), 281-296.

[4] J. Adámek: Categorical realization theory II : Nerode equivalences. Algebraische Modelle, Kategorien und Gruppoide (H.-J.Hoehnke ed.), Akademie-Verlag, Berlin 1979.

[5] J. Adámek, H. Ehrig, V. Trnková: An equivalence of system-theoretical and categorical notions. Kybernetika 16 (1980)

[6] M.A. Arbib, E.G. Manes: Machines in a category - an expository introduction. Lect. Notes Comp. Sci. 25, Springer-Verlag 1975.

[7] M.A. Arbib, E.G. Manes: Adjoint machines, state-behavior machines and duality. J. Pure Appl. Algebra 6 (1975), 313-344.

[8] B. Banaschewski, E. Nelson: Tensor products and bimorphisms. Canad. Math. Bull. 19 (1976), 385-402.

[9] V. Trnková: Automata and categories. Lect.N. Comp. Sci. 32, Springer-Verlag 1975, 138-152.

[10] V. Trnková, J. Adámek: Realization is not universal. Preprint , Heft 21, Weiterbildungszentrum TU Dresden (1977), 38-55.

[11] V. Trnková, J. Adámek, V. Koubek, J. Reiterman: Free algebras, input processes and free monads. Comment. Math. Univ. Carolinae 16 (1975), 339-351.

[12] V. Trnková: On minimal realizations of behavior maps in categorial automata theory. Comment. Math. Univ. Carolinae 15 (1974), 555-566.

SOME UNIVERSAL ALGEBRAIC AND MODEL THEORETIC RESULTS IN COMPUTER SCIENCE

H. **Andréka** and I. **Németi**
Mathematical Institute of the Hungarian Academy of Sciences
Budapest, Reáltanoda u. 13-15, H-1053, Hungary

Applications of universal algebra and model theory in theoretical computer science are so abundant nowadays that it is impossible to collect even the most important developments into a single paper. In our sequence of survey papers [2] we distinguished 11 main branches of directions of research in this field and in almost each of them there have been rather deep and surprising new results recently.

For a systematic survey of "universal algebra, model theory and categories in computer science" the reader is referred to [2] and Goguen [18]. Actually, [18] is more than a survey, it contains a concise exposition of how and why universal algebra is used in computer science. This exposition [18] is so fundamental that it is indispensable not only to read but to understand it (to understand at least the first 12 pages well) for anybody who intends to discuss or to form opinion on universal algebra applied in computer science. If somebody discusses universal algebra in computer science without knowing what [18] writes then he simply does not know what he is talking about. By an analogy, if in the prehistoric age somebody said that writing and calculating was a strong weapon then the other person might have thought that his partner intended to throw clay tablets to the heads of the attacking enemy. For this reason, [18] should be read before reading [2].

Now we turn to touch upon recent results in some of the 11 directions distinguished in the surveys [2]. Here we shall mention only a few of these 11 branches.

Direction 7 in [2] is universal algebraic theory of abstract data types. A good introduction to this branch is Zilles [36]. The basic tools here are such axiomatizable classes of algebraic systems or models in which free algebraic systems (initial ones) exist. Quasivarieties are most frequently used since they always have free algebras. It should be observed that classes definable by universally quantified Horn formulas do have free algebras and therefore they could and should replace quasivarieties in abstract data type theory. The problem to characterize those axiomatizable model classes in which free models exist is a common problem of data type theory, universal algebra, and model theory. Chapter 8 of the standard textbook of universal algebra Grätzer [19] contains results in this direction, see also §57 there. It was proved in [3], [0] that quasivarieties of partial algebras behave well and therefore abstract data type theory could be based on them. In [28] initial algebra semantics of context sensitive languages is treated on the basis of these results. Reichel, Hupbach [31a], [20], [21] and their collegues have done much to base computer science on partial algebra theory to satisfy a need that was recognized by several researchers. Lehmann-Pasztor [23], Pasztor [29], Pasztor [30], Burmeister [11] contain recent results in this direction.

It is very hard to separate universal algebra from model theory, see e.g. [24]. It appears that the subject of these two fields is the same, the difference is only in the kind of questions most frequently investigated, in where the emphasis is put, in style and method, but these distictions are not really clear. Accordingly, there is more and more model theory used in abstract data type theory. For example there is a school in München and Milano which uses a large amount of model theory to treat e.g. hierarchies of abstract data types in a very elegant way, see [9], [10], [34], [35]. Wirsing, Bertoni, Broy, Miglioli and others obtained greater freedom and flexibility for data type specification and use the category of models to unify several approaches, e.g. fixpoint theory, algebraic, category theoretic, and model theoretic or logical approaches [9], [10], [34], [35]. Wirsing et al [35] prove sufficient conditions for the existence of initial models for data types which go far beyond quasivarieties or even universally quantified classes.

Bergstra-Tiuryn-Tucker [8] applied a deep model theoretic result, the omitting types theorem, to solve a problem in the theory of program schemes raised by A.Meyer. So far nobody was able to eliminate the omitting types theorem from the proof of this result.

The investigation of behaviour of program schemes in axiomatizable classes of models is a field which became very active today. Some of the reasons for the importance of this field are that it can provide a simple and clear foundation for (i) structuring concepts in programming, (ii) stepwise refinement of abstract programs (see e.g. Dömölki [16]), (iii) portability, (iv) abstraction from irrelevant details (about the importance of abstraction in computer science, see Goguen [18]).

Definition 0: Let Th be any set of first order formulas of a fixed signature i.e. similarity type t . Mod(Th) denotes the class of all models of Th . Let p be a program-scheme of similarity type t . Let S(p) be a statement about p , i.e. S(p) is a property of p . Then Th $\models$ S(p) is defined to hold if in every $\underline{\underline{A}} \in$ Mod(Th) the program-scheme p has property S(p) .

A few years ago e.g. in the book of Manna only two extremes were investigated, either properties of programs in Mod($\emptyset$) or properties of programs in a single model $\underline{\underline{A}}$. Here Mod($\emptyset$) is the class of all possible models. Program properties in Mod($\emptyset$) are too general while properties in $\{\underline{\underline{A}}\}$ are too special to be of practical interest. Therefore emphasis has shifted to investigate program properties in classes of the form Mod(Th) for arbitrary first order theories Th . Burstall-Darlington [12] uses the Mod(Th) approach for program optimization succesfully, then the French school Courcelle-Guessarian [14] and their collegues adopted the Mod(Th) attitude to be fundamental and now it is quite widely accepted, e.g. in [8], [15], [17], [32], [33].

By using the extended omitting types theorem, Bergstra-Tiuryn-Tucker proved a strong connection between partial correctness and program equivalence in Mod(Th) for every complete theory Th (Thm.5.1 in [8]). This is only one example when heavy model theoretic or algebraic artillery was used in computer science. E.g. J.B.Paris and L.Csirmaz used recursively saturated models and ultraproducts to solve a problem of the authors about comparing methods of programverification [6]. Ultrapowers are used to check applicability of methods of program-verification [27], [32].

Adamek, Banaschewski, Koubek, Nelson have recently obtained results on Z-continuous algebras introduced by the ADJ group and the French free magma school independently as a basic tool for semantics of programming. A good survey of ω-continuous algebras is Goguen [18]. Some of the new results can be found in [25], [29]. Lehmann and Pasztor proved that the class of all ω-continuous algebras is nothing but a variety of partial algebras, see [23]. The impact of this result is that the existing machinery of universal algebra (of partial algebras) can be applied to a central tool of program-semantics while before this result one had to re-prove the results of universal algebra for these new structures before applying them. The universal algebraic variety representing the ω-continuous algebras has an infinitary operation. Thus by the Lehmann-Pasztor result infinitary algebras became important for computer science. A finitary logic for infinitary structures is developed in [1]. By using cylindric algebraic methods it was proved that a certain version of ultraproducts do work for these structures.

Cylindric algebras can be used to extend the theory of stepwise refinement of program specifications in the sense of Burstall-Goguen [13] and Dömölki [16] from specifications by equations to specifications by arbitrary first order theories because the category of all first order theories and theory morphisms is isomorphic to the category Lf_{ω} of cylindric algebras, see [1a]. Here the words "theory" and "theory morphism" are understood as defined in sections 2.1 and 2.3 of Dömölki [16]. The more useful part of this result is that all cylindric homomorphisms are theory morphisms and vice versa, especially if we consider the representation of these morphisms by base-homomorphisms which correspond to the semantic theory-morphisms operating on model classes outlined in sec.2.9 of [13].

Cylindric algebra theory was used by the authors and Sain to solve problems (some raised by Pratt) on dynamic algebras. Dynamic algebras are related to logics of programs the same way as Boolean algebras are related to classical logic. Dynamic algebras were introduced by Pratt [31] and Kozen [22] to study logics of programs. The connection between dynamic algebras and cylindric algebras was already pointed out by Pratt. Some applications of cylindric algebras to dynamic algebras can be found in [26], [7], [4]. For example, the class $\mathbf{I}$ Ds of dynamic set algebras is an infinitary quasivariety, therefore $\mathbf{I}$ Ds has free algebras, but $\mathbf{I}$ Ds is not first order axiomatizable. By using cylindric algebraic tools, Andréka and Sain characterized those relativizations of dynamic algebras which are homomorphisms and characterized direct decomposability of dynamic algebras. The latter appears to be a rather drastic tool in simplifying (decomposing) theories of programs i.e. dynamic theories. The idea of using and investigating relativizations originates from classical logic where it is commonly used for various purposes. In the present context relativizations can be proved to correspond to the finitely presentable surjective theory morphisms (interpretations) between dynamic theories.

REFERENCES

0. Andréka,H. Burmeister,P. Németi,I., Quasivarieties of Partial Algebras - A unifying approach towards a two-valued model theory for partial algebras. Preprint Nr.557, Technische Hochschule Darmstadt, Fachbereich Mathematics. July 1980.

1. Andréka,H. Gergely,T. Németi,I., On universal algebraic construction of logics. Studia Logica 36, 1977, Nr.1-2, pp.9-47.

1a. Andréka,H. Gergely,T. Németi,I. Sain,I., Theory morphisms, stepwise refinement of program specifications, representation of knowladge, and cylindric algebras. Preprint, June 1980.

1b. Andréka,H. Gergely,T. Németi,I., Investigations in Language Hierarchies. Preprint, Research Inst. for Applied Comp. Sci. 1980.

2. Andréka,H. Németi,I., Applications of universal algebra, model theory, and categories in computer science. Survey and bibliography, parts I-III. Part I in CL&CL - Comput. Ling. Comput. Lang. Vol.XIII (1979), Part II in CL&CL - Comput. Ling. Comput. Lang. Vol.XIV (1980), Part III as Preprint Math.Inst.Hung.Acad.Sci. 1980.

3. Andréka,H. Németi,I., Generalization of variety and quasivariety concept to partial algebras through category theory. Preprint Math.Inst.Hung.Acad.Sci. No.1976/5. Dissertationes Mathematicae (Rozprawy) No.204. In press.

4. Andréka,H. Németi,I., Every free algebra in the variety generated by the representable dynamic algebras is separable and representable. Theoretical Computer Science, to appear.

5. Andréka,H. Németi,I. Sain,I., Representable dynamic algebras are not first order axiomatizable and other considerations on the representation theory of dynamic algebras. Preprint, 1980.

6. Andréka,H. Németi,I. Sain,I., Henkin type model theory for first order dynamic logic. Preprint, 1980.

7. Andréka,H. Németi,I. Sain,I., Relativization of dynamic algebras a tool for decomposition of dynamic theories. Preprint, 1980.

8. Bergstra,J.A. Tiuryn,J. Tucker,J.V., Correctness theories and program equivalence. Mathematisch Centrum Preprint IW 119/79, Amsterdam 1979.

9. Bertoni,A. Mauri,G. Miglioli,P.A. Wirsing,M., On different approaches to abstract data types and the existence of recursive models. Bulletin of the European Association for Theoretical Computer Science EATCS 9, pp.47-57.

10. Broy,M. Wirsing,M., Algebraic definition of a functional programming language and its semantic models. Preprint Technische Universität München TUM-I8008, June 1980.

11. Burmeister,P., Partial algebras - Survey of a unifying approach towards a two-valued model theory for partial algebras. Preprint 1980.

12. Burstall,R. Darlington,J., A system which automatically improves programs. Proc. 3rd IJCAI, S.R.I. 1973, pp.537-542.

13. Burstall,R. Goguen,J.A., The semantics of CLEAR, a specification language. Abstract Software Specifications (1979 Copenhagen Winter School Proceedings) Ed.: D.Bjørner. Lecture Notes in Computer Science Vol.86, Springer Verlag Berlin, pp.292-333.

14. Courcelle,B. Guessarian,I., On some classes of interpretations. J. of Computer and System Sciences 17, 1978, pp.388-413.

15. Csirmaz,L., Programs and program verifications in a general setting. Theoretical Computer Science, to appear.

16. Dömölki,B., An example of hierarchical program specification. Abstract Software Specifications (1979 Copenhagen Winter School Proceedings) Ed.: D.Bjørner. Lecture Notes in Computer Science Vol.86, Springer Verlag Berlin, pp.333-353.

16a. Gergely,T., Algebraic Representation of Language Hierarchies. Preprint, Research Inst. for Applied Comp. Sci. 1981. Acta Cybernetica, to appear.

17. Gergely,T. Úry,L., Program behaviour specification through explicit time consideration. Information Processing 80. Ed.: S.H. Lavington. North Holland Publ. Co. IFIP, 1980, pp.107-111.

18. Goguen,J.A., Some ideas in algebraic semantics. Proceedings, 3rd IBM Symposium on Math.Foundations of Comp. Sci. (Kobe Japan).

19. Grätzer,G., Universal algebra. (Second Edition) Springer Verlag, New York, 1979.

20. Hupbach,U.L., Abstract implementation of abstract data type. MFCS'80 Proceedings of the 8th Symp. on Mathematical Foundations of Computer Science (Rydzina-Zamek, Poland 1980). Lecture Notes in Computer Science, Springer Verlag, 1980.

21. Hupbach,U.L., A uniform mathematical framework for initial algebraic specifications. Preprint VEB ROBOTRON Zentrum für Forschung und Technik, 1979.

22. Kozen,D., A representation theorem for models of *-free PDL. Report RC7864, IBM Research, Yorktown Heights, New York, 1979.

23. Lehmann,D. Pasztor,A., On a conjecture of Meseguer. Theoretical Computer Science, to appear.

24. Mal'cev,A.I., Algebraic systems. Springer Verlag Berlin, 1973.

25. Nelson,E., Z-continuous algebras. Preprint McMaster University, 1980.

26. Németi,I., Some constructions of cylindric algebra theory applied to dynamic algebras of programs. CL&CL - Comput. Ling. Comput. Lang. Vol.XIV, to appear.

27. Németi,I., Nonstandard runs of Floyd-provable programs. Submitted to Proc. Conf. Algorithmic Logic 1980 (Poznan).

28. Németi,I., Connections between cylindric algebras and initial algebra semantics of CF languages. Mathematical Logic in Computer Science (Proc.Coll. Salgótarján 1978) Colloq.Math.Soc.J.Bolyai Vol.26, North-Holland. Eds.: B.Dömölki, T.Gergely. pp.561-606.

29. Pasztor,A., Characterization of epis of continuous algebras. Proc.Coll. Categorical and algebraic methods in computer science and system theory (3rd Workshop) Dortmund, 1980.

30. Pasztor,A., Faktorisierungssysteme in der Kategorie der partiellen Algebren - Kennzeichnung von (Homo-)Morphismenklassen. Hochschul Sammlung Naturwissenschaften Mathematik Band 1, Hochschul Verlag, Freiburg, 1979.

31. Pratt,V.R., Dynamic algebras: examples, constructions, applications. Report MIT/LCS/TM-138. July 1979.

31a. Reichel,H., Initially restricting algebraic theories. MFCS'80. Proc. 8th Symp. on Mathematical Foundations of Computer Science (Rydzina-Zamek, Poland), Lecture Notes in Computer Science,

Springer Verlag 1980.

32. Sain,I., Ultraproducts in theoretical computer science. Seminar notes SZKI 1980, and Lecture notes of NJSZT summer school Siófok 1980.

33. Sain,I., There are general rules for specifying semantics: Observations on abstract model theory. CL&CL - Comput. Ling. Comput. Lang. Vol.XIII, 1979, pp.195-250.

34. Wirsing,M. Broy,M., Abstract data types as lattices of finitely generated models. MFCS'80, Proc. 8th Symp. on Mathematical Foundations of Computer Science (Rydzina-Zamek, Poland). Lecture Notes in Computer Science, Springer Verlag, 1980.

35. Wirsing,M. Pepper,P. Partsch,H. Dosch,W. Broy,M., On hierarchies of abstract data types. Preprint Technische Universität München TUM-I8007, May 1980.

36. Zilles,S.N., Introduction to data algebras. Abstract Software Specifications (1979 Copenhagen Winter School Proceedings), Ed.: D.Bjørner. Lecture Notes in Computer Science Vol.86, Springer Verlag Berlin, pp.248-273.

PROBABILISTIC ANALYSIS OF THE PERFORMANCE OF GREEDY STRATEGIES OVER DIFFERENT CLASSES OF COMBINATORIAL PROBLEMS

G. Ausiello
Istituto di Automatica, Universita di Roma
A. Marchetti-Spaccamela
IASI-CNR, Roma

M. Protasi
Istituto Matematico, Universita dell'Aquila

0. INTRODUCTION

Greedy algorithms are widely used in the solution of discrete optimization problems because despite their simple structure in many cases this kind of algorithms may achieve either the exact optimal solution (such as for problems on matroids) or a very good approximate one (at least for large classes of instances, such as in the case of knapsack).

In this paper we analyze the performance of greedy algorithms from a probabilistic point of view, based on the mathematical approach introduced by Erdős widely developed by Karp (see for example [3] and [6]). In particular we examine the behaviour of two different strategies of greedy algorithms on two different classes of problems (clique and knapsack problems). The results presented in this paper may be added to the results of [2,5,7] in order to obtain a complete picture of the behaviour of greedy algorithms with respect to two classes of problems which are known to have deeply different computational properties although both belonging to the class of NP-complete problems.

1. 'BLIND' AND 'SHORT-SIGHTED' GREEDY ALGORITHMS

A widely used approach for the study of theoretical and practical difficulty of solution of NP-complete problems is to analyze the corresponding algorithms from a probabilistic point of view. In particular it is interesting to observe the difference of behaviour of the same algorithmic strategy when applied to problems with different combinatorial properties. In the literature the probabilistic analysis of greedy algorithms has been carried on by several authors. In particular we will focus on two classes of results which show that, when applied to NP-complete problems, greedy strategies, in some cases, may provide either the exact solution or a very good approximate one (almost everywhere) while in other cases the quality of the approximation is limited also from a probabilistic point of view.

In [2] D'Atri shows that a simple algorithm based on greedy strategies achieves the optimal solution of the knapsack problem almost everywhere under natural assumptions on the input distribution. For several other optimization problems, instead, in [5,7] it has been shown that the performance of greedy algorithms is not so good because the solution which is achieved is far from the optimal by a factor of two. This result appears to be satisfactory if we consider that, in some cases (clique, graph colouring) no polynomial time algorithm is known to provide a finite ratio of approximation on all inputs; but, on the other side, the difference of behaviour with respect to the case of knapsack requires a deeper discussion.

The reasons for the differences appearing in the said results may be twofold:

A) the greedy algorithms considered in the two cases are indeed based on different strategies

B) the two classes of problems which are considered are already known to have deeply different combinatorial properties which are at the base of a different characterization of these problems also with respect to approximability by polynomial algorithms [4].

The two strategies differ in the sense that one of them (which we call "blind greedy" strategy) in order to go from an achieved feasible solution to a better one simply picks up a new item at random and adds it provided the obtained solution is still feasible. The other strategy instead (which we call "short sighted greedy" strategy) chooses the new item to be added to the solution on the base of some "quality" of the item itself(*).

The algorithms studied in [5,7] are indeed "blind greedy" algorithms while the algorithms used in [2] for the knapsack problem essentially is a "short sighted greedy" algorithm, where the intended quality of the item is the so-called density (=profit/occupancy) of an element.

In order to establish the impact of the strategy on the performance of the greedy algorithms, in the next paragraph, we explore the behaviour of a short sighted greedy algorithm in the case of the clique

---

(*) The term "short sighted" derives from the fact that on the enumeration tree of the problem this heuristic only takes into account the quality of the nearest nodes and does not consider the quality of deeper nodes.

problem and we realize that the same limitations of the known blind greedy algorithm occur. In par. 3, instead, the blind greedy technique is applied to the knapsack problem. In this case there is a neat difference between the result obtained here, which shows that in meaningful cases the blind algorithm cannot achieve the optimal solution, with respect to the good performance of the short sighted algorithm of [2].

The results presented in this paper are hence suggesting that the difference of behaviour of the greedy algorithms observed at the beginning of this paragraph is actually mainly due to the structural differences between clique and knapsack problems (hypothesis B). The fact that the results concerning clique and knapsack may, in fact, be extended to larger classes of problems (such as the class defined in [8]) confirms the hypothesis and puts in evidence that the difference with respect to full approximability and the difference of probabilistic behaviour under greedy algorithms are indeed related.

2. ANALYSIS OF GREEDY STRATEGIES FOR THE CLIQUE PROBLEM

In [1,5,7] the properties of the solutions of the clique problem and the analysis of the performance of simple versions of the greedy algorithm over random graphs were studied.

In these papers the following model of random graph was assumed and throughout our paper we will adopt the same approach.

If G_n is a *random graph* of n nodes then any edge $\{i,j\}$ has constant probability p to occur, $0 < p < 1$ (independently on the presence or absence of any other edge).

For this model the following properties were proved concerning the size of the maximum clique $\omega(G)$ (*clique number* of G).

THEOREM 2.1. [5]. Let G_n be a random graph of n nodes. The sequence of random variables $\omega_n = \omega(G_n)$ satisfies $\frac{\omega_n}{\log n} \rightarrow \frac{2}{\log 1/p}$ as $n \rightarrow \infty$ almost everywhere.

Note that this result was strenghtened in [1,7] but the given formulation is sufficient for our purposes.

The simple version of greedy algorithm that was analyzed by the mentioned authors is the following (*blind greedy* algorithm (BG1) for the clique problem):

$T := \emptyset$; $S :=$ set of nodes of G_n;
while $S \neq \emptyset$ *do*

1. pick up a node $i \in S$;

2. $T:=T \cup \{i\}$;
3. $S:=\{j | j \in S-\{i\} \wedge \{i,j\} \in E\}$ /E is the set of edges of G/

   end;
   output T

Despite of its simple nature and the poor performance with respect to worst case analysis, this algorithm has been shown to behave satisfactorily on random graphs:

THEOREM 2.2. [5]. Given a random graph G_n, Algorithm BG1 provides a solution T_n such that

$$\frac{T_n}{\log n} \to \frac{1}{\log 1/p} \text{ as } n \to \infty$$

almost everywhere.

On the other side by combining theorems 2.1 and 2.2 we obtain

THEOREM 2.3. [5]. Given a random graph G_n, Algorithms BG1 provides a solution T_n such that

$$\frac{\omega_n}{T_n} \to 2 \quad \text{as } n \to \infty$$

almost everywhere.

In order to develop the analysis discussed in par. 1 let us now consider a more sophisticated algorithm.

The algorithm SG1 is similar to algorithm BG1 except for step 1 of the iteration where at any stage the node with largest valence among the nodes candidates for the augmentation of the clique is now chosen.

Then this node is added to the clique and it is erased from the graph together with all nodes not adjacent to it.

This algorithm follows a short sighted strategy because the augmentation is carried on according to the quality (valence) of a node among the candidates. Differently from what one could expect this more clever strategy does not produce a better approximation then the one achieved by BG1.

THEOREM 2.4. Given a random graph G_n, Algorithm SG1 provides a solution T_n such that

$$\frac{\omega_n}{T_n} \to 2 \text{ as } n \to \infty$$

in probability.

PROOF. The upper bound stated in the theorem is based on the same analysis carried on in [5] for algorithm BG1.

In order to prove the lower bound we have to show that SG1 cannot provide, in probability, a complete maximal subgraph of size larger than $\frac{1}{2}\omega_n + o(\omega_n)$.

In fact we can show that after $d(n) = \frac{\log n}{\log 1/p} - \sqrt{\log\log n}$ steps the set S of candidate nodes has a size $o(\omega_n)$.

Let us define $k_o = n$ and k_i, $i = 1,2,\ldots,d$, the cardinality of S at stage i. Let be E_j, $j = 0,1,\ldots,d-1$, the event that $k_j \leq np^j$ and $k_i > np^i$, $i > j$ (if $k_i > np^i$ $i = 1,2,\ldots,d-1$ then E_o occurs).

Now we evaluate the probability that at stage d we have no more than $np^d \prod_{\ell=o}^{d-1} (1 + \frac{1}{(np^{\ell+1})^{1-\alpha}})$ nodes when $\alpha > 1/2$. We have

$$\text{(A)}\quad \text{Prob}(k_d \leq np^d \prod_{\ell=o}^{d-1} (1+\frac{1}{(np^{\ell+1})^{1-\alpha}})) = \sum_{j=o}^{d-1}\left[\text{Prob}(k_d \leq np^d \prod_{\ell=o}^{d-1} (1+\frac{1}{(np^{\ell+1})^{1-\alpha}}) \,|\, E_j\right) \cdot$$
$$\cdot \text{Prob}(E_j)\Big]$$

When E_o occurs we have the following bounds from the Chernoff's inequality on the tail of the binomial distribution

$$\text{Prob}(k_1 \leq np(1+\frac{1}{(np)^{1-\alpha}})) \geq 1 - n\exp(-(np)^{\beta/3}) \quad \text{where } \beta = 2\alpha - 1$$

$$\text{Prob}(k_2 \leq np^2(1+\frac{1}{(k_1p)^{1-\alpha}})(1+\frac{1}{(np)^{1-\alpha}})) \geq 1-k_1\exp(-k_1p)^{\beta/3})-k_2\exp(-(np)^{\beta/3})$$

Iteratively we have

$$\text{Prob}(k_i \leq np^i \prod_{\ell=o}^{i-1} (1 + \frac{1}{(k_\ell p)^{1-\alpha}})) \geq 1 - \sum_{\ell=o}^{i-1} k_\ell \exp(-(k_\ell p)^{\beta/3})$$

Since $k_i > np^i$ we have for n sufficiently large

$$\text{Prob}(k_i \leq np^i \prod_{\ell=o}^{i-1} (1 + \frac{1}{(np^{\ell+1})^{1-\alpha}})) \geq 1-\sum_{\ell=o}^{i-1} k_\ell \exp(-(k_\ell p)^{\beta/3}) \geq$$
$$\geq 1-\sum_{\ell=o}^{i-1} np^\ell \exp(-(np^{\ell+1})^{\beta/3})$$

If $j > 0$ we have that for $i > j$

$$\text{Prob}(k_i \leq np^j \cdot p^{i-j} \prod_{\ell=j}^{i-1} (1 + \frac{1}{(np^\ell)^{1-\alpha}})) \geq 1-\sum_{\ell=j}^{i-1} np^\ell \exp(-(np^{\ell+1})^{\beta/3})$$

So from (A) we have

$$\text{Prob}(k_d \leq np^d \prod_{\ell=o}^{d-1} (1 + \frac{1}{(np^{\ell+1})^{1-\alpha}})) \geq \sum_{j=o}^{d-1} \left\{ \left[1 - \sum_{\ell=j}^{d-1} np^{\ell} \exp(-(np^{\ell+1})^{\beta/3}) \right] \cdot \right.$$

$$\left. \cdot \text{Prob}(E_j) \right\} \geq 1 - \sum_{\ell=o}^{d-1} np^{\ell} \exp(-(np^{\ell+1})^{\beta/3})$$

Besides it can be shown that for $\alpha > 1/2$

$$\lim_{n \to \infty} \sum_{\ell=o}^{d-1} np^{\ell} \exp(-(np^{\ell+1})^{\beta/3}) = 0$$

Therefore after d stages we have that there are no more than

$np^d \prod_{\ell=1}^{d} (1 + \frac{1}{(np^{\ell})^{1-\alpha}})$ nodes in probability.

In order to evaluate this quantity we consider the logarithm

$$\log k_d \leq \log n + d \log p + \sum_{\ell=1}^{d} \log(1 + \frac{1}{(np^{\ell})^{1-\alpha}}) = \sum_{\ell-1}^{d} \log(1 + \frac{1}{(np^{\ell})^{1-\alpha}}) +$$

$$+ \log \frac{1}{p} \cdot \sqrt{\log\log n} \leq \sum_{\ell=1}^{d} \frac{1}{(np^{\ell})^{1-\alpha}} + \log \frac{1}{p} \sqrt{\log\log n} =$$

$$= \frac{1}{n^{1-\alpha}} \sum_{\ell=1}^{d} \frac{1}{(p^{1-\alpha})^{\ell}} + \log \frac{1}{p} \sqrt{\log\log n} = \frac{1}{n^{1-\alpha}} \frac{1-(p^{1-\alpha})^d}{1-p^{1-\alpha}} +$$

$$+ \log \frac{1}{p} \sqrt{\log\log n}$$

While n goes to infinity the first term of the last expression goes to 0. This means that after d stages the size of the set of nodes which may still be added to the clique by algorithm SG1 is at most $O(e^{\sqrt{\log\log n}})$. Therefore the size of the clique found by algorithm SG1, T_n, is bounded by $\frac{\log n}{\log 1/p} - \sqrt{\log\log n} + O(e^{\sqrt{\log\log n}})$.

QED

3. ANALYSIS OF GREEDY STRATEGIES FOR THE KNAPSACK PROBLEM

First of all, we recall the definition of the 0/1 Knapsack problem:

$$\text{Max} \sum_{i=1}^{n} p_i x_i$$

Subject to

$$\sum_{i=1}^{n} a_i x_i \leq b$$

where $(p_1, \ldots p_n, a_1, \ldots a_n, b)$ is a $(2n+1)$-tuple of positive integers and $x_i = 0$ or 1 for $i = 1,2,\ldots,n$.

In [2] D'Atri has studied the probabilistic behaviour of an approximate algorithm for the solution of the knapsack problem. The algorithm is essentially based on the shortsighted greedy strategy where the quality of the items is given by the ratio between profit (p_i) and occupancy (a_i). For this reason the list of items is assumed to be sorted in decreasing order, that is

$$\frac{p_i}{a_i} > \frac{p_j}{a_j} \quad \text{implies} \quad i < j;$$

If $\frac{p_i}{a_i} = \frac{p_j}{a_j}$ and $a_i < a_j$ then $i < j$.

Algorithm SG2:

T:=∅; S:= sorted list of items;

while S ≠ ∅ *do*

1. pick up smallest item k is S; S:=S-{k};

2. if $\sum_{i \in T} a_i + a_k \leq b$ *then* T:=T ∪ {k}

end;

output T.

In order to perform the probabilistic analysis of such algorithm it has been assumed that the coefficients a_i, p_i were uniformly distributed over the interval {1,...,c} for some c.

The result of the analysis is the following:

THEOREM 3.1. [2]. Algorithm SG2 provides an optimal solution of the knapsack problem almost everywhere.

As it has been announced in §1, we are interested in comparing the behaviour of various greedy strategies applied in the solution of the knapsack problem. Hence our further step is to consider the behaviour of an algorithm (BG2) based on a blind strategy, obtained by eliminating the preprocessing phase of sorting the items. This is equivalent to assuming that the items are chosen at random. The assumption on the distribution of the items is the same as for algorithm SG2.

Since in this model no hypothesis is made on the value of b, we will consider b as a parameter in the evaluation of the performance of the algorithm.

THEOREM 3.2. If $b > \frac{n(c+1)}{2}$ then algorithm BG2 provides a solution z_{BG2} such that

$$\frac{z^*}{z_{BG2}} \to 1 \quad \text{as} \quad n \to \infty \text{ almost everywhere}$$

where z* is the optimal solution of the problem.

PROOF. Let us first observe that the expected value of the total profit and the total occupancy of all items is $n\frac{(c+1)}{2}$. Hence if b is larger than such value all items can be expected to be put in the knapsack. Therefore, the thesis follows from the law of large numbers.

QED

On the other hand for any ε, if we choose $b < n/c^2(1-\varepsilon)$ it is not difficult to prove (by Chernoff's inequality) that be optimal solution is only given by those elements which have maximum profit c and minimum occupancy 1 almost everywhere. We will not consider anymore these cases because the optimal solution can be trivially found almost everywhere.

Hence the meaningful range of the values of b, which still has to be analyzed, is $n/c^2 \leq b \leq \frac{nc}{2}$.

Unfortunately, a complete analysis of the performance of the algorithm in this range is not yet fully developed in a rigorous way. Anyway we can prove the behaviour of the algorithm in some meaningful points corresponding to those instances such that the optimal solution may be obtained by considering all items with:

a) profit c and occupancy 1,

b) profit larger than or equal to $\left\lceil \frac{c}{2} \right\rceil$ and occupancy 1,

c) profit larger than or equal to the occupancy,

d) profit greater than 1 or occupancy smaller than or equal to $\left\lfloor \frac{c}{2} \right\rfloor$

THEOREM 3.3. Correspondingly to the said instances we have

a) $b = \left\lfloor \frac{n}{c^2} \right\rfloor \Rightarrow \frac{z^*}{z_{BG2}} \to c$ a.e.

b) $b = \left\lfloor \frac{n}{c^2} \right\rfloor \cdot \left\lceil \frac{c}{2} \right\rceil \Rightarrow \frac{z^*}{z_{BG2}} \leftrightarrow \frac{3}{4}c$ a.e.

c) $b = \left\lfloor \frac{n}{c^2} \right\rfloor \cdot (c(c+1)\ \frac{c+2}{6}) \Rightarrow \frac{z^*}{z_{BG2}} \to \frac{2c+1}{c+2}$ a.e.

d) $b = \left\lfloor \frac{n}{c^2} \right\rfloor \cdot (\frac{(c-1)(c+1)c}{2} + \frac{\left\lfloor \frac{c}{2} \right\rfloor (\left\lfloor \frac{c}{2} \right\rfloor +1)}{2}) \Rightarrow \frac{z^*}{z_{BG2}} \to$

$$\to \frac{c^2(c+1)-2\lfloor c/2\rfloor}{c(c^2-1)+\lfloor c/2\rfloor(\lfloor c/2\rfloor+1)} \quad \text{a.e.}$$

PROOF. Let us only examine case c); the other cases are treated essentially in the same way.

Let us first verify that in case c) the optimum is obtained by considering all items with profit larger then or equal to the occupancy. In fact if we choose such items we have that, by the law of large numbers, the expected value of the total occupancy is

$$\left\lfloor \frac{n}{c^2} \right\rfloor \cdot \sum_{i=1}^{c} (i\cdot(c-i+1)) = \left\lfloor \frac{n}{c^2} \right\rfloor \cdot (c(c+1)\frac{c+2}{6}) = b$$

and the expected profit is

$$\left\lfloor \frac{n}{c^2} \right\rfloor \cdot \sum_{i=1}^{c} i^2 = \left\lfloor \frac{n}{c^2} \right\rfloor \cdot \frac{1}{3}\, c(c+\frac{1}{2})(c+1)$$

Since, as we saw in theorem 3.2, the blind greedy algorithm achieves a solution z_{BG2} with expected value b, we obtain that

$$\frac{E(z^*)}{E(z_{BG2})} = \frac{2c+1}{c+2}$$

Almost everywhere convergence follows from the general properties of the law of large numbers.

QED

Theorem 3.3 provides enough evidence that the function $f(b) = \frac{z^*(b)}{z_{GB2}(b)}$ in the meaningful range $\frac{n}{c^2} \leq b \leq \frac{nc}{2}$, is nonincreasing and hence the behaviour of a blind greedy algorithm on the knapsack for such a range of values of b is indeed worse than the shortsighted algorithm used in [2].

REFERENCES

[1] B.BOLLOBAS, P.ERDŐS: *Cliques in random graphs*, Math. Proc. Cambridge Phil. Soc. Vol. 80 (1976).

[2] G.D'ATRI: *Probabilistic analysis of the knapsack problem*, Techn. Rep. n.7, g.d.r.n. 22 CNRS (1978).

[3] P.ERDŐS, J.SPENCER: *Probabilistic methods in combinatories*, Academic Press, N.Y. (1974).

[4] M.R.GAREY, D.S.JOHNSON: *Computers and intractability: a guide to the theory of NP-completeness*, Freeman, San Francisco (1979).

[5] G.R.GRIMMETT, C.J.H.Mc DIARMID: *On colouring random graphs*, Math.

Proc. Cambridge Phil. Soc. Vol. 77 (1975).

[6] R.M. KARP: *The probabilistic analysis of some combinatorial searc algorithms;* in J.F.Traub (ed.) "Algorithms and Complexity: New directions and recent results", Academic Press, N.Y. (1976)

[7] D.W.MATULA: *The largest clique size in a random graph,* Techn. Rep. CS76-03, Southern Metho. Univ. (1976).

[8] R.TERADA: *Polynomial time algorithms for NP-hard problems which are optimal or near optimal with probability one,* Techn. Rep. 79-351, Dep. Comp. Sc., Wisconsin Univ. (1979).

MODERATELY EXPONENTIAL BOUND FOR GRAPH ISOMORPHISM

László Babai
Dept. Algebra, Eötvös University
Budapest, Hungary H-1088

In this expository paper, we present recent major development concerning the complexity of graph isomorphism testing. The main result is a general $\exp(v^{1-c})$ bound, due to V.N. Zemlyachenko. The proof operates on the pioneering techniques of E.M. Luks which we briefly review here.

1. Brief history

The first significant complexity result in graph isomorphism testing was the linear time canonical labeling of trees, found by Hopcroft and Tarjan [16], and independently by V.N. Zemlyachenko [29]. Next came planar graphs, where again a linear time canonical labeling algorithm was found [17], [18]. As far as algorithms with proven efficiency concern, little has been done after this early work until quite recently when it became evident that bounds on certain parameters of a graph are likely to give enough structure that makes polynomial time isomorphism testing possible. Of the three examples I am going to give below, the first one can be viewed as a generalization of the work on planar graphs. As usual, P stands for polynomial time recognizable languages.

Theorem 1.1. Isomorphism of graphs of bounded genus $\in P$. /Filotti-Mayer [13], Lichtenstein [20] and G. Miller [23]./

The ideas in the other two examples go in another direction:

they exploit constraints on the structure of the automorphism groups of the graphs considered.

Theorem 1.2. Isomorphism of graphs with bounded multiplicities of eigenvalues $\in P$ /Babai-Grigoryev [8]/.

The major breakthrough was achieved by linking in depth the structure of permutation groups to the algorithmic problem of graph isomorphism:

Theorem 1.3. Isomorphism of graphs of bounded valence $\in P$ /Luks [22]/.

Although 1.3 seems by far the most substantial result in the subject, all attempts to derive either of 1.1, 1.2, 1.3 by modification of the techniques of the others have failed so far. A common generalization, involving perhaps boundedness of some more general parameter of graphs has yet to be seen. A common disturbing feature of the result 1.i $(1 \le i \le 3)$ is that the parameter itself goes into the exponent of the complexity estimate. Yet another, simpler result where this is not the case, is the following. A colored graph is a pair (X,f) where X is a graph and f is a map of the vertex set into some set of colors. Isomorphisms of colored graphs preserve colors by definition. The multiplicity of a color is the number of vertices with that color.

Theorem 1.4. Isomorphism of colored graphs with bounded color multiplicities can be determined in $O(v^4)$ steps /Babai [1]/.

Even this result seems independent of the above.

/The fact that this problem is in P follows immediately from 1.3./

As for the general case, these results give nothing. Let $T(v)$ denote the worst case complexity of the "best" graph isomorphism algorithm. /Throughout the paper, v denotes the number of vertices./ Brute force yields $T(v) < v!$. In general, for problems in NP there is usually a natural brute force bound, say $b(n)$ which is exponential, i.e. $\exp(n^{c_1}) < b(n) < \exp(n^{c_2})$ for certain positive constants $0 < c_1 < c_2$. I would like to propose to call a bound $t(n)$ moderately exponential if it reduces the logarithm of $b(n)$ by a positive fraction: $\log t(n) < (1-c)\log b(n)$ for some constant $c > 0$. In the case of graph isomorphism, a moderately exponential bound means $\exp(v^{1-c})$. Until very recently, no such bound was available. In contrast, very good bounds have been obtained for almost all graphs /$v^{\log v}$ by Zemlyachenko [30], polynomial time by Lipton [21] and Zemlyachenko [32], $O(v^2 \log v)$ by Karp [19], linear time /i.e. $O(v^2)$/ by Babai-Erdős-Selkow [7] and Babai-Kučera [9]/ and for the average complexity for all graphs /$v^{\log v}$ by Deo-Davis-Lord [12], linear time by Babai-Kučera [9]/. The fact is that extremely naive heuristics have an excellent performance on almost all graphs, hence these results again are irrelevant for the general worst case complexity. Results slightly better than brute force are c^v by Zemlyachenko [30] assuming the Ulam-Kelly reconstruction conjecture and the same bound by M. Goldberg [15] without such hypothesis. Previously, Zemlyachenko has obtained a hardly-better-than-brute-force bound $\exp(v \log\log v)$ [31], but combining the methods of that paper with those of Luks [22], Zemlyachenko now has announced what in my view is the first major result on isomorphism of general graphs: a moderatley exponential bound.

Theorem 1.5 /V.N. Zemlyachenko/. Graph isomorphism can be determined in $\exp(v^{1-c})$ steps where c is a positive constant.

One can prove 1.5 with $c = 1/3 + o(1)$. The main aim of this paper is to prove Theorem 1.5, together with a slight variation on Luks' theme. For $c = 1/4 + o(1)$, the proof presented is essentially self-contained. In order to obtain $c = 1/3 + o(1)$, some recent results of the present author are required [4], [5] /cf. the last section/. Otherwise, my role here is little more than that of a /travelling/ interpreter.

It may be worth pointing out that reaching an $\exp(v^{1/2+o(1)})$ bound still seems about as remote as before. For some important classes of highly symmetric graphs /strongly regular, primitive/ $\exp(v^{1/2+o(1)})$ has been reached [3], [4], but there is little indication of what the next step might be.

2. Naive vertex classification

We begin with describing a common heuristic procedure, central to Zemlyachenko's argument.

By a coloring of a v-set V we mean a map $f: V \to \{1,\ldots,v\}$ such that the actual range of f is an initial segment of $\{1,\ldots,v\}$. A coloring f_1 is a refinement of f if $f_1(x) \le f_1(y)$ implies $f(x) \le f(y)$ for all $x,y \in V$. The color-classes are the sets $C_i = f^{-1}(i)$. They form the partition Ker f.

The input of the naive refinement procedure is a graph with colored vertices. The output is a /not necessarily proper/ refinement of the coloring.

Let $X = (V,E)$ be a graph and $f: V \to \{1,\ldots,v\}$ a coloring. We refine the coloring as follows. For $x \in V$, let $k_i(x)$ denote the number of neighbors of x having color i. We assign the vector $g(x) = (f(x), k_1(x), \ldots, k_v(x))$ to x and sort the vertices of X according to the lexigraphic order of their corresponding vectors. Let f' be the coloring determined by this order: $f'(x) = f'(y)$ iff $g(x) = g(y)$ and $f'(x) < f'(y)$ iff $g(x) < g(y)$. With the additional stipulation that the range of f' be an initial segment of $\{1,\ldots,v\}$, f' is uniquely defined. Clearly, this refinement is canonical in the following sense: any map $V_1 \to V_2$ is an isomorphism of the colored graphs (X_i, f_i) $(i=1,2)$ iff it is an isomorphism of (X_i, f_i').

Let now $f_o = f$, $f_1 = f', \ldots, f_{i+1} = f'_i$. As there are no more than v-1 proper refinements possible, we have $f_i = f_{i+1} = \ldots = f_v$ for some i . Let us call f_v the stable refinement of f and denote it by $\overline{f}$. A coloring is stable if f=f' /hence $f=\overline{f}$/.

By a semiregular bipartite graph we mean a bipartite graph with given chromatic partition $V = V_1 \cup V_2$ /edges go between V_1 and V_2/ such that all vertices in the same class have equal valences.

For V_1, V_2 disjoint subsets of the vertex set of the graph $X = (V,E)$, let $X(V_1,V_2)$ denote the bipartite subgraph of X induced between V_1 and V_2 . Also, let $X(V_1)$ stand for the subgraph induced by V_1 .

Stable colorings are characterized by the following straightforward proposition.

Proposition 2.1. The coloring f of the graph X is stable if and only if all induced subgraphs $X(f^{-1}(i))$ are regular and all induced bipartite subgraphs $X(f^{-1}(i),f^{-1}(j))$ are semiregular.

While stable refinement is the simplest breadth-first trick in isomorphism testing, its depth-first counterpart consists of turning a vertex into a singleton color-class by assigning a new color to it. For f a /not necessarily stable/ coloring and $x \in V$, let f_x denote the obtained new coloring: $f_x^o(y) = f(y)$ for $y \neq x$ and $f_x^o(x) = \min(\{1,\ldots,v\} \setminus \{f(y):y \neq x\})$. Let f_x denote the stable refinement of f_x^o . We call f_x the stabilizer of x . /The term is borrowed from permutation groups./ The vertex x is a singleton under f_x . For $S = (x_1,\ldots,x_s)$ an ordered sequence of vertices we define the stabilizer of S to be $f_S = f_{x_1 \ldots x_s}$, obtained by repeated application of the above operation. The coloring f_S is stable and the vertices $x_1,\ldots,x_s$ form singleton color-classes.

3. Vertex-colored graphs with moderate color-valences

Let f be a coloring of the vertices of $X = (V,E)$. The <u>valence of vertex</u> x <u>in color</u> i is the number of neighbors of x in $C_i = f^{-1}(i)$. The <u>co-valence</u> of x in C_i is the corresponding valence in the complement of X. We say that X has <u>color-valence</u> $\leq d$ w.r. to f if for each color i and vertex x, either the valence or the co-valence of x in C_i is $\leq d$.

<u>Valence reduction lemma 3.1</u> /Zemlyachenko [31]/. Let the graph $X = (V,E)$ have color-valence $\leq d$ with respect to some stable coloring f. Then there exists a sequence $S = (x_1,\dots,x_k)$ of $k < 2v/d$ vertices such that X has valence $\leq d/2$ with respect to f_S.

<u>Proof</u>. Suppose $S(i) = (x_j : j < i)$ is given. If X has valence $\leq d/2$ w.r. to the stabilizer $f_{S(i)}$ then we declare $k = i-1$ and halt. Otherwise, there exists a vertex x_i and a color m_i such that both the valence and the co-valence of x_i to the color-class $C_i = f^{-1}_{S(i)}(m_i)$ are greater than $d/2$. Let N_i denote either the set of neighbors or the set of non-neighbors of x_i in C_i so that $|N_i| \leq |C_i|/2$. We have $d/2 < |N_i| \leq d$.

We claim that the sets $N_1,\dots,N_k$ are pairwise disjoint. From this claim we infer $kd/2 < |N_1|+\dots+|N_k| \leq v$ thus $k < 2v/d$ follows.

Suppose, by contradiction, that $N_j \cap N_i \neq \emptyset$ for some $j < i \leq k$. This implies $N_j \cap C_i \neq \emptyset$. Since x_j is a singleton class under $f_{S(i)}$, the set N_j is the union of classes and therefore $N_j \supseteq C_i$. Now we have $d/2 < |N_i| \leq |C_i|/2 \leq |N_j|/2 \leq d/2$, a contradiction, proving the lemma.

<u>Corollary 3.2.</u> Given d and any graph X, there exists a sequence $S = (x_1,\dots,x_k)$ of $k < 4v/d$ vertices such that X has color-valence $\leq d$ w.r. to f_S for any initial coloring f.

Proof. X has valence $\leq n/2$ w.r. to $\bar{f}$. Repeated application of 3.1 shows that there is a sequence S of less then $4+8+\ldots+2^s < 2^{s+1}$ vertices such that X has color-valence $\leq v/2^s$ w.r. to f_S . Choosing $v/2^s \leq d < v/2^{s-1}$, 3.2 follows. In fact, $k < (4v/d)-4$ vertices suffice.

Let $T(v,d)$ denote the worst case complexity of isomorphism of vertex-colored graphs of color-valence $\leq d$. $T(v) = T(v,v)$ is the complexity of the isomorphism problem for vertex colored graphs without any restriction.

Theorem 3.3 /V.N. Zemlyachenko [31]/. $T(v) < v^{4v/d}T(v,d)$.

Proof. There are less than v^k ways to select a sequence S of at most k distinct vertices. For each S , it takes $O(v^3)$ to calculate f_S . In order to test isomorphism of two colored graphs (X,f) and (Y,g) one finds /by exhaustive search/ a k-sequence S of vertices of X such that the color-valence of X w.r. to f_S is $\leq d$, and then one checks isomorphism of (X,f_S) with (Y,g_T) for all k-sequences T of vertices of Y . If Y has color-valence $\leq d$ w.r. to g_T , this isomorphism can be decided in $T(v,d)$ steps; otherwise $(Y,g_T) \not\cong \not\cong (X,f_S)$. By Cor. 3.1, one can choose k to be less than $(4v/d)-4$.

4. Further reductions

We consider isomorphism of colored graphs with color-valence $\leq d$. We need some simple polynomial time reductions. One may assume that every vertex has valence $\leq d$ in each color. As a matter of fact, we may restrict ourselves to stable colorings. Now, if C_i and C_j are color-classes and the number of edges between them is greater than $|C_i||C_j|/2$, we may switch to the complement of $X(C_i,C_j)$, retaining the edges in the rest of X . Similarly we may complement the induced subgraphs $X(C_i)$. In a pair of colored graphs to be tested for isomorphism, these operations can be carried out simultaneously or else, nonisomorphism is established. The result is that the valence of every vertex to any color-class is not greater than the corresponding co-valence hence it is $\leq d$.

It suffices to test isomorphism of connected components.

Finally let $\mathfrak{X}(v,d)$ denote the class of those connected colored graphs on v vertices each vertex of which has valence $\leq d$ in each color and which have a singleton color class.

We claim that the isomorphism problem for connected colored graphs having valence $\leq d$ in each color is polynomial time reducible to the determination of a set of generators of Aut X for members of $\mathfrak{X}(v,d)$. /There is no restriction on d here; as $v \to \infty$, d can be arbitrarily large /$3 \leq d < v$/./

As a matter of fact, let X and Y be the two connected colored graphs to be tested for isomorphism. Form their disjoint union. Select an edge $[x_1,x_2]$ of X and an edge $[y_1,y_2]$ of Y . Remove these edges. Introduce three new vertices x_o, y_o and z , assign one new color to x_o and y_o , and another new color to z . Join x_o to x_1, x_2 and z ; similarly y_o to y_1, y_2 and z . Let Z be the obtained colored graph. Clearly, an isomorphism $X \to Y$ taking the edge $[x_1,x_2]$ to $[y_1,y_2]$ exists iff x_o and y_o belong to the same orbit of Aut Z . We have to repeat this construction for every edge $[y_1,y_2]$, thus increasing execution time by a factor of $vd < v^2$ at most.

Next we make a simple observation about the automorphism groups in question.

Let $\mathcal{G}(d)$ denote the class of those groups G which possess a chain of subgroups $G = G_o \geq G_1 \geq \ldots \geq G_m = 1$ such that $|G_{i-1}:G_i| \leq d$, for all i .

Proposition 4.1. If $X \in \mathfrak{X}(v,d)$ then Aut X belongs to $\mathcal{G}(d)$.

Remark 4.2. This result goes back essentially to an observation by Tutte [26]. For graphs without colors it is explicitely stated in [10].

Proof. Let us number the elements of $V = \{x_1,\dots,x_v\}$ such that x_1 is the "root" /singleton color-class/, and for $i \geq 2$, each x_i is adjacent to at least one of the x_j, $j < i$. Let G_i be the pointwise stabilizer of the set $\{x_1,\dots,x_{i+1}\}$ in $G_o = \text{Aut } X$. Now, $|G_{i-1}:G_i|$ is the length of the orbit of x_{i+1} in G_{i-1}. If x_{i+1} belongs to color class C and some x_j $(j \leq i)$ is adjacent to x_{i+1} then the orbit of x_j consists of neighbors of x_j in C. There are at most d such vertices.

5. Reduction to G-automorphisms of colored sets

Let G be a permutation group acting on a set Ω of n elements and $h_1,h_2:\Omega \to \{1,2,\dots,n\}$ be two colorings. The set of G-isomorphisms of h_1 and h_2 is defined to be

$$\text{Iso}_G(h_1,h_2) = \{g \in G : h_2(g\alpha) = h_1\alpha \text{ for each } \alpha \in \Omega\} .$$

Setting $\text{Aut}_G(h) = \text{Iso}_G(h,h)$, we have either $\text{Iso}_G(h_1,h_2) = \emptyset$ or $\text{Iso}_G(h_1,h_2) = \text{Aut}_G(h_2)g_o$ where g_o is any particular member of $\text{Iso}_G(h_1,h_2)$. Thus if non-empty, $\text{Iso}_G(h_1,h_2)$ is a coset and can be represented by an element g_o and a set of generators of the group $\text{Aut}_G(h_2)$. It is such a representation that we seek when we want to "determine" $\text{Iso}_G(h_1,h_2)$. Luks has reduced the problem of determining Aut X for a connected rooted d-valent graph on v vertices to less than v instances of the problem of determining $\text{Aut}_G(h)$ for $n < v^d$ where $G \in \mathcal{G}(d)$. It looks like a bold idea to allow for such a drastic increase in problem size, but this idea has turned out to be an exceptionally successful one.

With some care, we can adapt Luks' method to handle the class

$\mathfrak{X}(v,d)$ of colored graphs as well.

Claim 5.1. For $(X,f)\in\mathfrak{X}(v,d)$, the determination of $\mathrm{Aut}(X,f)$ reduces to less than v problems $\mathrm{Aut}_G(h)$ for sets of size $n < v^d$ and groups $G\in\mathcal{G}(d)$.

Proof. Let $(X,f)\in\mathfrak{X}(v,d)$ where $X = (V,E)$. We may assume f is a stable coloring since $\mathrm{Aut}(X,f) = \mathrm{Aut}(X,\bar{f})$. Let x_1 be the root. We may assume that $f(x_1) = 1$ and for any $i > 1$ there is a j_i, $1 \le j_i \le i-1$ such that the color-classes C_i and C_{j_i} are adjacent. /Note that by Prop. 2.1, if a vertex in C_i is adjacent to some vertex in C_j then all of them are./ Let $C_s \ne \emptyset = C_{s+1}$. Let $Y = (V,F)$ be the subgraph of X containing, for every $i \ge 2$, all edges between C_i and C_{j_i} and no other edges. Clearly, $(Y,f)\in\mathfrak{X}(v,d)$ and $\mathrm{Aut}(Y,f) \ge \mathrm{Aut}(X,f)$.

First we show how to determine $\mathrm{Aut}(X,f)$ given $G = \mathrm{Aut}(Y,f)$. Let Ω be the set of $\binom{v}{2}$ pairs of vertices, colored red and blue according to whether a pair belongs to E or not. Let $\tilde{G}$ be the induced action of G on Ω. The induced action on Ω of $\mathrm{Aut}(X,f)$ is precisely the group of $\tilde{G}$-automorphisms of the 2-colored Ω. Clearly, $\tilde{G} \cong G \in \mathcal{G}(d)$.

Therefore it suffices to determine $\mathrm{Aut}(Y,f)$.

Let $V_i = C_1 \cup \ldots \cup C_i$ and let Y_i be the subgraph of Y induced by V_i. Set $G_i = \mathrm{Aut}(Y_i, f|V_i)$. We determine the groups G_i inductively. Denoting by H_i the restriction of G_{i+1} to V_i, it is a trivial matter to find generators of G_{i+1} given generators of H_i. /Note that there are no edges in F joining a pair of vertices of the same color./ So our task is to find generators for H_i, the subgroup of G_i consisting of those automorphisms of the colored Y_i which extend to Y_{i+1}.

Let Ω be the set of subsets of cardinality $\le d$ of V_i. For each $\alpha\in\Omega$, let $h(\alpha)$ denote the number of those $x \in C_{i+1}$ of which

α is the full neighborhood. /Clearly, $h(\alpha) = 0$ unless $\alpha \subseteq C_{j_{i+1}}$ ./ Let G_i and H_i denote the induced actions of G_i and H_i , resp., on Ω . We observe that H_i consists of the G_i -automorphisms of the colored set (Ω, h) /because the vertices in Y have $\le d$ "backward" neighbors/. We have $|\Omega| < v^d$ and $G_i \in \mathcal{G}(d)$ since $(Y_i, f|V_i) \in \mathfrak{X}(|V_i|, d)$.

6. Finding G-isomorphisms of colored sets

Using the notation of the previous section, our problem is now to determine $\mathrm{Iso}_G(h_1,h_2)$ where $G \in \mathcal{G}(d)$ is a group acting on a set Ω of n elements and h_1, h_2 are two colorings of this set. Let us call n the size of the problem. We briefly review Luks' elegant "divide and conquer" algorithm for this problem. We assume the reader is familiar with the first few pages of some text on permutation groups /e.g. [28]/.

If G is intransitive, let $\Omega_1,\dots,\Omega_k$ be its orbits. Let $\mathrm{Iso}_G^i(h_1,h_2) = \{g \in G : h_2(g\alpha) = h_1(\alpha)$ for each $\alpha \in \Omega_1 \cup \dots \cup \Omega_i\}$ $(i=0,\dots,k)$. By definition, $\mathrm{Iso}_G^0(h_1,h_2) = G$. Assuming $\mathrm{Iso}_G^i(h_1,h_2) =$ $= \mathrm{Aut}_G^i(h_2)g_i$ has already been determined, we have $\mathrm{Iso}_G^{i+1}(h_1,h_2) =$ $= \mathrm{Iso}_H^{i+1}(h_1^{g_i},h_2)g_i$ where $H = \mathrm{Aut}_G^i(h_2)$ where h^g is the coloring defined by $h^g(\alpha) = h(g^{-1}\alpha)$. Now, determining $\mathrm{Iso}_H^{i+1}(h_1^{g_i},h_2)$ is essentially a color-isomorphism problem on the set Ω_{i+1} . This way, the problem of size n has been sliced into pieces of sizes $n_i = |\Omega_i|$, $n=n_1+\dots+n_k$.

If G is transitive, let Ω_1 be a maximal block of imprimitivity of G . /If G itself is primitive, let $|\Omega_1| = 1$ ./ Let $\Omega_1,\dots,\Omega_k$ be the images under G of Ω_1 ; they form an invariant partition of G and G acts on them as a primitive permutation group. Let K denote the kernel of the resulting homomorphism $G \to \mathrm{Sym}(k)$, and let $g_1,\dots,g_m$ be a set of representatives of the cosets of K . The Ω_i are invariant under K and therefore, as before, a problem of the type $\mathrm{Iso}_K(h_1,h_2)$

reduces to k instances of a problem of size n/k. /The value of d remains unchanged./ On the other hand, $G = \bigcup_{j=1}^{m} g_j K$, hence $\mathrm{Iso}_G(h_1,h_2) = \bigcup_{j=1}^{m} g_i \mathrm{Iso}_K(h_1^{g_i},h_2)$.

Our conclusion is the following. Let $T_d(n)$ denote the worst case complexity of determining $\mathrm{Iso}_G(h_1,h_2)$ for $G \le \mathrm{Sym}(n)$, $G \in \mathcal{G}(d)$. /$\mathrm{Sym}(n)$ denotes the symmetric group of degree n./ Then either there is a partition of n, $n=n_1+\ldots+n_k$ $(k \ge 2)$ such that $T_d(n) \le$ $\le T_d(n_1) + \ldots + T_d(n_k)$, or there is a proper divisor k of n such that

$$T_d(n) \le m_d(k) k T_d(n/k)$$

where $m_d(k)$ denotes <u>the maximum order of a primitive permutation group of degree</u> k <u>belonging to</u> $\mathcal{G}(d)$.

Assuming that we have a bound

$$(*) \qquad m_d(k) \le k^{t(k,d)}$$

where $t(k,d)$ is a non-decreasing function of k for each d, a solution to the above recurrent inequalities is

$$T_d(n) < Cn^{1+t(n,d)}$$

/C is an absolute constant/. We have thus proved:

<u>Lemma 6.1.</u> If (*) holds, then for $(X,f) \in \mathfrak{X}(v,d)$, the group $\mathrm{Aut}(X,f)$ can be determined in $Cv^{d(1+t(v^d,d))}$ steps.

7. <u>Estimating the order of primitive groups</u>

<u>Theorem 7.1</u> [6]. There is a function $t(d)$ such that any primitive group $G \in \mathcal{G}(d)$ of degree n has order at most $n^{t(d)}$.

By Lemma 6.1, Luks' Theorem 1.3 now follows /slightly generalized, to colored graphs of valence $\le d$ in each color, d fixed/.

The difficulties in proving Theorem 7.1 occur when the orders of certain subgroups of GL(k,p) have to be estimated in terms of $n = p^k$, the number of vectors in the space on which GL(k,p) naturally acts.

Theorem 7.1 was not yet available at the time Luks wrote [22]. In a brilliant tour de force, he overcame the difficulty by inventing an additional algorithmic trick based on a result essentially due to Weir [27] and Suprunenko-Apatenok [25], which can conveniently be stated as follows:

Proposition 7.2. For r and p distinct primes, the order of a Sylow r-subgroup of GL(k,p) is less than n^2 where $n = p^k$.

Since clearly $|GL(k,p)| < n^{\log n}$, we do not face the difficulties related to the linear group when aiming at complexity estimates of the order $n^{\log n}$ and above; in particular, in completing the proof of Zemlyachenko's Theorem 1.5.

First, we have to find out more about the structure of the groups in $\mathcal{G}(d)$.

A section of a group G is a factor group H/K where $K \triangleleft H \leq G$.

We have already used the fact that if G belongs to $\mathcal{G}(d)$ then so do its sections. In particular, each composition factor L of G has a proper subgroup of index $\leq d$. Representing L on the cosets of this subgroup, it follows that L is isomorphic to a subgroup of Sym(d). In particular

$$(**) \qquad |L| < d! < d^d, \quad \text{and} \quad |\mathrm{Aut}\, L| < |L|^{\log|L|} < d^{d^2 \log d}.$$

/log is meant to be base 2./

In what follows we use more detailed but still elementary knowledge about primitive groups. Sections 2 to 4 of Cameron [11] contain more than enough information.

The socle of a group is the product of its minimal normal subgroups. If G is a primitive group of degree n then its socle is the direct product $T_1 \times \dots \times T_s$ of isomorphic simple groups. There are two cases. If the socle is abelian then G is a subgroup of the affine linear group $AGL(k,p)$ where $n = p^k$ hence $|G| < n^{1+\log n}$. In the case of nonabelian socle, G is a subgroup of the wreath product $(\mathrm{Aut}\ T)$ wr S_s where $T \cong T_i$ hence $|G| \le |\mathrm{Aut}\ T|^s s!$. On the other hand, it follows from what Cameron calls the O'Nan-Scott Theorem [11, p.5], that $n \ge 5^s$ in this case. Consequently,

$$|G| < |\mathrm{Aut}\ T|^{\log n} n^{\log\log n} .$$

In view of the estimates (**), we infer

$$|G| < n^{d^2 \log^2 d + \log n}$$

By Lemma 6.1, we conclude that:

Theorem 7.3. Isomorphism of colored graphs with color-valence $\le d$ can be tested in $T(v,d) < v^{d^2(\log v + d \log d)}$ steps.

We remark, that in [4] and [5], the following result is proved:

Theorem 7.4. If G is a primitive group of degree n, other than the symmetric or the alternating group, then $|G| < \exp(4\sqrt{n}\log^2 n)$ provided $n > n_0$.

From this, an improvement of (**) follows. Representing L on the cosets of a maximal subgroup of index $\le d$ we see that L is isomorphic to a primitive group of degree $d' \le d$, hence either $L \cong \mathrm{Alt}(d')$ and therefore $|\mathrm{Aut}\ L| = d'! < d^d$, or $|L| < \exp(c\sqrt{d}\log^2 d)$ hence $|\mathrm{Aut}\ L| < |L|^{\log |L|} < \exp(cd \log^4 d)$. /$c$ is an absolute constant./ This implies

Theorem 7.3'. $T(v,d) < v^{d^2(c \log^4 d + \log v)}$.

8. Proof of the moderately exponential bound

All we have to do now is to combine 7.3 /7.3', resp./ with 3.3 and choose the value of d appropriately. With $d = v^{1/3}$ and using 7.3', we obtain that graph isomorphism can be tested in $T(v) < \exp(v^{2/3+o(1)})$ steps. /The weaker bound 7.3 suffices for the moderately exponential claim: setting $d = v^{1/4}$, we obtain $T(v) < \exp(v^{3/4+o(1)})$./

Acknowledgement. My debt to E.M. Luks and to V.N. Zemlyachenko is implicit. I wish to express my gratitude to Dept. Math., Bucknell University, Pa. and to the Leningrad Branch of the Math. Inst. of the Acad. Sci. USSR for their hospitality during my short visits in 1980 and 1981, resp.

References

1. L. Babai, Monte-Carlo algorithms in graph isomorphism testing, preprint 1979.

2. —, Isomorphism testing and symmetry of graphs, in: Ann. Discr. Math. 8 /1980/ 101-109.

3. —, On the complexity of canonical labelling of strongly regular graphs, SIAM J. Comp. 9 /1980/ 212-216.

4. —, On the order of uniprimitive permutation groups, Ann. Math. 109 /1981/.

5. —, On the order of doubly transitive permutation groups, submitted

6. —, P.J. Cameron, P.P. Pálfy, On the order of primitive permutation groups with bounded nonabelian composition factors, in preparation.

7. —, P. Erdős, S.M. Selkow, Random graph isomorphism, SIAM J. Comp. 9 /1980/ 628-635.

8. —, D.Yu. Grigoryev, Isomorphism testing for graphs with bounded eigenvalue multiplicities, in preparation.

9. —, L. Kučera, Canonical labelling of graphs in linear average time, Proc. 20th IEEE FOCS Symp. /1979/ 39-46.

10. —, L. Lovász, Permutation groups and almost regular graphs, Studia Sci. Math. Hungar. 8 /1973/ 141-150.

11. P.J. Cameron, Finite permutation groups and finite simple groups, Bull. Lond. Math. Soc. 13 /1981/ 1-22.

12. N. Deo, J.M. Davis, R.E. Lord, A new algorithm for digraph isomorphism, BIT 17 /1977/ 16-30.

13. I.S. Filotti, J.N. Mayer, A polynomial-time algorithm for determining the isomorphism of graphs of fixed genus, Proc. 12th ACM Symp. Th. Comp. /1980/ 236-243.

14. M.L. Furst, J.E. Hopcroft, E.M. Luks, Polynomial time algorithms for permutation groups, Proc. 21st IEEE FOCS Symp. /1980/ 36-41.

15. M.K. Goldberg, A nonfactorial algorithm for testing isomorphism of two graphs, preprint 1981.

16. J.E. Hopcroft, R.E. Tarjan, Efficient planarity testing, J. ACM 21 /1974/ 549-568.

17. —, —, Isomorphism of planar graphs, in Complexity of Computations /Miller, Thatcher, eds./ Plenum Press, N.Y. 1972, 143-150.

18. —, J. Wong, Linear time algorithm for isomorphism of planar graphs, Proc. Sixth ACM Symp. Th. Comp. /1974/ 172-184.

19. R.M. Karp, Probabilistic analysis of a canonical numbering algorithm for graphs, Proc. Symp. Pure Math. 34, AMS 1979, 365-378.

20. D. Lichtenstein, Isomorphism for graphs embeddable on the projective plane, Proc. 12th ACM Symp. Th. Comp. /1980/ 218-224.

21. R.M. Lipton, The beacon set approach to graph isomorphism, SIAM J. Comp. 9 /1980/.

22. E.M. Luks, Isomorphism of graphs of bounded valence can be tested in polynomial time, Proc. 21st IEEE FOCS Symp. /1980/ 42-49.

23. G.L. Miller, Isomorphism testing for graphs of bounded genus, Proc. 12th ACM Symp. Th. Comp. /1980/ 225-235.

24. P.P. Pálfy, On the order of primitive solvable groups, J. Algebra, submitted.

25. D.A. Suprunenko, R.F. Apatenok, Nilpotent irreducible groups of matrices over a finite field /Russian/, Doklady Akad. Nauk Belorus. SSR 5 /1961/ 535-537.

26. W.T. Tutte, A family of cubical graphs, Proc. Cambr. Phil. Soc. 43 /1947/ 459-474

27. A.J. Weir, Sylow p-subgroups of the classical groups over finite fields with characteristic prime to p, Proc. A.M.S. 6 /1955/ 529-533.

28. H. Wielandt, Finite Permutation groups, Acad. Press, London 1964.

29. V.N. Zemlyachenko, Canonical numbering of trees /Russian/, Proc. Seminar on Comb. Anal. at Moscow State Univ., Moscow 1970.

30. —, On algorithms of graph identification /Russian/, Questions of Cybernetics 15, Proc. 2nd All-Union Seminar on Combinat. Math., Moscow /1975/ 33-41.

31. —, Determination of isomorphism of graphs /Russian/, in Mathem. Problems of Modelling Complex Objects, Karelian Branch of the Acad. Sci. USSR, Petrozavodsk 1979, 57-64.

32. —, Polynomial algorithm of identification of almost all graphs /Russian/, Karelian Branch of the Acad. Sci. USSR, in print.

AN ALGEBRAIC DEFINITION OF ATTRIBUTED TRANSFORMATIONS

by Miklós Bartha
Department of Computer Science
University of Szeged
Szeged, Aradi v. tere 1., H-6720

Abstract

A general concept of attributed transformation is introduced. It is shown that the domain of attributed tree transformations is a regular tree language, and an alternative proof is given for the decidability of the K-visit property of deterministic attributed tree transducers. Finally some closure properties are investigated concerning the composition of attributed tree transformations.

Introduction

The purpose of this paper is to present an algebraic treatment of attributed transformations. We assume a knowledge of magmoids [1,2] and rational algebraic theories, e.g. [3]. We mostly follow these works in notation, too. That is, if Σ is a finite ranked alphabet, the free projective (decomposable) magmoid generated by Σ is denoted by $T(\Sigma)$ ($\widetilde{T}(\Sigma)$, respectively). The magmoid of torsions or base morphisms is denoted by Θ. $R(\Sigma)$ denotes the free rational algebraic theory generated by Σ, and $\mathrm{Rec}(\Sigma)$ denotes the smallest rational algebraic theory in $PT(\Sigma)$ that contains $P_F T(\Sigma)$ as a submagmoid (or sub-algebraic theory, because $P_F T(\Sigma)$ is projective). It is natural that we use tensor product: $\otimes$ as a basic operation, and source-tupling: $\langle\!\langle \ldots \rangle\!\rangle$ as a derived one. Iteration is denoted by $^{+}$.

Definition 1. Let R be a rational algebraic theory, $k \geq 1$, $l \geq 0$ integers. Define $R(k,l) = (\{R(k,l)^{p}_{q} \mid p,q \geq 0\}, \cdot, \otimes, e, e_0)$ to be the following structure:

(i) $R(k,l)^{p}_{q} = R^{k\cdot p + l\cdot q}_{k\cdot q + l\cdot p}$.

(ii) If $a \in R(k,l)^{p}_{q}$, $b \in R(k,l)^{q}_{r}$, then

$a.b = \langle\!\langle \mu^{k\cdot p}, \nu_{l\cdot r} \rangle\!\rangle . \langle\!\langle a.\theta_{p,q,r}, b.\psi_{p,q,r} \rangle\!\rangle^{+}$, where

μ^{n}_{m} ($=\mu^{n}$ if m is understood) $= id_n \otimes 0_m \in \Theta^{n}_{n+m}$,

ν^{n}_{m} ($=\nu_{m}$ if n is understood) $= 0_n \otimes id_m \in \Theta^{m}_{n+m}$,

$\theta_{p,q,r} = \nu^{k\cdot p + l\cdot q}_{k\cdot q} \otimes \nu^{(k+l)\cdot r}_{l\cdot p}$,

$\psi_{p,q,r} = 0_{k\cdot p} \otimes \langle\!\langle \nu^{(k+l)\cdot q + l\cdot r}_{k\cdot r}, \mu^{l\cdot q}_{k\cdot q + (k+l)\cdot r} \rangle\!\rangle \otimes 0_{l\cdot p}$.

See also the figure below:

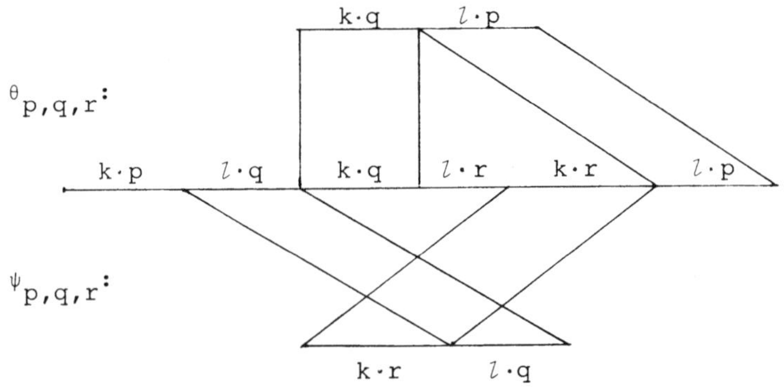

(iii) If $a \in R(k,l)^{p_1}_{q_1}$, $b \in R(k,l)^{p_2}_{q_2}$, then

$$a \otimes b = \langle \mu^{k\cdot p_1}_{l\cdot q_1} \otimes \mu^{k\cdot p_2}_{l\cdot q_2}, \nu^{k\cdot p_1}_{l\cdot q_1} \otimes \nu^{k\cdot p_2}_{l\cdot q_2} \rangle . (a \otimes b) . \langle \mu^{k\cdot q_1}_{l\cdot p_1} \otimes \mu^{k\cdot q_2}_{l\cdot p_2}, \nu^{k\cdot q_1}_{l\cdot p_1} \otimes \nu^{k\cdot q_2}_{l\cdot p_2} \rangle^{-1}.$$

(iv) $e = id_{k+l}$, $e_0 = 0_0$.

(We shall never add any distinctive mark to the sign of the operations when working in different magmoids in the same time, because only one interpretation is reasonable anywhere in the context.)

Theorem 2. $R(k,l)$ is a magmoid.

All the requirements can be proved by a computation in a free rational algebraic theory generated by an appropriate finite ranked alphabet.

Let $\chi : R \to R'$ be a homomorphism between rational algebraic theories. Clearly, χ defines a homomorphism $\chi(k,l) : R(k,l) \to R'(k,l)$, and so the operator (k,l) becomes a functor.

Definition 3. An attributed transducer is a 6-tuple: $\mathfrak{A} = (\Sigma, R, k, l, h, S)$, where

(i) Σ is a finite ranked alphabet, $S \notin \Sigma$,

(ii) R is a rational algebraic theory, $k \geq 1$, $l \geq 0$ are integers,

(iii) $h : \Sigma_S \to DR(k,l)$ is a ranked alphabet map, where $\Sigma_S = \Sigma \cup \{S\}$ with S having rank 1, and $h(S) = a \otimes 0_l$ for some $a \in R^{k+l}_k$. We say that $h(S)$ is a synthesizer.

$\tau_{\mathfrak{A}} : \tilde{T}(\Sigma)^1_0 \to R^1_0$, the transformation induced by $\mathfrak{A}$ is the following function: $\tau_{\mathfrak{A}}(t) = a$, where $\pi^1_k . h(S(t)) = a \otimes 0_l$. It is clear that $\tau_{\mathfrak{A}}(t)$ is uniquely determined by this imlicit form. (As it is usual, we denoted the unique homomorphic extension of h also by h.)

Definition 4. An attributed tree transducer is a 6-tuple: $\mathfrak{A}=(\Sigma,\Delta,k,l,h,S)$, where Σ,k,l and S are as in the previous definition, Δ is a finite ranked alphabet, $h:\Sigma_S\to P_FT(\Delta)$ is such that $h((\Sigma_S)_n)\subseteq P_FT(\Delta)^{k+l\cdot n}_{k\cdot n+l}$ and $h(S)\in P_FT(\Delta)^{k+l}_{k}$.

To define the transformation $\tau_{\mathfrak{A}}$, consider the attributed transducer $\mathfrak{B}=(\Sigma,\mathrm{Rec}(\Delta),k,l,h,S)$. $\mathfrak{B}$ is correct, since $P_FT(\Delta)\subseteq\mathrm{Rec}(\Delta)$ and h(S) is a synthesizer. Now, $\tau_{\mathfrak{A}}=\{\langle t,u\rangle \mid t\in\tilde{T}(\Sigma)^1_0,\ u\in\tau_{\mathfrak{B}}(t)\}$. $a\in PT(\Delta)^p_q$ is called deterministic (completely defined) if for each $i\in[p]=\{1,\dots,p\}$, $\pi^i_p\cdot a$ contains at most (at least) one element. Let a be deterministic. Then a is linear if none of the variables $x_1,\dots,x_q$ occur more than once in a. We say that $\mathfrak{A}$ is deterministic (completely defined, deterministic and linear) if for all $\sigma\in\Sigma_S$, $h(\sigma)$ is deterministic (completely defined, deterministic and linear, respectively).

Definition 4 might be interpreted as follows. Let $t\in\tilde{T}(\Sigma)^1$, α a node in t having some label $\sigma\in\Sigma_n$. A component of $h(\sigma)$ describes how to compute the value of a synthesized attribute at α (the first k components), or an inherited attribute at an immediate descendant of α (the last $l\cdot n$ components) as a function (polynomial) of the synthesized attributes at the immediate descendants (the variables $x_1,\dots,x_{k\cdot n}$) and the inherited attributes at α itself (the variables $x_{k\cdot n+1},\dots,x_{k\cdot n+l}$). The role of the synthesizer h(S) is to produce the final result of the computation.

It will be convenient to identify the nodes of a tree $t\in\tilde{T}(\Sigma)^1_q$ with the set $\mathrm{nds}(t)\subseteq\mathbb{N}^*\times(\Sigma\cup X_q)$, and the leaves of t with $\mathrm{lvs}(t)\subseteq\mathbb{N}^*\times X_q$ as follows:

(i) if $t=x_1$, then $\mathrm{nds}(t)=\mathrm{lvs}(t)=\{\langle\lambda,x_1\rangle\}$,

(ii) if $t=t_0\cdot(id_{p-1}\otimes\sigma(x_1,\dots,x_n)\otimes id_{q-p})$ with $t_0\in\tilde{T}(\Sigma)^1_q$, $q\geq1$, $p\in[q]$, $n\geq0$, $\sigma\in\Sigma_n$, then $\mathrm{nds}(t)=\bigcup_{i=1}^{5}V_i$, where

$V_1=\{\langle w,x_j\rangle \mid j\in[p-1] \text{ and } \langle w,x_j\rangle\in\mathrm{lvs}(t_0)\}$,

$V_2=\{\langle w,x_j\rangle \mid j\geq p+n \text{ and } \langle w,x_{j-n+1}\rangle\in\mathrm{lvs}(t_0)\}$,

$V_3=\{\langle wj,x_{p+j-1}\rangle \mid j\in[n] \text{ and } \langle w,x_p\rangle\in\mathrm{lvs}(t_0)\}$,

$V_4=\mathrm{nds}(t_0)\setminus\mathrm{lvs}(t_0)$,

$V_5=\{\langle w,\sigma\rangle\}$, where $\langle w,x_p\rangle\in\mathrm{lvs}(t_0)$;

$\mathrm{lvs}(t)=V_1\cup V_2\cup V_3$.

It is easy to verify that nds(t) and lvs(t) are uniquely defined by the above construction, and for each $w\in\mathbb{N}^*$ there exists at most one $\alpha\in\mathrm{nds}(t)$ having w as first component. Clearly, $\|\mathrm{nds}(t)\|=r(t)$, the number of nodes in t.

Let $\mathfrak{A}=(\Sigma,\Delta,k,l,h,S)$ be an attributed tree transducer, fixed in the rest of the paper. Let $t\in\tilde{T}(\Sigma)^1_q$, $Z_t=\{x(\alpha,i),y(\alpha,m) \mid \alpha\in\mathrm{nds}(t),i\in[k],m\in[l]\}$ a set of variable symbols. Construct a system $E_{t,h}$ of nondeterministic

Δ-equations over the variables Z_t as follows:

$$E_{t,h}=\{E_{x,h}(\alpha,i)\mid \alpha\in nds(t)\setminus lvs(t),\ i\in[k]\}\cup$$
$$\cup\{E_{y,h}(\alpha,m)\mid \alpha\in nds(t)\setminus\{<\lambda,root(t)>\},m\in[l]\},$$

where

(i) If $\alpha=<w,\sigma>$ with $\sigma\in\Sigma_n$ and

$$h(\sigma)=<T_1,\ldots,T_k,\ Q_1,\ldots,Q_{l\cdot n}>, \qquad (1)$$

then the equation $E_x(\alpha,i)$ is of the form:

$$x(\alpha,i)=T_i[x_{k\cdot(r-1)+p}\leftarrow x(\alpha_r,p),x_{k\cdot n+s}\leftarrow y(\alpha,s)\mid p\in[k],\ r\in[n],\ s\in[l]],$$

where $\leftarrow$ denotes variable substitution, $\alpha_r\in nds(t)$ is the unique node having wr as first component. (We omitted the index h, which is fixed.)

(ii) If $\alpha=<wj,a>$ with $a\in\Sigma\cup X_q$, then consider the unique node $\bar\alpha=<w,\sigma>$, where $\sigma\in\Sigma_n$, $n\geq j$, and the nodes $\bar\alpha_r$, $r\in[n]$. (Naturally $\bar\alpha_j=\alpha$.) Let $h(\sigma)$ be as (1) above. Then the equation $E_y(\alpha,m)$ looks as:

$$y(\alpha,m)=Q_{l\cdot(j-1)+m}[x_{k\cdot(r-1)+p}\leftarrow x(\bar\alpha_r,p),\ x_{k\cdot n+s}\leftarrow y(\alpha,s)\mid p\in[k],\ r\in[n],\ s\in[l]].$$

The variables:

$$Z_t^1=\{x(\alpha,i)\mid \alpha\in lvs(t),\ i\in[k]\}\cup\{y(<\lambda,root(t)>,m)\mid m\in[l]\}$$

do not occur on the left-hand side of these equations, so they are considered as parameters. On the other hand, the variables:

$$Z_t^2=\{x(<\lambda,root(t)>,i)\mid i\in[k]\}\cup\{y(\alpha,m)\mid \alpha\in lvs(t),m\in[l]\}$$

do not occur on the right-hand side of the equations. It can be proved that for any $t\in\tilde T(\Sigma)_0^1$, $\tau_{\mathfrak{A}}(t)$ equals to the $x(<\lambda,S>,1)$ component of $E^+_{S(t)}$. This result links our work to [4], where the same technic was used to define the semantics of attribute grammars.

Now we turn our attention to the domain of $\tau_{\mathfrak{A}}$, that is the set $D\tau_{\mathfrak{A}}=\{t\in\tilde T(\Sigma)_0^1\mid$ for some $u\in T(\Delta)_0^1 <t,u>\in\tau_{\mathfrak{A}}\}$. Let $G(k,l)$ be the following finite set:

$G(k,l)=\{(G;V_{1,1},V_{1,2},V_{2,1},V_{2,2})\mid G=(V,E)$ is a directed acyclic bipartite graph, and

(i) $V=V_1\cup V_2$, $V=[k+l]$, $V_1=[k]$, $V_2=V\setminus V_1$, $E=E_1\cup E_2$, $dom(E_1)\subseteq V_1$, $dom(E_2)\subseteq V_2$,

(ii) $V_1=V_{1,1}\cup V_{1,2}$, $V_{1,1}\cap V_{1,2}=\emptyset$; $V_2=V_{2,1}\cup V_{2,2}$, $V_{2,1}\cap V_{2,2}=\emptyset$,

(iii) for each $j\in V_{2,1}$ there exists an $i\in V_{1,1}$ such that $<i,j>\in E_1$ and the vertices $V_{1,2}\cup V_{2,2}$ are isolated.$\}$. We construct a finite top-down tree automaton $\mathcal{B}$, that operates nondeterministically on $\tilde T(\Sigma)^1$ with states $A=G(k,l)$. The fact that, starting from state a_0, $\mathcal{B}$ is able to reach the vector of states $<a_1,\ldots,a_q>$ on input $t\in\tilde T(\Sigma)_q^1$ will be denoted by $a_0t\vdash^*_{\mathcal{B}}t(a_1,\ldots,a_q)$. If for some $\sigma\in\Sigma_q$ $t=\sigma(x_1,\ldots,x_q)$, we simply write $a_0\sigma\vdash_{\mathcal{B}}\sigma(a_1,\ldots,a_q)$.

Let $\sigma\in(\Sigma_S)_n$, $h(\sigma)=<T_1,\ldots,T_{k+l\cdot n}>$, $I_\sigma=\{i\in[k+l\cdot n]\mid T_i=\emptyset\}$. The set of alternatives of σ is:

$A[\sigma]=\{<t_1,\ldots,t_{k+l\cdot n}>|$ if $i\in I_\sigma$, then $t_i=\perp$, else $t_i\in T_i\}$.
We say that $c\in A[S]$ realizes the initial state $a=(G_c;V^c_{1,1},\ldots,V^c_{2,2})$ if the following conditions are satisfied. If $c=<t_1,\ldots,t_{k+l}>$, then

(a) if $j>k$ and $<j,i>\in E^c_2$, then x_i occurs in t_j, and conversely, if $j\in V^c_{2,1}$, $i\in[k]$ and x_i occurs in t_j, then $j\vdash^+_{G_c} i$ (where $\vdash^+_{G_c}$ denotes the transitive closure of $\vdash_{G_c}=E_c$);

(b) $V^c_{1,1}\supseteq Q=\{i\in[k]|x_i$ occurs in $t_1\}$, and for each $i\in V^c_{1,1}\backslash Q$ there exists an $i'\in Q$ such that $i'\vdash^+_{G_c} i$;

(c) $V^c_{2,2}\supseteq\{j>k|j\in I_S\}$.

Define the set of initial states of $\mathfrak{S}$ as $A_0=\{a\in A|a$ is realized by some $c\in A[S]\}$.

Let $n\geq 0$, $\sigma\in\Sigma_n$, $a_0,\ldots,a_n\in A$, $a_m=(G_m;V^m_{1,1},\ldots,V^m_{2,2})$ for each $0\leq m\leq n$, and $c=<t_1,\ldots,t_{k+l\cdot n}>\in A[\sigma]$. Construct the graph $G[c,a_0,\ldots,a_n]$ by adding the edges $E[c,a_0,\ldots,a_n]$ to the disjoint union of graphs G_m, $0\leq m\leq n$. An edge $<<i,m_1>,<j,m_2>>\in E[c,a_0,\ldots,a_n]$ if and only if one of the following conditions is satisfied:

(i) $m_1=m_2=0$, $i\in V^0_{1,1}$, $j>k$ and $x_{k\cdot n+(j-k)}$ occurs in t_i;

(ii) $m_1=0$, $m_2\geq 1$, $i\in V^0_{1,1}$, $j\leq k$ and $x_{k\cdot(m_2-1)+j}$ occurs in t_i;

(iii) $m_1\geq 1$, $m_2=0$, $i\in V^{m_1}_{2,1}$, $j>k$ and $x_{k\cdot n+(j-k)}$ occurs in $t_{k+l\cdot(m_1-1)+(i-k)}$;

(iv) $m_1\geq 1$, $m_2\geq 1$, $i\in V^{m_1}_{2,1}$, $j\leq k$ and $x_{k\cdot(m_2-1)+j}$ occurs in $t_{k+l\cdot(m_1-1)+(i-k)}$.

We say that c realizes the transition $a_0\sigma\vdash_{\mathfrak{S}}\sigma(a_1,\ldots,a_n)$ if $G[c,a_0,\ldots,a_n]$ is acyclic, and the following conditions are satisfied. (The mark $[c,a_0,\ldots,a_n]$ will be omitted from the right of G and E)

(A) If $i\in I_\sigma$, then $i\in V^0_{1,2}\cup(\bigcup_{m=1}^{n} V^m_{2,2})$.

(B) If $<<i,m_1>,<j,m_2>>\in E$, then $j\in V^{m_2}_{1,1}\cup V^{m_2}_{2,1}$.

(C) If $<i,j>\in E^0_1$, then $<i,0>\vdash^+_G <j,0>$ on such a path that consits of edes $E\cup(\bigcup_{s=1}^{n} E^s_1)\cup E^0_2$ only.

(D) If $<i,0>\vdash^+_G <j,0>$ for some $i\in V^0_{1,1}$, $j\in V^0_{2,2}$, then $i\vdash^+_{G_0} j$.

(E) If $<j,i>\in E^m_2$ and $m\geq 1$, then $<j,m>\vdash^+_G <i,m>$ on a path consisting of edges $E\cup(\bigcup_{s=1}^{n} E^s_1)\cup E^0_2$ only.

(F) If $<j,m>\vdash^+_G <i,m>$ for some $m\geq 1$, $j\in V^m_{2,1}$, $i\in V^m_{1,1}$, then $j\vdash^+_{G_m} i$.

(G) For each $m \geq 1$, $i \in V^m_{1,1}$, there exists an $i' \in V^O_{1,1}$ such that $<i',O> \vdash^+_G <i,m>$ on a path consisting of edges $E \cup (\bigcup_{s=1}^{n} E^s_1)$ only.

Now for each $\sigma \in \Sigma_n$, $a_0 \sigma \vdash_{\mathcal{B}} \sigma(a_1,\ldots,a_n)$ if and only if this transition is realized by some $c \in A[\sigma]$.

Let $q \geq 0$, $t \in \tilde{T}(\Sigma)^1_q$. A deterministic part of $E_{S(t)}$ can be chosen as follows. Replace the equations of the form $z = \emptyset$ by $z = z$, then for each $z \in Z_{S(t)} \setminus Z^1_{S(t)}$, replace the right-hand side of the equation $z = T_z$ by an arbitrary $t_z \in T_z$. Further on $DE_{S(t)}$ will always denote a deterministic part of $E_{S(t)}$. For each $z \in Z_{S(t)} \setminus Z^1_{S(t)}$, $\pi(z).E^+_{S(t)} \neq \emptyset$ if and only if there exists a $DE_{S(t)}$ such that $\pi(z).DE^+_{S(t)} \neq \emptyset$. ($\pi(z).$ means the selection of the component z.) Let $\vdash_{DE_{S(t)}}$ denote the dependence relation among the variables $Z_{S(t)}$ in a deterministic part of $E_{S(t)}$, that is, $z_1 \vdash_{DE_{S(t)}} z_2$ if and only if z_2 occurs in t_{z_1}. It is clear that $\pi(z).DE_{S(t)} \neq \emptyset$ if and only if $z \not\vdash^+_{DE_{S(t)}} z$, and $z \vdash^+_{DE_{S(t)}} z'$ implies $z' \not\vdash^+_{DE_{S(t)}} z'$.

For each $n \in [l]$ take a new symbol γ_n, and construct the ranked alphabet $\Gamma = \bigcup_{n=1}^{l} \Gamma_n$ with $\Gamma_n = \{\gamma_n\}$. Let $q \geq 0$, $t \in \tilde{T}(\Sigma)^1_q$, $a_1,\ldots,a_q \in A$, $a_j = (G_j; V^j_{1,1},\ldots,V^j_{2,2})$ for each $j \in [q]$. By $E_t[a_1,\ldots,a_q]$ we mean the following system of equations:

$$E_t[a_1,\ldots,a_q] = \{x(<w,x_j>,i) = \gamma_n(y(<w,x_j>,m_1),\ldots,y(<w,x_j>,m_n)) \mid j \in [q], <w,x_j> \in lvs(S(t)),\ i \in [k] \text{ and } m_1,\ldots,m_n \text{ are all the possible values of such an } m \text{ for which } <i,k+m> \in E^j_1\}.$$

Lemma 5. Let $q \geq 0$, $t \in \tilde{T}(\Sigma)^1_q$, $a_1,\ldots,a_q \in A$ and for each $j \in [q]$ $a_j = (G_j; V^j_{1,1},\ldots,V^j_{2,2})$. There exists an $a \in A_0$ for which $a\, t \vdash^*_{\mathcal{B}} t(a_1,\ldots,a_q)$ if and only if a $DE_{S(t)}$ can be chosen such that:

(i) $\pi(x(<\lambda,S>,1)).(DE_{S(t)} \cup E_t[a_1,\ldots,a_q])^+ \neq \emptyset$,

(ii) for each $j \in [q]$, $<w,x_j> \in lvs(S(t))$, $i \in [k]$, $x(<\lambda,S>,1) \vdash^+ x(<w,x_j>,i)$ holds in $DE_{S(t)} \cup E_t[a_1,\ldots,a_q]$ if and only if $i \in V^j_{1,1}$,

(iii) for each $m \in [l]$,
$y(<w,x_j>,m) \vdash^+ x(<w,x_j>,i)$ if and only if $m+k \vdash^+_{G_j} i$.

Proof.

Only if: If $t = x_1$, then $a = a_1 \in A_0$. In this case $E_{S(t)}$ is the same as $h(S)$, written in the form of equations, so (i), (ii) and (iii) follow from the conditions (a), (b) and (c) that must hold for $a \in A_0$. Let $q \geq 1$, $p \in [q]$,

$n\geq 0$, $\sigma\in\Sigma_n$, $t_0\in\tilde{T}(\Sigma)^1_q$ and $t=t_0\cdot(id_{p-1}\otimes\sigma(x_1,\ldots,x_n)\otimes id_{q-p})$. If a $t\overset{+}{\vdash_{\mathcal{L}}}$ $\overset{+}{\vdash_{\mathcal{L}}} t(a^1,\ldots,a^{p-1}, a_1,\ldots,a_n,a^{p+1},\ldots,a^q)$, then there exists an $a_0\in A$ such that a $t_0\overset{*}{\vdash_{\mathcal{L}}} t_0(a^1,\ldots,a^{p-1},a_0,a^{p+1},\ldots,a^q)$ and $a_0\sigma\vdash_{\mathcal{L}}\sigma(a_1,\ldots,a_n)$. Suppose the Only if part is true for t_0 and states $a^1,\ldots,a^{p-1},a_0,$ $a^{p+1},\ldots,a^q$, and the transition $a_0\sigma\vdash_{\mathcal{L}}\sigma(a_1,\ldots,a_n)$ is realized by $c=\langle t_1,\ldots,t_{k+l\cdot n}\rangle\in A[\sigma]$. Then there exists an appropriate $DE_{S(t_0)}$ satisfying the three conditions. For each $i\in[k]$, $m\in[l]$, replace the variables $x(\langle w,x_p\rangle,i)$ and $y(\langle w,x_p\rangle,m)$ in $DE_{S(t_0)}$ by $x(\langle w,\sigma\rangle,i)$ and $y(\langle w,\sigma\rangle,m)$, respectively, and add the set of equations:

$$\{x(\langle w,\sigma\rangle,i)=t_i[x_{k\cdot(j-1)+r}\leftarrow x(\langle wj,x_{j+p-1}\rangle,r),x_{k\cdot n+s}\leftarrow y(\langle w,\sigma\rangle,s),$$
$$\bot\leftarrow x(\langle w,\sigma\rangle,i)\mid j\in[n],r\in[k],s\in[l]]\mid i\in[k]\}\cup\{y(\langle wj,x_{j+p-1}\rangle,m)=$$
$$=t_{k+l\cdot(j-1)+m}[x_{k\cdot(u-1)+r}\leftarrow x(\langle wu,x_{u+p-1}\rangle,r),x_{k\cdot n+s}\leftarrow y(\langle w,\sigma\rangle,s),$$
$$\bot\leftarrow y(\langle wj,x_{j+p-1}\rangle,m)\mid u\in[n],r\in[k],s\in[l]]\mid j\in[n],m\in[l]\}$$

to obtain $DE_{S(t)}$. By the inductive hypothesis and condition (A) imposed on the transitions of $\mathcal{L}$, if for some $z\in Z_{S(t)}$ $x(\langle\lambda,S\rangle,1)\overset{+}{\vdash} z$ holds in $DE_{S(t)}\cup E_t[a^1,\ldots,a^{p-1},a_1,\ldots,a^{p-1},a_1,\ldots,a_n,a^{p+1},\ldots,a^q]$, then $z\not\vdash z$. Suppose $z\overset{+}{\vdash} z$. Then, it can easily be seen that it is possible to choose $z=x(\langle w,\sigma\rangle,i)$ for some $i\in[k]$. This assumption, however, would lead to a contradiction using the inductive hypothesis and condition (D), so (i) is proved. The Only if part of (ii) is a consequence of condition (B). If $p\leq j<p+n$, then the If part of (ii) follows from (G), else - i.e. if $j\in[p-1]$ or $p+n\leq j\leq q-1+n$ - it follows from (C). (iii) is a consequence of (E) and (F).

<u>If:</u> The case $t=x_1$ is again trivial. Let $t=t_0\cdot(id_{p-1}\otimes\sigma(x_1,\ldots,x_n)\otimes id_{q-p})$ as above, and suppose the If part is true for t_0 and any appropriate states $b_1,\ldots,b_q$. Let $DE_{S(t)}$ and states $a^1,\ldots,a^{p-1},a_1,\ldots,a_n,a^{p+1},\ldots$ $\ldots,a^q$ satisfy (i), (ii) and (iii). Split $DE_{S(t)}$ into $DE_{S(t_0)}$ and a part that can be derived from $c=\langle t_1,\ldots,t_{k+l\cdot n}\rangle\in A[\sigma]$. Let $a_0=(G_0;$ $V^0_{1,1},\ldots,V^0_{2,2})$ be the following state:

$i\in V^0_{1,1}$ if and only if $x(\langle\lambda,S\rangle,1)\overset{+}{\vdash} x(\langle w,\sigma\rangle,i)$ holds in $DE_{S(t)}\cup$ $\cup E_t[a^1,\ldots,a^{p-1},a_1,\ldots,a_n,a^{p+1},\ldots,a^q]$, where w is the first component of the node $\langle w,x_p\rangle$ in $S(t_0)$;

$\langle i,j\rangle\in E^0_1$ if and only if $i\in V^0_{1,1}$ and $x(\langle w,\sigma\rangle,i)\overset{+}{\vdash} y(\langle w,\sigma\rangle,j-k)$, $V^0_{2,1}=\{j\mid$ for some $i\in V^0_{1,1}\langle i,j\rangle\in E^0_1\}$;

$\langle j,i\rangle\in E^0_2$ if and only if $j\in V^0_{2,1}$ and $y(\langle w,\sigma\rangle,j-k)\overset{+}{\vdash} x(\langle w,\sigma\rangle,i)$.

It is clear that $DE_{S(t_0)}$ and states $a^1,\ldots,a^{p-1},a_0,a^{p+1},\ldots,a^q$ satisfy (i), (ii) and (iii), hence, by the inductive hypothesis a $t_0\overset{*}{\vdash_{\mathcal{L}}}$

$\vdash_{\mathcal{B}}^{*} t_0(a^1,\dots,a^{p-1},a_0,a^{p+1},\dots,a^q)$ for some $a\in A_0$. On the other hand, it can easily be checked that $a_0\sigma\vdash_{\mathcal{B}}\sigma(a_1,\dots,a_n)$ is realized by c, so we are through.

Taking q=0 in the lemma we get

Theorem 6. The domain of attributed tree transformations is a regular tree language.

However, Lemma 5 is worth some further considerations. It can be seen that Lemma 5 remains valid if we require the states of $\mathcal{B}$ not contain any redundant edges. (an edge $\langle i,j\rangle$ is redundant if there is another path from i to j containing more than one edge.) Let $\mathfrak{A}$ be deterministic, and suppose the states of $\mathcal{B}$ satisfy the above additional requirement. The following statement can be proved by a bottom-up type induction combined with Lemma 5.

Proposition 7. Let $t\in D\tau_{\mathfrak{A}}$, $t=t_0.u$ with $t_0\in\tilde{T}(\Sigma)_1^1$. There exists a unique $a\in A$ such that for some $a_0\in A_0$ we have $a_0t_0\vdash_{\mathcal{B}}^{*} t_0(a)$ and $a\,u\vdash_{\mathcal{B}}^{+} u$. This unique $a=(G;V_{1,1},\dots,V_{2,2})$ is the following: $V_{1,1}\cup V_{2,1}\cong Z_\alpha=\{z\in Z_{S(t)}\mid$ the "node" index of z is $\alpha=\text{root}(u)$ and $x(\langle\lambda,S\rangle,1)\vdash_{DE_{S(t)}}^{+} z\}$, and $\vdash_G^{+}\cong$ $\cong\vdash_{DE_{S(t)}}^{+}/Z_\alpha$. (Obviously, $DE_{S(t)}$ is unique in this case.)

As an application of Proposition 7 we show how to decide the K-visit property of deterministic attributed tree transducers. (Alternative proofs can be derived from [5] and [6].) Let $t\in D\tau_{\mathfrak{A}}$, $\alpha=\langle w,\sigma\rangle\in nds(t)$. Proposition 7 shows that the state $a=(G_\alpha;V^\alpha_{1,1},\dots,V^\alpha_{2,2})$ in which $\mathcal{B}$ passes through α during the recognition of t is uniquely determined, and it describes the dependence relation among the useful attributes of α. If p is a path in G_α ($p\in\text{path}(G_\alpha)$), then let $v_p=\|\{i\in V^\alpha_{1,1}\mid p$ passes through i and $\pi^i_{k+l\cdot n}.h(\sigma)\neq 0_{k\cdot n}\otimes a$ for any $a\in T(\Delta)^1_l\}\|$. Let $v_\alpha=\max\{v_p\mid p\in$ $\in\text{path}(G_\alpha)\}$. v_α shows how much times we must "enter" the subtree below α to ask for the value of certain attributes. (Supposing an optimal, maximally paralleled evaluation of the useful attributes.) Define $v_{\mathfrak{A}}=\max\{v_\alpha\mid\alpha\in nds(t)$ for some $t\in D\tau_{\mathfrak{A}}\}$. It is easy to give an algorithm that computes $v_{\mathfrak{A}}$, and obviously, $\mathfrak{A}$ is K-visit if and only if $v_{\mathfrak{A}}\leq K$. Moreover, it follows from the construction that: if $l<k$, then $v_{\mathfrak{A}}\leq l+1$, else $v_{\mathfrak{A}}\leq k$.

Again, let $\mathfrak{A}$ be deterministic, and consider the unique $\varepsilon:R(\Delta)\to$ $\to\text{Rec}(\Delta)$ homomorphic extension of the ranked alphabet map ε, by which $\varepsilon(\delta)=\{\delta(x_1,\dots,x_n)\}$ for $\delta\in\Delta_n$. For every $n\geq 0$, $\sigma\in(\Sigma_S)_n$ let $h'(\sigma)\in T(\Delta_\perp)$ such that $\varepsilon(h'(\sigma))=h(\sigma)$. $\mathfrak{A}$ can be characterized by the attributed transducer $\mathfrak{A}'=(\Sigma,R(\Delta),k,l,h',S)$ as follows:

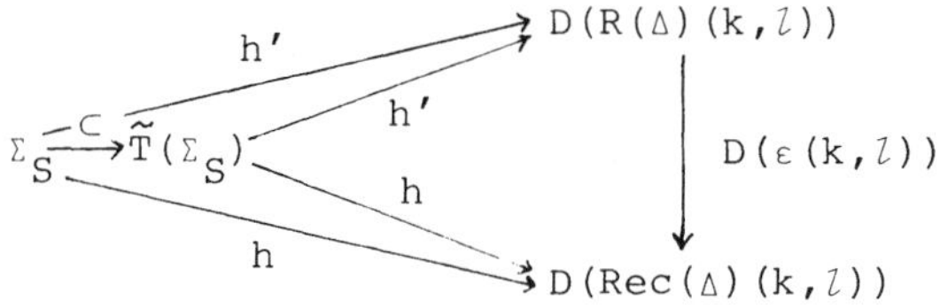

That's why we always consider a deterministic attributed tree transducer from Σ into Δ as an attributed transducer from Σ into $R(\Delta)$. (keeping in mind the homomorphism ε.) A deterministic and completely defined attributed tree transducer is called noncircular, if $h(S(t))\in T(\Delta)$ for any $t\in\tilde{T}(\Sigma)$[1].

<u>Theorem 8.</u> Let $\mathfrak{A}=(\Sigma,R(\Delta),k,l,h,S)$ be a deterministic, completely defined and noncircular attributed tree transducer, $\phi:T(\Delta)\to k'\text{-dil}\ T(\Gamma)$ a deterministic and completely defined top-down tree transformation. Then $\tau_{\mathfrak{A}}\circ\phi_{[1]}$ is a deterministic, completely defined and noncircular attributed tree transformation.

<u>Proof.</u> $\tau_{\mathfrak{A}}\circ\phi_{[1]}=\tau_{\mathfrak{B}}$, where $\mathfrak{B}=(\Sigma,R(\Gamma),k\cdot k',l\cdot k',h',S)$ and h' is defined by the following two diagrams:

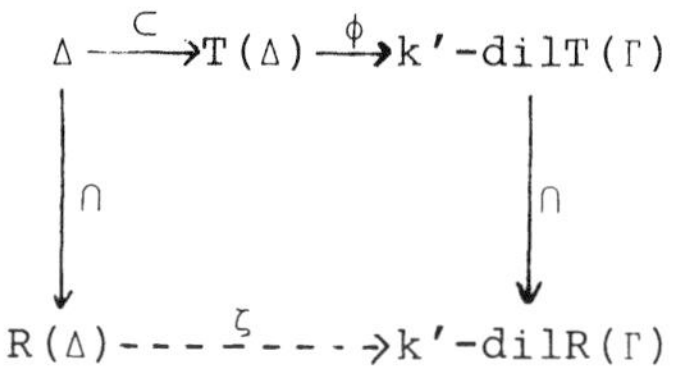

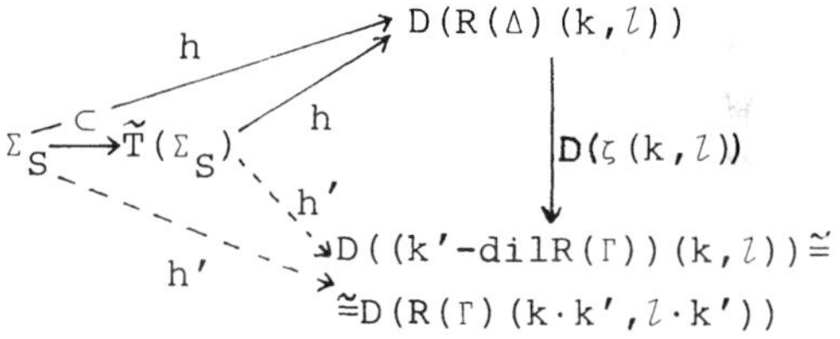

The following theorem can be proved similarly.

<u>Theorem 9.</u> Let $\mathfrak{A}$ be as in the previous theorem, $\phi:T(\Delta)\to k'\text{-dil}P_F T(\Gamma)$ an arbitrary top-down tree transformation. Then $\tau_{\mathfrak{A}}\circ\phi_{[1]}$ is an attributed tree transformation.

If ϕ is nondeleting in Theorems 8 and 9, then the conditions: completely defined and noncircular can be dropped.

The following theorem will be proved in a forthcoming paper.

<u>Theorem 10.</u> The class of all deterministic, completely defined and linear attributed tree transformations as well as the class of all deterministic and linear attributed tree transformations is closed under composition.

References

[1] Arnold, A., Dauchet, M., Theorie des magmoides, Preliminary work to the authors' theses, Univ. de Lille, France, 1977.

[2] Dauchet, M., Transductions de forets, bimorphismes de magmoides, These, Univ. de Lille, France, 1977.

[3] Wright, J.B., Thatcher, J.W., Wagner, E.G., Goguen, J.A., Rational algebraic theories and fixed-point solutions, 17-th IEEE Symposium on Foundations of Computing, Houston, 1976, pp. 147-158.

[4] Chirica, L.M., Martin, D.F., An order-algebraic Definition of Knuthian Semantics, Math. Systems Theory, v. 13, 1979, pp. 1-27.

[5] Riis, H., Skyum, S., k-Visit attribute grammars, DAIMI PB-121, Aarhus University, Denmark, 1980.

[6] Fülöp, Z., Attribute grammars and attributed tree transducers, 15-th National Scientific Conference for Students, Budapest, Hungary, 1981.

Analogies of PAL and COPY

Franz-Josef Brandenburg

Institut für Informatik, Universität Bonn
Wegelerstr. 6, 5300 Bonn, Federal Republic of Germany

Abstract:

The LIFO or pushdown principle and the FIFO or queue principle are compared in the framework of language theory. To this effect, languages which characteristically describe these principles are studied comparing the least cones or semiAFLs containing them. Although the classes of languages so obtained are different and often are incomparable, very interesting analogies are established between LIFO type languages and FIFO type languages. Thus our investigations show both, the common and the contrasting properties of LIFO and FIFO structures.

Introduction:

The LIFO or pushdown principle "last-in, first-out" is modelled in its simplest form by the language PAL = $\{w\bar{w} \mid w \in \{a,b\}^*,\ \bar{w}^R \in \{\bar{a},\bar{b}\}^*$ is the reversal of the barred copy of $w\}$. PAL describes the essential part of the behaviour of a one-turn pushdown automaton, which in terms of language theory means that PAL is a cone or a semiAFL generator of the class of languages recognized by nondeterministic one-turn pushdown automata. Accordingly, the FIFO or queue principle "first-in, first-out" in its simplest form is modelled by the language COPY = $\{w\bar{w} \mid w \in \{a,b\}^*,\ \bar{w} \in \{\bar{a},\bar{b}\}^*$ is the barred copy of $w\}$, and COPY describes the behaviour of a single reset machine [4]. Both languages have in common that they consist of pairs of equal strings, which are distinguished by bars on the second components, and are composed in reverse for PAL and straight for COPY. The storage structures modelled by PAL and COPY, respectively, make one full check for comparison and operate in a LIFO and in a FIFO manner.

The canonical generalizations of these storage structures are a pushdown stack and a queue, respectively. These allow many piece-by-piece checks for comparison. Their characteristic languages are the Dyck set $D_2'^*$ and FIFO, respectively. In common to LIFO and FIFO type storage structures are counters, and the characteristic languages are now $\{a^n\bar{a}^n \mid n \geq o\}$ and the Dyck sets over one pair of matching symbols. The various relationships between all these languages are summarized in Figure 1.

Generalizing to multiple storage, which in terms of algebraic language theory are the equivalent concepts of real time multitape machines, intersections of languages, or the shuffle of characteristic languages, the distinctive character of LIFO and FIFO type languages disappears somewhat. Surprisingly, using intersection one obtains finite hierarchies of three classes each, whose limits are MULTI-RESET [4], BNP [5], and NTIME(n) [3], respectively. We show that these classes can be defined by several combinations of our basic languages from above, which means that the storage structures modelled by these languages are equivalent when used with nondeterministic real time acceptors. On the other hand, MULTI-RESET $\subseteq$ BNP $\subseteq$ NTIME(n). Thus under intersection closure LIFO type languages dominate FIFO type languages. More relations between the languages so obtained are shown in Figure 2.

Clearly this paper studies some popular and very important languages as part of algebraic language theory. But it also contributes to a better understanding of some fundamental data structures, namely, a pushdown stack, which operates LIFO, and a queue, which operates FIFO, and some structures made up by their combination.

Preliminaries:

For the sake of brevity and clearness we restrict ourselves to the discussion of languages and their comparison, and we omit the interpretation of our results for the storage structures modelled by the languages. We assume that the reader is familiar with the basic concepts from language theory and refer to [1,11,17].

A cone (called trio in [11]) is a class of languages containing a nonempty language and closed under the operations of inverse homomorphism, nonerasing homomorphism, and intersection with regular sets. A semiAFL (AFL) is a cone closed under union (product, star). For a language L let $\mathcal{M}(L)$ denote the least cone containing L. $\mathcal{M}(L)$ is a principal cone with generator L, and L is called a characteristic language of $\mathcal{M}(L)$. Notice that $\mathcal{M}(L)$ is closed under union, and thus is a principal semiAFL, and that each language in $\mathcal{M}(L)$ can be represented as $g(h^{-1}(L) \cap R)$, where h (g) is a (nonerasing) homomorphism, and R is a regular set. Let $\mathcal{M}_\cap(L)$ denote the least intersection closed cone containing L.

DEFINITION: For languages L_1 and L_2, L_1 is called weaker than L_2, denoted $L_1 \leq L_2$, if $\mathcal{M}(L_1) \subseteq \mathcal{M}(L_2)$. L_1 and L_2 are equivalent, $L_1 \equiv L_2$, if $L_1 \leq L_2$ and $L_2 \leq L_1$, and L_1 and L_2 are incomparable, if neither is weaker than the other.

Notice that "$\leq$" induces a partial ordering on non-equivalent languages, and it is of particular interest (and widely unsolved) to determine the greatest lower bound of certain natural languages.

Here we compare languages in terms of cones or equivalently semiAFLs. A comparison based on other sets of operations can be done as well. We belief that cones provide a good insight into characteristic properties of languages and provide useful tools for an exact analysis. E.g., we are independent of certain variations of the languages, such as the choice of the alphabets, and we heavily make use thereof.

Let $\{a,\bar{a},b,\bar{b}\}$ be a fixed alphabet which is used in this paper. Here bars are used to indicate pairs of matching symbols, such as $(a,\bar{a})$ and $(b,\bar{b})$. The barring of symbols canonically extends to strings, such that $\bar{w} = \bar{a}_1\bar{a}_2\ldots\bar{a}_n$ is the barred copy of $w = a_1a_2\ldots a_n$. When it is appropriate we use k-fold bars on symbols and strings. Note that the fixed alphabet is no real restriction, since we are dealing with cones.

Next we introduce the basic languages of this paper.

NOTATION:

PAL $= \{w\bar{w}^R \mid w \in \{a,b\}^*,\ \bar{w}^R$ is the reversal of the barred copy of $w\}$.

COPY $= \{w\bar{w} \mid w \in \{a,b\}^*,\ \bar{w}$ is the barred copy of $w\}$.

$D_2'^*$ denotes the one-sided Dyck set over $\{a,\bar{a},b,\bar{b}\}$ (see [1,2,17]).

D_2^* denotes the two-sided Dyck set over $\{a,\bar{a},b,\bar{b}\}$ (see [1,2,17]).

$D_1'^*$ denotes the one-sided Dyck set over $\{a,\bar{a}\}$ (see [1,2,17]).

D_1^* denotes the two-sided Dyck set over $\{a,\bar{a}\}$ (see [1,2,17]).

C $= \{a^n\bar{a}^n \mid n \geq 0\}$.

E $= \{w \in \{a,\bar{a},b,\bar{b}\}^* \mid w = u_1v_1u_2v_2\ldots u_nv_n,\ u_i \in \{a,b\}^*,\ v_i \in \{\bar{a},\bar{b}\}^*,$
$\bar{u}_1\bar{u}_2\ldots\bar{u}_n = v_1v_2\ldots v_n\}$ (see [6,8,10]).

FIFO $= \{w \in E \mid w = u_1v_1\ldots u_nv_n$ with $u_i \in \{a,b\}^*,\ v_i \in \{\bar{a},\bar{b}\}^*$, and for $i=1,\ldots,n$ $|u_1\ldots u_i| \geq |v_1\ldots v_i|\}$ (see [6,8,24]).

Finally we define two related operations on strings and languages.

DEFINITION: For strings x and y define the shuffle of x and y by
$\text{shuf}(x,y) = \{u_1v_1u_2v_2\ldots u_nv_n \mid x = u_1u_2\ldots u_n$ and $y = v_1v_2\ldots v_n\}$. See e.g. [11]. The shuffle operation naturally extends to languages, denoted by $⧢(L_1,L_2,\ldots,L_n)$.

For a string $x\bar{y}$ with $x \in \{a,b\}^*$ and $\bar{y} \in \{\bar{a},\bar{b}\}^*$ define the merge of $x\bar{y}$ by $\text{merge}(x\bar{y}) = \text{shuf}(x,\bar{y})$, and extend the merge operation to languages $L \subseteq \{a,b\}^*\{\bar{a},\bar{b}\}^*$, denoted by $\mathfrak{M}(L)$.

The merge operation has been introduced in [6]. It can be extended to strings and languages, whose segments are unbarred, barred, double barred, etc. When applied to COPY the merge operation functions as a diagonal of $⧢(\{a,b\}^*,\{\bar{a},\bar{b}\}^*)$.

Comparison of PAL and COPY:

We first recall some characterizations of the cones generated by PAL and COPY.

THEOREM 1: For a language L the following are equivalent:

(i) L is in M(PAL).

(ii) L is accepted (in real time) by a pushdown automaton making one reversal.

(iii) L is generated by a linear context-free grammar.

(iv) There is a (nonerasing) homomorphism h and a regular set R such that $L = \{h(w)\, h(\bar{w})^R \mid w \in R\}$.

(v) There is a regular set $R \subseteq \Sigma^* \times \Sigma^*$ such that $L = \{xy^R \mid (x,y) \in R\}$.

M(PAL) is known as the class of linear context-free languages.

THEOREM 2: For a language L the following are equivalent:

(i) L is in M(COPY).

(ii) L is accepted (in real time) by a single reset machine (see [4]).

(iii) L is generated by a single reset grammar (see [18]) or by an equal matrix grammar of order two (see [22]).

(iv) There is a (nonerasing) homomorphism h and a regular set R such that $L = \{h(w)\, h(\bar{w}) \mid w \in R\}$.

(v) There is a regular set $R \subseteq \Sigma^* \times \Sigma^*$ such that $L = \{xy \mid (x,y) \in R\}$.

M(COPY) is the class of single reset languages from [4].

Further similarities of M(PAL) and M(COPY) are the closure under union, homomorphism, and reversal, the nonclosure under product, star, intersection, and complementation, the decidability of the membership, emptiness, and finiteness problems, and an intercalation theorem (pumping lemma) of the same appearance, which follows easily from (iv) and the pumping lemma for regular sets. Besides these similarities PAL and COPY are essentially different.

THEOREM 3: PAL and COPY are incomparable.

What is the common structure of PAL and COPY? What is their greatest lower bound with respect to "$\leq$". Obviously, $\text{PAL} \cap \text{COPY} = \{w\bar{w} \mid w = w^R,\ w \in \{a,b\}^*\}$ is not appropriate here, since it is neither a single reset language, nor a linear context-free language. Consider the language C, which is the characteristic language of the one-reversal one counter languages. $C = \text{PAL} \cap \{a,\bar{a}\}^* = \text{COPY} \cap \{a,\bar{a}\}^*$, which implies that $M(C) \subseteq M(\text{PAL}) \cap M(\text{COPY})$. However, we do not know whether $M(\text{PAL}) \cap M(\text{COPY}) \subseteq M(C)$. Only a few results of this form have appeared in the literature. They would be valuable contributions to language theory and would be a fundamental improvement of our knowledge of certain classes of languages.

Generalizations of PAL and COPY: $D_2'^*$ *and FIFO:*

The context-free languages are the canonical generalizations of the linear context-free languages. Thus PAL generalizes, e.g., to $D_2'^*$, which is a characteristic language of the context-free languages. In fact, $D_2'^*$ can be obtained from the variation $PAL_c = \{wc c\bar{w}^R \mid w \in \{a,b\}^*\}$ of PAL be nested iterated substitution (see [15]), substituting c by $PAL_c \cup \{e\}$ and d by {d}, otherwise. This method corresponds to the generation of $D_2'^*$ by a context-free grammar with productions $S \to aSS\bar{a}$, $S \to bSS\bar{b}$, and $S \to e$.

In a different way the Dyck sets, E and FIFO can be characterized by congruence relations or cancellations on the free monoid $\{a,\bar{a},b,\bar{b}\}^*$, and restricted applications of these relations to the center defines PAL and COPY, respectively. It is well-known (see, e.g., [1,17]) that the one-sided Dyck sets $D_2'^*$ and $D_1'^*$ are the sets of strings congruent with the empty string according to the least congruence relations that contain $a\bar{a} \sim b\bar{b} \sim e$ and $a\bar{a} \sim e$, respectively. Similarly, the two-sided Dyck sets D_2^* and D_1^* are defined by the symmetric relations $a\bar{a} \sim_s \bar{a}a \sim_s b\bar{b} \sim_s \bar{b}b \sim_s e$ and $a\bar{a} \sim_s \bar{a}a \sim_s e$, respectively.

For the definition of E and FIFO by congruence relations consider strings w_1 and w_2 in $\{a,\bar{a},b,\bar{b}\}^*$. Let $w_1 \simeq w_2$, if $w_1 = ax\bar{a}y$ or $w_1 = bx\bar{b}y$, $w_2 = xy$ and $x \in \{a,b\}^*$, and let $w_1 \simeq_s w_2$, if $w_1 \simeq w_2$ or if $w_1 = \bar{a}uav$ or $w_1 = \bar{b}ubv$, $w_2 = uv$ and $u \in \{\bar{a},\bar{b}\}^*$. Thus the relations "$\simeq$" and "$\simeq_s$" cancel the first occurrences of {a,b} and $\{\bar{a},\bar{b}\}$, provided they are matching counterparts from $(a,\bar{a})$ and $(b,\bar{b})$. Notice that $ab\bar{b}\bar{a} \sim e$, but $ab\bar{b}\bar{a} \not\simeq_s e$, and $ab\bar{a}\bar{b} \simeq e$, but $ab\bar{a}\bar{b} \not\sim_s e$. These strings distinguish "$\sim$" and "$\simeq$".

<u>LEMMA</u>:

$$D_2'^* = \{w \in \{a,\bar{a},b,\bar{b}\}^* \mid w \sim e\}, \qquad D_2^* = \{w \in \{a,\bar{a},b,\bar{b}\}^* \mid w \sim_s e\},$$

$$FIFO = \{w \in \{a,\bar{a},b,\bar{b}\}^* \mid w \simeq e\}, \qquad E = \{w \in \{a,\bar{a},b,\bar{b}\}^* \mid w \simeq_s e\},$$

$$D_1'^* = \{w \in \{a,\bar{a}\}^* \mid w \sim e\} \quad \text{and} \quad D_1^* = \{w \in \{a,\bar{a}\}^* \mid w \sim_s e\}$$

$$= \{w \in \{a,\bar{a}\}^* \mid w \simeq e\} \qquad\qquad = \{w \in \{a,\bar{a}\}^* \mid w \simeq_s e\}.$$

Furthermore, $E = \mathfrak{M}(COPY)$ and $D_1^* = \mathfrak{M}(C)$.

The language E has first appeared in [10], where it is called the "complete twin shuffle language". E is an equality set (see [6,10,21]), i.e., $E = \{w \mid g(w) = h(w)\}$, where g,h are homomorphisms with $g(a) = h(\bar{a}) = a$, $g(b) = h(\bar{b}) = b$, and g,h erase, otherwise. In fact, E is the hardest equality set. (See [6,10] for more discussions).

For a language L define $\min(L) = \{w \in L \mid w = x,y,\ x \in L \text{ and } x \neq e \text{ implies } y = e\}$. It has been shown in [2] that $\min(D_2'^*) \equiv \min(D_2^*) \equiv D_2'^*$, and $\min(D_1'^*) \equiv \min(D_1^*) \equiv D_1'^*$. Thus the min operation preserves equivalence among these Dyck sets. min(E) is a minimal equality set in the sense of [9], and from results in [6,7] we obtain that $\min(E) \equiv \min(FIFO) \equiv FIFO$.

The similarities and dissimilarities of $D_2'^*$ and FIFO are stated in the following theorems, the proof of which can be found in [1,6,8,24].

<u>THEOREM 4</u>: For a language L the following are equivalent:

(i) L is in $M(D_2'^*)$.

(ii) L is in $M(D_2^*)$.

(iii) L is in the least (full) AFL generated by $D_2'^*$.

(iv) L is accepted (in real time) by a pushdown automaton.

$M(D_2'^*)$ is the well-known class of <u>context-free languages</u>.

<u>THEOREM 5</u>: For a language L the following are equivalent:

(i) L is in M(FIFO).

(ii) L is in M(E).

(iii) L is in the least AFL generated by FIFO.

(iv) L is accepted in real time by a Post machine. (A Post machine is a non-deterministic acceptor, whose storage tape is a queue. See [6,24,25]).

M(FIFO) is the class of simple <u>Post languages</u> from [6,24].

Notice that both $M(D_2'^*)$ and M(FIFO) are AFLs. M(FIFO), however is not a full AFL, since languages in M(FIFO) can model computations of Turing machines. Furthermore, COPY is not context-free and PAL is not a simple Post languages, which follows from results in [7,23], and solves an open problem in [24].

<u>THEOREM 6</u>: Each pair of languages from $\{D_2'^*, \text{PAL}\} \times \{\text{FIFO}, \text{COPY}\}$ is incomparable.

<u>THEOREM 7</u>: The context-free languages $M(D_2'^*)$ are closed under homomorphism. The closure of M(FIFO) under (erasing) homomorphism is the class of all recursively enumerable sets.

Figure 1 summarizes <u>all</u> equivalence, strictly weaker, and incomparability relations among the languages shown and introduced so far. Let $R = \{a,b\}^* \{\bar{a},\bar{b}\}^*$ and $S = \{a,\bar{a}\}^*$, and let $L \underset{A}{\rightarrow} L'$ indicate that $L' = L \cap A$ with $A = R$ or $A = S$.

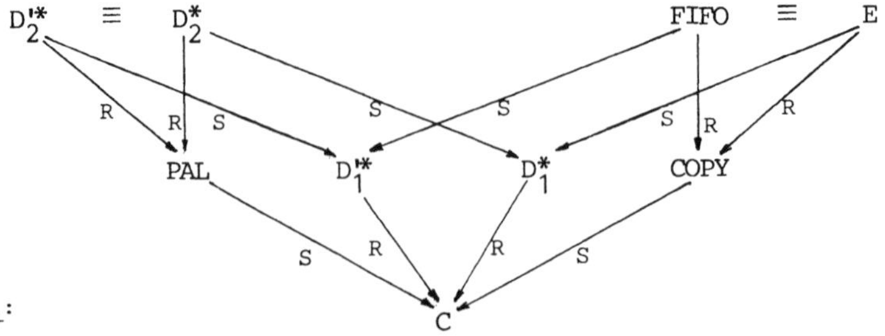

<u>FIGURE 1</u>:

Shuffle languages of PAL, COPY, D_2' *and FIFO:*

We now turn to intersection closed cones, which contain the languages from above. It is a well-known fact from AFL theory [11] that intersection closure corresponds to multitape machines, or to the shuffle of characteristic languages.

Let MULTI-RESET, BNP, and NTIME(n) denote the least intersection closed cones containing COPY, PAL, and $D_2'^*$, respectively. These classes of languages have been introduced and studied in [3-5], where it has been shown that they have many properties in common. In particular, the intersection of only three basic languages is necessary for their specification, i.e., these classes equal the least cones containing the intersection of three languages from the basic cones M(COPY), M(PAL), and $M(D_2'^*)$, respectively. Hence, MULTI-RESET, BNP, and NTIME(n) are principal cones (or equivalently principal semiAFLs), and generators are e.g., the threefold shuffles of COPY, PAL, and $D_2'^*$, resp., with themselves. Furthermore, these classes of languages are closed under linear erasing homomorphism (see e.g. [11]), and define a hierarchy, whose strictness is unknown.

These facts are summarized in the next theorem.

THEOREM 8:

(i) MULTI-RESET = $M_\cap$(COPY) = $M(⧢$(COPY, COPY, COPY)). †

(ii) BNP = $M_\cap$(PAL) = $M(⧢$(PAL, PAL, PAL)).

(iii) NTIME(n) = $M_\cap(D_2'^*)$ = $M(⧢(D_2'^*, D_2'^*, D_2'^*))$.

and there is a hierarchy

(iv) MULTI-RESET $\subseteq$ BNP $\subseteq$ NTIME(n).

Now the question arises, what $M_\cap$(FIFO) is? Speaking in terms of the related storage structures, it is easily seen that a Post tape can be simulated in linear time by two reset tapes, which implies that FIFO is in MULTI-RESET. Conversely, two Post tapes can simulate any number of Post or reset tapes. Thus we obtain the fact, that in the context of intersection closed cones, FIFO is no more powerful than COPY. This result has been established in [6], where it has been shown, too, that merging two copies of strings defines a generator of MULTI-RESET. Thus we can state.

THEOREM 9: $M_\cap$(COPY) = $M_\cap$(FIFO) = $M(⧢$(FIFO, FIFO))

$= M(\mathfrak{m}(\{w\bar{w}\bar{\bar{w}} \mid w \in \{a,b\}^*\}))$.

---

† When we use the shuffle operation we assume that the shuffled languages are defined over disjoint alphabets.

These results show that our basic languages PAL, COPY, $D_2'^*$, and FIFO are comparable, when the least cones containing them are considered, which are closed under intersection, or which contain at least the shuffle of three languages. Then COPY and FIFO are equivalent, they are weaker than PAL, and PAL is weaker than $D_2'^*$. It is unknown, whether these relations are strict and it has been conjectured in [4,5] that they are strict.

Next we weaken the standard characteristic languages of the classes MULTI-RESET, BNP, and NTIME(n) obtained from Theorem 8 and use FIFO for shuffles of languages. Using techniques from [4,6] it can be shown that two occurrences of COPY, PAL, or $D_2'^*$ can be replaced by FIFO without harm. In fact the language *COPY from [4] can be used, but this language is inadequate here.

<u>THEOREM 1o</u>: Each of the following languages is a characteristic language (cone generator) of MULTI-RESET, BNP, NTIME(n), respectively.

(i) ⧢(COPY, COPY, COPY) $\equiv$ ⧢(FIFO,COPY).

(ii) ⧢(PAL, PAL, PAL) $\equiv$ ⧢(FIFO,PAL) $\equiv$ ⧢(COPY, COPY, PAL) $\equiv$ ⧢(PAL,PAL,COPY).

(iii) ⧢($D_2'^*$, $D_2'^*$, $D_2'^*$) $\equiv$ ⧢(FIFO,$D_2'^*$) $\equiv$ ⧢(COPY, COPY, $D_2'^*$) $\equiv$ ⧢(PAL, PAL, $D_2'^*$) $\equiv$ ⧢(COPY, PAL, $D_2'^*$).

Next we consider the two-fold shuffles of COPY, PAL, and $D_2'^*$. The class M(⧢($D_2'^*$, $D_2'^*$)) equals the family of real time list storage languages from [14]. It is unknown, whether the real time list storage languages are properly included in NTIME(n), i.e., whether ⧢($D_2'^*$, $D_2'^*$) is stricly weaker than ⧢($D_2'^*$, $D_2'^*$, $D_2'^*$). In fact, the class of languages accepted by nondeterministic real time deque automata, whose storage tape is a deque (see [19,2o]), lies between the real time list storage languages and NTIME(n) (and contains FIFO).

The least cones containing ⧢(PAL, PAL) and ⧢(COPY, COPY), respectively, have deeply been studied in [18], where it has been shown, that e.g., the four-fold products PAL^4 and $COPY^4$ are not contained in these cones. Using the results from [18] we obtain the following facts.

<u>THEOREM 11</u>: ⧢(COPY, COPY) is strictly weaker than ⧢(COPY, COPY, COPY).
⧢(COPY, COPY) is incomparable with PAL, $D_2'^*$, FIFO.
⧢(PAL, PAL) is strictly weaker than ⧢(PAL, PAL, PAL).
⧢(PAL, PAL) is incomparable with $D_2'^*$, FIFO, ⧢(COPY, COPY).
⧢(PAL, PAL) is strictly stronger ($\geq$ and $\not\equiv$) than COPY.

Figure 2 summarizes our results on the comparison of languages obtained by the shuffle of the basic languages PAL, COPY, $D_2'^*$, and FIFO. It shows <u>all</u> nonequivalent languages that can be defined in this way. (Nevertheless, some of the languages shown may be equivalent). Weaker and strictly weaker relations among languages are indicated by single and double arrows, respectively.

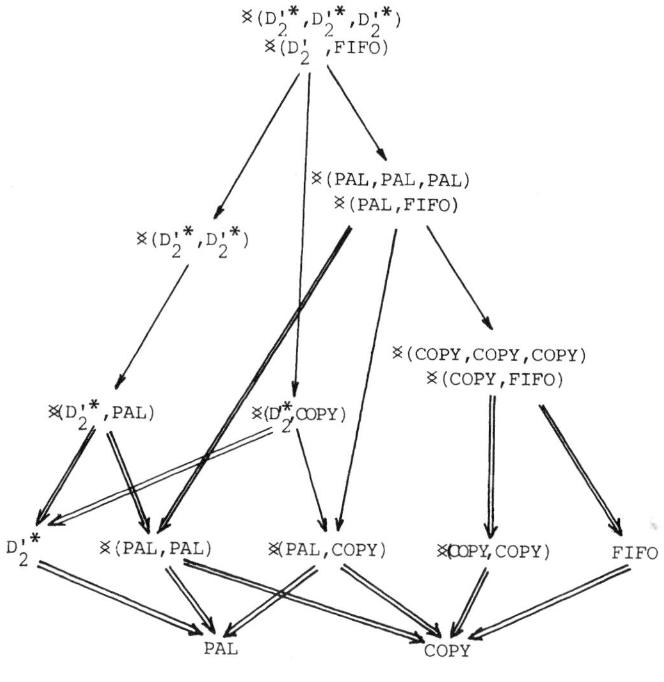

<u>FIGURE 2</u>:

<u>References</u>:

1. J. Berstel, <u>Transductions and Context-Free Languages</u>. Teubner, Stuttgart, 1979.
2. L. Boasson, Two iteration theorems for some families of languages, J. Comput. System Sciences 7, 583-596 (1973).
3. R.V. Book, S.A. Greibach, Quasi-realtime languages. Math. Systems Theory 4, 97-111 (1970).
4. R.V. Book, S.A. Greibach, and C. Wrathall, Reset machines. J. Comput. System Sciences 19, 256-276 (1979).
5. R.V. Book, M. Nivat, and M. Paterson, Reversal-bounded acceptors and intersections of linear languages. SIAM J. Computing 3, 283-295 (1974).
6. F.J. Brandenburg, Multiple equality sets and Post machines. J. Comput. System Sciences 21, 292-316 (1980).
7. F.J. Brandenburg, Three write heads are as good as k. Math. Systems Theory 14, 1-12 (1981).
8. F.J. Brandenburg, Analogies of certain families of languages arising from PAL and COPY. Report, University of Braunschweig (1980).

9. K. Culik II, A purely homomorphic characterization of recursively enumerable sets. J. Assoc. Comput. Mach. 26, 345-350 (1979).

10. J. Engelfriet and G. Rozenberg, Fixed point languages, equality languages and representation of recursively enumerable languages. J. Assoc. Comput. Mach. 27, 499-518 (1980).

11. S. Ginsburg, Algebraic and Automata Theoretic Properties of Formal Languages. North-Holland, Amsterdam 1975.

12. S. Ginsburg and S.A. Greibach, Principal AFL. J. Comput. System Sciences 4, 308-338 (1970).

13. S. Ginsburg and S.A. Greibach, Multitape AFA. J. Assoc. Comput. Mach. 16, 193-221 (1972).

14. S. Ginsburg and H.A. Harrison, One-way nondeterministic real-time list storage languages. J. Assoc. Comput. Mach. 15, 428-446 (1968).

15. S.A. Greibach, Full AFLs and nested iterated substitution. Inform. Control 16, 7-35 (1970).

16. S.A. Greibach, Remarks on blind and partially blind multicounter machines. Theoret. Comput. Science 7, 311-324 (1978).

17. M.A. Harrison, Introduction to Formal Language Theory. Addison-Wesley, Reading 1978.

18. R.B. Hull, Containment between intersection families of linear and reset languages. Ph. D. Thesis, Berkeley (1979).

19. D.E. Knuth, The Art of Computer Programming, Vol 1, Fundamental Algorithms. Addison-Wesley, Reading 1967.

20. S.R. Kosaraju, Real-time simulation of concatenable double-ended queues by double-ended queues. Proc. 11 ACM Symposium Theory of Computing, 346-351 (1979).

21. A. Salomaa, Equality sets for homomorphisms of free monoids. Acta Cybernetica 4, 127-139 (1978).

22. R. Siromoney, On equal matrix languages. Inform. Control 14, 135-151 (1969).

23. I. H. Sudborough, One-way multihead writing finite automata. Inform. Control 30, 1-20 (1976).

24. B. Vauquelin and P. Franchi-Zannettacci, Automates à file. Theoret. Comput. Science 11, 221-225 (1980).

25. R. Vollmar, Über einen Automaten mit Pufferspeicherung. Computing 5, 57-7o (197o).

QUASI-EQUATIONAL LOGIC FOR PARTIAL ALGEBRAS

PETER BURMEISTER

Partial algebras occur quite often in connection with the foundations of Computer Science. Even many-sorted algebras are quite often partial ones at the first glance and only afterwards they are usually completed (as many-sorted algebras) by so called "error values".
In this note we want to show, that partial algebras can quite easily be treated by model theoretic methods as structures independently from any possible completion. It is one in a sequence of papers with the subtitle:
"A unifying approach towards a two-valued model theory for partial algebras" (cf. [ABN 80] and [B 81]).
We start here by considering special axiomatizable classes of partial algebras in which free (partial) algebras still exist with respect to any set, and universal solutions of arbitrary partial algebras exist whenever this structure allows any homomorphism into at least one object of the class at all. All the results which we will state can be proved without using the Axiom of Choice (cf. [ABN 80]).
We will restrict our considerations to the one-sorted case in order to keep the representation simpler; but everything can easily be transferred to many-sorted partial algebras as long as all sorts of the structures under consideration are supposed to be nonvoid (cf. [GoM 81]) for many-sorted algebras with possibly void sorts).
One concern of this note is also, to provide a common background for all the so called *notions of validity of equations in partial algebras* which are floating around, and to show that all but one are implications of very special structure in each case.

1. Motivation:

When dealing with algebraic structures, the notions of *terms* and *equality* between terms play a very important rôle, and for a model theoretic approach they are fundamental. But so far at least algebraists could not decide for the most appropriate concept of equality in connection with partial algebras. The reason is that as soon as one wants to say something about equality of terms in connection with partial algebras, one has also to say something about the existence of the interpretation of the terms involved, and that is the point where the "algebraic" approaches differ, as we shall make more precise later (cf. [S/ 68], [Hö 73] or [J 78], also [Ed 73]). But also the approaches proposed so far by logicians (at least those of which we know) are not satisfactory as far as they immediately use a three-valued logic, whence they cannot express the non-existence of the interpretation of a term in the object language (cf. [E 69] and the references there). But it is often quite useful to be able to formulate axioms which express just the non-existence of the interpretation of a term. Since this is possible in the approach which we are going to propose (and for some other logical reasons), and since this is done through the semantics of our notion of equality, we want to speak about *existence-equality* (briefly: *E-equality*) and *existence-equations* (briefly: *E-equations*) in order to stress this new quality of our semantics. For the same reason we shall use the symbol " $\overset{e}{=}$ " instead of " = ", in order to indicate E-equality.

2. The language:

Actually we use the same language with terms and equality which one usually considers in connection with total algebras of some given type. The only difference is the notation "E-equality" and the symbol "$\overset{e}{=}$":

2.1. Definition of terms. Let Ω be any set, the set of *operation symbols*, and $\Delta := (n_f)_{f\in\Omega}$ the corresponding *type*, i.e. a family of natural numbers, such that n_f is the *arity* of the operation symbol f. Moreover let V be any denumerable set of *variables*. Then *terms of type* Δ *with variables in* V are defined as follows:

(i) Each variable $v\in V$ is a term.

(ii) If $f\in\Omega$ is any operation symbol, and if $t_0,\dots,t_{n_f-1}$ are terms, then $ft_0\dots t_{n_f-1}$ is a term.

(iii) Only such sequences are terms, which are formed according to (i) and (ii). We denote the set of all terms by T.

2.2. Definition of formulas. Let Ω, Δ, V and T be given as in Definition 2.1. A *formula corresponding to the set* T *of terms* is then defined as follows:

(i) If t and t' are any terms, then the E-equation $t\overset{e}{=}t'$ is a formula.

(ii) If Φ and Ψ are any formulas, then $\sim\Phi$, $(\Phi\wedge\Psi)$, $(\Phi\vee\Psi)$, $(\Phi\Rightarrow\Psi)$, and $(\Phi\Leftrightarrow\Psi)$ are formulas.

(iii) If Φ is any formula, and if v is any variable, then $(\forall v)\Phi$, and $(\exists v)\Phi$ are formulas.

(iv) Formulas are only such sequences which are formed according to the rules (i), (ii) and (iii), above.

3. The semantics:

3.1. Definition of the models (partial algebras). Our models are *partial algebras* of type Δ, i.e. ordered pairs $\underline{A}:= (A,(f^{\underline{A}})_{f\in\Omega})$, where each $f^{\underline{A}}$ is a *partial operation* $f^{\underline{A}}:\text{dom } f^{\underline{A}} \to A$, such that $\text{dom } f^{\underline{A}} \subseteq A^{n_f}$. If we have $\text{dom } f^{\underline{A}}=A^{n_f}$, then we call $f^{\underline{A}}$ a *total operation* on A; and if all operations of $\underline{A}$ are total, then we call $\underline{A}$ a *total algebra of type* Δ. The set A is called the *carrier set* of the partial algebra $\underline{A}$.

Basic, but not new (cf. [E 69]), for our approach is the definition of the interpretation of terms with respect to an *evaluation* of the variables, i.e. with respect to a mapping $\mu:V \to A$, where A is the carrier set of some partial algebra $\underline{A}$. (Note that we keep - from now on - the type Δ fixed, and do not mention it in every case explicitely.)

3.2. Definition of the interpretation of terms. Let $\underline{A}$ be any partial algebra of the given type, and let $\mu:V \to A$ be any evaluation of the variables in the carrier set of $\underline{A}$. Then we define recursively, what it means, that *a term* t *is interpreted in* $\underline{A}$ *with respect to (under) the evaluation* μ (or: *that* μ *interprets* t *in* $\underline{A}$). If this interpretation exists, then we will denote it by $J^{\mu}(t)$:

(i) For every variable $v\in V$ its interpretation under μ in $\underline{A}$ exists, and it is given by $J^{\mu}(v):=\mu(v)$.

(ii) Let f be any operation symbol, and let t_i be any terms such that their interpretations $J^{\mu}(t_i) =: a_i$ exist in $\underline{A}$ for $0\leq i\leq n_f-1$. Moreover, assume that $(a_0,\ldots,a_{n_f-1})\in\text{dom } f^{\underline{A}}$; then the interpretation of the term $ft_0\ldots t_{n_f-1}$ exists in $\underline{A}$ with respect to μ, and it is given by

$$J^{\mu}(ft_0\ldots t_{n_f-1}) := f^{\underline{A}}(a_0,\ldots,a_{n_f-1}) = f^{\underline{A}}(J^{\mu}(t_0),\ldots,J^{\mu}(t_{n_f-1})).$$

(iii) A term $t\in T$ is interpreted in $\underline{A}$ with respect to the evaluation μ iff this is done according to (i) and (ii). Note that we have used for the interpretation of the operation symbols the *standard interpretation* $J^{\mu}(f) = f^{\underline{A}}$, which is induced by the notation for $\underline{A}$.

The real difference of our approach to a model theory for partial algebras in comparison with earlier ones lies in the following definition of the satisfaction of formulas, especially as far as negation is concerned.

3.3. Definition of the satisfaction of a formula. Using the same assumptions and notations as in Definition 3.2 we define recursively what it means that *a formula* Φ *is satisfied in a partial algebra* $\underline{A}$ *with respect to an evaluation* μ *of the variables in* V (briefly: $\underline{A} \models \Phi[\mu]$):

(i) Let t and t' be any two terms, then:

$\underline{A} \models t \stackrel{e}{=} t' [\mu]$ iff $J^{\mu}(t)$ and $J^{\mu}(t')$ both exist and are equal, i.e. iff $J^{\mu}(t) = J^{\mu}(t')$ (where, by writing this equation, in the metalanguage we tacitly assume that both sides are defined).

(ii) Let Φ and Ψ be any two formulas, then:

$\underline{A} \models \sim\Phi[\mu]$ iff it is not true that $\underline{A} \models \Phi[\mu]$.

$\underline{A} \models (\Phi\wedge\Psi)[\mu]$ iff $\underline{A} \models \Phi[\mu]$ and $\underline{A} \models \Psi[\mu]$.

$\underline{A} \models (\Phi\vee\Psi)[\mu]$ iff $\underline{A} \models \Phi[\mu]$ or $\underline{A} \models \Psi[\mu]$.

$\underline{A} \models (\Phi\Rightarrow\Psi)[\mu]$ iff $\underline{A} \models \Phi[\mu]$ implies that $\underline{A} \models \Psi[\mu]$.

$\underline{A} \models (\Phi\Leftrightarrow\Psi)[\mu]$ iff $\underline{A} \models \Phi[\mu]$ is equivalent to $\underline{A} \models \Psi[\mu]$.

(iii) Let Φ be any formula, and let v be any variable, then:

$\underline{A} \models (\forall v)\Phi[\mu]$ iff for every $a \in A$ there holds $\underline{A} \models \Phi[\mu_v^a]$.

$\underline{A} \models (\exists v)\Phi[\mu]$ iff there exists $a \in A$ such that $\underline{A} \models \Phi[\mu_v^a]$.

Here $\mu_v^a : V \to A$ is the evaluation defined by

$$\mu_v^a(w) := \begin{cases} \mu(w) & \text{, if } w \neq v \text{ ,} \\ a & \text{, if } w = v \text{ .} \end{cases}$$

(iv) Finally we say that *a formula* Φ *holds in* $\underline{A}$ *(is valid in* $\underline{A}$*)* (briefly: $\underline{A} \models \Phi$), iff $\underline{A} \models \Phi[\mu]$ for every evaluation $\mu : V \to A$.

3.4. Remarks.

(i) The language presented above together with its semantics seems to form a solid basis for model theoretic research in connection with partial algebras. It allows to talk about the existence or nonexistence of the interpretation of a term within the object language. For instance $\underline{A} \models t \stackrel{e}{=} t[\mu]$ just means that *the term* t *is interpreted in* $\underline{A}$ *under* μ, i.e. that $J^{\mu}(t)$ exists, since all the other statements occurring on the right hand of 3.3.(i) are trivially satisfied. Similarly $\underline{A} \models \sim t \stackrel{e}{=} t' [\mu]$ means that either at least one of $J^{\mu}(t)$ or $J^{\mu}(t')$ does not exist, or both interpretations exist but are not equal. Hence $\underline{A} \models \sim t \stackrel{e}{=} t[\mu]$ means that the interpretation of t does not exist in $\underline{A}$ under μ. Thus we are able to *forbid* through first order axioms the existence of the interpretation of a term in a given model.

(ii) The approach of EBBINGHAUS in [E 69], which uses three truth values, is weaker than our approach, namely, when we denote satisfaction in the sense of EBBINGHAUS by " $\models_E$ ", we get for instance:

$$\underline{A} \models_E {\sim} t=t' \, [\mu] \quad \text{iff} \quad \underline{A} \models (t \stackrel{e}{=} t \wedge t' \stackrel{e}{=} t' \wedge {\sim} t \stackrel{e}{=} t') \, [\mu],$$

and this procedure can be extended to other formulas, which contain negations.

(iii) The *notions of validity of equations in partial algebras* correspond to the consideration of implications of very special structure in our langage, for instance (cf. HÖFT [Hö 73]):
Weak validity of $t=t'$ corresponds to the validity of the implication $t \stackrel{e}{=} t \wedge t' \stackrel{e}{=} t' \Rightarrow t \stackrel{e}{=} t'$;
and *validity of* $t \stackrel{e}{=} t'$ *in the sense of SŁOMINSKI* [SŁ 68] corresponds to the validity of the two implications $t \stackrel{e}{=} t \Rightarrow t \stackrel{e}{=} t'$ and $t' \stackrel{e}{=} t' \Rightarrow t \stackrel{e}{=} t'$ (note that we omitted brackets when possible).
In this way all but the *strong validity* of [SŁ 68] can be handled.

(iv) The implications which one meets in connection with "notions of validity" are all of the following special form:

$$t_1 \stackrel{e}{=} t_1 \wedge \ldots \wedge t_n \stackrel{e}{=} t_n \Rightarrow t_0 \stackrel{e}{=} t_0' \quad (\text{or} \quad \bigwedge_{i=1}^{n} t_i \stackrel{e}{=} t_i \Rightarrow t_0 \stackrel{e}{=} t_0' \;),$$

i.e. in the premise we have only what we will call *term-existence statements (TE-statements)*. Implications of this form will be called *ECE-equations* (briefly for *existentially conditioned existence-equations*). We have the feeling that these axioms fill the place that equations hold in the case of total algebras.

(v) In analogy to MAL'CEV (cf. [Mal 73]) we call implications of the form $\bigwedge_{i=1}^{n} t_i \stackrel{e}{=} t_i' \Rightarrow t_0 \stackrel{e}{=} t_0'$ *QE-equations* (briefly for *quasi-existence-equations*). Moreover, it seems that in connection with partial algebras we have to include in the axioms which are of fundamental importance the *forbidden terms*, i.e. axioms of the form ${\sim} t \stackrel{e}{=} t$. All the axioms which we have mentioned in this and the previous remark, together with E-equations, are special universal Horn-formulas (cf. [ChK 73]).
Before we now turn to some results we would like to recall that a QE-equation is satisfied whenever the satisfaction of the E-equations occurring in the premise implies that the E-equation of the conclusion is satisfied.

(vi) Since in the partial algebra with empty carrier set every formula is valid (since there are no evaluations) we exclude it from all the following considerations.

4. Some BIRKHOFF-type results.

4.1. Some basic notions and notation. Let $\underset{\sim}{P}^{\Delta}$ designate the class of all partial algebras of a given type Δ, and let $\underset{\sim}{K}$ be any subclass. Moreover let F be any set of formulas. Then we denote by Mod F the subclass of $\underset{\sim}{P}^{\Delta}$ which contains all those $\underset{\sim}{P}^{\Delta}$-algebras in which every formula from F is valid. Moreover we denote by Eeq $\underset{\sim}{K}$, ECEeq $\underset{\sim}{K}$, QEeq $\underset{\sim}{K}$ the set of all E-equations, ECE-equations and QE-equations, respectively, which are valid in every $\underset{\sim}{K}$-algebra.
Let $P\underset{\sim}{K}$ designate the class of all *cartesian products* of $\underset{\sim}{K}$-algebras (where the partial operations are defined on a sequence of elements of the product iff they are defined for every "component-sequence"), and $P_{+}\underset{\sim}{K}$ means that only nonempty index sets are allowed. By $P^{r}\underset{\sim}{K}$ we understand the class of all *reduced products of* $\underset{\sim}{K}$*-algebras*, i.e. the carrier sets are quotients of cartesian products of the carrier sets of $\underset{\sim}{K}$-algebras with respect to a "filter congruence" and an operation is defined on a sequence, iff the set of indices for which the operation is defined on the corresponding component sequence belongs to the filter; $P^{r}_{+}\underset{\sim}{K}$ means that we allow only nonempty index sets (cf. [ABN 80] or [C 65]). $S\underset{\sim}{K}$ will designate the class of all *subalgebras* (with the induced structure on closed subsets) of $\underset{\sim}{K}$-algebras, and $I\underset{\sim}{K}$ will designate the class of all *isomorphic copies* of $\underset{\sim}{K}$-algebras.
Finally we shall need some operators concerning homomorphic images: A *homomorphism* $\varphi:\underline{A} \to \underline{B}$ is a mapping $\varphi:A \to B$ such that for every fundamental operation symbol $f \in \Omega$ and for every sequence $\underline{a} \in A^{n_f}$ there holds: $\underline{a} \in \text{dom } f^{\underline{A}}$ implies $\varphi \circ \underline{a} \in \text{dom } f^{\underline{B}}$ and $\varphi(f^{\underline{A}}(\underline{a})) = f^{\underline{B}}(\varphi \circ \underline{a})$. If, in addition, there holds that $\varphi \circ \underline{a} \in \text{dom } f^{\underline{B}}$ implies $\underline{a} \in \text{dom } f^{\underline{A}}$, then we call the homomorphism *closed*. We call $\underline{B}$ a *(closed) homomorphic image of* $\underline{A}$ iff there exists a (closed and) surjective homomorphism from $\underline{A}$ onto $\underline{B}$. By $H\underset{\sim}{K}$ and $H_{c}\underset{\sim}{K}$ we denote the classes of all homomorphic images and closed homomorphic images of $\underset{\sim}{K}$-algebras, respectively.

4.2. Theorem. It can be proved in ZERMELO-FRAENKEL set theory without the Axiom of Choice that for any class $\underset{\sim}{K} \subseteq \underset{\sim}{P}^{\Delta}$ there holds:

(i) Mod Eeq $\underset{\sim}{K}$ $= HSP\,\underset{\sim}{K}$,

(ii) Mod ECEeq $\underset{\sim}{K}$ $= H_{c}SP^{r}\,\underset{\sim}{K}$,

(iii) Mod QEeq $\underset{\sim}{K}$ $= ISP^{r}\,\underset{\sim}{K}$.

4.3 Remark. A proof which avoids the Axiom of Choice has been given so far only for 4.2.(iii) in [ABN 80], but the methods and ideas there show how one can modify the proofs of 4.2.(i) in [B 71] or [Hö 73] in order to avoid the Axiom of Choice, and also how one can attack a proof of 4.2.(ii). Note that the category theoretical proofs of the above results which are contained in [AN 76] or [NSa 77] do not avoid the Axiom of Choice, since they assume the existence of enough projective objects in special subcategories; but the projectivity of the partial PEANO-algebras which one has to prove then is equivalent to the Axiom of Choice (cf. [ABN 80], Proposition 5.2.2).

Let HEeq $\underset{\sim}{K}$ designate the set of all *universal HORN-formulas* which are valid in each $\underset{\sim}{K}$-algebra, i.e. besides QE-equations we consider also negations of conjunctions of E-equations; and let $EC\tilde{E}eq\ \underset{\sim}{K}$ be analogously defined in connection with ECE-equations and negations of (single) TE-statements.

<u>4.4. Theorem</u>. It can be proved in ZERMELO-FRAENKEL set theory without the Axiom of Choice that for any class $\underset{\sim}{K} \subseteq \underset{\sim}{P}^{\Delta}$ there holds:

(i) $\quad \text{Mod HEeq}\ \underset{\sim}{K} = ISP^{r}_{+}\ \underset{\sim}{K}$;

(ii) $\quad \text{Mod } EC\tilde{E}eq\ \underset{\sim}{K} = \begin{cases} H_{c}SP^{r}_{+}\ \underset{\sim}{K} & \text{, if there is a } t\in T \text{ with } \underset{\sim}{K} \models \sim t \overset{e}{=} t \text{ ,} \\ H_{c}SP^{r}\ \underset{\sim}{K} & \text{, otherwise.} \end{cases}$

<u>4.5. Remark</u>. As we have already mentioned at the beginning, in the axiomatizable classes considered in 4.2 and 4.4 there exist free partial algebras on every set, whenever these classes contain at least one object with at least two elements; the structure of these free objects is described in [B 70], they behave very similar to total free algebras, since all induced term operations in them, which are defined for an injective sequence of the basis, are total, and each element of the free partial algebra is uniquely representable with respect to such induced total term operations and sequences with elements only from the basis. If $\underset{\sim}{K} \subseteq \underset{\sim}{P}^{\Delta}$ is closed with respect to the operators S and P, then $\underset{\sim}{K}$-universal $\underset{\sim}{K}$-solutions exist for every $\underset{\sim}{P}^{\Delta}$-algebra, but if we have only closedness with respect to S and P_{+}, then $\underset{\sim}{K}$-universal $\underset{\sim}{K}$-solutions exist only for such $\underset{\sim}{P}^{\Delta}$-algebras which allow at least one homomorphism into some $\underset{\sim}{K}$-algebra (cf. [ABN 80]).
In [AN 76], [NSa 77] and [B 81] some additional BIRKHOFF-type results in connection with quasi-equational logic for partial algebras are listed.

<u>5. Closed sets of QE-equations</u>:

We give here a set theoretical and algebraic description of closed sets of QE-equations which seems also to be new for total algebras. For simplicity of the representation we use the fact that commutativity of the conjunction is derivable from the logical axioms which hold in every first order language.

<u>5.1. Some notation</u>. Let $\mathcal{S}_f$ designate the set of all finite subsets of $T\times T$, where T is our set of terms. We consider T in a natural way as a total algebra of type Δ, and then we denote it by $\underline{T}$. Let QEE:= $\mathcal{S}_f\times(T\times T)$, and associate with each QE-equation $\bigwedge_{i=1}^{n} t_i \overset{e}{=} t_i' \Rightarrow t \overset{e}{=} t'$ the element $(\{(t_i,t_i') \mid 1\leq i\leq n\},(t,t'))$ of QEE. In what follows we will make no difference between a QE-equation and its associated element from QEE (see the remark above). (Note that (t,t') and $(\emptyset,(t,t'))$ both represent the same E-equation, in the first case as a subformula of some QE-equation, and in the second case standing for itself, but we hope that this will not cause any confusion.)

Let F be any set of QE-equations, then we introduce

$\downarrow F := \{t\in T \mid t$ is a subterm of a term t' occurring in F$\}$, and

$\downarrow_V F := V \cup \downarrow F$.

$\underline{\downarrow F}$ and $\underline{\downarrow_V F}$ shall indicate that we consider these sets as relative subalgebras of $\underline{T}$ (note that $\underline{\downarrow F}$ is always generated by $V\cap\downarrow F$, and $\underline{\downarrow_V F}$ is generated by V).

If $Q \subseteq \mathcal{S}_f\times(T\times T)$ represents any set of QE-equations, and if $\Phi\in\mathcal{S}_f$ is any finite subset of $T\times T$, then

$Q(\Phi) := \{(t,t')\in T\times T \mid (\Phi,(t,t'))\in Q\}$ designates the set of all conclusions of QE-equations which are represented in Q, and the premise of which is represented by Φ.

<u>5.2. Theorem</u>. Let $Q \subseteq \mathcal{S}_f\times(T\times T) = \text{QEE}$ be any subset, then the following statements are equivalent:

(i) Q = QEeq Mod Q.

(ii) Q has the following properties (I1), (I2), (I3) and (I4) for every $\Phi,\Phi'\in\mathcal{S}_f$:

- (I1) $\underline{\downarrow Q}$ is generated by V as a relative subalgebra of $\underline{T}$.
- (I2) $Q(\Phi)$ is a closed congruence relation on $\underline{\downarrow Q(\Phi)}$ (i.e. the natural homomorphism from $\underline{\downarrow Q(\Phi)}$ onto its factor algebra $\underline{\downarrow Q(\Phi)}/Q(\Phi)$ is closed).
- (I3) $\Phi \subseteq Q(\Phi)$.
- (I4) For every homomorphism $\varphi:\underline{\downarrow_V \Phi}\to\underline{\downarrow Q(\Phi')}$ which satisfies $(\varphi\times\varphi)[\Phi]\subseteq Q(\Phi')$, there exists a homomorphic extension $\bar\varphi:\underline{\downarrow Q(\Phi)}\to\underline{\downarrow Q(\Phi')}$ satisfying $(\bar\varphi\times\bar\varphi)[Q(\Phi)]\subseteq Q(\Phi')$.

<u>5.3. Remarks</u>.

(i) The proof of the above theorem is contained in [ABN 80]. As for the results stated earlier we do not need the Axiom of Choice.

(ii) If we replace in Theorem 5.2 the set S_δ by the set of all finite subsets of T or by the empty set, then the corresponding theorems characterize the closed sets of ECE-equations or of E-equations, respectively. Especially in the case of E-equations the condition (I4) becomes much more familiar, since it can then be reformulated as (note that $\emptyset$ means the empty set):

$(I4)^E$ For every mapping $\varphi: V \to \underline{\downarrow Q(\emptyset)}$ there exists a homomorphic extension $\bar{\varphi}: \underline{\downarrow Q(\emptyset)} \to \underline{\downarrow Q(\emptyset)}$, satisfying

$(\bar{\varphi}\times\bar{\varphi})[Q(\emptyset)] \subseteq Q(\emptyset)$.

But this means nothing else than that V freely generates $\underline{\downarrow Q(\emptyset)}$ and that $Q(\emptyset)$ is a fully invariant (and by (I2) closed) congruence relation of $\underline{\downarrow Q(\emptyset)}$.

(iii) Theorem 5.2 can also be restricted to total algebras and quasi-equations. Then $\underline{\downarrow Q(\Phi)}$ and $\underline{\downarrow Q(\Phi')}$ can everywhere be replaced by $\underline{T}$, and condition (I2) can be replaced by the statement that $Q(\Phi)$ is a congruence relation on $\underline{T}$, while (I1) becomes superfluous.

(iv) All the results stated in this note can easily be transferred to many-sorted partial algebras as long as in each model each sort is kept nonempty (in this case one has to be careful with the process of generation, since only such families of subsets which induce in each sort in a natural way a nonempty subset really generate substructures).

(v) The results of this note (except for the statements concerning the Axiom of Choice) can also - with the necessary precautions - be transferred to infinitary types.

(vi) More details - but still no proofs - can be found in [B 81]. Some of the proofs of results mentioned here and in [B 81] will possibly first be published in a planned monograph on partial algebras.

R E F E R E N C E S :

ABN 80 H. Andréka, P. Burmeister, I. Németi. Quasivarieties of partial algebras - A unifying approach towards a two-valued model theory for partial algebras. Preprint Nr. 557, Technische Hochschule Darmstadt, Fachbereich Mathematik, 1980.

AN 76 H. Andréka, I. Németi. Generalization of variety and quasivariety concept to partial algebras through category theory. MKI Budapest, Preprint 1976/5; to appear in Dissertationes Mathematicae No. 204.

B 70 P. Burmeister. Free partial algebras. J. reine u. angewandte Math. 241, 1970, pp. 75 - 86.

B 71 P. Burmeister. Primitive Klassen partieller Algebren. Habilitationsschrift, Bonn, 1971.

B 81 P. Burmeister. Partial algebras - Survey of a unifying approach towards a two-valued model theory for partial algebras. Preprint Nr. 582, Technische Hochschule Darmstadt, Fachbereich Mathematik, 1981.

ChK 73 C. C. Chang, H. J. Keisler. Model Theory.North-Holland, 1973.

C 65 P. M. Cohn. Universal Algebra. Harper & Row, 1965.

E 69 H.-D. Ebbinghaus. Über eine Prädikatenlogik mit partiell definierten Prädikaten und Funktionen. Archiv f. math. Logik 12, 1969, pp. 39 - 53.

Ed 73 G. A. Edgar. The class of topological spaces is equationally definable. Algebra universalis 3, 1973, pp. 139 - 146.

GoM 81 J. A. Goguen, J. Meseguer.Completeness of many-sorted equational logic. Manuscript 1981.

Hö 73 H. Höft. Weak and strong equations in partial algebras. Algebra universalis 3, 1973, pp. 203 - 215.

J 78 R. John. Gültigkeitsbegriffe für Gleichungen in partiellen Algebren. Math. Zeitschrift 159, 1978, pp. 25 - 35.

Mal 73 A. I. Mal'cev. Algebraic Systems. Springer-Verlag, 1973.

NSa 77 I. Németi, I. Sain. Cone-implicational subcategories and some Birkhoff-type theorems. Contributions to Universal Algebra (Proc. Coll. Esztergom 1977) Colloq. Math. Soc. J. Bolyai, North Holland, to appear.

Sł 68 J. Słomiński . Peano-algebras and quasi-algebras. Dissertationes Mathematicae (Rozprawy Mat.) 62, 1968.

Address of the author:

Fachbereich Mathematik, AG 1
Technische Hochschule
Schloßgartenstr. 7
D-6100 Darmstadt
Fed. Rep. of Germany

HOMOGENEITY AND COMPLETENESS

B. Csákány
Bolyai Institute, József Attila University
6720 Szeged, Hungary

1. Homogeneity. In the everyday usage homogeneity means to be formed of parts of the same kind. This loose notion can be made more exact in concrete instances. E.g., a mathematical object A may be called homogeneous if each element of the underlying set A of A has the same relevant properties. After Felix Klein, it is natural to call a property P relevant if it is invariant under Aut A (the group of automorphisms of A). We can imagine P as a partition of A (induced by the mapping of A into a full system of mutually excluding attributes). Then P is invariant with respect to Aut A iff the partition consisting of orbits of Aut A is a refinement of P. Thus the above heuristic notion of mathematical homogeneity of A expresses the requirement that Aut A has a unique orbit, i.e. the automorphism group of A is transitive.

Here we used elements of A for "parts" of A. However, we may also consider pairs of elements of A, subsets of A, etc. as parts of A. Now we cannot require transitivity because the full permutation group S_A of A does not act transitively neither on $A \times A$, nor on the power set of A; thus, it remains to require that Aut A acts as transitively as S_A does.

On the other hand, we used Aut A for characterizing "relevant" properties. However, invariance with respect to other related structures of A may also be considered; the most natural example is Part A, the inverse semigroup of partial automorphisms (i.e., isomorphisms between subobjects) of A. The caution made in the preceding paragraph is valid for this case, too.

Now, combining various choices of "parts" and "relevancy" we obtain various notions of homogeneity. Pairs, triples, etc. combined with automorphisms provide the properties of having doubly, triply, etc. transitive automorphism group. Even n-tuples with $n = |A|$ together with automorphisms furnish a meaningful case: every permutation of A is an automorphism of A. Apparently, this is the highest degree of homogeneity which may be reached using automorphisms; it may be further heightened if we apply partial automorphisms.

Here we have at least two possibilities of generalization. Our notions of homogeneity require Aut A to be sufficiently large for acting as transitively as possible on some set derived from A. We may postulate the size of Aut A as a measure of homogeneity of A even if this size may not be measured by transitivity properties. E.g., primitivity of the automorphism group of A may also be considered as a kind of homogeneity. (Primitivity of permutation groups lies between transitivity and double transitivity; however, it is easy to see that there is no set $\tau \subseteq A \times A$ such that a subgroup of S_A is primitive iff it is transitive on τ.) Another way of generalization is to deal with objects whose underlying set A is equipped with some structure and use the automorphisms of this structure instead of S_A.

Homogeneity occurs in various mathematical contexts, e.g. in the study of combinatorial, algebraic, and topological systems. A classical appearance of homogeneity in the mathematical background of computer science is John von Neumann's cellular model of self-reproduction: its underlying object is an infinite planar graph whose vertices are labelled with finite automata, and the main result is that self-reproduction is possible in such a model even under the very restrictive assumption that this object is homogeneous in the elements-and-automorphisms sense ([18], pp. 103-108).

In what follows we give a survey of examples and results concerning homogeneity of algebraic objects related to multiple-valued logics: finite sets provided with finitary operations (or relations), i.e. finite algebras (or relational algebras). Next we present some knowledge on homogeneous algebras under several usual assumptions. Then we give a report on recent research concerning the influence of homogeneity upon completeness. Here completeness of an algebra A means that all possible operations on A may be expressed from the operations of A in some standard way, e.g., using projections and superposition (this kind of completeness is often called *primality*), or using also constants besides projections and superposition (*functional completeness*). The main observation is that while homogeneity keeps back finite algebras from being primal, it encourages them to be functionally complete.

2. Examples. Traditional finite algebras are not homogeneous: no group (and hence no ring), further, no finite semilattice (and hence no finite lattice) has a transitive automorphism group. Stein [27] proved that a quasigroup with transitive automorphism group can be defined on a finite set M iff $|M| \not\equiv 2 \pmod 4$. He also studied quasigroups with doubly transitive automorphism groups; a typical example is a vector space over a finite field K with an operation of form $x \circ y =$

$= ax + by$ $(a,b\in K)$ where $a + b = 1$. Obviously, no quasigroup with at least four elements has triply transitive automorphism group. Babai and Pastijn [1] have shown that a finite semigroup has a transitive automorphism group iff it is a rectangular band (i.e. the direct product of a left zero semigroup and a right zero semigroup, the last ones being the only semigroups with n-fold transitive automorphism groups for $n \geq 2$).

As for finite algebras in general, it is customary to consider them up to equivalence (two algebras are equivalent if their underlying sets are the same and their operations are mutually derivable from each other, i.e., they can be expressed from each other using projections and superposition). In the case of two-element underlying set, an algebra is homogeneous (in any meaningful sense) iff its operations are self-dual Boolean functions. Then Post's classification shows that there are 7 homogeneous self-dual algebras (see, e.g. [12], pp. 72-76).

Demetrovics, Hannák and Marčenkov [4][16] gave a partial description of three-element algebras with transitive automorphism groups. They have found a countable set of such algebras and conjecture that also the set of all such algebras is countable. Three-element algebras with doubly (hence triply) transitive automorphism groups were determined by Marčenkov [14]; their number is 7. For arbitrary n, n-element algebras whose automorphism groups are n-fold transitive (i.e., all permutations are automorphisms) were introduced by Marczewski [17] under the name "homogeneous algebras". He determined all the possible operations of such algebras and the entire research in this area is based on his work. For $n \geq 4$, we have found $2n-2$ n-element algebras whose all permutations are automorphisms [3]; recently, Marčenkov determined all of them [15]. Their total number equals $4n-3$ for $n > 4$ and 14 for $n = 4$.

In order to reproduce Marczewski's basic observation, we recall the notion of the *pattern*, due to Quackenbush [24]. Two k-tuples over A are said to be of the same pattern if they belong to the same orbit of S_A, acting on A^k. An operation f on A is called a *pattern function* if $f(a_1,\ldots,a_k) = a_i$ with $1 \leq i \leq k$ so that i depends upon the pattern of $\langle a_1,\ldots,a_k\rangle$ only. Now, following Marczewski, $\mathrm{Aut}\langle A;g\rangle$ equals S_A iff for any $a_1,\ldots,a_k\in A$, g behaves as a pattern function with the possible exception that $g(a_1,\ldots,a_k)$ can also be the *unique* element of $A\setminus\{a_1,\ldots,a_k\}$, and in any case $\langle a_1,\ldots,a_k, g(a_1,\ldots,a_k)\rangle$ and $\langle b_1,\ldots,b_k, g(b_1,\ldots,b_k)\rangle$ are of the same pattern if $\langle a_1,\ldots,a_k\rangle$ and $\langle b_1,\ldots,b_k\rangle$ do so.

The most popular pattern function is *the* (ternary) *discriminator* t, defined on any set by the rule: $t(x,y,z) = z$ if $x = y$, and $t(x,y,z) = x$ otherwise. The book [32] (including a list of references of 180 items) is devoted to the study of the discriminator. On any set, every pattern function is derivable from the discriminator [8]. Probably, the first appearance of any pattern function was the normal transform n [9]: $n(x,y,u,v) = u$ if $x = y$, and $n(x,y,u,v) = v$ otherwise. The operations t and n are mutually derivable from each other ([32], p. 12). Further useful pattern functions are *the dual discriminator* d [7] ($d(x,y,z) = x$ if $x = y$, and $d(x,y,z) = z$ otherwise), and *the near-trivial functions* l_n [17],[3] whose introduction is meaningful on any at least n element set: $l_n(x_1,\ldots,x_n) = x_1$ if $x_1,\ldots,x_n$ are pairwise distinct, and $l_n(x_1,\ldots,x_n) = x_n$ otherwise. Note that the notion of pattern function fits naturally into the above system of homogeneities: $\mathrm{Part}\langle A;f\rangle$ is the full inverse semigroup of partial permutations of A iff f is a pattern function.

As a sample of a non-pattern function which may serve as an operation of a homogeneous algebra in Marczewski's sense, take the $(n-1)$-ary operation d_n, definable on the n element set A as follows [15]: $d_n(a_1,\ldots,a_{n-1}) = a_n$ if $\{a_1,\ldots,a_{n-1},a_n\} = A$, and $d_n(a_1,\ldots,a_{n-1}) = d(a_1,a_2,a_3)$ otherwise. Another example is the three-element field with the operation $x \circ y = 2x + 2y$; this is also the unique quasigroup with triply transitive automorphism group.

Universal algebra provides a lot of algebras with transitive, doubly transitive, etc. automorphism groups. The study of such algebras was initiated by Grätzer [10] (cf. Problem 20 in [11]). *Diagonal algebras* were introduced and studied by Płonka [21]; their underlying set is a product of $n(\geq 2)$ sets, and their (n-ary) operation is defined by the law that the i-th entry of the result is equal to the i-th entry of the i-th variable (in the case $n = 2$, this gives rectangular bands). Diagonal algebras have transitive automorphism groups. *Affine spaces* arise from vector spaces by appointing all linear functions of form $a_1x_1 + \ldots + a_nx_n$ with $a_1 + \ldots + a_n = 1$ to basic operations. An affine space A over a $q(>2)$ element finite field F is equivalent to the quasigroup $\langle A;\circ\rangle$ with $x \circ y = rx + (1-r)y$ where r is a generator of the multiplicative group of F (see, e.g.[29]). As we have seen, such quasigroups have doubly transitive automorphism groups; thus, affine spaces over at least three element finite fields do so, too. Finally, affine spaces over the two element field have triply transitive automorphism groups.

For relational algebras, the requirement of having k-fold transitive automorphism group is very strong. For example, on an at least two element set A there exist only two non-trivial binary relations ρ with $\mathrm{Aut}\langle A;\rho\rangle$ doubly transitive, namely the equality and the inequality. It is easy to extend this simple remark to the general case; in particular, a relational algebra $\langle A;\rho\rangle$ is homogeneous in Marczewski's sense iff, for any $a_1,\ldots,a_n,b_1,\ldots,b_n \in A$, $\rho(a_1,\ldots,a_n)$ implies $\rho(b_1,\ldots,b_n)$ whenever $\langle a_1,\ldots,a_n\rangle$ and $\langle b_1,\ldots,b_n\rangle$ are of the same pattern.

In order to obtain a special kind of generalized homogeneous algebras, take a relational algebra $\langle A;\rho\rangle$ and consider an operation f on A such that $\mathrm{Part}\langle A;f\rangle \supseteq \mathrm{Part}\langle A;\rho\rangle$. If ρ is the equality relation, our assumption means that f is a pattern function. In the general case, we may call f a ρ-pattern function. E.g., let ρ be a linear order on A; then the n-ary maximum and minimum functions are ρ-pattern functions.

3. Completeness. If $\mathrm{Aut}\langle A;F\rangle$ is transitive, than for any g derivable from the set of operations F, $\mathrm{Aut}\langle A;g\rangle$ ($\supseteq \mathrm{Aut}\langle A;F\rangle$) is also transitive. Hence a finite algebra with transitive automorphism group cannot be primal. Moreover, a beautiful theorem of Rousseau [26] asserts that a finite algebra of form $\langle A;f\rangle$ is primal iff it has no non-trivial congruences, proper subalgebras, and non-identical automorphisms. Thus homogeneity seems to be a property opposite to primality.

However, homogeneity does not exclude functional completeness (remind that an algebra is functionally complete iff the inclusion of all constant operations into its operation makes it primal). Werner [31] proved that algebras $\langle A;t\rangle$ with t the discriminator are functionally complete. Quackenbush [23] proved that the triangle is functionally complete (*the triangle* is an algebra of form $\langle \{a,b,c\};\circ\rangle$ where $\circ$ is commutative, idempotent, and satisfies $a\circ b = b$, $b\circ c = c$, $c\circ a = a$; thus, the automorphism group of the triangle is transitive). Fried and Pixley [7] proved that algebras $\langle A;d\rangle$ with $|A| \geq 3$ and d the dual discriminator are also functionally complete.

Inspirited by these results, we proved [2] that if a non-trivial*) finite algebra is homogeneous in Marczewski's sense then it is functionally complete, except it is equivalent to one of the following six algebras: a two element set with (1) the negation; (2) the ternary majority function (which coincides, in this case, with the dual discriminator);

*) An operation is called non-trivial if it is not a projection and an algebra is non-trivial if it has a non-trivial operation.

(3) the ternary minority function; (4) the negation of the ternary minority function; further (5) the three element homogeneous quasigroup; and (6) the four element affine space over the two element field. In particular, algebras of form $\langle A;f\rangle$ where f is a non-trivial pattern function are functionally complete, except $|A| = 2$ and f is the ternary majority or minority function. This contains the mentioned theorems of Werner and Fried - Pixley.

Szabó and Á. Szendrei [28] extended this result proving that if a non-trivial finite homogeneous algebra with at least four element underlying set has triply transitive automorphism group then it is either functionally complete or equivalent to an affine space over the two element field. For an infinite version of this theorem, see [13]. Further, P[3], Szabó and Á. Szendrei [19] proved that if an at least three element non-trivial finite algebra has doubly transitive automorphism group, then it is either functionally complete or equivalent to an affine space over a finite field. The next result in this direction was obtained by Demetrovics, Hannák and Rónyai [5]: for any odd prime p, a p element algebra A with transitive automorphism group is either functionally complete or linear (this means that the operations of A are of form $a_1x_1 + \ldots + a_nx_n + b$ with $a_1,\ldots,a_n,b\in A$ and suitably chosen field structure on A). This contains the mentioned theorem of Quackenbush.

The final result improving all the preceding theorems was proven also by P[3], Szabó and Á. Szendrei [20]: if a non-trivial finite algebra has primitive automorphism group then it is functionally complete, except it is equivalent to one of the following algebras: (1) A^k with $k > 1$ where A is a functionally complete algebra with primitive automorphism group of composite order; (2) an affine space over a finite field; (3) a two element set with the ternary majority function; and a prime element set A with (4) $x+a$, (5) $x-y+z+a$, where addition is meant in an Abelian group $A = [a]$. This result is a final one, indeed, as for any imprimitive permutation group G on a set A there exists a functionally incomplete algebra on A whose automorphism group is G.

The rationale of these functional completeness results is the following. By Rosenberg's well-known completeness criterion [25], a finite algebra A is primal iff every member of a well-defined list of relations on A is violated by some operation of A. A lot of permutations of A and all the non-trivial unary relations are included in the list. Assuming that A has a non-identical automorphism it is easy to deduce that either there is a permutation in the list respected by the operations of A or there is a unary relation respected by them. (This is why an algebra with a non-identical automorphism cannot be primal.) However, the constant operations violate every permutation, and every non-trivial unary relation is violated by some constant. Hence the operations of A have to take care of the remainder part of Rosenberg's relations only. It turns out that respecting anyone of these relations is not compatible with a large automorphism group. Note that in several proofs of the theorems listed above no explicit use of Rosenberg's criterion is made (see, e.g., [2],[6] and [19]).

There are also partial results concerning completeness of generalized homogeneous algebras, mentioned at the end of the preceding section, Vármonostory [30] proved, e.g., that a finite algebra $\langle A;f\rangle$ is functionally complete whenever f arises from the discriminator by using a linear order or a permutation instead of equality in its definition. Finally, we remark that Pöschel [22] proved the exact counterpart of the theorem in [2]: if a non-trivial finite relational algebra is homogeneous in Marczewski's sense then it is relationally complete (in a natural and well-defined sense), except it essentially coincides with one of five particular (at most four element) relational algebras.

References

[1] L. Babai - F. Pastijn, On semigroups with high symmetry, *Simon Stevin*, 52(1978), 73-84.

[2] B. Csákány, Homogeneous algebras are functionally complete, *Algebra Universalis*, 11(1980), 149-158.

[3] B. Csákány - T. Gavalcová, Finite homogeneous algebras, *Acta Sci. Math. (Szeged)*, 42(1980), 57-65.

[4] J. Demetrovics - L. Hannák - S.S. Marčenkov, Some remarks on the structure of P_3, *C.R. Math. Rep. Acad. Sci. Canada*, 2(1980), 215-219.

[5] J. Demetrovics - L. Hannák - L. Rónyai, Prime-element algebras with transitive automorphism groups, *C.R. Math. Rep. Acad. Sci. Canada*, 3(1981), 19-22.

[6] E. Fried - H.K. Kaiser - L. Márki, An elementary way for polynomial interpolation in universal algebras, *Algebra Universalis*, to appear.

[7] E. Fried - A.F. Pixley, The dual discriminator function in universal algebra, *Acta Sci. Math. (Szeged)*, 41(1979), 83-100.

[8] B. Ganter - J. Płonka - H. Werner, Homogeneous algebras are simple, *Fund. Math.*, 79(1973), 217-220.

[9] M.I. Gould - G. Grätzer, Boolean extensions and normal subdirect powers of finite universal algebras, *Math. Z.*, 99(1967), 16-25.

[10] G. Grätzer, A theorem on doubly transitive permutation groups with application to universal algebras, *Fund. Math.*, 53(1963), 25-41.

[11] G. Grätzer, *Universal Algebra*, Van Nostrand, Princeton, 1968.

[12] S.W. Jablonski - G.P. Gawrilow - W.B. Kudrjawzew, *Boolesche Funktionen und Postsche Klassen*, Akademie-Verlag, Berlin, 1970.

[13] H.K. Kaiser - L. Márki, Remarks on a paper of L. Szabó and Á. Szendrei, *Acta Sci. Math.*, 42(1980), 95-98.

[14] S.S. Marčenkov, On closed classes of self-dual functions of many-valued logic (in Russian), *Problemy Kibernet*, 36(1979), 5-22.

[15] S.S. Marčenkov, On homogeneous algebras (in Russian), *Dokl. Akad. Nauk SSSR*, 256(1981), 787-790.

[16] S.S. Marčenkov - J. Demetrovics - L. Hannák, On closed classes of self-dual functions in P_3 (in Russian), *Diskret. Analiz*, 34(1980), 38-73.

[17] E. Marczewski, Homogeneous algebras and homogeneous operations, *Fund. Math.*, 56(1964), 81-103.

[18] J. von Neumann, *Theory of Self-Reproducing Automata*, University of Illinois Press, Urbana and London, 1966.

[19] P.P. Pálfy - L. Szabó - Á. Szendrei, Algebras with doubly transitive automorphism groups, in: *Coll. Math. Soc. J. Bolyai, Vol.* 28., *Finite Algebra and Multiple-valued Logic*, North-Holland Publ. Co., 1981, pp. 521-535.

[20] P.P. Pálfy - L. Szabó - Á. Szendrei, Automorphism groups and functional completeness, *Algebra Universalis*, to appear.

[21] J. Płonka, Diagonal algebras, *Fund. Math.*, 58(1966), 309-321.

[22] R. Pöschel, Homogeneous relational algebras are relationally complete, in: *Coll. Math. Soc. J. Bolyai, Vol.* 28., *Finite Algebra and Multiple-valued Logic*, North-Holland Publ. Co., 1981, pp. 587-601.

[23] R.W. Quackenbush, The tringle is functionally complete, *Algebra Universalis*, 2(1972), 128.

[24] R.W. Quackenbush, Some classes of idempotent functions and their compositions, *Colloq. Math.*, 29(1974), 71-81.

[25] I.G. Rosenberg, Completeness properties of multiple-valued logic algebras, in: *Computer Science and Multiple-valued Logic*, North-Holland Publ. Co., 1977, pp. 144-186.

[26] G. Rousseau, Completeness in finite algebras with a single operation, *Proc. Amer. Math. Soc.*, 18(1967), 1009-1013.

[27] S.K. Stein, Homogeneous quasigroups, *Pacific J. Math.*, 14(1964), 1091-1102.

[28] L. Szabó - Á. Szendrei, Almost all algebras with triply transitive automorphism groups are functionally complete, *Acta Sci. Math. (Szeged)*, 41(1979), 391-402.

[29] Á. Szendrei, On the arity of affine modules, *Colloq. Math.*, 38 (1977), 1-4.

[30] E. Vármonostory, Relational pattern functions, in: *Coll. Math. Soc. J. Bolyai Vol. 28., Finite Algebra and Multiple-valued Logic*, North-Holland Publ. Co., 1981, pp. 753-758.

[31] H. Werner, Eine Characterisierung funktional vollständiger Algebren, *Arch. Math.*, 21(1970), 381-385.

[32] H. Werner, *Discriminator-Algebras*, Akademie-Verlag, Berlin, 1978.

On the Error Correcting Power of Pluralism in Inductive Inference†

Preliminary Report

Robert P. Daley
Department of Computer Science
University of Pittsburgh
Pittsburgh, Pennsylvania 15260

In this paper we show that it is always possible to reduce errors for some forms of inductive inference by increasing the number of machines involved in the inference process. Moreover, we obtain precise bounds for the number of machines required to eliminate all errors. The type of inference we consider here was originally defined by Barzdin [1] (called GN^{∞} inference), and later independently by Case [2] (called BC inference) who expanded the definition to include the inference of programs which are allowed a finite number of errors (called BC^m inference). This latter type of inference has been studied extensively by Smith [2,3]. We use here the definitions and notations from Case and Smith.

We say that an inductive inference machine M BC^m *identifies* a total function f (written $f \in BC^m(M)$) if and only if when M is succesively fed the graph of f as input it outputs over time a sequence of programs $p_0, p_1, \ldots$ such that $(\overset{\infty}{\forall} n)[\phi_{p_n} =^m f]$, where $g =^m h$ means that g and h disagree at at most m places. Thus, when M is presented a function f it is permitted to change its mind infinitely often so long as eventually each of its conjectures contain at most m errors. We define the following classes,

$$BC^m = \{S \mid (\exists M)[S \subseteq BC^m(M)]\},$$

$$BC^m(M_1, M_2, \ldots, M_n) = \bigcup_{i=1}^{n} BC^m(M_i),$$

$$C(n, BC^m) = \{S \mid (\exists M_1, M_2, \ldots, M_n)[S \subseteq BC^m(M_1, M_2, \ldots, M_n)]\}.$$

Thus, a class of functions S belongs to $C(n, BC^m)$ if and only if there are n inductive inference machines $M_1, M_2, \ldots, M_n$ such that for each $f \in S$ one of them BC^m identifies f.

We now introduce notation which will be used throughout the paper. If T is a finite set, then $\# T$ denotes the cardinality of T. If f and g are functions, then $g(x) \downarrow\neq f(x)$ means that $g(x)$ is defined and unequal to $f(x)$, and $f \subseteq g$ means that f is a subfunction of g, i.e., $(\forall x \in \mathbf{dom}\ g)[g(x) = f(x)]$. We assume that ϕ_0 is the everywhere undefined function, and Φ_i denotes the computational complexity of ϕ_i. We will always use σ and τ to denote finite functions whose domains consist of an initial segment of the natural numbers. Finally, we introduce some special programs which will be used below. If σ is a finite function, then $\sigma \circ (x, v)$ is the finite function which agrees with σ except at x where it has the value v,

$$\sigma \circ (x, v)(y) = \begin{cases} v, & \text{if } y = x, \\ \sigma(y), & \text{otherwise.} \end{cases}$$

If σ is a finite function, and S is a set of programs, then the program $\langle \sigma : S \rangle$ is defined as follows:

$$\phi_{\langle \sigma : S \rangle}(x) = \begin{cases} \sigma(x), & \text{if } x \in \mathbf{dom}\ \sigma, \\ \phi_p(x), & \text{where } \Phi_p(x) = \mathbf{min}\{\Phi_q(x) \mid q \in S\}. \end{cases}$$

Thus, given x the program $\langle \sigma : S \rangle$ looks up x in the finite table σ and gives as output the corresponding value, and if x is not present in the table then it runs all programs in the set S on x and gives as output the output of the first program which halts.

From Smith [2] one easily obtains that $C(n+1, BC^0) \not\subseteq C(n, BC^m)$ for all m, so that it is not possible to trade errors for machines. We show now that it is always possible to trade machines for errors.

Theorem 1: $(\forall M)(\exists M_0, M_1)[BC^m(M) \subseteq BC^0(M_0) \cup BC^{m-1}(M_1)]$.

Proof: Suppose $f \in BC^m(M)$ and let $\{\sigma_i\}$ be the sequence of finite subfunctions of f which are fed to M and let $p_i = M(\sigma_i)$. Then every such f belongs to one of the following two cases:

Case 0: $(\overset{\infty}{\forall} i)[\phi_{p_i} \subseteq f]$.

Case 1: $(\overset{\infty}{\exists} i)(\exists x)[\phi_{p_i} \downarrow\neq f(x)]$.

Case 0 represents the situation where eventually all errors are errors of omission, and Case 1 represents the situation where infinitely many errors are errors of commission, i.e., mistakes. We now define machines M_0 and M_1 which will deal with Case 0 and Case 1 respectively.

M_0: Given σ as input, let $\sigma_0 \subset \sigma_1 \subset \ldots \subset \sigma_k = \sigma$ be the sequence of proper subfunctions of σ and let $q_i = M(\sigma_i)$. Then $M_0(\sigma) = \langle\sigma:\{\hat{q}_1,\ldots,\hat{q}_k\}\rangle$, where

$$\hat{q}_i = \begin{cases} 0, & \text{if } (\exists x \in \mathbf{dom}\ \sigma)(\exists t \le k)[\Phi_{q_i}(x) \le t \text{ and } \phi_{q_i}(x) \neq \sigma(x)], \\ q_i, & \text{otherwise.} \end{cases}$$

M_1: Given σ as input, M_1 requests further input until σ' and τ' are given such that $\sigma \subseteq \sigma' \subset \tau'$ and $\phi_{M(\sigma')}(x) \neq \tau'(x)$ for some $x \in \mathbf{dom}\ \tau'$. Then $M_1(\sigma) = \langle\tau':\{M(\sigma')\}\rangle$.

Returning to the given $f \in BC^m(M)$ suppose that Case 0 applies to f. Then, since all but finitely many p_k have at most m errors of omission, there exists an integer N so large that

1) $(\forall k \ge N)[\phi_{p_k} \subseteq f]$,

2) $(\forall k < N)[\phi_{p_k} \not\subseteq f \Rightarrow (\exists x \in \mathbf{dom}\ \sigma_N)[\Phi_{p_k}(x) \le N \text{ and } \phi_{p_k}(x) \neq f(x)]]$,

3) $(\exists k_0 < N)[\phi_{p_{k_0}} \subseteq f \text{ and } (\forall x)[\phi_{p_{k_0}}(x)\uparrow \Rightarrow x \in \mathbf{dom}\ \sigma_N]]$.

Then for all $\sigma_k \supseteq \sigma_N$, $M_0(\sigma_k) = \langle\sigma_k:S\rangle$ is such that $p_{k_0} = \hat{p}_{k_0} \in S$, and if $p_j \not\subseteq f$ and $\hat{p}_j \in S$ then $\phi_{\hat{p}_j} = \emptyset$. Therefore, for all x either $x \in \textbf{dom}\ \sigma_N$ or $\phi_{p_{k_0}}(x) = f(x)$ and for all $\hat{p}_j \in S$ if $\phi_{\hat{p}_j}(x)\downarrow$ then $\phi_{\hat{p}_j}(x) = f(x)$, so that $\phi_{M_0(\sigma_k)}(x) = f(x)$. Hence, $f \in BC^0(M_0)$.

Suppose now that Case 1 applies to f. Let N be so large that $(\forall k \geq N)[\phi_{p_k} =^m f]$. Then for any $\sigma_N \subseteq \sigma_k \subset f$ there exist $\sigma_k \subseteq \sigma_i \subset \sigma_j \subset f$ such that $\phi_{p_i}(x) \neq \sigma_j(x) = f(x)$ for some $x \in \textbf{dom}\ \sigma_j$. Then $M_1(\sigma_k) = \langle \sigma_j : \{p_i\} \rangle$ is such that

$$\phi_{M_1(\sigma_k)}(x) = \begin{cases} f(x), & \text{if } x \in \textbf{dom}\ \sigma_j, \\ \phi_{p_i}(x), & \text{otherwise.} \end{cases}$$

Since $\phi_{p_i} =^m f$, we have $\phi_{M_1(\sigma_k)} =^{m-1} f$, and hence $f \in BC^{m-1}(M_1)$.

■

By iterating the procedure given in Theorem 1 to the machine M_1 we obtain,

Theorem 2: $C(1, BC^m) \subseteq C(k+1, BC^{m-k})$ *for* $0 \leq k \leq m$.

From the remarks preceding Theorem 1 we have

Corollary 3: $C(1, BC^m) \subset C(k+1, BC^{m-k})$ *for* $0 \leq k \leq m$.

We also clearly have,

Corollary 4: $C(n, BC^m) \subset C(n \times (k+1), BC^{m-k})$ *for* $0 \leq k \leq m$.

The following corollary to Theorem 2 gives an upper bound on the number of machines required to eliminate all errors.

Theorem 5: $C(n, BC^m) \subset C(n \times (m+1), BC^0)$.

We now show that this upper bound is the least upper bound. We define the following class of sets of functions,

$$S_n^m = \{S \mid (\forall f \in S)(\exists k)[0 \le k < n \text{ and } (\overset{\infty}{\forall} x)[x \equiv k \textbf{ mod } n \Rightarrow \phi_{f(x)} =^m f]]\}.$$

Theorem 6: $S_n^m \in C(n, BC^m) - C(n \times (m+1) - 1, BC^0)$.

Proof: To see that $S_n^m \in C(n, BC^m)$ we define n machines M_k for $0 \le k < n$ as follows:

$$M_k(\sigma) = \sigma(\textbf{max}\{x \in \textbf{dom } \sigma \mid x \equiv k \textbf{ mod } n\}).$$

Let $N = n \times (m+1)$. We now give a construction which will produce for any set of $N-1$ machines $M_1, \ldots, M_{N-1}$ a total recursive function $f \in S_n^m$ such that $f \notin BC^0(M_1, \ldots, M_{N-1})$. Let $T_0 = \{M_1, \ldots, M_{N-1}\}$. The construction is an elaboration of the construction given by Case and Smith in [2] to show that $BC^m \subset BC^{m+1}$. The construction consists of a process which can create subprocesses (which in turn can create their own subprocesses, etc.), and it is the goal of these subprocesses to extend the finite portion of f thus far constructed in such a way that a certain number of the machines will make an error. The maximum depth of nesting of processes will be N, and each process will occur at some level of nesting from level 1 (the deepest) to level N. The success of this construction lies in the fact that there are N alternative ways of extending any finite segment of f and only $N-1$ machines which must be fooled. The main process is the only level N process and there may be infinitely many processes created at all other levels. However, at any given moment only one process at each level can be active. A process is initially everywhere undefined until it is created, and it remains active from the time it is created until it either returns to or is recalled by its parent process, from which point onward it duplicates the actions of its parent process. We denote the main process by p_N and for any $1 \leq i < N$, $p_{i,1}, p_{i,2}, \ldots$ will denote the sequence of level i processes which are created during the construction. Also, $\phi_{p_{i,s}}$ will denote the function defined by process $p_{i,s}$, and at any point in the construction $\tilde{\phi}_{p_{i,s}}$ will denote the finite segment of $\phi_{p_{i,s}}$ thus far defined. We use $p_{i,?}$ to denote the currenly active level i process. The construction is so designed that if $\phi_{p_{i,s}}$ is total and $\phi_{p_{i-1,t}}$ is any process created by it, then $\phi_{p_{i-1,t}} =^1 \phi_{p_{i,s}}$. Because of this and because f is so constructed that if $x \equiv k \textbf{ mod } n$ then $f(x) = p_{k \times (m+1)+1,?}$, it follows that $f \in S_n^m$. The goal of any level i process is to find an extension of the finite segment of the parent process's function thus far defined, which will cause i of the $N-1$ machines to make an error. When this goal is achieved it returns to its parent process with the set of i machines which have been fooled and the extension which fools them.

We give now the description of these processes -- how they behave both when they are created (and hence activated) and when they are recalled. In the following **create**$(i;\sigma)$ is a system call which creates the next level i process $p_{i,?}$ whose function $\phi_{p_{i,?}}$ is initialized to σ; **return**(σ,T) is a system call which returns the current process to its parent with the finite segment σ and the set of machines T; **returns**$(i;\sigma,T)$ is a predicate which states that the current level i process has returned with σ and T; and **recall**(i) is a system call which recalls the current level i subprocess.

Main process p_N:

> **repeat forever** {
>
> > **create**$(N-1;\tilde{\phi}_{p_N})$;
> >
> > **when returns**$(N-1;\sigma,T)$ $\tilde{\phi}_{p_N} \leftarrow \sigma$;

}

Level 1 process $p_{1,s}$: (*after* **create**$(_1;\sigma)$)

create an alias $P_{1,s}$; (i.e., $\phi_{P_{1,s}} \equiv \phi_{p_{1,s}}$)

$\tilde{\phi}_{p_{1,s}} \leftarrow \sigma$;

$S \leftarrow \{\sigma\}$;

repeat {

$y \leftarrow y+1$;

$\tilde{\phi}_{p_{1,s}}(y) \leftarrow p_{k\times(m+1)+1,?}$, where $y \equiv k$ **mod** n;

$S \leftarrow S \cup \{\tilde{\phi}_{p_{1,s}}\}$;

} **until** { $(\exists\tau\in S)(\exists M\in T_0)(\exists x \le y)$ **such that**

$[x \equiv 0$ **mod** n and $x \notin$ **dom** τ and $\Phi_{M_i(\tau)}(x) \le y]$

}

$v \leftarrow$ **min**$(\{p_{1,?}, P_{1,?}\} - \{\phi_{M_i(\tau)}(x)\})$;

return$(\tilde{\phi}_{p_{1,s}} \circ (x,v), \{M_i\})$;

Level i *process* $p_{i,s}$ $(1 < i < N)$: (*after* **create**$(_i;\sigma)$)

$\tilde{\phi}_{p_{i,s}} \leftarrow \sigma$;

create$(i-1;\sigma)$;

when **returns**$(i-1;\sigma',T')$ {

$\tilde{\phi}_{p_{i,s}} \leftarrow \sigma'$;

$S \leftarrow \{\sigma'\}$;

$T \leftarrow T'$;

}

repeat {

create$(i-1;\tilde{\phi}_{p_{i,s}})$;

when **returns**$(i-1;\sigma'',T'')$ {

$\tilde{\phi}_{p_{i,s}} \leftarrow \sigma''$;

$S \leftarrow S \cup \{\sigma''\}$;

$T \leftarrow T \cup T''$;

}

} **until** {

either { $(\exists\tau\in S)(\exists M\in T_0-T)(\exists x\in \textbf{dom}\ \tilde{\phi}_{p_{i,s}})$ **such that**

$[x \equiv 0$ **mod** n and $x \notin$ **dom** τ and $\phi_{M(\tau)}(x)\in$ **dom** $\tilde{\phi}_{p_{i,s}}]$

} **or** $\#T \geq i$

}

if $\#T \geq i$ **return**$(\tilde{\phi}_{p_{i,s}}, T)$;

else {

recall$(i-1)$;

when returns$(i-1;\sigma''',\emptyset)$ $\tilde{\phi}_{p_{i,s}} \leftarrow \sigma'''$;

$v \leftarrow$ **min**$(\{p_{1,?}, P_{1,?}\} - \{\phi_{M(\tau)}(x)\})$;

return$(\tilde{\phi}_{p_{i,s}} \circ (x,v), T\cup\{M\})$;

}

Process $p_{i,s}$: (*after* **recall**(i))

recall$(i-1)$; (only if $i > 1$)

when returns$(i-1;\sigma,\emptyset)$ $\tilde{\phi}_{p_{i,s}} \leftarrow \sigma$;

return$(\tilde{\phi}_{p_{i,s}}, \emptyset)$;

We illustrate the correctness of the construction by considering the case where $n = 2$ and $m = 1$. In this case there are 4 levels of processes, with p_4 being the main process, and $p_{1,?}$ being defined while it is active by,

$$\phi_{p_{1,?}}(y) = \begin{cases} p_{1,?}, & \text{if } y \text{ is even,} \\ p_{3,?}, & \text{if } y \text{ is odd.} \end{cases}$$

We consider the following four cases:

<u>Case 1:</u> Only finitely many level 1 processes are ever created. Let $p_{1,s}$ be the last level 1 process created. Since $p_{1,s}$ never returns, $\phi_{p_{1,s}}$ is total and $\phi_{p_{1,s}}(x) = p_{1,s}$ for all but finitely many even x. Let $f = \phi_{p_{1,s}}$, then clearly $f\in S_2^1$. Also, since $p_{1,s}$ never returns, for infinitely many $\sigma \subset \phi_{p_{1,s}}$ we have $\phi_{M_i(\sigma)}(x)\uparrow$ for

$i = 1,2,3$ for all but finitely many even x, so that $f \notin BC^0(M_1,M_2,M_3)$.

Case 2: There are infinitely many level 1 processes created, but only finitely many level 2 processes are ever created. Let $p_{2,s}$ be the last level 2 process created. Then $\phi_{p_{2,s}}$ is total, so let $f = \phi_{p_{2,s}}$. Now for all but finitely many even x, $f(x) = p_{1,t}$, where $p_{1,t}$ is some process created by $p_{2,s}$. When $p_{1,t}$ returns to $p_{2,s}$, $\tilde{\phi}_{p_{2,s}}$ is made equal to $\tilde{\phi}_{p_{1,t}}$ except for at most one even value x. Thus $\phi_{p_{1,t}} =^1 \phi_{p_{2,s}}$ and $f \in S^1_2$. Since $\phi_{p_{2,s}}$ never returns, $\#T < 2$ and since at least one level 1 process returns to $p_{2,s}$, $\#T \geq 1$ so that $\#T = 1$. Suppose $T = \{M_1\}$. Since infinitely many level 1 processes return with σ and $\{M_1\}$, where $\sigma(x) \neq \phi_{M_1(\tau)}(x)$ for some $\tau \subset \sigma$, and upon return $\tilde{\phi}_{p_{2,s}} \leftarrow \sigma$, we see that for infinitely many $\sigma \subset f$ there is an even x such that $\phi_{M_1(\sigma)}(x) \neq f(x)$ and so $f \notin BC^0(M_1)$. Moreover, since $p_{2,s}$ never returns, for infintely many $\sigma \subset f$ for all even $x \notin \mathbf{dom}\ \sigma$ $\phi_{M_2(\sigma)}(x)\uparrow$ and $\phi_{M_1(\sigma)}(x)\uparrow$. Therefore, $f \notin BC^0(M_2,M_3)$ also.

Case 3: There are infinitely many level 2 (and hence level 1) processes created, but only finitely many level 3 processes are ever created. Let $p_{3,s}$ be the last level 3 process created and let $f = \phi_{p_{3,s}}$. As in Case 2 above it can be easily shown that $f \notin BC^0(M_1,M_2,M_3)$. Now since $p_{3,s}$ is the last level 3 process created, from some point onward it is the active level 3 process, so that $f(x) = p_{3,s}$ for all but finitely many odd integers x. Therefore, $f \in S^1_2$.

Case 4: There are infinitely many level 3 processes created. Then ϕ_{p_4} is total, so let $f = \phi_{p_4}$. Each time a level 3 process returns to p_4 it returns σ and $\{M_1,M_2,M_3\}$ where σ is such that there exist distinct x_1, x_2, x_3 and τ_1, τ_2, τ_3 such that $\tau_i \subset \sigma$ and $\phi_{M_i(\tau_i)}(x_i) \neq \sigma(x_i) = f(x_i)$, for $i = 1,2,3$. Therefore, $f \notin BC^0(M_1, M_2, M_3)$. Furthermore, each subprocess $p_{3,s}$ of p_4 satisfies $\phi_{p_{3,s}} =^1 \phi_{p_4}$ so that $\phi_{f(x)} =^1 f$ for all but finitely many odd values x and so $f \in S^1_2$. ∎

We therefore see that for each n, m, and $0 \leq k \leq m$ that there exists a (least) integer $N_{n,m,k}$ such that $n < N_{n,m,k} < n \times (k+1)$ and that $C(n, BC^m) \subset C(N_{n,m,k}, BC^{m-n})$. It is an open question at this time what the precise value of $N_{n,m,k}$ is for $m \neq k$.

Acknowledgement: I am indebted to Carl Smith for bringing this interesting problem to my attention and for preliminary discussions on the problem. I have also learned from him that he has independently proved that $C(n, BC^m) - C(n \times m, BC^0) \neq \emptyset$.

References

1. Barzdin, J., "Two theorems on the limiting synthesis of functions", *Theory of Algorithms and Programs*, Latvian State University (1974), 82-88.
2. Case, J., and Smith, C., Anomaly hierarchies of mechanized inductive inference", Proceedings of the 10th Symposium on the Theory of Computing (1978), 314-319.
3. Smith, C., "The power of parallelism for automatic program synthesis", Technical Report CSD TR 350 (1980), Department of Computer Science, Purdue University.

---

†) This work was supported by NSF grant MCS-8017332.

EQUALITY LANGUAGES AND LANGUAGE FAMILIES

Jürgen Dassow
Technological University Otto von Guericke
Department of Mathematics and Physics
Magdeburg
German Democratic Republic

In the last years some papers have been published on so-called equality languages and fixed point languages which are defined by

$$EQ(h_1,h_2) = \{w : w \in X^*,\ h_1(w) = h_2(w)\} ,$$
$$FP(h) = \{w : w \in X^*,\ h(w) = w\} ,$$

where X is an alphabet, h, h_1, and h_2 are homomorphisms on X^*. These languages are often used to characterize language families (for instance see /BB/, /Br/, /Cu/, /CD/, /ER2/). The most results of this type are of the following form: For each recursively enumerable language L, there are homomorphisms h, h_1, h_2, and a regular set R such that

$$L = h(EQ(h_1,h_2) \cap R) .$$

Therefore it seems to be of interest to study the language families

$$EQ(\mathcal{L}) = \{EQ(h_1,h_2) \cap L : L \in \mathcal{L},\ h_1,h_2 \text{ homomorphisms}\} ,$$

where $\mathcal{L}$ is a language family.

This paper is devoted to a first study of these families. We shall prove some closure properties of $EQ(\mathcal{L})$ and some conditions which ensure that $\mathcal{L} = EQ(\mathcal{L})$ holds. Using these results we shall give further characterizations of some language families.

The results will be formulated for AFL's, i.e. language families which are closed under union, product, catenation closure, λ-free homomorphisms, inverse homomorphisms, and intersections with regular languages, or pre-AFL's, i.e. families closed under marked product L_1cL_2, marked catenation closure $(Lc)^+$ (where c is not contained in the alphabet), inverse homomorphisms, and intersections with regular languages, and union with $\{\lambda\}$, although they are sometimes also valid under weaker closure requirements (for instance for semi AFL's). An AFL is called full iff it is closed with respect to arbitrary homomorphisms. We assume that the reader is familiar with the rudiments of formal languages and AFL theory (see /S1/).

For an arbitrary homomorphism h and each language $L \in \mathcal{L}$ we have

$$EQ(h,h) \cap L = L$$

and thus

$$\mathcal{L} \subseteq EQ(\mathcal{L}) .$$

The replication c_1 and the reverse replication c_2 of a language L are defined by

$$c_1(L) = \{wcw : w \in L\} ,$$
$$c_2(L) = \{wcw^R : w \in L\} ,$$

where c is a new letter and $(x_1x_2 \ldots x_n)^R = x_nx_{n-1} \ldots x_1$. For an alphabet X, let

$$X' = \{x' : x \in X\} , X \cap X' = \emptyset .$$

We put

$$u(y) = \begin{cases} x & \text{if } y = x' \in X' \\ \lambda & \text{if } y \in X \end{cases} ,$$

$$u'(y) = \begin{cases} y & \text{if } y \in X \\ \lambda & \text{otherwise} \end{cases} .$$

Lemma 1 Let $\mathcal{L}$ be a full AFL closed under intersection. Then each of the following conditions implies $\mathcal{L} = EQ(\mathcal{L})$:
i) $EQ(u,u') \in \mathcal{L}$,
ii) $\mathcal{L}$ is closed under replication,
iii) $\mathcal{L}$ is closed under reverse replication.

Proof i) By /ER2/, for any homomorphisms h_1 and h_2, there is a homomorphism h such that

$$EQ(h_1,h_2) = h^{-1}(EQ(u,u')) ,$$

and therefore

$$EQ(h_1,h_2) \cap L = h^{-1}(EQ(u,u')) \cap L$$

for all $L \in \mathcal{L}$.

ii) We consider the generalized sequential machine A_{h_1,h_2} (for a general definition of these machines see e.g. /GGH/) given by

$$A_{h_1,h_2} = (Z,\ X \cup \{c\},\ Y,\ g,\ f,\ z_o)$$

where

$$Z = \{z_o,\ z_1\}\ ,\quad Y = h_1(X) \cup h_2(X) \cup \{c\}\ ,$$

$$g(z_o,x) = \begin{cases} z_o & \text{if } x \in X \\ z_1 & \text{if } x = c \end{cases},\qquad g(z_1,x) = z_1 \quad \text{for } x \in X \cup \{c\},$$

$$f(z_o,x) = \begin{cases} h_1(x) & \text{if } x \in X \\ c & \text{if } x = c \end{cases},\qquad f(z_1,x) = \begin{cases} h_2(x) & \text{if } x \in X \\ c & \text{if } x = c \end{cases}.$$

Further, let $A = (Z,\ X \cup \{c\},\ X,\ g,\ f',\ z_o)$ with

$$f'(z_o,x) = \begin{cases} x & \text{if } x \in X \\ \lambda & \text{otherwise} \end{cases},\qquad f'(z_1,x) = \lambda \quad \text{for } x \in X \cup \{c\}.$$

Then

$$EQ(h_1,h_2) \cap L = A(A^{-1}_{h_1,h_2}(c_1(Y \setminus \{c\})^*) \cap c_1(X^*)) \cap L\ .$$

This implies $EQ(\mathcal{L}) \subseteq \mathcal{L}$.

iii) We have

$$c_2(c_2(L)) = \{wcw^Rc'wcw^R\ :\ w \in L\}\ ,$$

and it is easy to construct a gsm-mapping B such that

$$c_1(L) = B(c_2(c_2(L)))\ .$$

Let REG, CS, R, RE denote the families of regular, context sensitive, recursive, and recursively enumerable languages, respectively.

<u>Corollary 2</u> $EQ(CS) = CS,\quad EQ(R) = R,\quad EQ(RE) = RE$.

<u>Proof</u> Salomaa (/S2/) proved $EQ(u,u')$ CS. Since Lemma 1 i) holds for AFL's, too, the assertion follows.

<u>Lemma 3</u> i) RE is the smallest full (semi) AFL with $\mathcal{L} = EQ(\mathcal{L})$.

ii) RE is the smallest full (semi) AFL closed under intersection and replication.

<u>Proof</u> i) By the above result, for $L \in RE$, we have

$L = h(EQ(h_1,h_2) \cap R)$ for some homomorphisms h, h_1, h_2, and $R \in REG$. Hence any full (semi) AFL $\mathcal{L}$ with $\mathcal{L} = EQ(\mathcal{L})$ contains RE. Now the statement follows from Corollary 2.
ii) follows by Lemma 1 and i).

Remark In /Bo/ Book proved that RE is the smallest full semi AFL which is closed under intersection and homomorphic replication. Since c_1 is a special homomorphic replication and there are homomorphic replications which cannot be simulated by iterated applications of c_1, Lemma 3 is a slight improvement of Book's result. We also believe that our proof is easier.

Lemma 4 Let $\mathcal{L}$ be a full AFL with $\mathcal{L} = EQ(\mathcal{L})$, then $\mathcal{L}$ is closed under replication and intersection.

Proof replication: Let $X' = \{x' : x \in X\}$ be a primed version of X, $X' \cap X = \emptyset$. For $w = x_1x_2 \ldots x_n \in X^*$ and $L \subseteq X^*$ we put $w' = x_1'x_2' \ldots x_n'$ and $L' = \{w' : w \in L\}$, respectively. Then $LcL' \in \mathcal{L}$ and therefore

$$L_1 = \{wcw' : w \in L\} = EQ(u,u') \cap LcL' \in EQ(\mathcal{L}) = \mathcal{L} ,$$

where $u(c) = u'(c) = \lambda$. Let u_1 be the homomorphism defined by

$$u_1(y) = \begin{cases} u(y) & \text{for } y \in X' \\ u'(y) & \text{for } y \in X \\ c & \text{for } y = c . \end{cases}$$

Then $c_1(L) = u_1(L_1)$.
intersection: Let $L_1, L_2 \in \mathcal{L}$. Then we have

$$L'' = \{wcw' : w \in L_1 \cap L_2\} = EQ(u,u') \cap L_1cL_2' \in EQ(\mathcal{L}) = \mathcal{L} ,$$

and clearly, we can construct a generalized sequential machine which maps L'' on $L_1 \cap L_2$.

Combining Lemma 1 and Lemma 4 we obtain

Theorem 5 A full AFL $\mathcal{L}$ is closed under intersection and replication if and only if $\mathcal{L} = EQ(\mathcal{L})$.

Lemma 6 Let $\mathcal{L}$ be an AFL.
i) Then $EQ(\mathcal{L})$ is closed under inverse homomorphisms and intersection with regular languages.

ii) The following statements are equivalent:

a) $\mathcal{L} = EQ(\mathcal{L})$,

b) $EQ(\mathcal{L})$ is closed under union,

c) $EQ(\mathcal{L})$ is closed under homomorphisms,

d) $EQ(\mathcal{L})$ is closed under product.

Proof i) follows by

$$(EQ(h_1,h_2) \cap L) \cap R = EQ(h_1,h_2) \cap (L \cap R),$$

$$h^{-1}(EQ(h_1,h_2) \cap L) = EQ(hh_1,hh_2) \cap h^{-1}(L) .$$

ii) Obviously, a) implies b), c), and d).

b) $\Rightarrow$ a) Assume that $EQ(\mathcal{L}) \setminus \mathcal{L} \neq \emptyset$, $L \in EQ(\mathcal{L}) \setminus \mathcal{L}$. Further let V be the set of letters which occur in words of L. Clearly, $V \in EQ(\mathcal{L})$. We prove $L \cup V \notin EQ(\mathcal{L})$. Assume

$$L \cup V = EQ(h_1,h_2) \cap L_1, \quad L_1 \in \mathcal{L} .$$

Because $L \subseteq V^*$, we have $L_1 \subseteq V^*$. Since $h_1(x) = h_2(x)$ for $x \in V$, we obtain $V^* \subseteq EQ(h_1,h_2)$. Therefore

$$L \cup V = L_1 \cap V^* = L_1 .$$

Thus $L = L_1 \setminus (V \setminus L)$. This contradicts $L \notin \mathcal{L}$.

c) $\Rightarrow$ a) Let L, L_1, h_1, h_2, V be as in b) $\Rightarrow$ a). Let $V' = \{x' : x \in V\}$. We put

$$h_i'(y) = \begin{cases} h_i(y) & \text{if } y \in V \\ y & \text{if } y \in V' \end{cases} , \quad i = 1,2,$$

and obtain

$$EQ(h_1',h_2') \cap (L_1 \cup V') = L \cup V' \in EQ(\mathcal{L}) .$$

Now let h be the homomorphism

$$h(y) = \begin{cases} x & \text{if } y = x \in V \\ x & \text{if } y = x' \in V' . \end{cases}$$

Then

$$h(L \cup V') = L \cup V \in EQ(\mathcal{L}) .$$

d) $\Rightarrow$ a) can be proved analogously.

Lemma 7 If $\{a^n b^n : n \geq 0\} \in EQ(h_1,h_2)$, then $h_1 = h_2$ or $h_1(a), h_2(a), h_1(b), h_2(b) \in v^*$ for some word v.

Proof Let $h_1 \neq h_2$. Then $h_1(a) = w_1^{\alpha}$, $h_2(a) = w_1^{\beta}$, $h_1(b) = w_2^{\gamma}$, $h_2(b) = w_2^{\delta}$ by /ER1/, Lemma 2. Without loss of generality we can assume that $\alpha < \beta$. Then $\delta < \gamma$. Since $w_1^{\alpha n} w_2^{\beta n} = w_1^{\gamma n} w_2^{\delta n}$, we obtain $w_2^{(\gamma-\delta)n} = w_1^{(\beta-\alpha)n}$, and thus $w_1, w_2 \in v^*$ by /ER1/, Lemma 2.

Now we are in a position to add a further closure result.

Lemma 8 EQ(REG) is not closed under catenation closure.

Proof Obviously, $S = \{a^n b^n : n \geq 0\} = EQ(h_1,h_2) \cap a^* b^*$ where

$$h_1(a) = a^2, \quad h_1(b) = a,$$
$$h_2(a) = a, \quad h_2(b) = a^2.$$

We prove $S^+ \notin EQ(REG)$. Assume the contrary, i.e.

$$S^+ = EQ(g_1,g_2) \cap L, \quad L \in REG.$$

Since $L \in REG$ and $a^n b^n a^m b^m \in L$ for all n, m, we can prove by usual methods of automata theory that there are numbers r and s such that

$$a^n b^{n+kr} a^{m+ls} b^m \in L \quad \text{for all } k, l \in \mathbb{N},\ n \geq n_0,\ m \geq m_0.$$

By Lemma 7 we have to regard the following two cases.
Case 1. $g_1 = g_2$. Then $EQ(g_1,g_2) \cap L = L$ and hence $S^+ \in REG$. This is a contradiction.
Case 2. $g_1(a) = v^i$, $g_2(a) = v^x$, $g_1(b) = v^j$, $g_2(b) = v^y$. Without loss of generality we assume $i > x$. We put $l = r(y-j)$ and $k = s(i-x)$ and obtain

$$g_1(a^n b^{n+kr} a^{m+ls} b^m) = v^{i(n+m+sr(y-j))+j(n+m+rs(i-x))},$$

$$g_2(a^n b^{n+kr} a^{m+ls} b^m) = v^{x(n+m+rs(y-j))+y(n+m+rs(i-x))}.$$

Since $a^n b^n a^m b^m \in EQ(g_1,g_2) \cap L$, we have

$$i(n+m) + j(n+m) = x(n+m) + y(n+m),$$

and now it is easy to check that

$$g_1(a^n b^{n+kr} a^{m+ls} b^m) = g_2(a^n b^{n+kr} a^{m+ls} b^m).$$

This contradicts the structure of S^+.

Theorem 9 Let $\mathcal{L}$ be a pre-AFL and let $\mathcal{L}_1$ be the smallest full AFL which contains $\mathcal{L}$ and which is closed under intersection and replication. Then $L \in \mathcal{L}_1$ if and only if

$$L = h(EQ(h_1,h_1') \cap EQ(h_2,h_2') \cap \ldots \cap EQ(h_n,h_n') \cap L')$$

for some $L' \in \mathcal{L}$ and some homomorphisms $h, h_1, h_1', \ldots, h_n, h_n'$.

Proof Let $H_\lambda(\mathcal{S})$ and $H(\mathcal{S})$ denote the closure of $\mathcal{S}$ under λ-free and arbitrary homomorphisms, respectively.
First we prove that $H_\lambda(EQ(\mathcal{L}))$ is a pre-AFL. We have

$$h(EQ(h_1,h_2) \cap L) \cap R = h(EQ(h_1,h_2) \cap L \cap h^{-1}(R))$$
$$= h(EQ(h_1,h_2) \cap L')$$

where $L' = L \cap h^{-1}(R) \in \mathcal{L}$. Further

$$h_2^{-1}h_1(L) = h_3(h_4^{-1}(L) \cap R)$$

for some $R \in REG$ and some homomorphisms h_3, h_4, where h_3 is λ-free, by /GGH/, page 43/44. If $L \in EQ(\mathcal{L})$, then $h_2^{-1}h_1(L) \in H_\lambda(EQ(\mathcal{L}))$ by Lemma 6 i).
Now let $L = EQ(h_1,h_2) \cap L'$. Then $(Lc)^+ = EQ(h_1',h_2') \cap (L'c)^+$ where

$$h_i'(x) = \begin{cases} h_i(x) & \text{if } x \in X \\ c & \text{if } x = c \end{cases}, \quad i = 1,2.$$

If h is an arbitrary λ-free homomorphism, then we put

$$h'(x) = \begin{cases} h(x) & \text{if } x \quad X \\ c & \text{if } x = c \end{cases}$$

and obtain $h'(Lc)^+) = (h(L)c)^+$. This proves the closure under marked catenation closure.
Now let

$$L_i = h_i(EQ(h_i',h_i'') \cap L_i'), \quad i = 1,2.$$

Without loss of generality we can assume that L_1', h_1, h_1', h_1'', and L_2', h_2, h_2', h_2'' are defined over disjoint alphabets X_1 and X_2. Then let

$$h_3(x) = \begin{cases} h_1(x) & \text{if } x \in X_1 \\ h_2(x) & \text{if } x \in X_2 \\ c & \text{if } x = c, \end{cases}$$

and analogously we define h_3' and h_3''. Then

$$L_1cL_2 = h_3(EQ(h_3',h_3'') \cap L_1'cL_2') .$$

The closure under union with $\{\lambda\}$ is obvious.
We define

$$EQ^1(\mathcal{L}) = EQ(\mathcal{L}),$$

$$EQ^{n+1}(\mathcal{L}) = EQ(EQ^n(\mathcal{L}))$$

and

$$EQ^+(\mathcal{L}) = \bigcup_{n \geq 1} EQ^n(\mathcal{L}) .$$

As above we can prove that $H_\lambda(EQ^+(\mathcal{L}))$ is a pre-AFL. Therefore $\mathcal{L}'' = H(EQ^+(\mathcal{L}))$ is a full AFL by /GGH/. Let $L \in EQ(\mathcal{L}'')$, i.e.

$$L = EQ(h_o,h_o') \cap h(EQ(h_1,h_1') \cap EQ(h_2,h_2') \cap \ldots \cap EQ(h_n,h_n') \cap L')$$

for some homomorphisms h, h_o,h_o', $h_1,h_1',\ldots,$ h_n,h_n', and some $L' \in \mathcal{L}$.
Then

$$L = h(EQ(hh_o,hh_o') \cap EQ(h_1,h_1') \cap EQ(h_2,h_2') \cap \ldots \cap EQ(h_n,h_n') \cap L').$$

Hence we have $L \in \mathcal{L}''$, i.e. $\mathcal{L}'' = EQ(\mathcal{L}'')$. By Theorem 5 $\mathcal{L}''$ is closed under intersection and replication. Clearly, it is minimal with respect to these closure properties.

Finally we prove a result on the effect of the restriction to λ-free homomorphisms.
By $EQP(\mathcal{L})$ we denote the set of all languages of the form $EQ(h_1,h_2) \cap L$ with $L \in \mathcal{L}$ and λ-free homomorphisms h_1 and h_2.

<u>Lemma 10</u> $EQP(REG) \subsetneq EQ(REG)$.

<u>Proof</u> Let

$$h_1(a) = a^2,\ h_1(b) = a,\ h_1(c) = c^2,\ h_1(d) = c,\ h_1(e) = \lambda,$$

$$h_2(a) = a,\ h_2(b) = a^2,\ h_2(c) = c,\ h_2(d) = c^2,\ h_2(e) = \lambda,$$

$$h(a) = a,\ h(b) = b,\ h(c) = c,\ h(d) = d,\ h(e) = \lambda .$$

Then

$$L = EQ(h_1,h_2) \cap h^{-1}(a^*b^*c^*d^*) = h^{-1}(\{a^nb^nc^md^m : n, m \geq 0\}).$$

Hence $L \in EQ(REG)$. We prove $L \notin EQP(REG)$.
Assume $L = EQ(g_1,g_2) \cap L_1$ where $L_1 \in REG$ and g_1 and g_2 are λ-free homomorphisms. By Lemma 7, g_1 and g_2 agree on $\{a,b\}$ or $g_1(a), g_1(b), g_2(a), g_2(b) \in v^*$ for some v. Again, the first case is impossible because

$$\{a^nb^n : n \geq 0\} = L \cap a^*b^* = EQ(g_1,g_2) \cap L_1 \cap a^*b^* = L_1 \cap a^*b^* \in REG$$

does not hold. Consider the words $u_n = a^ne^nb^n$. We have $u_n \in L$ and $g_1(u_n) = g_2(u_n)$. If $g_i(e) = w_i$, $i = 1,2$, we conclude that

$$g_1(u_n) = v^{in}w_1{}^nv^{jn}, \quad g_2(u_n) = v^{xn}w_2{}^nv^{yn},$$

where we have used the same powers as in the proof of Lemma 8. Without loss of generality we have $i > x$ and therefore

$$w_2{}^nv^{yn} = v^{(i-x)n}w_1{}^nv^{jn} \quad \text{and} \quad v^{in}w_1{}^n = v^{xn}w_2{}^nv^{(y-j)n}.$$

By /ER1/, Lemma 2, hence $w_1, w_2, v \in v_1{}^*$ for some word v_1, i.e. $g_1(a), g_2(a), g_1(b), g_2(b), g_1(e), g_2(e) \in v_1{}^*$.
The same arguments can be applied to c,d. Therefore $g_i(z) \in v_2{}^*$ for $z \in \{c,d,e\}$. Thus $g_i(x) \in w^*$ for some word w, $i = 1,2$, $x \in \{a,b,c,d,e\}$. Now we can construct a contradiction as in Lemma 8.

If we consider the fixed point languages and the associated language family

$$FP(\mathcal{L}) = \{FP(h) \cap L : L \in \mathcal{L}, \text{ h homomorphism}\},$$

the situation is very simple. We have

<u>Theorem 11</u> For any AFL $\mathcal{L}$, $FP(\mathcal{L}) = \mathcal{L}$.

<u>Proof</u> $FP(h)$ is a regular language (see /ER1/).

References

/Bo/ R.V.Book, Simple representations of classes of languages. J. Assoc. Comput. Mach. 25, 23-31 (1978).

/BB/ R.V.Book and F.-J.Brandenburg, Equality sets and complexity classes. Techn. report, University of Santa Barbara, 1979.

/Br/ F.-J.Brandenburg, Multiple equality sets and Post machines. Techn. report, University of Santa Barbara, 1979.

/Cu/ K.Culik II, A purely homomorphic characterization of recursively enumerable sets. J. Assoc. Comput. Mach. 26, 345-350 (1979).

/CD/ K.Culik II and N.D.Diamond, A homomorphic characterization of time and space complexity classes of languages. Techn. report, University of Waterloo, 1979.

/ER1/ J.Engelfriet and G.Rozenberg, Equality languages and fixed point languages. Information and Control 43, 20-49 (1979)

/ER2/ J.Engelfriet and G.Rozenberg, Fixed point languages, equality languages and representations of recursively enumerable languages. J. Assoc. Comput. Mach., to appear.

/GGH/ S.Ginsburg, S.Greibach, and J.Hopcroft, Studies in Abstract Families of Languages. Memoirs of the Amer. Math. Soc., No. 87, 1969.

/S1/ A.Salomaa, Formale Sprachen. Spriger-Verlag, 1978.

/S2/ A.Salomaa, Equality sets for homomorphisms of free monoids. Acta Cybernetica 4, 127-139 (1978).

EXTREMAL COMBINATORIAL PROBLEMS IN RELATIONAL DATA BASE

by

J. DEMETROVICS
Computer and Automation Institute
Hungarian Academy of Sciences

G.O.H. KATONA
Mathematical Institute of the Hungarian
Academy of Sciences

1. INTRODUCTION

One of the possible models of a data base is the following "relational" model introduced by Codd [1]. In a data base there are several different data of several different individuals. E.g. a data base may contain the name, the place of birth, the date of birth, and so on ... of different persons. The possible data are called *attributes* while the whole of the data of one individual is its *record*. In the above example the name, the place of birth, the date of birth are attributes. The whole of data of one person is a record. It is rather natural to describe this system by a matrix whose rows and columns correspond to the records and attributes, respectively. The entries of the jth row or jth attribute can be chosen from a set D_j where these sets are not necessarily disjoint. However, for our investigations the choice of sets D_j plays no role, thus we may assume the D_j's are equal to the set of non-negative integers. That is, we shall simply consider $m\times n$ matrices with non-negative entries. As two different individuals should have different records we may assume that the rows of the matrix are different.

For a fixed $m\times n$ matrix M, a set $K\subset\{1,\dots,n\}$ of columns is called a *key* iff the rows of the submatrix $M(K)$ determined by the given columns are different. That is, K is a key iff the attribute corresponding to K *uniquely* determine the records (individual) in the sense that given any values of these attributes there is at most one record having these values. A key K is called *minimal* iff there is no different key K' satisfying $K'\subset K$.

Let $A\subset B\subset\{1,\dots,n\}$. We say (write) that A implies B $(A\to B)$ when two rows of $M(A)$ are equal iff the corresponding rows of $M(B)$ are also equal. In other words, A implies B when the values of the attributes in A uniquely determine the values of the attributes in B. (That is, A is a "key" in B in a certain sense.) If $A\to B$ then the pair (A,B) is called a *functional dependency* or simply a *depen-*

dency. A dependency (A,B) is called *basic* iff

1) $A \neq B$
2) there is no $A' \subset A$, $A' \neq A$ such that $A' \to B$
3) there is no $B' \supset B$, $B' \neq B$, such that $A \to B'$.

There are many natural extremal problems concerning the above concepts. The present paper gives a *survey* of the not very numerous results of this type. However, the main air is to call the reader's attention to these applicable and mathematically interesting problems.

The next section determines the maximal number of minimal keys when the number n of attributes is given. Section 3 gives lower and upper estimates on the maximal number of basic dependencies. In the last section there are some initial results in the determination of the smallest number m of rows such that there is a data base ($m \times n$ matrix) in which the family of minimal keys is given beforehand.

2. MAXIMUM NUMBER OF MINIMAL KEYS

The keys play an important role in data bases. The records can be uniquely found by them. Of course, it is worth-while to consider the minimal ones, only. It is quite natural to ask at most how many minimal keys can exist.

Theorem 1 [3]. *The maximal number of minimal keys in a data base with n attributes is*

$$(1) \qquad \binom{n}{\left[\frac{n}{2}\right]}$$

Proof. The minimal keys K are subsets of $\{1,\dots,n\}$ and do not contain each other. Sperner's well-known theorem [2] states that such a family can not contain more than $\binom{n}{\left[\frac{n}{2}\right]}$ members.

We will now construct an $m \times n$ matrix (with $m = \binom{n}{\left[\frac{n}{2}\right] - 1} + 1$) having $\binom{n}{\left[\frac{n}{2}\right]}$ minimal keys.

The first row of M consists of nothing but 1's. The other rows contain $\left[\frac{n}{2}\right] - 1$ 1's in all possible ways while the remaining entries of the i-th row are i's $\left(2 \leq i \leq \binom{n}{\left[\frac{n}{2}\right] - 1} + 1\right)$. For $n = 4$, see the matrix below:

$$\begin{matrix} 1 & 1 & 1 & 1 \\ 1 & 2 & 2 & 2 \\ 3 & 1 & 3 & 3 \\ 4 & 4 & 1 & 4 \\ 5 & 5 & 5 & 1 \end{matrix} .$$

If we chose $\left[\frac{n}{2}\right]$ places in a row then we find there either only 1's or at least one number i different from 1. Therefore the row i is uniquely determined. Any K with $|K| = \left[\frac{n}{2}\right]$ is a key. On the other hand, it is easy to see that no set K with $|K| < \left[\frac{n}{2}\right]$ can be a key, the first row coincides with another one in $M(K)$. The proof is complete.

Let us remark that a stronger statement is proved in [3]. There is a matrix M for any prescribed family of keys if it has the Sperner-property.

3. MAXIMUM NUMBER OF BASIC DEPENDENCIES

All dependencies trivially follow from the basic dependencies. Therefore their number can be considered the complexity of the data base. Thus our aim given in the title of the section is in fact equivalent to the problem of finding the most complex data base.

Let $N(n)$ denote the maximum number of basic dependencies in a data base with n attributes (or of a matrix with n columns).

It is easy to construct amatrix in which the basic dependencies are of the form $A \to A \cup \{x\}$ where x is a fixed attribute. That is, $2^{n-1} \leq N(n)$. However, the real value is greater:

Theorem 2 [4]

$$(2) \qquad 2^n \left(1 - \frac{4}{\log_2 e} \frac{\log_2 \log_2 n}{\log_2 n} ((1 + o(1))\right) \leq N(n) \leq 2^n \left(1 - \frac{1}{n+1}\right)$$

Sketch of the proof (It can be found in [4]) *Lower estimate.* Let $q_1, q_2, \dots, q_k$ be positive integers satisfying $\sum_{i=1}^{k} q_i = n$. The $(\binom{q_i}{2}+1) \times q_i$ matrix Q_i is defined in the following way. The first row of Q_i consists of nothing but 1's. The other rows contain $q_i - 2$ 1's in all the possible ways. In the rest of the places we put i's in the ith row $(2 \leq i \leq \binom{q_i}{2} + 1)$. The rows of the matrix $M = Q_1 \times \ldots \times Q_k$ are all the $\prod_{i=1}^{k} (\binom{q_i}{2} + 1)$ possible combinations of the rows of $Q_1, \ldots, Q_k$. (More precisely, we put together an arbitrary row of Q_1, then an arbitrary row of Q_2 e.t.c., in this order.) It follows from $\sum_{i=1}^{k} q_i = n$ that M has n columns. We show the construction for $n = 5$, $k = 2$, $q_1 = 3$, $q_2 = 2$:

1	1	1	:	1	1
1	1	1	:	2	2
1	2	2	:	1	1
1	2	2	:	2	2
3	1	3	:	1	1
3	1	3	:	2	2
4	4	1	:	1	1
4	4	1	:	2	2 .

Let Q_i^* denote the set of indices of columns of Q_i in M, that is $Q_i^* = \{q_1 + \ldots + q_{i-1} + 1, \ldots, q_1 + \ldots + q_i\}$. One can see that the basic dependencies (A,B) in M are those satisfying

$$\min(q_i - |A \cap Q_i^*|) = 1$$

$$B = A \cup \bigcup_{|A \cap Q_i^*| = q_i - 1} Q_i^* .$$

Their number is

$$(3) \qquad \prod_{i=1}^{k} (2^{q_i} - 1) - \prod_{i=1}^{k} (2^{q_i} - q_i - 1).$$

This give a lower estimate on $N(n)$. In the above example $(n=5,\ q_1=3,\ q_2=3)$ this estimate gives $17 \leq N(5)$

For large n we choose first

$$q = q(n) = \left[\log n - \log \left(\frac{1}{\log e}(\log \log n - \log \log \log n - \log \log e - 1)\right)\right]$$

where $\log$ means the logarithm of base 2. Then $k = k(n)$ and $r = r(n)$ are defined by $n = q(n)\, k(n) + r(n)$ $\quad 0 \leq r(n) < q(n)$.

Finally choose

$$q_1 = q_2 = \ldots = q_r = q + 1$$

$$q_{r+1} = \ldots = q_k = q.$$

With this choice (3) gives the left hand side of (2) after some long but elementary calculations.

Upper estimate. Let H be the family of sets A having a pair B such that (A,B) is a basic dependency. It can be seen that the set C satisfying $A \subset C \subset B$ $|C| = |A| + 1$ is not an element of H. Such a C can be obtained from at most n different sets A only, consequently for at least $|H|/n$ sets C $C \notin H$, implying $|H| + \frac{|H|}{n} \leq 2^n$. This is equivalent to the right hand side of (2). The theorem is proved.

That is, it remains an *Open question*: what is the proper second term of $N(n)$?

4. SMALLEST NUMBER OF INDIVIDUALS REALIZING A GIVEN FAMILY OF MINIMAL KEYS

Suppose a family F of subsets of n-element sets is given and it is known that the data base has exactly these sets as minimal keys. It is useful to know something about the number of rows of the data base. This section offers some results along this line.

Let F satisfy the Sperner-property, that is, $A,B \in F$ $A \neq B$ implies $A \not\subset B$. Then $s(F)$ denotes the minimum number m of the rows of an $n \times m$ matrix M in which the set of minimal keys is F.

<u>*Theorem 3.*</u> [5] *For any* $n > 0$ *there is an* F *satisfying*

(4) $$s(F) \geq \frac{1}{n^2} \binom{n}{\left[\frac{n}{2}\right]} .$$

Proof. Let $s(n) = \max_F s(F)$, where F runs over the families satisfying the Sperner-property. We say that M *realizes* F if the set of minimal keys of M is F. Then any F can be realized by an $m \times n$ M where $m \leq s(n)$. Write completely new and different integers into the $(m+1)$th, $(m+2)$th,...,$s(n)$th row. The new enlarged $s(n) \times n$ matrix realizes the same F. Consequently any F can be realized by an $s(n) \times n$ matrix. Furthermore, we may suppose that the entries of these matrices are $1, 2, \ldots, s(n)$, only.

Indeed, if the integers found in the first column of M are $i_1 < i_2 < \ldots i_r$ $(1 \leq r \leq s(n))$ then let us make the changes $i_j \to j$. This change does not change the family of minimal keys. Thus, following the same procedure with all the columns independently, we finally arrive to a $s(n) \times n$ matrix containing only $1, \ldots, s(n)$ as entries. The number of such matrices is $s(n)^{n\, s(n)}$, therefore we have

(5) $$s(n)^{n\, s(n)} > \text{number of } F\text{'s}.$$

Any family of some $\left[\frac{n}{2}\right]$-element sets has the Sperner-property, consequently,

(6) $$s(n)^{n\, s(n)} > 2^{\binom{n}{[\frac{n}{2}]}}$$

follows from (5). (6) results in (4). The theorem is proved.

Now we will try to determine $s(F)$ for some simple particular F's. Let F_k^n denote the family of k-element subsets of an n-set. First an easy lemma.

Lemma 1. *For any* $0 < k \leq n$

(7) $$\binom{s(F_k^n)}{2} \geq \binom{n}{k-1}$$

holds.

Proof. Suppose that an $m \times n$ matrix M realizes F_k^n. For any $(k-1)$-element set A of columns there is a pair of rows in which these columns have identical entries. Moreover for different A's these pairs of rows must be different. Hence $\binom{m}{2} \geq \binom{n}{k-1}$ and (7) follow. The lemma is proved.

From this lemma $s(F_1^n) \geq 2$ follows. The construction

$$\begin{matrix} 0 & 0 & \dots & 0 \\ 1 & 1 & \dots & 1 \end{matrix}$$

shows that $s(F_1^n) = 2$.

Let us determine $s(F_2^n)$. If an $m \times n$ matrix realizes F_2^n then by the above lemma

(8) $$n \leq \binom{m}{2}$$

holds. On the other hand, if (8) is satisfied we are able to construct an $m \times n$ matrix M realizing F_2^n. M contains two 0's in each column. Of course the pair of places is different for different columns. The other entries of the ith row will be i's $(1 \leq i \leq m)$. Consequently $s(F_2^n)$ is the smallest integer m satisfying (8).

Let us apply lemma 1 for $k = n-1$. We obtain $s(F_{n-1}^n) \geq n$. On the other hand $m = n$ is enough for the construction:

$$\begin{matrix} 1 & 0 & \dots & 0 \\ 0 & 1 & \dots & 0 \\ \vdots & & & \\ 0 & 0 & \dots & 1 \end{matrix} ,$$

this leads to $s(F_{n-1}^n) = n$.

The determination of $s(F_n^n)$ needs a slightly harder consideration. If M is an $m \times n$ matrix, let $G(M)$ denote the graph whose vertices are the rows of M, two vertices are connected with an edge iff the set A of places where the two rows are equal is non-empty. The edge is *labelled* by A.

Lemma 2. *Let M be a matrix and let $A_1, \dots, A_r$ be the labels along a circuit of $G(M)$. Then*

(9) $$\bigcap_{\substack{i=1\\ i\neq j}}^{r} A_i - A_j = \emptyset$$

Proof. Suppose that, in the contrary, (9) is non-empty, that is, there is a column, say the uth one, which is an element of all A_i but A_j. Let the vertices of the circuit be $k_1,\ldots,k_r$ in such a way that the edge (k_i, k_{i-1}) is labelled by A_i $(1 \leq i < r)$ and the edge (k_r, k_1) is labelled by A_r. From $u \in A_{j+1}$ it follows that the k_{j+1}th and k_{j+2}th entries in the uth column are equal. The same holds for the k_{j+2}th and k_{j+3}th entries, etc. Consequently, the k_{j+1}th, k_{j+2}th,...,k_rth, k_1th,...,k_jth entries in the uth column are all equal. This leads to $u \in A_j$ contradicting the assumption. The proof is complete.

Now we are able to determine $s(F_n^n)$. Suppose that the $m \times n$ matrix M realizes F_n^n. An $(n-1)$-element subset A of $\{1,\ldots,n\}$ is not a key therefore there must be an edge in $G(M)$ labelled with A. Consequently $G(M)$ has n different edges labelled with the $(n-1)$-element subsets of $\{1,\ldots,n\}$. These edges cannot form a circuit because the $(n-1)$-element subsets cannot satisfy (9), the lemma is applicable. $G(M)$ has at least $n+1$ vertices, $m \geq n+1$. The following $(n+1) \times n$ matrix realizes F_n^n

$$\begin{matrix} 0 & 0 & \ldots & 0 \\ 1 & 0 & \ldots & 0 \\ 0 & 1 & \ldots & 0 \\ \vdots & & & \\ 0 & 0 & \ldots & 1 \end{matrix} .$$

We summarize our moderate results in a theorem.

Theorem 4

$$s(F_1^n) = 2, \qquad s(F_2^n) = \left[\frac{1+\sqrt{1+4n}}{2} \right]$$

$$s(F_n^n) = n+1 , \qquad s(F_{n-1}^n) = n.$$

In the case of $k = 3$ lemma 1 gives $s(F_3^n) \geq n$, again. There are problems, however, with the construction of an $n \times n$ matrix realizing F_3^n. Checking the proof of lemma 1 we can see that equality can stand in (7) iff for any pair of rows there are exactly two equal entries. In fact a column is a partition $P_i = (A_{i1}, \ldots, A_{ir_i})$ on n elements (the set of rows) $(1 \leq i \leq n)$. These partitions have to satisfy the following conditions:

1) For any pair x,y of elements of the n-element set X there are exactly two sets A_{ip} and A_{jq} containing both of them $(i \neq j)$

2) For any given $i \neq j$ there is a unique pair p,q such that $|A_{ip} \cap A_{jq}| = 2$. $|A_{ip} \cap A_{jq}| < 2$ holds for the other pairs.

Condition 1) implies

$$\sum_{j=1}^{r_i} \binom{|A_{ij}|}{2} = n - 1 \tag{10}$$

for any $1 \leq i \leq n$. If $n = 3$ or 6 (10) has no solution, therefore $s(F_3^n) > n$ for these values. However, for the values of form $n = 3k + 1$ there is always a solution $|A_{i1}| = \ldots = |A_{ik}| = 3$, $|A_{i,k+1}| = 1$. This suggests the following conjecture.

<u>Conjecture 1.</u> There is a system of 3-element subsets of a $3k + 1$ -element set satisfying the following conditions 1) Any pair of elements is contained in exactly two 3-sets, 2) the family of 3-sets can be divided into subfamilies where a subfamily consists of k disjoint 3-sets. 3) Exactly one pair of members of two different subclasses meet in 2 elements.

The problem of *resolvable Steiner systems* is very closely related. This problem is solved in [6]. We were able to construct them for $n = 4$ and 7 (using the Fano-geometry):

0	0	0	1
0	0	1	0
0	1	0	0
1	0	0	0

2	1	1	1	0	0	0
0	2	1	0	1	1	0
0	0	2	1	0	1	1
0	1	0	2	1	0	1
1	0	1	0	2	0	1
1	0	0	1	1	2	0
1	1	0	0	0	1	2

Conjecture 1 implies $s(F_3^n) = n$ for infinitely many n. However we state

Conjecture 2 $s(F_3^n) = n \quad (n \geq 7)$.

We state the next lemma without proof.

Lemma 3.

$$s(F_k^{n+1}) \leq s(F_k^n) + s(F_{k-1}^n) .$$

This implies $s(F_3^n) \leq c\, n^{3/2}$, a fairly weak upper bound.

REFERENCES

[1] CODD, E.F., A relational model of data for large shared data banks, Comm. ACM 13(1970) 377-387.

[2] SPERNER, E., Ein Satz über Untermengen einer endlichen Menge, Math. Z. 27(1928) 544-548.

[3] DEMETROVICS, J., On the equivalence of candidate keys with Sperner Systems, Acta Cybernetica 4 (1979) 247-252.

[4] BÉKÉSSY, J., DEMETROVICS, J., HANNÁK, K., FRANKL, P. and KATONA, G.O.H., On the number of maximal dependencies in a data base relation of fixed order, Disc. Math. 30(1980) 83-88.

[5] DEMETROVICS, J. and GYEPESI, GY., On the functional dependency and some generalization of it (to appear in Acta Cybernetica).

[6] RAY-CHAUDHURI, D.K., and WILSON, R.M., Solution of Kirkman's school girl problem, Proc. Symp. in Pure Math., Combinatorics, Am. Math. Soc. 19(1971) 187-204.

Specifying algebraic data types by domain equations

H.-D. Ehrich

Abt.Informatik,Univ. Dortmund, PF 500500,D-4600 Dortmund 50

ABSTRACT - The paper provides the theoretical foundation for a new algebraic specification method, using parameterized specifications and algebraic domain equations, an algebraic analogon to the domain equations used in Scott's theory of data types. The main result is that algebraic domain equations always have an initial solution. Also, a parametric version of algebraic domain equations is investigated. In either case, there is a simple syntactic solution method.

1. INTRODUCTION

Scott's order theoretic approach to data types (SC 71 ff) offers a characteristic specification method: there are given basic types like BOOL, INT, etc., and there is a fixed set of type constructors like sum, product, function space, and powerdomain. Data types are then defined by so-called domain equations that may be recursive, e.g. (cf. SC 72a, SP 77) $N \cong N+1$ (natural numbers), $L \cong (L \times L)+D$ (list structures over D), $F \cong [F \longrightarrow F]$ (a pure λ-calculus model), $S \cong S \times D+1$ (stacks over D), etc.

In contrast, the algebraic approach to data types has hardly more to offer than explicit specifications: one has to give all sorts, all operations, and all axioms that describe the behaviour of the operations (cf. ADJ 78). Some caution is required when comparing the order-theoretic and algebraic approaches to data types, since they do not model the same aspects and serve somewhat different purposes. It may, however, be fruitful to apply ideas from one approach to the further development of the other one. The papers of Lehmann and Smyth (LS 77) and Kanda (KA 78) represent different attempts to do this.

Recently, an algebraic analogon to type constructors, called parameterized data types, has been investigated (BG 77, EH 79, EL 79a+b, BG 80, EKTWW 80a+b). So, in a sense, algebraic data types can be specified by expressions involving parameterized data types applied to actual parameters. The question naturally arises whether a meaningful and useful analogon to recursive domain equations can be established in the algebraic framework. This paper shows that the answer is positive.

Our approach is inspired by the categorical versions of recursive domain equations due to Wand (WA 75), Lehmann (LE 76), and Smyth and Plotkin (SP 77). Thus, fixpoints of functors play an important role. However, we do not need ω-continuity, and we have a simple syntactic method for solving domain equations. The solutions are algebraic data types, i.e. many-sorted algebras with all their operations.

2. FUNDAMENTAL NOTIONS

A signature is a quadruple $\Sigma=\langle S,\Omega$, arity, sort$\rangle$, where S and Ω are sets of sorts and operators, respectively, and arity:$\Omega \longrightarrow S^*$, sort:$\Omega \longrightarrow S$ are mappings. We will write $\Sigma=\langle S,\Omega\rangle$ for short, assuming tacitly the existence of the arity and sort mappings. A signature morphism $f:\Sigma \longrightarrow \Sigma'$ is a pair of mappings $f=\langle f_s:S \longrightarrow S', f_\omega:\Omega \longrightarrow \Omega'\rangle$ such that arity (ωf_ω)=arity$(\omega)f_s$ and sort (ωf_ω)=sort$(\omega)f_s$. For convenience, we often omit the index, writing f for f_s or f_ω.

Algebras are interpretations of signatures: a Σ-algebra A is an S-indexed family of sets, $\{s_A\}$, the carrier of A, together with an Ω-indexed family of mappings, $\{\omega_A:\text{arity}(\omega)_A \longrightarrow \text{sort}(\omega)_A\}$, the operations of A (if $x=s_1s_2\ldots s_n\in S^*$, x_A denotes the cartesian product $s_{1,A}\times\ldots\times s_{n,A}$). A Σ-algebra morphism $\varphi:A \longrightarrow B$ is an S-indexed family of mappings $\varphi_s:s_A \longrightarrow s_B$ such that, for each operator $\omega\in\Omega$ with arity x and sort s, we have $\omega_A\varphi_s=\varphi_x\omega_B$. Here, $\varphi_x=\varphi_{s_1}\times\ldots\times\varphi_{s_n}$ if $x=s_1\ldots s_n$. The class of all Σ-algebras with all Σ-algebra morphisms forms a category Σ-alg. It is well known that Σ-alg has an initial algebra I_Σ, having a unique morphism to any other algebra in Σ-alg.

If $f:\Sigma \longrightarrow \Sigma'$ is a signature morphism, there is a corresponding forgetful functor f-alg:Σ'-alg $\longrightarrow$ Σ-alg sending each Σ'-algebra B to that Σ-algebra A such that $s_A=(sf)_B$ and $\omega_A=(\omega f)_B$.

Let $\Sigma=\langle S,\Omega\rangle$ be a signature . A Σ-equation is a triple $\langle X, \tau_1, \tau_2\rangle$ where X is an S-indexed family of sets (of variables), and τ_1,τ_2 are terms over X and Ω of the same sort, called the sort of the equation. A Σ-algebra A satisfies a Σ-equation $\langle X,\tau_1,\tau_2\rangle$ iff the formula $\forall X:\tau_1=\tau_2$ is true when interpreted in A in the obvious way.

A specification is a pair $\underline{D}=\langle\Sigma,E\rangle$ where Σ is a signature and E is an S-sorted set of Σ-equations. If $\Sigma=\langle S,\Omega\rangle$, we sometimes write $\underline{D}=\langle S,\Omega, E\rangle$ instead of $\underline{D}=\langle\Sigma,E\rangle$. A Σ-algebra satisfies a specification $\underline{D}=\langle\Sigma,E\rangle$ iff it satisfies each equation in E. Σ-algebras satisfying a specification $\underline{D}$ will be called $\underline{D}$-algebras. The full subcategory of Σ-alg consisting of all $\underline{D}$-algebras is denoted by $\underline{D}$-alg. $\underline{D}$-alg, too, has an initial algebra, denoted by I_D.

A specification morphism $f:\underline{D} \longrightarrow \underline{D}'$ is a signature morphism $f:\Sigma \longrightarrow \Sigma'$ such that the corresponding forgetful functor f-alg:Σ'-alg $\longrightarrow$ Σ-alg sends each $\underline{D}'$-algebra to a $\underline{D}$-algebra. Thus, a specification morphism defines a forgetful functor also denoted by f-alg, but with $\underline{D}'$-alg and $\underline{D}$-alg as domain and range, respectively, obtained by restricting f-alg to $\underline{D}'$-alg. The class of all specifications together with all specification morphisms forms a category denoted by spec.

It is not difficult to show that spec has all colimits, i.e. spec is cocomplete. We will use especially pushouts and coequalizers. A pushout is a square like that in figure 2.1(a) such that, whenever there are morphisms g_1, g_2 such that $f_1g_1=f_2g_2$, there is exactly one h such that $f_3h=g_1$ and $f_4h=g_2$. Given f_1, f_2, there are spec morphisms f_3, f_4 such that they form a pushout. We say that

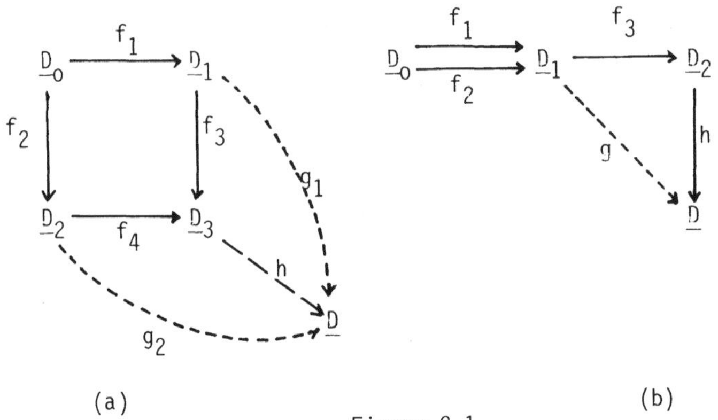

Figure 2.1

(f_3, f_4) is the pushout of f_1 and f_2. Analogously, a coequalizer is a diagram like that in figure 2.1(b) (without broken lines) such that, whenever there is a g such that $f_1g=f_2g$, then there is exactly one h such that $f_3h=g$. Given f_1 and f_2 as shown in figure 2.1(b), there is always a spec morphism f_3 forming a coequalizer. We say that f_3 is the coequalizer of f_1 and f_2.

Let $f:D \longrightarrow D'$ be a specification morphism. A functor $F:D\text{-alg} \longrightarrow D'\text{-alg}$ is called strongly persistent with respect to f iff F.f-alg is the identity on D-alg. The following lemma is proven in EKTWW 80a+b:

Extension lemma: Let the pushout in figure 2.1(a) be given. Let $F_1:D_0\text{-alg} \longrightarrow D_1\text{-alg}$ be strongly persistent wrt f_1. Then there is exactly one functor $F_4:D_2\text{-alg} \longrightarrow D_3\text{-alg}$, called the extension of F_1 via f_2, that is strongly persistent wrt f_4 and satisfies $f_2\text{-alg}.F_1 = F_4.f_3\text{-alg}$.

Associated with a specification morphism $f:D \longrightarrow D'$, there is a functor f-free: $D\text{-alg} \longrightarrow D'\text{-alg}$ that is left adjoint to f-alg. f-free sends each D-algebra A to the free D'-algebra over A (wrt f). Also from EKTWW 80a+b, we have the following:

Extension lemma supplement: If, in addition, $F_1 \cong f_1\text{-free}$, then we have $F_4 \cong f_4\text{-free}$.

A parameterized specification is an injective spec morphism $p:X \longrightarrow XP$. X is the formal parameter of p. A parameterized data type is a pair (p,P) where $p:X \longrightarrow XP$ is a parameterized specification, and $P:X\text{-alg} \longrightarrow XP\text{-alg}$ is a functor. (p,P) is called strongly persistent iff P is strongly persistent wrt p. We call p strongly persistent iff (p,p-free) has this property. In the case of strong persistency, parameter passing works as follows. Given $p:\underline{X} \longrightarrow \underline{XP}$, an actual parameter for p is a pair $(f,\underline{D})$ where $\underline{D}$ is a specification and $f:\underline{X} \longrightarrow \underline{D}$ is a spec morphism. Let (p',f') be the pushout of p and f (cf. fig. 2.2). Let P' be the unique extension of P via f.

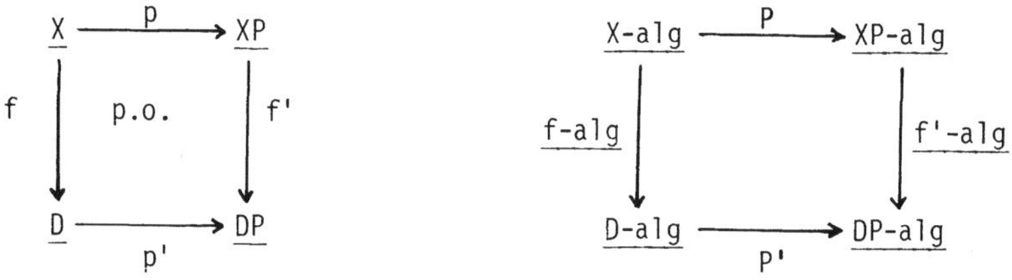

(a) (b)

Figure 2.2

Then (p',P') is a strongly persistent parameterized data type, also called the extension of (p,P) via f. Now, each actual parameter $\underline{D}$-algebra A is sent to AP'. In this way, (p,P) works as a data type constructor, operating on various actual parameter algebras and preserving their structure.

Parameter passing works in the same way when the actual parameter is again parametric, i.e. $\underline{D}=\underline{YP}$ of a parametric specification $\hat{p}:\underline{Y} \longrightarrow \underline{YP}$. In this case, an (actual parameter) parametric data type $(\hat{p},\hat{P})$ is sent to the parametric data type $(\hat{p}p',\hat{P}P')$ where p',P' are obtained as above.

Example 2.1: The "n-product with constant"

$$\underline{X}_1 \times \ldots \times \underline{X}_n + \underline{1}$$

is the embedding $p:\underline{X} \hookrightarrow \underline{XP}$, where $\underline{X}$ and $\underline{XP}$ are defined as follows $(i=1,\ldots n)$.

	$\underline{X}$	rest of $\underline{XP}$
sorts	$X_1,\ldots,X_n$,	P
ops	$\bar{x}_i: \ \longrightarrow X_i$	$\bar{p}: \ \longrightarrow P$
	$\equiv_i : X_i \times X_i \longrightarrow BOOL$	$<\_,\ldots,\_>: X_1 \times \ldots \times X_n \longrightarrow P$
		$\_(i) \ : P \longrightarrow X_i$
		$\equiv \ : P \times P \longrightarrow BOOL$
eqs	$\bar{x}_i \equiv_i \bar{x}_i$ = true	$\bar{p}(i) = \bar{x}_i$
		$<x_1,\ldots,x_n> (i) = x_i$
		$\bar{p} \equiv \bar{p}$ = true
		$\bar{p} \equiv <x_1,\ldots,x_n>$ = false
		$<x_1,\ldots,x_n> \equiv \bar{p}$ = false
		$<x_1,\ldots,x_n> \equiv <x_1,\ldots,x_n> = x_1 \equiv x_1 \ \ldots \ x_n \equiv x_n$

Consider (p,p-free) for n=2. Let D be a specification of the integers (sort INT) and booleans (sort BOOL). Let $f:X \longrightarrow D$ be defined by $X_1 \longmapsto$ BOOL, $X_2 \longmapsto$ INT, $\bar{x}_1 \longmapsto$ false, $\bar{x}_2 \longmapsto 0$, $\equiv_i \longmapsto$ identity on BOOL or INT, respectively, i=1,2. Then, the initial boolean-integer algebra (D-algebra) is sent to the algebra of (boolean,integer)-pairs with one additional constant, all other operations of XP, and all boolean and integer operations on the components retained. The specification of this algebra is DP, the pushout object of p and f. It is obtained from XP by substituting BOOL for X_1, false for $\bar{x}_1$, INT for X_2, 0 for $\bar{x}_2$, and the respective identities for $\equiv_i$. Therefore, a suggestive notation for DP is BOOL×INT + 1.

3. ALGEBRAIC DOMAIN EQUATIONS

In the order theoretic approach to data types, parameterized data types can be viewed as functors, domain equations as endofunctors, and their solutions as fixpoints of functors (WA 75, LE 76, SP 77).

In the algebraic approach, parameterized data types are essentially functors, too, but most often they are not endofunctors. Typically, the signatures of actual parameter and resultant algebras are different. In order to get endofunctors, we define algebraic domain equations to consist of a parameterized data type and a functor in the reverse direction. We restrict ourselves to free and strongly persistent parameterizations of the form (p,p-free) and to algebraic reverse functors of the form e-alg for some spec morphism e.

Definition 3.1: An algebraic domain equation is a pair of spec morphisms (p,e), $p,e:X \longrightarrow XP$, such that p is a strongly persistent parameterized specification.

Let P=p-free, $\bar{P}$=p-alg, and $\bar{E}$=e-alg. There are two endofunctors, namely $P\bar{E}$ on X-alg and $\bar{E}P$ on XP-alg. A fixpoint is an object that is sent to an isomorphic one. It is immediate to see that the fixpoints of $P\bar{E}$ and $\bar{E}P$ are very closely related: A is a fixpoint of $P\bar{E}$ iff AP is a fixpoint of $\bar{E}P$, and vice versa.

For the definition of what we mean by a solution of an algebraic domain equation, we make use of the following result. Let (q,Q) be the coequalizer in spec of p and e,

$$X \underset{e}{\overset{p}{\rightrightarrows}} XP \xrightarrow{\;q\;} Q \quad ,$$

and let $\bar{Q}$=q-alg.

Theorem 3.2: If B is a fixpoint of $\bar{E}P$, then there is a unique (up to isomorphism) Q-algebra C such that $B=C\bar{Q}$.

It is convenient not to take fixpoints of $P\bar{E}$ or $\bar{E}P$ as solutions, but these uniquely associated Q-algebras.

Definition 3.3: A solution of an algebraic domain equation (p,e) is a Q-algebra C such that $C\bar{Q}$ is a fixpoint of $\bar{E}P$.

The main result can now be stated as follows.

Theorem 3.4: The initial $\underline{Q}$-algebra I_Q is a solution of (p,e).

The proof is rather involved and requires some more technical machinery to be developed. It will be published elsewhere. Clearly, I_Q is an initial solution, i.e. it is initial in the full subcategory of $\underline{Q}$-alg of all solutions of (p,e). A specification of I_Q is $\underline{Q}$, the coequalizer object of p and e. There is a simple construction for $\underline{Q}$, given p and e, based on the coequalizer construction in set applied to sorts and operators.

Example 3.5: Consider the n-product with constant from example 2.1 for n=1. Let

$$\underline{X} \overset{p}{\underset{e}{\rightrightarrows}} \underline{X} + \underline{1}$$

be defined by taking p as in example 2.1, and e sending sort X_1 to P, $\bar{x}_1$ to $\bar{p}$, and $\equiv_1$ to $\equiv$. Then the solution is I_Q where $\underline{Q}$ is the following specification obtained from $\underline{X} + \underline{1}$ by identifying X_1 and P, $\bar{x}_1$ and $\bar{p}$, and $\equiv_1$ and $\equiv$. For convenience, we rename P by N, $\bar{p}$ by 0, <_> by succ, and _(1) by pred.

sorts N

ops
$0: \to N$
$succ: N \to N$
$pred: N \to N$
$\equiv : N \times N \to BOOL$

eqs
$pred(0) = 0$
$pred(succ(n)) = n$
$0 \equiv 0 = true$
$0 \equiv succ(n) = false$
$succ(n) \equiv 0 = false$
$succ(n) \equiv succ(m) = n \equiv m$

This is a specification of the natural numbers.

Example 3.6: Let the above specification be $\underline{N}$. Let $\underline{X}\times\underline{N} + \underline{1}$ be the parametric specification obtained from $\underline{X}_1\times\underline{X}_2 + \underline{1}$ (example 2.1) by parameterized parameter passing with actual parameter $(f,\hat{p})$, where $\hat{p}:\underline{X} \hookrightarrow \underline{X} + \underline{N}$ and f sends X_1 to X (forgetting the index 1), X_2 to N, $\bar{x}_2$ to 0, and $\equiv_2$ to $\equiv$. Then, the algebraic domain equation

$$\underline{X} \overset{p}{\underset{e}{\rightrightarrows}} \underline{X}\times\underline{N} + \underline{1}$$

has stacks as solutions, specified by the following specification (with obvious renamings):

$\underline{\text{sorts}}$ S,N

$\underline{\text{ops}}$ empty: $\rightarrow$ S

push : S $\times$ N $\rightarrow$ S

pop : S $\rightarrow$ S

top : S $\rightarrow$ N

= : S$\times$S $\rightarrow$ BOOL

... (ops from N)

$\underline{\text{eqs}}$ pop(empty) = empty

top(empty)= 0

pop(push(s,n))= s

top(push(s,n))= n

... (eqs from N and eqs for $\equiv$)

In a similar way, we get trees with a natural number attached to each node as solutions of the domain equation $\underline{X} = \underline{X} \times \underline{X}\times\underline{N} + \underline{1}$, where = denotes an appropriate pair (p,e) of morphisms, etc.

4. PARAMETERIZED ALGEBRAIC DOMAIN EQUATIONS

Our theory so far gives algebraic data types as solutions of algebraic domain equations. It is natural to ask whether we can get parameterized data types as solutions of parameterized algebraic domain equations by a similar method of implicit specification and syntactic solution. This works indeed, if we proceed as follows.

Definition 4.1: A parameterized algebraic domain equation is a triple (r;p,e) of spec morphisms,

$$\underline{Y} \xrightarrow{\ r\ } \underline{XY} \overset{p}{\underset{e}{\rightrightarrows}} \underline{XYP}$$

such that (1) (p,e) is an algebraic domain equation, (2) r is a parameterized specification, and (3) rp=re. (r;p,e) is called strongly persistent iff r has this property.

Let (f,$\underline{D}$) be an actual parameter for r. Then, using the mechanism of parameter passing as defined in the last section, we can construct an algebraic domain equation (p',e'), the (f,D)-instance of (r;p,e), as follows (cf. figure 4.1):

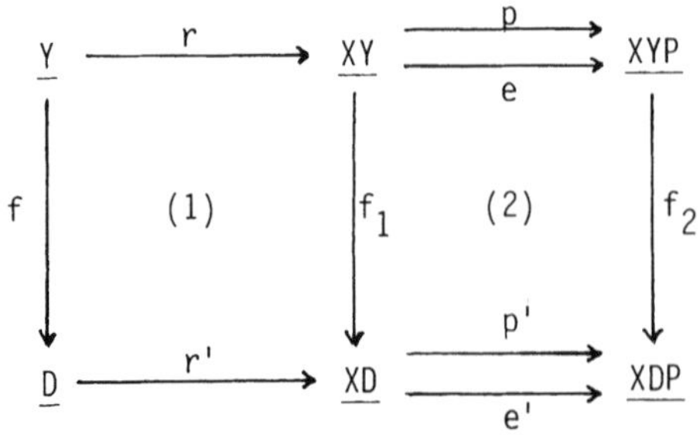

Figure 4.1

1) Let r', $\underline{XD}$,f_1 be such that (1) is a pushout.
2) Let p',$\underline{XDP}$,f_2 be such that (2) is a pushout wrt p,p'.
3) Define e' by r'e' = r'p' and f_1e'=ef_2. (e' is well defined since r' and f_1 are jointly surjective, and we easily prove that rf_1e' = fr'e'.)

Definition 4.2: A solution of a parameterized algebraic domain equation (r;p,e) is a parameterized data type (s,S) such that, for each actual parameter (f,$\underline{D}$) of r, (s,S) sends I_D via f to the initial solution of the (f,$\underline{D}$)-instance (p',e') of (r;p,e).

Of course, r and s must have the same source $\underline{Y}$. Our main theorem 3.4 now extends to the parametric case as follows.

Theorem 4.3: Let (r;p,e) be a strongly persistent parameterized algebraic domain equation, and let (q,$\underline{YQ}$) be the coequalizer of p and e. Then, if s=rpq, (s,s-free) is a solution of (r;p,e).

Again, the proof is a little bit lengthy and will be published elsewhere. Examples of implicit parameteric specifications are $\underline{X}$=$\underline{X}$×$\underline{Y}$ + $\underline{1}$ (stacks over Y), $\underline{X}$=$\underline{X}$×$\underline{X}$×$\underline{Y}$ + $\underline{1}$ (trees over Y), etc. Here, = denotes a pair (p,e) of morphisms that are defined like those in example 3.6.

5. CONCLUSIONS

The theoretical results presented here provide a sound and consistent semantics for a new algebraic specification method using parameterized specifications and algebraic domain equations. The feasibility and usefulness of this method for the development of specification methods and specification languages should be subject to further study.

Another possible area of application is the algebraic semantics of programming languages. In denotional semantics, domain equations are used extensively to specify the syntactic and semantic domains. Our theory can provide algebraic interpretations for them. There is, however, one difficulty: we get only "finitary" solutions, for example (initial) algebras of finite sets or finite functions. The central semantic domains of environments and states usually are finitary, so there seems to be no problem. It is, however, not quite clear how to cope with cases like procedure parameters. In particular, we cannot obtain a model for λ-calculus with our method, like Scott's reflexive domain (SC 72b). For these and similar cases, an extension of our theory to continuous algebras is necessary. This is subject to further study.

REFERENCES

ADJ 77 Goguen,J.A./Thatcher,J.W./Wagner,E.G./Wright,J.B.: Initial Algebra Semantics and Continuous Algebras. Journal ACM 24,(1977), 68-95

ADJ 78 Goguen,J.A./ Thatcher,J.W./Wagner,E.G.: An Initial Algebra Approach to the Specification, Correctness, and Implementation of Abstract Data Types.Current Trends in Programming Methodology,Vol IV (R.T. Yeh,ed.).Prentice Hall, Englewood Cliffs 1978, 80-149

BG 77 Burstall,R.M./Goguen,J.A.: Putting Theories Together to Make Specifications. Proc. 5th Int. Joint Conf. on Artificial Intelligence, MIT, Cambridge (Mass.), 1977

BG 80 Burstall,R.M./Goguen,J.A.: The Semantics of CLEAR, a Specification Language. Proc. 1979 Copenhagen Winter School on Abstract Software Specifications (D. Bjørner,ed.). LNCS 86, Springer-Verlag, Berlin 1980, 292-331

EH 79 Ehrich,H.-D.: On The Theory of Specification, Implementation, and Parameterization of Abstract Data Types. Bericht Nr. 82/79, Abtlg. Informatik,Univ. Dortmund 1979 (also to appear in Journal ACM)

EL 79a Ehrich,H.-D./Lohberger,V.G.: Parametric Specification of Abstract Data Types, Parameter Substitution, and Graph Replacements. Graphs, Data Structures, Algorithms (M.Nagl/H.-J. Schneider,eds.). Applied Computer Science 13,Hanser Verlag, München 1979, 169-182

EL 79b Ehrich,H.-D./Lohberger,V.G.: Constructing Specifications of Abstract Data Types by Replacements. Proc. Int. Workshop on Graph Grammars and Their Application to Computer Science and Biology (V.Claus/ H.Ehrig/G.Rozenberg,eds.).LNCS 73,Springer-Verlag,Berlin 1979,180-191

EKTWW 80a Ehrig,H./Kreowski,H.-J./Thatcher,J.W./Wagner,E.G./Wright,J.B.: Parameterized Data Types in Algebraic Specification Languages. Proc.7th ICALP (J.W.deBakker/J.van Leeuwen,eds.) LNCS 85, Springer-Verlag, Berlin 1980, 157-168

EKTWW 80b Ehrig,H./Kreowski,H.-J./Thatcher,J.W./Wagner,E.G./Wright,J.B.: Parameter Passing in Algebraic Specification Languages. Internal Report, FB 20 TU Berlin, 1980

KA 78 Kanda,A.: Data Types as Initial Algebras: a Unification of Scottery and ADJery. Proc. 19th FOCS 1978, 221-230

LE 76 Lehmann,D.J.: Categories for Fixpoint Semantics. Proc.7th FOCS 1976, 122-126

LS 77 Lehmann,D.J./Smyth,M.B.:Data Types. Proc 18th FOCS 1977, 7-12

SC 71 Scott,D.S.:The Lattice of Flow Diagrams.Proc.Symp. on Semantics of Algorithmic Languages (E.Engeler,ed.).LNM 188, Springer-Verlag, Berlin 1971, 311-372

SC 72a Scott,D.S.: Lattice Theory,Data Types and Semantics.Formal Semanctics of Algorithmic Languages (R.Rustin,ed.).Prentice Hall, Englewood Cliffs 1972, 65-106

SC 72b Scott,D.S.: Continuous Lattices. Toposes, Algebraic Geometry and Logic (F.W.Lawvere,ed.).LNM 274,Springer-Verlag,Berlin 1972,97-136

SC 76 Scott,D.S.: Data Types as Lattices. SIAM Journal of Computing 5 (1976), 522-587

SP 77 Smyth,M.B./Plotkin,G.D.: The Category-Theoretic Solution of Recursive Domain Equations. Proc. 18th FOCS 1977, 13-17

WA 75 Wand,M.: On the Recursive Specification of Data Types. Proc. 1st Int. Coll. on Category Theory Applied to Computation and Control (E.G. Manes,ed.). LNCS 25, Springer-Verlag, Berlin 1975, 214-217

AN AXIOMATIZATION OF REGULAR FORESTS IN THE LANGUAGE OF ALGEBRAIC THEORIES WITH ITERATION

by Z. Ésik
University of Szeged
6720 Szeged, Aradi v. tere 1, Hungary

Introduction

Let Σ be an arbitrary ranked alphabet. We denote by

$$Reg_{\Sigma} = (Reg_{\Sigma}(n,p); \cdot, \langle\,\rangle, \pi_p^i, \subseteq)$$

the rational algebraic theory (cf. [8], [9]) of all regular forests (cf. [7]) of finite Σ-trees on the variables $x_1, x_2 \ldots$. Reg_{Σ} can be viewed as an algebra of type $(\cdot, \langle\,\rangle, \pi_p^i, {}^+)$, as well: define $T^+ \in Reg_{\Sigma}(n,p)$ - i.e. $T^+ : n \to p$ - as the least fixed point of the mapping $T\langle -, 1_p\rangle$ for each $T : n \to n+p$.

Now let $\vee$ be a binary operational symbol not in Σ and consider $P_{\Sigma \cup \{\vee\}}$, the absolutely free algebra of type $(\cdot, \langle\,\rangle, \pi_p^i, {}^+)$ generated by $\Sigma \cup \{\vee\}$. There exists a unique homomorphism $\|\ \| : P_{\Sigma\cup\{\vee\}} \to Reg_{\Sigma}$ extending $\Sigma\cup\{\vee\} \hookrightarrow Reg_{\Sigma}$, where the latter correspondence is given by $\sigma \mapsto (r; \{\sigma(x_1,\ldots,x_r)\}) \in Reg_{\Sigma}(1,r)$ if $\sigma \in \Sigma_r$, and $\vee \mapsto (2; \{x_1, x_2\}) \in Reg_{\Sigma}(1,2)$. This homomorphism $\|\ \|$, which is in fact an epimorphism, induces a congruence relation $\Theta_{\|\ \|}$ in $P_{\Sigma\cup\{\vee\}}$. The problem we are going to solve concerns with $\Theta_{\|\ \|}$ and can be formulated as follows: find a generating system of $\Theta_{\|\ \|}$.

Let $I_{\Sigma\cup\{\vee\}} = (I_{\Sigma\cup\{\vee\}}(n,p); \cdot, \langle\,\rangle, \pi_p^i, {}^+)$ denote the algebra obtained by introducing the operator ${}^+$ into the free rational algebraic theory (cf. [8]) generated by $\Sigma\cup\{\vee\}$ with the insertion $\Sigma\cup\{\vee\} \hookrightarrow I_{\Sigma\cup\{\vee\}}$. There is a unique way to factor the epimorphism $\|\ \|$ through $I_{\Sigma\cup\{\vee\}}$ in such a manner that the following diagram becomes commutative:

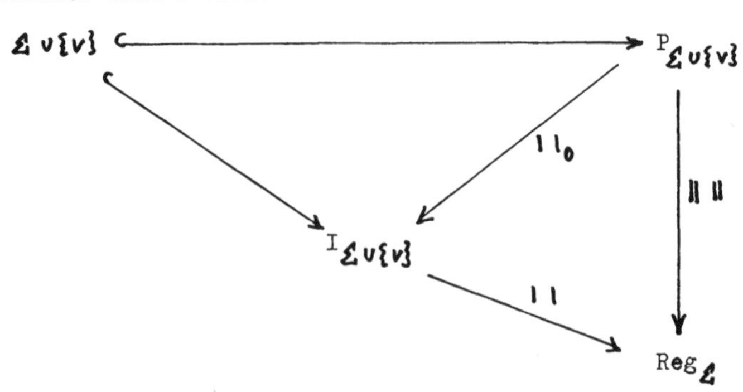

Denote by $\Theta_{|\ |_0}$ and $\Theta_{|\ |}$ the congruence relations induced by the epimorphisms $|\ |_0$ and $|\ |$, respectively. The commutativity of the previous diagram allows us to seek for a generating system of $\Theta_{\|\ \|}$ in the form $E_0 \cup E_1$ where E_0 generates $\Theta_{|\ |_0}$

and $|E_1|_0$ generates Θ_{11}.

Relations Θ_{11_0} is a completely invariant congruence relation and a generating system E_0 of Θ_{11_0} has been already found in [3] by giving a base of identities of the equational class generated by the class of rational algebraic theories (being considered as algebras of type $(\cdot, \langle\,\rangle, \pi^i_p, {}^+)$). This equational class is called the class of generalized iterative theories in [3]. In [4] E_0 is expressed with scalar iteration. Hence, our problem reduces to the following one: find a generating system E of Θ_{11}, i.e. such that $\Theta_E = \Theta_{11}$.

The result

In this section we give a presentation of system E. But first some definitions are needed in connection with theory $I_{\Sigma\cup\{\vee\}}$.

Let $m \geq 1$, $a = \langle a_1,\dots,a_m\rangle : m \to n+p$. Then a is called canonical provided for every $i \in [m] = \{1,\dots,m\}$ there exists a finite family $\{\sigma_j \mid j \in I\} \subseteq \Sigma - \Sigma_0$ together with base morphisms (cf. [2]) $\alpha_j : r_j \to n+p$, as well as $\Delta \subseteq \Sigma_0$ and $J \subseteq [p]$, such that for a_i we have $a_i = \bigvee_{j\in I} \sigma_j \alpha_j \vee \bigvee_{\sigma\in\Delta} \sigma + 0_{n+p} \vee \bigvee_{j\in J} \pi^{n+j}_{n+p}$ [1],

and for each $j \in I$ α_j is of the form $\alpha_j = \langle \pi^{i_1}_{n+p},\dots,\pi^{i_r}_{n+p}\rangle$ for some integers $r, i_1,\dots,i_r$ with $r > 0$, $i_1,\dots,i_r \in [n]$.

Now let $a = \langle a_1,\dots,a_n\rangle : n \to (n+p)+q$ be canonical. We say that a satisfies the ($*$ - p)-condition, provided a_1 has the form $a_1 = \bigvee_{i\in I} \sigma_i \alpha_i$, where $\sigma_i \in \Sigma_{r_i}$ and $r_i > 0$ for any $i \in I$, furthermore, one of the next two conditions are valid for each $\alpha_i = \langle \pi^{i_1}_{n+p+q},\dots,\pi^{i_r}_{n+p+q}\rangle$:

(i) there is an index $j \in [r]$ such that either $i_j = 1$ or $n < i_j \leq n+p$,

(ii) there is an index $j \in [r]$ and a base morphism $\alpha : n-1 \to n-1$ such that $1 < i_j \leq n$ and α - being considered as a mapping of $[n-1]$ into itself - is bijective, moreover, $\alpha(1) = i_j - 1$ and $\alpha\langle a_2,\dots,a_n\rangle(\beta + 1_{p+q})(\alpha^{-1} + 1_{p+q+1})$ satisfies the ($*$ - p+1)-condition for $\beta = \langle 0_{n-1} + 1, 1_{n-1} + 0_1\rangle$.

Next let $a = \langle a_1,\dots,a_n\rangle : n \to n+p$ be canonical. We say that a has the empty-set-property if $a_1 = \bigvee_{i\in I} \sigma_i \alpha_i$ with $\sigma_i \in \Sigma - \Sigma_0$ $(i \in I)$, and either (i) or (ii) is valid for each

$$\alpha_i = \langle \pi^{i_1}_{n+p},\dots,\pi^{i_r}_{n+p}\rangle :$$

(i) there exists $j \in [r]$ with $i_j = 1$,

(ii) there is an index $j \in [r]$ and a bijective base morphism $\alpha : n-1 \to n-1$

---

1) If $b_1,\dots,b_k : 1 \to q$ then $\bigvee_{i=1}^{k} b_i$ is used to denote $\vee\langle\dots\vee\langle\vee\langle b_1,b_2\rangle, b_3\rangle\dots,b_k\rangle$. If $k = 0$ this is meant as $\perp_{1,q} = \pi^{1^+}_{1+q}$.

such that $\alpha(1) = i_j - 1$ and $\alpha\langle a_2,\dots,a_n\rangle(\beta + 1_p)(\alpha^{-1} + 1_{p+1})$ satisfies the $(*-1)$-condition, where $\beta = \langle 0_{n-1} + 1_1, 1_{n-1} + 0_1\rangle$.

In what follows we shall define an element $P(a) : m \to m+p$, where $m = 2^n$, for each canonical $a : n \to n+p$.

Let $\alpha : [m] \to P([n]) = \{X \mid X \subseteq [n]\}$ be an arbitrary fixed bijective mapping and assume that $a = \langle a_1,\dots,a_n\rangle$ with

$$a_i = \bigvee_{j\in I_i} \sigma_j \alpha_j \vee \bigvee_{\sigma\in\Delta_i} \sigma + 0_{n+p} \vee \bigvee_{j\in J_i} \pi^{n+j}_{n+p} \qquad (i\in[n]).$$

Set

$$\Sigma_a = \{\sigma(X_1,\dots,X_r) = X \mid \sigma\in\Sigma - \Sigma_0,\ X_1,\dots,X_r \subseteq [n],\ X =$$
$$= \{i \mid \forall i_1\in X_1,\dots,i_r\in X_r\ \exists_j\in I_i,\ \sigma_j = \sigma,\ \alpha_j = \langle \pi^{i_1}_{n+p},\dots,\pi^{i_r}_{n+p}\rangle\}\},$$

$$\Delta_a = \{\sigma = X \mid \sigma\in\Sigma_0,\ X = \{i \mid \sigma\in\Delta_i\}\},$$

$$\pi_a = \{j = X \mid j\in[p],\ X = \{i \mid j\in J_i\}\}.$$

Then $P(a) = \langle b_1,\dots,b_m\rangle$ is defined by

$$b_i = \bigvee(\sigma\langle \pi^{i_1}_{m+p},\dots,\pi^{i_r}_{m+p}\rangle \mid \sigma(\alpha(i_1),\dots,\alpha(i_r)) = \alpha(i)\in\Sigma_a) \vee$$
$$\vee \bigvee(\sigma + 0_{m+p} \mid \sigma = \alpha(i)\in\Delta_a) \vee \bigvee(\pi^{m+j}_{m+p} \mid j = \alpha(i)\in\pi_a) \quad (i\in[m]).$$

The last definition we are going to use in the enumeration of E is the following one. Let $a = \langle a_1,\dots,a_n\rangle : n \to n+p$ and $b = \langle b_1,\dots,b_m\rangle : m \to m+p$ be canonical,

$$a_i = \bigvee_{j\in I_i} \sigma_j\alpha_j \vee \bigvee_{\sigma\in\Delta_i} \sigma + 0_{n+p} \vee \bigvee_{j\in J_i} \pi^{n+j}_{n+p} \qquad (i\in[n]),$$

$$b_i = \bigvee_{j\in I'_i} \sigma'_j\alpha'_j \vee \bigvee_{\sigma\in\Delta'_i} \sigma + 0_{m+p} \vee \bigvee_{j\in J'_i} \pi^{m+j}_{m+p} \qquad (i\in[m]).$$

b is called a <u>homomorphic image</u> of a if there is a base morphism $\rho : n \to m$ which is surjective (being considered as a mapping), and satisfies all three conditions listed here for each $i\in[n]$:

(i) for any $j\in I_i$ there exists $k\in I'_{\rho(i)}$ with $\sigma_j = \sigma'_k$ and $\alpha_j(\rho + 1_p) = \alpha'_k$ and, conversely, for any $j\in I'_{\rho(i)}$, $r > 0$, $i_1,\dots,i_r\in[n]$ if $\alpha'_j = \langle \pi^{\rho(i_1)}_{m+p},\dots,\pi^{\rho(i_r)}_{m+p}\rangle$ then there exist $i_0\in[n]$ and $k\in I_{i_0}$ such that $\rho(i_0) = \rho(i)$, $\sigma_k = \sigma'_j$, $\alpha_k(\rho + 1_p) = \alpha'_j$,

(ii) $\Delta'_{\rho(i)} = \bigcup(\Delta_j \mid \rho(j) = i)$,

(iii) $J'_{\rho(i)} = \bigcup(J_j \mid \rho(j) = i)$.

Now we are ready to introduce the system E. E consists of the following pairs (written as equalities), from (A_1) to (A_9):

(A_1) $\vee\langle\vee\langle f,g\rangle ,h\rangle = \vee\langle f, \vee\langle g,h\rangle\rangle$,

(A_2) $\vee\langle f,g\rangle = \vee\langle g,f\rangle$,

(A_3) $\vee\langle f,f\rangle = f$,

(A_4) $\vee\langle f, \perp_{1,p}\rangle = f$, where $f,g,h:1\to p$,

(A_5) $\sigma\langle f_1,\dots,f_{i-1}, \perp_{1,p}, f_{i+1},\dots,f_r\rangle = \perp_{1,p}$ if $\sigma\in\Sigma_r$, $r > 0$, $i\in[r]$, $f_1,\dots,f_{i-1}, f_{i+1},\dots, f_r:1\to p$,

(A_6) $(\vee\langle\pi^1_{1+p}, f\rangle)^+ = f^+$ where $f : 1\to 1+p$,

(A_7) $\pi^1_n a^+ = \perp_{1,p}$ if $a : n\to n+p$ is canonical and has the empty-set--property,

(A_8) $\pi^1_n a^+ = \vee(\pi^i_m P(a)^+ \mid 1\in\alpha(i))$ if $a : n\to n+p$ is canonical, $m = 2^n$ and $\alpha : [m]\to P([n])$ denotes the bijection fixed in the construction of $P(a)$,

(A_9) $\vee(\pi^i_n a^+ \mid \rho(i) = 1) = \pi^i_m b^+$ if $a : n\to n+p$ and $b : m\to m+p$ are canonical and b is a homomorphic image of a under $\rho : n\to m$.

With the above definition of E we have

<u>Theorem 1</u> $\Theta_E = \Theta_{||}$.

In case of unary Σ, i.e. $\Sigma = \Sigma_1$, we have found a simpler version of system E. In order to present this we modify the construction of $P(a)$.

Let $a = \langle a_1,\dots,a_n\rangle : n\to n+p$, $n \geq 1$. Assume that for every $i\in[n]$ there exist $\Delta_i \subseteq \Sigma$, $J_i \subseteq [p]$ with $a_i = \bigvee_{\sigma\in\Delta_i}\sigma a^i_\sigma \vee \bigvee_{j\in J_i}\pi^{n+j}_{n+p}$ where for each $\sigma\in\Delta_i$ $a^i_\sigma : 1\to n+p$ a morphism of type $\vee(\pi^i_{n+p} | i\in I_\sigma)$ for some $I_\sigma \subseteq [n]$. Now let $m = 2^n$ and fix an arbitrary bijective mapping $\alpha: [m]\to P([n])$ with $\alpha(1) = [1]$. Then $P(a) = \langle b_1,\dots,b_m\rangle : m\to m+p$ is determined by

$$b_i = \bigvee_{\sigma\in\Delta}\sigma b_\sigma \vee \bigvee_{j\in J}\pi^{m+j}_{m+p}$$

if and only if $\Delta = \bigcup(\Delta_j \mid j\in\alpha(i))$, $b_\sigma = \vee(a^j_\sigma \mid j\in\alpha(i))$ and $J = \bigcup(J_j \mid j\in\alpha(i))$.

<u>Theorem 2</u> If Σ is unary then the system E' consisting of $(A_1),\dots,(A_8)$ and (B) - with the above new definition of $P(a)$ - constitutes a generating system of $\Theta_{||}$. (B) is given by

(B) $f\cdot \vee\langle g,h\rangle = \vee\langle fg,fh\rangle$, $f : 1\to 1$, $g,h : 1\to p$.

Preliminary lemmas

Let I be an arbitrary generalized iterative algebraic theory and Θ a congruence relation of I. Define $\bar{\Theta}$ in the following way: for $a : n \to n+p$ and $b : m \to m+p$, $a \bar{\Theta} b$ if and only if $n \leq m$ and $a^{+} \Theta (1_n + 0_m) b^{+}$. The next two statements can be proved by short computations.

Lemma 1 Let $a : n \to n+m+p$, $b : m \to n+m+p$, $a' : n' \to n'+m+p$, $b' : m' \to n+m'+p$. Assume that $a \bar{\Theta} a'$ and $\bar{b} \bar{\Theta} \bar{b}'$ where $\bar{b} = b(\langle 0_m + 1_n, 1_m + 0_n \rangle + 1_p)$, $\bar{b}' = b'(\langle 0_{m'} + 1_n \; 1_{m'} + 0_n \rangle + 1_p)$. Let $\alpha = 1_{n'+m} + 0_{m'-m} + 1_p$, $\beta = 1_n + 0_{n'-n} + 1_{m'+p}$. Then there exists a bijective base morphism $\gamma : n' + m' \to n' + m'$ with $\langle a,b \rangle \bar{\Theta} \gamma \langle a'\alpha , b'\beta \rangle (\gamma^{-1} + 1_p)$.

Lemma 2 Let $n \geq 1$, $i \in [n]$ and $a = \langle a_1, \ldots, a_n \rangle : n \to n+p$, $b : 1 \to p$. Assume that $\pi_n^i a^{+} \Theta b$. Then also $a^{+} \Theta c^{+}$, i.e. $a \bar{\Theta} c$, where

$c = \langle a_1, \ldots, a_{i-1}, 0_n + b, a_{i+1}, \ldots, a_n \rangle$.

Additional lemmas

Now we return to theory $I_{\Sigma \cup \{v\}}$, and make some preparations for proving Theorem 1.

Lemma 3 If $a : n \to (n+p)+q$ is canonical then a has the $(* - p)$-property if and only if every element of $|\pi_n^1 a^{+}|$ contains one of the variables $x_1, \ldots, x_p$.

Proof By induction on n. The case $n = 1$ is obvious. The induction step follows by $|\pi_n^1 a^{+}| = (|\pi_n^1 a| \langle 1_1 + 0_{p+q}, |((0_1 + 1_{n-1}) a (\beta + 1_{p+q}))^{+}|, 0_1 + 1_{p+q})^{+}$, where $\beta = \langle 0_{n-1} + 1_1, 1_{n-1} + 0_1 \rangle$.

Lemma 4 Let $a : n \to n+p$ be canonical. Then a has the empty-set-property if and only if $|\pi_n^1 a^{+}|$ is void, i.e. $|\pi_n^1 a^{+}| = \emptyset_{1,p}$.

Proof By Lemma 3.

Lemma 5 (Consistency) $\Theta_E \subseteq \Theta_{||}$.

Proof We have to show that $E \subseteq \Theta_{||}$. Concerning $(A_1), \ldots, (A_6)$ it is quite obvious. $(A_7) \subseteq \Theta_{||}$ follows by Lemma 4. In order to prove that $(A_8) \cup (A_9) \subseteq \Theta_{||}$ observe that each canonical $a : n \to n+p$ can be viewed as a nondeterministic tree-automaton with set of states $[n]$ and final state 1. Thus the construction of $P(a)$ from a becomes the construction of a powerset-automaton (cf. [7]). This validiates $(A_8) \subseteq \Theta_{||}$. The fact that $b : m \to m+p$ is a homomorphic image of a expresses that b, as a nondeterministic tree automaton is a homomorphic image of the nondeterministic tree automaton a. This proves $(A_9) \subseteq \Theta_{||}$.

Lemma 6 For arbitrary $f : 1 \to p$ there exists a canonical element $a : n \to n+p$ with $f \Theta_E \pi_n^1 a^{+}$.

Proof We know that $I_{\Sigma \cup \{v\}}$ contains $T_{\Sigma \cup \{v,\perp\}}$ as a subtheory. $T_{\Sigma \cup \{v,\perp\}}$ is the free algebraic theory generated by the ranked alphabet $\Sigma \cup \{v,\perp\}$ where

$\perp$ is a new nullary operational symbol, $\perp = \pi_1^{1^+}$. Let us call an element $b : k \to \ell+p \in T_{\Sigma \cup \{\vee,\perp\}}$ primitive, if $k \geq 1$ and for every $i \in [k]$ $\pi_k^i b = \pi_{\ell+p}^{\ell+j}$ for some $j \in [p]$ or $\pi_k^i b = \sigma\alpha$ for some $\sigma \in (\Sigma \cup \{\vee,\perp\})_r$ and $\alpha = \langle \pi_{\ell+p}^{i_1}, \dots, \pi_{\ell+p}^{i_r} \rangle$ with $i_1,\dots,i_r \in [\ell]$. (A somewhat different notion of primitiveness is used in [2].) For arbitrary $f : 1 \to p \in I_{\Sigma \cup \{\vee\}}$ there exists a primitive $b : m \to m+p$ such that $f = \pi_m^1 b^+$. Therefore, Lemma 5 follows by the following statement. For any primitive $a : n \to (n+p)+q$ there is a canonical $b : m \to (m+p)+q$ with $a \,\Theta_E\, b$. This can be verified by induction on n, using $(A_1),\dots,(A_6)$ and Lemma 1.

Lemma 7 If $a : n \to n+p$ is canonical and $|\pi_n^1 a^+| \neq \emptyset_{1,p}$ then there exists a canonical $b : m \to m+p$ such that $b \,\overline{\Theta}_E\, a$ and $|\pi_m^i b^+| \neq \emptyset_{1,p}$ for every $i \in [m]$.

Proof By identities $(A_1),\dots,(A_5)$ and Lemma 2.

The proof of Theorem 1

$\Theta_E \subseteq \Theta$ is valid by Lemma 5. Conversely, it is enough to show that $f \Theta g$ implies $f \Theta_E g$ for $f,g : 1 \to p$. By Lemma 6, there exist canonical $a_1 : n_1 \to n_1+p$ and $b_1 : m_1 \to m_1+p$ with $f \,\Theta_E\, \pi_{n_1}^1 a_1^+$ and $g \,\Theta_E\, \pi_{m_1}^1 b_1^+$. Two cases arise. If $|f| = \emptyset_{1,p}$ then, by Lemma 4, $\pi_{n_1}^1 a_1^+ \,\Theta_E \perp_{1,p}$, and similarly, $\pi_{m_1}^1 b_1^+ \,\Theta_E \perp_{1,p}$. Hence, $f \Theta_E g$. If $|f| \neq \emptyset_{1,p}$ then $f \,\Theta_E\, \vee(\pi_{n_2}^i a_2^+ \mid i \in I)$ holds for $a_2 = P(a_1) : n_2 \to n_2+p$ and a suitable $I \subseteq [n_2]$. From a_2, by repeated applications of Lemma 7 and (A_4), one can obtain $a_3 : n_3 \to n_3+p$ and $J \subseteq [n_3]$ with $f \,\Theta_E\, \vee(\pi_{n_3}^i a_3^+ \mid i \in J)$ and $|\pi_{n_3}^i \;| \neq \emptyset_{1,p}$ $(i \in [n_3])$. Omit those components of a_3 which are "unreachable" from the components corresponding to the set J. In this way we get $a_4 : n_4 \to n_4+p$ and $K \subseteq [n_4]$ with $f \,\Theta_E\, \vee(\pi_{n_4}^i a_4^+ \mid i \in K)$. Starting with b_1, similar construction will produce $b_4 : m_4 \to m_4+p$ and $L \subseteq [m_4]$. One can view a_4 and b_4 as deterministic tree automata with set of final states K and L, resp. By the minimalizing process of deterministic tree-automata (cf. [1]) - which can be extended to the partial case - we obtain that a_4 and b_4 have a common homomorphic image $c : k \to k+p$. From this $f \Theta_E g$ follows by first applying (A_9) and then (A_1), (A_2) and (A_3).

Conclusion

An axiom system for regular expressions of regular forests has been already introduced in [5], on the basis of [6]. By Malcev's lemma a new axiom system can be obtained from our systems E_0 and E as well. This axiom system differs from the system in [5] in two respects. First, we have used a distinct collection of operations to build up "regular expressions". (In this language there are more polynomials than in the language of type $(\cup, \emptyset, x_i, \cdot x_i, *^{x}i.)$) Secondly, the operations with the equality sign and substitution are the only rules of inference in the system obtained here.

(A_6) is superfluous in E, it is contained in (A_7). (A_8) and (A_9) are extremely strong, it would be interesting to replace them by simpler axioms. (E.g. (*) $\sigma\langle f_1,\ldots, \vee\langle f_i,g\rangle, f_{i+1},\ldots,f_r\rangle = \vee\langle\sigma\langle f_1,\ldots,f_r\rangle, \sigma\langle f_1,\ldots,f_{i-1}, g, f_{i+1},\ldots,f_r\rangle\rangle$ follows by (A_8).)

Finally, let us mention another possible characterization of relation θ_{11}. Namely, one can show that θ_{11} is the weakest rational congruence relation (cf. [8]) of $I_{\Sigma\cup\{\vee\}}$ containing $(A_1),\ldots,(A_5)$ and (*).

References

[1] Brainerd, W. S., The minimalization of tree automata, Information and Control 13 (1968), 484-491.

[2] Elgot, C. C., S. L. Bloom, R. Tindell, On the algebraic structure of rooted trees, JCSS 16 (1978), 362-399.

[3] Ésik, Z., Identities in iterative and rational algebraic theories, Computational Linguistics and Computer Languages, to appear.

[4] Ésik, Z., On generalized iterative algebraic theories, Computational Linguistics and Computer Languages, submitted for publication.

[5] Ito, T. and S. Ando, A complete axiom system of superregular expressions, Proc. IFIP Congress 74, 661-665.

[6] Salomaa, A., Two complete axiom systems for the algebra of regular events, JACM 13 (1966), 158-169.

[7] Thatcher, J. W. and B. Wright, Generalized finite automata theory with an application to a decision problem of second-order logic, Math. Syst. Theory 2 (1968), 57-81.

[8] Wagner, E. G., J. B. Wright, J. A. Goguen and J. W. Thatcher, Some fundamentals of order-algebraic semantics, Lecture Notes in Computer Science 45 (1976), 151-168.

[9] Wright, J. B., J. W. Thatcher, E. G. Wagner and J. A. Gouen, Rational Algebraic theories and fixed-point solutions, Proc. 17th IEEE Symp. on Foundations of Computing, Houston, 1976, 147-158.

FAST RECOGNITION OF RINGS AND LATTICES

P. Goralčík, A. Goralčíková, V. Koubek
Charles University
Faculty of Mathematics and Physics
Sokolovská 83, 186 00 Praha, ČSSR

V. Rödl
Technical University
Faculty of Nuclear Engineering
Husova 5, 110 00 Praha, ČSSR

Both rings and lattices are interesting classes of algebraic structures with two binary operations (or of bigroupoids, as we shortly call such structures). One can, therefore, ask how to recognize a ring or a lattice in the class of all finite bigroupoids. More precisely, are there algorithms, effectuated on a RAM, which need less than $O(n^3)$ time to decide about a bigroupoid of size n (presented by the two tables of its operations) whether it is a ring or a lattice?

It is our aim to prove that there are algorithms which do the task for rings in $O(n^2)$ time and for lattices in $O(n^{5/2})$ time.

The first time complexity estimate cannot be improved and is closely related to the $O(n^2)$ recognition of groups; the second may not be final and arises from algorithms involving the reduct (Hasse diagram) of the lattice order whose size is shown to be $O(n^{3/2})$.

1. Recognition of rings and associative rings

A bigroupoid $(X,+,.)$, with the operations called addition and multiplication, respectively, is a ring if $(X,+)$ is an abelian group and the multiplication is both left and right distributive with respect to the addition:

(D) $\quad a.(x+y) = a.x + a.y \quad , \quad (x+y).a = x.a + y.a$

A ring $(X,+,.)$ is an associative ring if, moreover, $(X,.)$ is a semigroup.

Given a bigroupoid $(X,+,.)$ of size $|X| = n$, it is possible to decide in $O(n^2)$ time whether $(X,+)$ is an abelian group (cf. [2], [4], [5]), so we can further assume that under addition our bigroupoid is an abelian group.

Assume we know some base $b_1,\ldots,b_m$ of $(X,+)$, i.e. a list of elements of $(X,+)$ establishing a group isomorphism

$$Z_{n_1}\times\cdots\times Z_{n_m} \longrightarrow (X,+):\ (i_1,\dots,i_m)\longmapsto \sum_{k=1}^{m} i_k b_k$$

where Z_{n_k} denotes the additive group of integers modulo n_k, the order of b_k in $(X,+)$. As usual, we identify elements of Z_{n_k} with integers $0,1,\dots,n_k-1$. Distributivity (D) is clearly equivalent to

$$(D')\qquad a\left(\sum_{k=1}^{m} i_k b_k\right) = \sum_{k=1}^{m} i_k(ab_k)\ ,\quad \left(\sum_{k=1}^{m} i_k b_k\right)a = \sum_{k=1}^{m} i_k(b_k a)$$

for all $a\in X$, $(i_1,\dots,i_m)\in Z_{n_1}\times\cdots\times Z_{n_m}$.

We shall show that, in this form involving a base of $(X,+)$, the distributivity can be verified in $O(n^2)$ time.

Considering m-tuples from $Z_{n_1}\times\cdots\times Z_{n_m}$ as m-positional encodings of integers i, $0\le i<n$, by the injection

$$(i_1,\dots,i_m)\longmapsto i = i_1 + n_1 i_2 + n_1 n_2 i_3 + \dots + n_1\cdots n_{m-1} i_m\ ,$$

we can speak of $(i_1,\dots,i_m)$ as the i-th m-tuple. The verification of (D′) proceeds from m-tuple to m-tuple in this linear order.

Let us call $(i_1,\dots,i_m)$ an m-tuple of r-th kind if

$$i_1 = n_1-1\ ,\ \dots\ ,\ i_{r-1} = n_{r-1}-1\ ,\ i_r < n_r-1\ .$$

This means exactly that the next $(i+1)$-th m-tuple is

$$(j_1,\dots,j_m) = (0,\dots,0,i_r+1,i_{r+1},\dots,i_m) = (i_1,\dots,i_m) + (\underbrace{1,\dots,1}_{r},0,\dots,0)\ .$$

Assume that (D′) holds for a given $a\in X$ and the i-th m-tuple $(i_1,\dots,i_m)$ of r-th kind. Then checking of (D′) for the $(i+1)$-th m-tuple amounts to checking that

$$a\left(\sum_{k=1}^{m} j_k b_k\right) = a\left(\sum_{k=1}^{m} i_k b_k\right) + \sum_{k=1}^{r} ab_k\ ,$$

$$\left(\sum_{k=1}^{m} j_k b_k\right)a = \left(\sum_{k=1}^{m} i_k b_k\right)a + \sum_{k=1}^{r} b_k a\ ,$$

which requires time proportional to r. There are $(n_r-1)n_{r+1}\cdots n_m$ m-tuples of r-th kind, therefore the checking of (D′) for a single element $a\in X$ requires time proportional to

$$\sum_{r=1}^{m} r(n_r-1)n_{r+1}\cdots n_m = \sum_{r=1}^{m} n_r\cdots n_m - m \le n(1+\tfrac{1}{2}+\tfrac{1}{4}+\dots)\le 2n$$

It follows that (D′) can be verified in $O(n^2)$ time.

Knowing now that $(X,+)$ is an abelian group with base $b_1,\dots,b_m$ and that distributivity (D) holds for $(X,+,.)$, then in order to prove that the ring is associative it is enough to check for associativity of multiplication only triples $b_i(b_jb_k) = (b_ib_j)b_k$ for $i,j,k=1,\dots,m$. Since $2^m \leqslant n$, this test is carried out in $O(\log^3 n)$ time.

The elementary theory of abelian groups (cf. [1]) yields the following inductive construction of a base of $G = (X,+)$:

I. choose for b_1 an arbitrary element of maximal order;

II. if $b_1,\dots,b_k$ have been constructed choose for b_{k+1} an arbitrary element of maximal order in $X \setminus \langle b_1,\dots,b_k\rangle$, where $\langle b_1,\dots,b_k\rangle$ denotes the subgroup generated by $b_1,\dots,b_k$, such that $\langle b_1,\dots,b_k\rangle \cap \langle b_{k+1}\rangle = 0$.

As a preliminary step we make a list of all cyclic subgroups of G, descending by their orders, in the form of an $n{\times}n$-table with rows, corresponding to elements $x \in X$, filled by $x,2x,3x,\dots,kx=0$ in the first k places, where k is the order of x, and the remaining $n-k$ places blank. This preliminary step takes $O(n^2)$ time. We now start the construction of a base. We take the generator b_1 of the first group in the list for the first member of the base and strike off the list all cyclic groups generated by elements of $\langle b_1\rangle$. Then we search, from the beginning, the remaining list until we find b_2 - the generator of the first cyclic group in the list which meets $\langle b_1\rangle$ trivially. We generate $\langle b_1,b_2\rangle$, strike off the list all groups generated by elements of $\langle b_1,b_2\rangle$, look for the first b_3 in the list with $\langle b_1,b_2\rangle \cap \langle b_3\rangle = 0$, and so on.

It is clear that we go only once through each entry of the table constructed in the preliminary step, except for the rows corresponding to $b_1,\dots,b_m$ which are used twice - the first time to find out that $\langle b_k\rangle$ meets $\langle b_1,\dots,b_{k-1}\rangle$ trivially, the second to generate $\langle b_1,\dots,b_k\rangle$. It follows that a base can be constructed in $O(n^2)$ time, thus we have proved

<u>THEOREM 1.</u> There exists an algorithm deciding in $O(n^2)$ time whether a given bigroupoid $(X,+,.)$ of size n is a ring or an associative ring.

2. Recognition of lattices through the associated digraphs

A bigroupoid $(X, \vee, \wedge)$, with the two operations called join and meet, respectively, is a lattice if both $(X, \vee)$ and $(X, \wedge)$ is a commutative idempotent semigroup, i.e. a semilattice, and the two operations are linked by the absorption laws

$$x \vee (x \wedge y) = x \ , \quad x \wedge (x \vee y) = x \ .$$

The definition itself suggests an almost obvious reduction of the recognition of a lattice to that of two semilattices linked by the absorption laws. Since the verification of the latter is $O(n^2)$ time task (and one cannot expect any faster than $O(n^2)$ time recognition of a lattice) we see that the recognition of a lattice among bigroupoids can be reduced to the at least as complex recognition of a semilattice among groupoids.

A semilattice $(X, \vee)$ determines, and is completely determined by, its natural semilattice order

$$x \leq y \quad \text{iff} \quad x \vee y = y \ ,$$

giving rise to a special digraph (= directed graph) $(X, \leq)$, its upper semilattice poset, a partially ordered set in which supremum $\sup(x,y)$ exists for every couple of elements of X and $\sup(x,y) = x \vee y$.

Imitating the passage from a semilattice to its semilattice poset, we assign to an arbitrary groupoid $(X, \vee)$ a digraph (X,E), $E \subseteq X \times X$, denoted by $Dg(X, \vee)$, determined by

$$(x,y) \in E \quad \text{iff} \quad x \vee y = y \ ,$$

and called the associated digraph.

The associated digraph $Dg(X, \vee)$ is clearly obtainable from $(X, \vee)$ in $O(n^2)$ time. Since $(X, \vee)$ is a semilattice iff $Dg(X, \vee)$ is a semilattice poset in which $\sup(x,y) = x \vee y$, we have the recognition of a semilattice among groupoids reduced to the at least as complex recognition of a semilattice poset among digraphs.

This is an important reduction step since it brings into play graph-theoretical algorithms. Let us recall, for the reader's convenience, some known concepts and facts. More details and proofs can be found in [3].

Let $G = (X,E)$ be a digraph. G is called reflexive if $(x,x) \in E$ for all $x \in X$. Given a subset $Y \subseteq X$, an element $y \in Y$ is said to be maximal in Y if for every $z \in Y$ such that $(y,z) \in E$, $y = z$.

Y is upper closed if for every $x \in X$ and every $y \in Y$ such that $(y,x) \in E$ it is $x \in Y$. G is called acyclic if it does not contain a cycle of length ≥ 2, or equivalently, if every non-empty subset $Y \subseteq X$ has a maximal element.

The transitive closure $Clos(G) = (X,C)$ is a digraph defined by $(x,y) \in C$ iff there is a directed path of positive length from x to y. The transitive reduct is defined only for an acyclic digraph G, as the least digraph $Red(G) = (X,R)$ such that $Clos(Red(G)) = Clos(G)$.

A poset is a reflexive, acyclic, and transitive digraph G. In this case the reduct $Red(G) = (X,R)$ is also called the Hasse diagram of G and R the relation of covering in G. An upper semilattice poset is a poset $G = (X,E)$ in which all suprema exist ($(x,y) \in E$ must be interpreted as " x is smaller or equal to y "). A lattice poset is a poset which has both all suprema and all infima. In our finite case we can say equivalently that a lattice poset is a semilattice poset with the least element.

Given a digraph $G = (X,E)$, with $|X| = n$, $|E| = e$, one can use a dual version of the level-decomposition algorithm Lev from [3] to construct, in $O(n+e)$ time, a sequence $H(0), H(1), \ldots, H(n-1)$ of disjoint subsets of X such that $H(0)$ is the set of all maximal elements of X and $H(k)$ is the set of all maximal elements of $X \setminus \bigcup_{i=0}^{k-1} H(i)$ for all $k=1,\ldots,n-1$. Clearly, G is acyclic iff $\bigcup_{i=0}^{n-1} H(i) = X$. Thus for an acyclic digraph G the non-empty members $H(0), \ldots, H(s)$ of the sequence form a partition of X. Listing $H(0)$ in an arbitrary order, then adjoining a list of elements of $H(1)$, and so on, we make the vertices X of G into a sequence $x_1,\ldots,x_n$ such that initial segments $\{x_1,\ldots,x_k\}$ are upper closed for every $k=1,\ldots,n$.

If G is acyclic with $Red(G) = (X,R)$, $|R| = r$, then algorithm Recl introduced in [3] constructs both $Clos(G)$ and $Red(G)$ simultaneously in $O(n(r+1))$ time. It proceeds in $O(r)$ time steps, each producing a batch of at most n new edges for $Clos(G)$ and $Red(G)$.

By a mere combination of the above facts we get

STATEMENT 1. There exists an algorithm which for an arbitrary digraph $G = (X,E)$ decides in $O(n+e)$ time whether G is acyclic; if so then it in $O(n+e)$ time orders X into a sequence $x_1,\ldots,x_n$ in such a way that $\{x_1,\ldots,x_k\}$ is upper closed for every $k=1,\ldots,n$, next in $O(n+nr)$ time constructs $Red(G) = (X,R)$, tabulates the sets $xR = \{y \in X \mid (x,y) \in R\}$, $x \in X$, and decides whether G is a poset.

Next we show that at no additional time cost we can supplement the algorithm of Statement 1 so as to also decide whether G is an upper semilattice poset, as well as to compute and tabulate all the suprema. The following lemma gives the basis for such an algorithm.

LEMMA 1. Let $G = (X,E)$ be a poset with $Red(G) = (X,R)$, and let Y be a non-void upper closed subset of X such that any two elements of Y have a supremum. Let further y be an element of $X \setminus Y$ such that $Y \cup \{y\}$ is again an upper closed subset of X. Then, for any $z \in Y$, there exists $\sup(z,y)$ iff the set $\{\sup(u,z) \mid u \in yR\}$ has the least element w, in which case $\sup(z,y) = w$.

Proof. If, for some $z \in Y$, there exists $\sup(y,z)$ then $\sup(u,z) \geqslant \sup(y,z)$ for all $u \in yR$. Since $\sup(y,z) > y$, there is $u_0 \in yR$ such that $\sup(y,z) \geqslant u_0$. Evidently, $\sup(y,z) \geqslant z$, hence $\sup(y,z) \geqslant \sup(u_0,z)$ and the assertion follows.

If, conversely, for some $z \in Y$, there exists $u_0 \in yR$ such that $w = \sup(u_0,z)$ is the least element of $\{\sup(u,z) \mid u \in yR\}$ then clearly $w \geqslant z$ and $w \geqslant y$. Let v be an element of X such that $v \geqslant z$ and $v \geqslant y$. Then $v \in Y$ and there exists $u \in yR$ such that $v \geqslant u$, and, since $v \geqslant \sup(u,z) \geqslant w$, we conclude that $\sup(y,z) = w$.

Suppose we have a poset $G = (X,E)$ with X arranged into a sequence $x_1,\dots,x_n$ such that $\{x_1,\dots,x_k\}$ is upper closed for every $k=1,\dots,n$. Then by Lemma 1, the decision about existence and the computation of $\sup(x_i,x_k)$, for $i < k$, takes an amount of time proportional to $|x_kR|$. Thus we have proved

STATEMENT 2. There exists an algorithm which, for an arbitrary acyclic digraph $G = (X,E)$ with $Red(G) = (X,R)$, $|R| = r$, decides in $O(n+nr)$ time whether G is an upper semilattice poset.

Let us summarize what has been so far achieved. Given a groupoid $(X, \vee)$ of size n, we pass in $O(n^2)$ time to the associated digraph $G = Dg(X, \vee)$ and decide in $O(n^2)$ time if it is acyclic; if so then we decide in $O(n+nr)$ time if it is an upper semilattice poset, and, if so, tabulate all the suprema. Finally we go back to $(X, \vee)$ and compare in $O(n^2)$ time $x \vee y$ with $\sup(x,y)$ to make sure they coincide. If they do then we know that $(X, \vee)$ is a semilattice; if any of the stages just described fails then $(X, \vee)$ is not semilattice.

This $O(n^2+nr)$ time recognition procedure may also be considered as an indirect associativity test for a commutative idempotent groupoid $(X,\vee)$, with a recourse to its associated digraph. The latter can be used to this end in a much more straightforward manner, based on the following lemma.

LEMMA 2. A groupoid $(X,\vee)$ is a semilattice iff the following holds:

(1) $Dg(X,\vee) = (X,E)$ is a poset,

(2) $x \vee x = x$, $x \vee y = y \vee x$, $x \vee (x \vee y) = (x \vee y) \vee y = x \vee y$ for all $x,y \in X$,

(3) $(a \vee b) \vee x = a \vee (b \vee x)$ for all $(b,a) \in R$ and $x \in X$, where R comes from the reduct $(X,R) = Red(X,E)$.

Proof. Clearly the conditions are necessary. Assume they are fulfilled. If we show that

(4) $(x \vee y) \vee z = x \vee (y \vee z)$ for all $x,y,z \in X$,

we shall be done. Condition (3) makes (4) hold for all $(y,x) \in R$, $z \in X$. Assume that (4) has already been proved for all $(y,x) \in R^k$, $z \in X$, and let $(u,x) \in R^{k+1}$. Then there is $y \in X$ such that $(y,x) \in R^k$, $(u,y) \in R$, and, by induction hypothesis,

$$x \vee (u \vee z) = (x \vee y) \vee (u \vee z) = x \vee (y \vee (u \vee z)),$$
$$(y \vee u) \vee z = y \vee (u \vee z).$$

Using this and (2), (3), we get

$$(x \vee u) \vee z = x \vee z = (x \vee y) \vee z = x \vee (y \vee z) = x \vee ((y \vee u) \vee z) =$$
$$= x \vee (y \vee (u \vee z)) = x \vee (u \vee z).$$

We have proved (4) for all $(x,y) \in E$, $z \in X$. Let now $x,y,z \in X$ be arbitrary. It is

$$(x \vee y) \vee z = ((x \vee y) \vee y) \vee z = (x \vee y) \vee (y \vee z),$$

and, using commutativity,

$$x \vee (y \vee z) = x \vee (y \vee (y \vee z)) = (x \vee y) \vee (y \vee z).$$

Unfortunately, similarly as in Lemma 1, also here the number of edges in the reduct intervenes in such a way that the recognition of a semilattice based on Lemma 2 has again time complexity $O(n^2+nr)$.

If one has no better estimate than $r \leq n^2$ for the number of edges in the reduct of a semilattice poset then the above time complexity assessment becomes $O(n^3)$, which is trivial. However, we can do better.

STATEMENT 3. Let $h(n)$ denote the maximum number of edges in the reduct of a semilattice poset of size n. Then it holds

$$\tfrac{1}{2}2^{1/2}n^{3/2}(1 + o(1)) \leq h(n) \leq \tfrac{1}{3}(2n)^{3/2}(1 + o(1))$$

Proof. We first prove the upper bound. Let (X,R) be the reduct of a (semi)lattice poset of size n. We may assume that $X = \{1,2,\dots,n\}$ and that R is part of the natural ordering of X, i.e. $(i,j) \in R$ implies $i < j$.

Let X_k denote the set of all $i \in X$ such that $(i,k) \in R$. Then, as there are no four vertices i,j,k,ℓ such that all edges (i,k), (i,ℓ), (j,k), (j,ℓ) belong to R, we clearly have $|X_k \cap X_\ell| \leq 1$ for $k \neq \ell$. Thus, if we put $|X_k| = x_k$, we have

$$\sum_{j=1}^{k} \binom{x_j+1}{2} \leq \binom{k}{2} \qquad \text{for all } k=1,\dots,n,$$

hence

$$\sum_{j=1}^{k} x_j^2 \leq k^2 \qquad \text{for all } k=1,\dots,n.$$

Therefore, by a routine calculation we get

$$|R| = \sum_{k=1}^{n} x_k \leq \sum_{k=1}^{n} (2k-1)^{1/2} < \int_{1}^{n+1} (2x)^{1/2}dx < \tfrac{1}{3}2^{3/2}(n+1)^{3/2}.$$

To prove the lower bound, consider a projective plane $P(p)$ with $m = p^2+p+1$ points, for a prime p. Its lattice $LP(p)$ of subspaces consists of m points plus m lines plus the smallest and the biggest element, the reduct of the lattice poset of this lattice has $n=2m+2$ vertices and

$$2(p^2+p+1) + (p^2+p+1)\binom{p+1}{2} = \tfrac{1}{2}2^{1/2}n^{3/2}(1 + o(1)) \quad \text{edges.}$$

If n is arbitrary then take the largest prime p with $2(p^2+p+1) + 2 \leq n$ and, if necessary, add $n - 2(p^2+p+1) - 2$ points to the reduct of $LP(p)$ and connect each of them by two edges to the top and to the bottom of the reduct. Theresulting graph is the reduct of a lattice poset of size n. One can show, using the well known theorem on distribution of primes, that it has

$$\tfrac{1}{2}2^{1/2}n^{3/2}(1 + o(1)) \quad \text{edges.}$$

REMARK. We could not improve the upper bound but we conjecture that

$$h(n) = (\frac{1}{2}2^{1/2} + o(1))n^{3/2}$$

If we modify in the algorithm of Statement 2 the construction of Red(G), so as to reject G if after some $O(n^2)$ time step completing a new batch of edges for Red(G) the number of reduced edges so far constructed excedes $(n+1)^{3/2}$, we get an algorithm which makes true the following theorem.

THEOREM 2. There is an algorithm deciding in $O(n^{5/2})$ time whether a digraph $G = (X,E)$ of size $|X| = n$ is a semilattice poset and thereby also, through the associated digraph $Dg(X, \vee)$, whether a bigroupoid $(X, \vee, \wedge)$ of size n is a lattice.

REFERENCES

[1] G. Birkhoff, S. Mac Lane, Algebra, The Macmillan Comp., N. Y., 1967.

[2] A. Goralčíková, P. Goralčík, V. Koubek, Testing of Properties of Finite Algebras, Proceedings of ICALP´80, Springer-Verlag 1979, 273-281.

[3] A. Goralčíková, V. Koubek, A Reduct and Closure Algorithm for Graphs, Proceedings of MFCS´79, Springer-Verlag 1979, 301-307.

[4] R. E. Tarjan, Determining whether a Groupoid is a Group, Inf. Proc. Letters 1(1972), 120-124.

[5] J. Vuillemin, Comment verifier si une boucle est un groupe, Technical Report, Université de Paris-Sud, 1976, 1-12.

A DEFINITION OF THE P = NP-PROBLEM IN CATEGORIES

H.Huwig
Informatik II
Postfach 500 500
Universität Dortmund

In [Ha,Ho] Hartmanis and Hopcroft posed the question wether the P = NP?-problem is independent from a formal set theory, let's say Zermelo-Fraenkel,ZF, for example. Precisely this means, that neither $ZF \vdash \neg(P = NP)$ nor $ZF \vdash P = NP$ holds. In this paper we propose to look at categories different from Set in order to find modells M in which $M \models P = NP$ or $M \models (P \neq NP)$ holds, because then the converse result cannot proved in any theory, T, for which M is a modell. We detect that in the category of commutative, idempotent monoids, CIS, we have P = NP. Of course any theory of CIS will be considerable weaker as ZF. But to have an example: such a theory will probably have enough expressive power to carry out the minimization theory of automata in it (and to get the exspected results). In this sense independence results should enjoy some interest - even if $ZF \vdash \neg(P = NP)$ can be shown -, becouse they give some insight in the difficulty of the proof. Our result has to be seen in contrast to Lipton's result, showing that there is a modell of a fragment of arithmetic + 'P = NP'. Our natural numbers fullfile all the Peano axioms; especially they allow the definition of all primitive recursive functions, while Lipton's numbers have only a few arithmetical functions. On the other hand he can use a first order theory, while up to now we have to introduce some second order statements.

In this paper "category" means symmetrical monoidal and monoidal closed category, (The neutral element of the tensorproduct is denoted by 1 and structural isomorphins are denoted by α,λ,ρ and ν respectively.) Moreover we assume, that our categories have pullbacks and a coequalizer of kernel pair - mono factorization. Pullbacks allow to define subobjects by intersection - given $L \overset{i}{\rightarrowtail} X$ and $M \overset{j}{\rightarrowtail} X$, define $L \cap M$ to be their pullback - and also to define subobjects as inverse images: given $L \overset{i}{\rightarrowtail} X$ and $f\colon Y \to X$, $L f^{-1}$ is the subobject we get by pulling back i along f. Image factorization allows for existential quantification: Given $L \overset{i}{\rightarrowtail} X \oplus Y$ and $\pi_0\colon X \oplus Y \to X$, we get the object $L\exists\pi_0$ by factorization of $i\pi_0$:

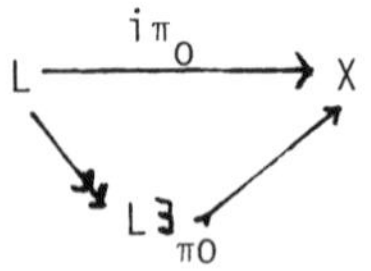

In Set, if $\pi_0\colon X \times Y \to X$ denotes the projection, we have:

$$L\exists\pi_0 = \{x \mid x \in X\ \exists\, z \in L : z\pi_0 = x\}$$
$$= \{x \mid x \in X\ \exists\, y \in Y : (x,y) \in L\}.$$

The key to a theory of primitive recursive computability is yield by the following definition due to Lawvere:

Def 1: A category satisfies the axiom of infinity or has a natural number object (NNO) <=>

There exists an object N equipped with two morphisms $1 \overset{o}{\to} N \overset{s}{\to} N$, such that for any other object A with morphisms $1 \overset{a}{\to} A \overset{s'}{\to} A$, there is a unique $a^{\#}: N \to A$ with $oa^{\#} = a$ and $sa^{\#} = a^{\#}s'$.

Theorem 1 (Freyd): In a category with NNO, morphisms can be defined by primitive iteration, i.e: Given $f: X \to Y$ and $g: Y \to Y$, there exists a unique $f^{\#}: X \oplus N \to Y$ with $\alpha^{-1}(X \oplus o)\ f^{\#} = f$ and $(X \oplus s)\ f^{\#} = f^{\#} g$.

As a consequence of this theorem, each primitive recursive function can be defined in a category with NNO. For example we may define $s_i = \lambda x.\ 2x + i + 1$, $i \varepsilon \{o,1\}$, by requiring $os_i = os^{i+1}$ and $ss_i = s_i s^2$.

Theorem 2 If in a category with NNO $1 \overset{o}{\to} N \overset{s_o}{\to} N \overset{s_1}{\to} N$ is an initial o-2-algebra, morphisms can be defined by syntactic iteration, i.e.: Given $f: X \to Y$ and $g_i: Y \to Y$, $i \varepsilon \{o,1\}$, there exists a unique $f^{\#}: X \oplus N \to Y$ with $\alpha^{-1}(X \oplus o)\ f^{\#} = f$, $s_o f^{\#} = f^{\#} g_o$ and $s_1 f^{\#} = f^{\#} g_1$.

Example: The assumption of the above theorem holds as soon as the category is cartesian closed. Moreover it holds for any monoidal closed category, which is algebraic over Set.

Remark 1: In a category with NNO

$$N \oplus N \xrightarrow{N \oplus \delta} N \oplus (N \oplus N) \overset{\sim}{\to} (N \oplus N) \oplus N \xrightarrow{+\oplus N} N \oplus N$$

defines "the" order relation on N. We denote this relation by $\geq \overset{r}{\to} N \oplus N$.

Note that δ, the diagonalization, and + can be defined by primitive iteration. The same statement holds for all morphisms of substitution, i.e. projections, interchanging of variables and so on. Under the assumption of theorem 2 we have a simple definition of logarithm: $lg_2: N \to N$ fullfiles $o\ lg_2 = o$ and

$$s_o lg_2 = lg_2 s = s_1\ lg_2.$$

Def 2: If B is a category with NNO, we define the category of polynomials in B, N[x] - B, to be the smallest monoidal subcategory of B, containing all "numbers" $1 \xrightarrow{p} N$ from B, all morphisms of substitution, the addition and the multiplication.

If moreover N is an initial o - 2 - algebra in B, we define the category of polynomial time computable morphisms in B, Poltime - B, to be the smallest monoidal subcategory of B containing all numbers $1 \xrightarrow{p} N$, the morphisms s_0 and s_1 and being closed under polynomial time bounded iteration.

The last statement says, that $f^{*}: X \oplus N \to Y$ (X and Y a finite product of copies of 1 and N) is in Poltime provided, there are morphisms $f: X \to Y$ and $g_i: Y \to Y$, $i \in \{o,1\}$, in Poltime-B and a $p: N \to N$ in N|x|-B, with

$$\text{i) } \alpha^{-1}(X \oplus o)f^{*} = f^{*},\ s_0 f^{*} = f^{*} g_0 \text{ and } s_1 f^{*} = f^{*} g_1$$

and ii) $\max_{X \oplus N} lg_2\, p \geq f^{*} \max_Y lg_2$ ($\max_X$ has the usual iductive definition starting from a primitive recursive definition of max: $N \oplus N \to N$. The condition $f \geq g$ can be expressed by the requirement, that the diagonal factors though the pullback of $\geq \xrightarrow{r} N \oplus N$ along $f \oplus g$).

In order to define polynomial time recognizable languages, we have to choose one language to be polynomial time recognizable and then to express the others as invers images of the prechoosen one. This is nothing else but the trace, which the use of accepting states in a machine oriented definition leaves in our abstract setting. Reminging our aim, that we want the learn something about "the" P = NP problem, one has to choose only those subobjects, which intuitively are polynomial time computable.

Def 3: Let B be a category with a NNO, which is also an initial o - 2 - algebra, and $C \xrightarrow{j} N$ a subobject of N. $L \xrightarrow{i} N$ is in deterministic polynomial time recognizable with respect to C <==>

There exists $f: N \to N \in$ Poltime - B, such that $L = Cf^{-1}$. $L \rightarrowtail^{i} N$ is in nondeterministic polynomial time recognizable with respect to C <=>

There exists $f: N \oplus N \to N \in$ Poltime-B and $p \in N[x]$-B, such that $L \xrightarrow{i} N = (Cf^{-1} \cap R)\exists\pi_0$, where R denotes the pullback of $\geq \xrightarrow{r} N \oplus N$ along $lg_2 p \oplus lg_2$.

We denote the discrete categories of these objects by D-Poltime$_C$-B respectively ND-Poltime$_C$-B. This definition is due to Scott [Sco].

Example 1: Choosing B = Set and $C \xrightarrow{i} \mathbb{N} = 1 \xrightarrow{o} \mathbb{N}$, we get the usual clases of deterministic and nondeterministic polynomial time recognizable languages. This remains true, if we replace $1 \xrightarrow{o} \mathbb{N}$ by any deterministic polynomial-timelanguage, except $\emptyset$ and $\mathbb{N}$.

Example 2: A strange example of a category with P = NP is yield by the free category over the graph $\underset{A}{\bullet} \to \underset{B}{\bullet}$. In fact this category is even cartesian closed and B is a NNO and a initial o-2-algebra. But B is at the same time a terminal object in that category, hence the NNO in that category fails to fullfile the axiom of nontriviality: $\neg(1 \cong N)$.

Example 2 shows, that in the absence of the power set axiom we have to check the Peano axioms, in order to get natural examples with P = NP.

Example 3: A nontrivial example of a category with P = NP is provided by CIS, the category of commutative, idempotent semigroups.

We will discuss this example in some detail now. Note that CIS is a symmetrical monoidal and monoidal closed category with respect to the usual tensorproduct, $\_\otimes\_$. The right adjoint of $\_\otimes H$ is given by $G \to G^H$, the set of all homomorphisms from H to G. The functor "free commutative idempotent semigroup", F:Set → CIS, which takes a set X to the set of all its finite nonempty subsets with $\_\cup\_$ as binary operation, sends products in Set to tensorproducts in CIS. Thus it is easy to check, that $1F \xrightarrow{oF} \mathbb{N}F \xrightarrow{sF} \mathbb{N}F$, which will be denoted by $1 \xrightarrow{o} N \xrightarrow{s} N$, is a natural number object in CIS. The remaining Peano Axiom then are: a) the induction principle, each o-s-closed subobject of N is the whole of N, b) Nontriviality, $\neg(1 \cong N)$ and c) s is a mono. These are all easy to check. For c) remark, that s in Set, is even a coretraction and each functor preserves coretractions. From theorem one we know, that we can define morphisms by primitive iteration in CIS. Moreover, because of the nice behaviour of F mentioned above, we have: if f is primitive recursive in Set, so is fF in CIS. We will use the same names for these morphisms, thus the addition in CIS is denoted by $+ : N \otimes N \to N$, which now takes a finite subset S of $\mathbb{N} \times \mathbb{N}$ to the set $\{s + t \mid (s,t) \in S\}$.

The definition of "the" order relation in Remark 1 yields, the subobject $R \rightarrowtail N \otimes N$ consting of all those finite subsets S of $\mathbb{N} \times \mathbb{N}$ with: $(s,t) \in S \Rightarrow s \geq t$. We may even check that R is indeed a order relation in the rather formal sense, that we have. $\Delta \subset R$, $R^2 \subset R$ and $R \cap R^{op} = \Delta$ in the appropriate arrow theoretic formulation of these axioms. Hence we will write $\geq \to N \otimes N$, instead of $R \rightarrowtail N \otimes N$. Again if $f: \mathbb{N} \to \mathbb{N}$ is pointwise less or equal to $g: \mathbb{N} \to \mathbb{N}$ in Set, $f \leq g$, then we have $fF \leq gF$ in CIS too. of course $fF \leq gF$ in CIS has the obvious arrow theoretic meaning. Since CIS has all primitive recursive functions, the definition of $s_i : N \to N$ with $s_i = \lambda x.\ 2x + (i+1)$ is obvious and an easy diagramm chase shows, that $1 \xrightarrow{o} N \xrightarrow{s_0} N \xrightarrow{s_1} N$ is indead an initial 0-2-algebra in CIS. As a consequence of theorem 2 we may define morphisms in CIS by syntactic iteration and the definition of polynomials in CIS and polynomial-time-computable morphisms in CIS is just the specialisation of definition 2 to CIS.

The above remarks on the order relations make sure that we have:

Lemma 1: If f is polynomial time computable in Set, so is fF in CIS.

Of course there are some more polynomialtime computable morphisms in CIS. The reason for this is provided by the so called nonstandard natural numbers, which we can't get by an iterated application of the successor to zero. An example of such a number is

$$\begin{array}{l} \underline{1} \xrightarrow{\{0,1\}} N \\ 0 \longmapsto \{0,1\}. \end{array}$$

Just these nonstandards allow for the construction of efficient algorithms in CIS. This fact is used in Lipton's modell theoretic considerations too . In order to apply definition 3 in our concrete situation, we chose the Heyting-complement of the positive numbers, which is the subobject $C \rightarrow N$ consisting of all those finite $S \subset \mathbb{N}$ with $0 \in S$. Since this definition yields the subobject $\{0\} \subset \mathbb{N}$ in Set, it meets the requirement we posed on the definition of "languages".

Lemma 2: If f and g are polynomial time computable in CIS, so is $f \cup g$.

Proof: α, the homomoprhic extension of α' is polynomial time computable in CIS because
$\alpha = \alpha' F$, where $\alpha' : (\mathbb{N} \times \mathbb{N}) \times \mathbb{N} \rightarrow \mathbb{N}$

$$(x,y,z) \mapsto \begin{cases} x & \text{if} \quad z = 0 \\ y & \text{if} \quad z \neq 0 \end{cases}$$

and α' is polynomial time computable. $f \cup g$ may be computed from f and g using only parallel execution and composition as indicated in the following diagram:

$$\underbrace{N \xrightarrow{\delta} N \otimes N \cong (N \otimes N) \otimes \underline{1} \xrightarrow{(f \otimes g)\otimes\{0,1\}} (N \otimes N) \otimes N \xrightarrow{\alpha} N}_{f \cup g} ;\blacksquare$$

Lemma 3: The morphism

$$\begin{array}{l} \phi : N \rightarrow N \\ \quad S \rightarrow \{s' \mid \exists\, s \in S : s\ lg_2 \geq s'\ lg_2\} \end{array}$$

is in Poltime-CIS

Proof: We define $\lambda x.2x + \{1,2\}$ using the diagram:

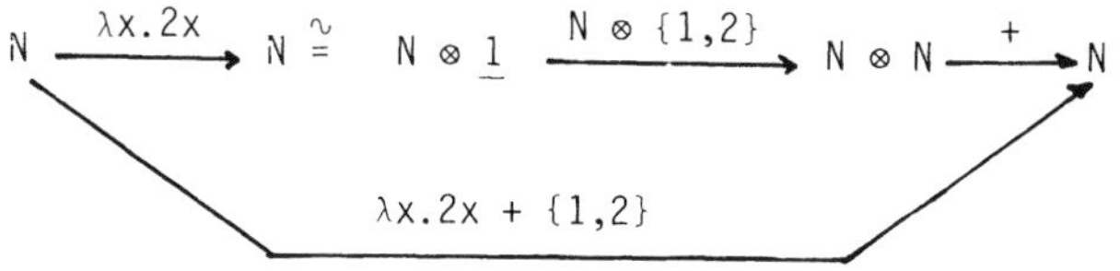

Clearly $\lambda x.2x + \{1,2\}$ is in Poltime-CIS because it is build up - using composition and parallel execution only - from morphisms which are polynomial time computable in CIS either because of Lemma 1 or because they are (nonstandard) numbers. From Lemma 2 it follows that $\psi = N \cup \lambda x.2x + \{1,2\}$ is in Poltime-CIS.

If $s \varepsilon N$ we have

$$S\psi = \{x \mid x \varepsilon \mathbb{N} \wedge \exists s \varepsilon S: x = s \vee x = 2s + 1 \vee x = 2s + 2\}.$$

Now ϕ is the solution of the iteration problem

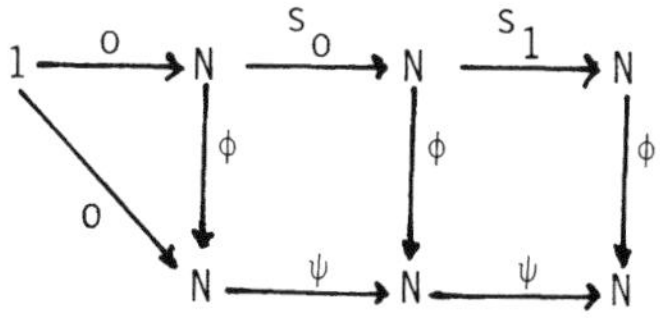

and it is easy to check, that the logarithmical growth of ϕ is bounded by $N: N \to N$.

Theorem 3: D-Poltime-CIS = ND-Poltime-CIS.

Proof: For each $p \varepsilon N[x]$ -CIS there are polynomials $p_0, \ldots, p_{r-1}$ in Set such that $p = \bigcup_{\rho \varepsilon \underline{r}} p_\rho$.

Clearly $v_p: \mathbb{N} \to \mathbb{N}$

$x \to \min\{x p_\rho \mid \rho \varepsilon \underline{r}\}$ is polynomial time computable in Set. From Lemma 1 we know, that

$u_p: \mathbb{N} \to N$

$S \to \{s' \mid \exists s \varepsilon S \ \ s' = \min\{s p_\rho \mid \rho \varepsilon \underline{r}\}\}$

is polynomialtime computable in CIS. As in Set we may show that $(\lambda x.2^x \dot{-} 1)\, lg_2 = N$ and that

$N \xrightarrow{lg_2} N \xrightarrow{f} N \xrightarrow{\lambda x.2^x \dot{-} 1} N$ is polynomial time computable if f is. (Remark that this does not imply, that $\lambda x.2^x \dot{-} 1$ is polynomialtime computable)

Define u'_p via the diagram

$$N \xrightarrow{lg_2} N \xrightarrow{u_p} N \xrightarrow{\lambda x\,.\,2^x \dot{-} 1} N$$ and ϕ_p via

$$N \xrightarrow{\delta} N \otimes N \xrightarrow{N \otimes (u'_p \phi)} N \otimes N$$

Let $L \rightarrowtail^{i} N$ be a nondeterministic polynomial time recognizable object in CIS, then there is f: $N \otimes N \to N$ from Poltime-CIS and $p \varepsilon$ N[x]-CIS, such that L is composed exactly of all those $S \varepsilon N$ for which there is $S' \varepsilon Cf^{-1} \cap R$ with $S = S' \exists \pi_0$. But such a S' will exists iff $0 \varepsilon S\phi_p f$. Thus the inverse image of $C \rightarrowtail^{j} N$ under the morphism

$$N \xrightarrow{\phi_p} N \otimes N \xrightarrow{f} N$$

defines the same language and hence ND-Poltime-CIS is contained in D-Poltime-CIS. The converse is easy for given $N \xrightarrow{f'} N \varepsilon$ Poltime-CIS just consider the morphism $N \otimes N \xrightarrow{N \otimes t} N \otimes \underline{1} \xrightarrow{\sim} N \xrightarrow{f} N$ and the polynomial $N \xrightarrow{t} 1 \xrightarrow{0} N$. (t denotes the morphism $t{:}N \to 1$, $S \mapsto o$ which is easily seen to be polynomial time computable).

We finish this paper with a short discussion of the relation between the P = NP-problem in CIS and that in Set. Intuitively the theorem says, that the first order theory of a symmetrical monoidal and monoidal closed category with a NNO, finite limits, finite colimits and a generator is not powerful to solve the P = NP-problem in the exspected way. To make this precise we will replace the second order definitions of Poltime-B and N[x]-B by first order one's in a forth-coming paper. An example of a theorem which can be proved in this theory is the minimization of automaton. Moreover a good deal of standard constructions of abstract algebra can be carried out in that theory with the exspected results. Set theory is an extension of our theory: one has to add the power set axiom, the axiom, that tensorproduct is just the cartesian product, and some very specialized axioms [Jo]. The last two axioms would tourn our theory in the theory of a topos with NNO. There are very many such topoi and the detection of a topos with P = NP would considerably improve the possibility, that the conjuncture of Hartmanis and Hopcroft is true.

REFERENCES

[BGS] Baker,T. & Gill, J. & Solovay, R. "Relativization of the P = NP question". SIAM Journal of Computing 4, 1975

[CO] Cook,S. "The Complexity of Theorem Proving Procedures", 3rd ACM Symposium on Theory of Computing, 1971, Seite 151-158

[G] Grzegorczyk,A. "Some classes of recursive functions", Rozprawy Matematyczne, Warszawa 1953

[F] Fachini,E. and Maggiolo-Schettini, "A Hirarchy of Primitive Recursive Sequence Funtions", Informatique theoretique, Nr. 13, 1979

[Ha,Ho] Hartmanis,J. & Hopcroft J. "Independence results in Computer Science SIGACT Newsletter 8, 1976

[Hu] Huwig,H., "Beziehungen zwischen beschränkter syntaktischer und beschränkter primitiver Rekursion", Dissertation, Berichtsreihe der Abteilung Informatik an der Universität Dortmund 25/76

[Hu] Huwig,H., "Das P = NP-Problem in der Kategorie der kommutativen idempotenten Halbgruppen", Habilitationsschrift, eingereicht bei der Universität Dortmund, 1980

[HWI] von Henke,F., & Indermark, K. & Weihrauch K. "Hierarchies of primitive recursive wordfunctions and transductions defined by automata", Automata, Languages and Programming, 1972, North-Holland Publishing Company, Amsterdam -London

[Li] Lipton,R.: "On the consistency of P = NP and fragments of arithmetic" FCT'79, Akademie Verlag Berlin

[Ma] Makkai, M & Reyes,G., "First order category logic" Springer Verlag, 1977

[Ra] Radziszowski,S. "Programmability and P = NP-Conjecture" Proceedings of FCT '77, Springer Verlag

[Sco] Scott,D."Some Definitional Suggestions for Automata Theory",JCSS 1,1967

[To] Thompson,D. "Subrecursiveness: Machine-Independent Notions of Computability in Restricted Time and Storage", Mathematical Systems Theory, Vol. 6, No. 1, 1972.

[Jo] Johnstone,P.T., "Topos Theory", Academic Press, 1977

GENERATING GRAPH LANGUAGES USING HYPERGRAPH GRAMMARS

D. Janssens
Department of Mathematics
University of Antwerp, UIA
Universiteitsplein 1
B-2610 Wilrijk
Belgium

G. Rozenberg
Institute of Applied Mathematics
and Computer Science
University of Leiden
Wassenaarseweg 80
Leiden
The Netherlands

INTRODUCTION

As documented e.g. in [N2] and [CER] (in particular in [E] and [N1]) the theory of graph grammars is a well-motivated research area. However, this theory is much poorer than that of the classical "string" grammars. This is due not only to the fact that the subject is intrinsically more difficult (a graph is a more complicated structure than a string) but also to the fact that the number of people working in this area is considerably smaller than the number of people working on string grammars.

In the present state of the theory, new approaches to defining graph languages are still needed (as well as in depth research of old approaches). In particular, we do not have yet the class of graph grammars (and languages) which would correspond to finite automata (and regular languages). What we mean by this correspondence is that (1) one would like to have a very "natural" device to generate nontrivial languages (as a finite automaton is), and (2) one would like to have a class of graph languages that would be as essential for the theory of graph languages as the regular languages are for string languages.

In this paper we present an attempt to provide a solution of (1) above. Our approach is methodologically quite analogous to that of defining languages by transition graphs (of finite automata). The notion of a graph is a natural generalization of the notion of a string and in finite automata theory one uses <u>one</u> (transition) graph to define a set of strings (its language). The notion of a hypergraph generalizes the notion of a graph. In our approach we will use <u>one</u> hypergraph (equipped with an additional graph structure) to define a set of graphs (its language).

The aim of this paper is to introduce some basic notions and formalisms concerning our approach, to illustrate it by examples and to compare several classes of graph-generating systems that we introduce.

The reader can find all formal definitions and proofs in the full version of this paper (available as a technical report at the University of Antwerp).

I. PRELIMINARIES

(1) A hypergraph is a system $H = (V,E,f)$ where V is a finite nonempty set, called the set of nodes of H, E is a finite set, called the set of edges of H and f is an injective function from E into $P(V)$ such that $\bigcup_{e \in E} f(e) = V$; f is called the edge function of H. If H is a hypergraph, then the set of nodes of H, the set of edges of H and the edge function of H will be denoted by V_H, E_H and f_H respectively.

(2) Let $H = (V,E,f)$ be a hypergraph. By int H we denote the set $\{X \mid X \neq \emptyset$ and there exist distinct edges e, $\bar{e}$ in E such that $f(e) \cap f(\bar{e}) = X\}$ (int H is the set of intersections of H).

(3) If X denotes a class of graph-generating systems (e.g. H systems, FDH systems, etc.), then by $L(X)$ we will denote the set of languages L for which there exists a X system G such that $L = L(G)$. ($L(G)$ is defined in the sequel.)

Gluing graphs is a very basic operation for our paper. Formally it is defined as follows. (A similar operation was used already before in graph grammars, see e.g. [E] and [N1] in [CER]).

Definition 1.1. Let A,B and H be graphs and let I be a discrete graph. Let f and g be injective homomorphisms from I into A and into B respectively. Then we say that H is the gluing of A and B along I by f and g if H is isomorphic to the graph (V,E) constructed as follows.
Let $\bar{A}$ and $\bar{B}$ be graphs, isomorphic to A and B such that $V_{\bar{A}} \cap V_{\bar{B}} = \emptyset$ and let h_A and h_B be the corresponding isomorphisms from A into $\bar{A}$ and from B into $\bar{B}$ respectively. Let $\bar{f} = h_A \circ f$ and $\bar{g} = h_B \circ g$. Then define

$$V = V_{\bar{A}} \cup (V_{\bar{B}} \setminus \mathrm{Im}(\bar{g})),$$

$$\begin{aligned} E = E_{\bar{A}} &\cup \{\{x,y\} \mid \{x,y\} \in E_{\bar{B}} \text{ and } x,y \notin \mathrm{Im}(\bar{g})\} \\ &\cup \{\{\bar{f} \circ \bar{g}^{-1}(x), y\} \mid \{x,y\} \in E_{\bar{B}} \text{ and } x \in \mathrm{Im}(\bar{g}),\ y \notin \mathrm{Im}(\bar{g})\} \\ &\cup \{\{\bar{f} \circ \bar{g}^{-1}(x),\ \bar{f} \circ \bar{g}^{-1}(y)\} \mid \{x,y\} \in E_{\bar{B}} \text{ and } x,y \in \mathrm{Im}(\bar{g})\}. \end{aligned}$$

Let h be an isomorphism from the graph (V,E) into H, let $\hat{f} = h \circ h_A$, and let $\hat{g}$ be defined by $\hat{g}(x) = h \circ h_B$ for $x \in V_B \setminus \text{Im}(y)$ and $\hat{g}(x) = \hat{f} \circ f \circ g^{-1}(x)$ for $x \in \text{Im}(g)$. Then $\hat{f}$ and $\hat{g}$ are isomorphisms from A and B respectively into subgraphs of H. $\hat{f}$ and $\hat{g}$ will be called the natural injections of A into H and B into H respectively. □

In the rest of this paper we will frequently use the set of nodes of I instead of the graph I itself. The homomorphisms f and g of the above definition are then simply injective functions from a finite nonempty set into V_A and V_B.

II. HYPERGRAPH SYSTEMS

In this section a grammatical device to define graph languages is introduced - it is very basic for this paper. It is based on hypergraphs. Given a hypergraph one first imposes on its node set an additional graph structure. Hence to each edge corresponds a subgraph of this structure. Then one uses these subgraphs as elementary "blocks" ("letters") to build graphs. The way that these elementary blocks are glued together is controlled through the structure of the given hypergraph (the way its edges intersect). Formally such a construct is defined as follows.

Definition 2.1.

A hypergraph system (abbreviated H system) is a system $G = (H,\Gamma,e_{in})$ where H is a hypergraph, Γ is a set of multisets of the form $\{x,y\}$ with $x,y \in V_H$ and e_{in} is an element of E_H; e_{in} is called the initial edge of G. □

The hypergraph system $G = (H,\Gamma,e_{in})$ defines a graph-language in the following way : one starts with the graph H_{in} (corresponding to e_{in}). Then one may choose an edge e_1 such that $f_H(e_{in}) \cap f_H(e_1) \neq \emptyset$, and one can glue H_{in} to the graph H_1 (corresponding to the edge e_1) along the intersection $f_H(e_{in}) \cap f_H(e_1)$, obtaining a graph of the language. Iterating this process one can again choose an edge e_2 with $f_H(e_2) \cap f_H(e_1) \neq \emptyset$ and glue the corresponding graph H_2 to the previously obtained graph along $f_H(e_1) \cap f_H(e_2)$, etc. The so obtained set of graphs is called the language of G and denoted by L(G).

When deriving a graph in the language of a hypergraph system, the way that the "building blocks" are glued together is determined by the way that edges in the hypergraph intersect. Hence it is natural to consider a system also based on a hypergraph, in which rather than to follow "consecutive" edges and glue them according to their intersections, one "follows" the intersections themselves.

Definition 2.2. An Intersection-based Hypergraph system (abbreviated IH system) is a system $G = (H,\Gamma,u_{in})$ where H is a hypergraph, Γ is a set of multisets of the form $\{x,y\}$ with $x,y \in V_H$ and u_{in} is an element of int H; u_{in} is called the initial intersection of G. □

The language L(G) of an IH system $G = (H,\Gamma,u_{in})$ is obtained as follows : first one chooses an edge e_0 of H such that $u_{in} \subseteq f_H(e_0)$. The graph H_0 corresponding to the edge e_0 is in the language. Then one may choose an edge e_1 such that $u_{in} = f_H(e_0) \cap f_H(e_1)$, glue the graph H_1 (corresponding to e_1) to H_0 along u_{in} and choose an element $u_1 \in$ int(H) such that there exists an edge $\tilde{e} \in E_H$ with $\tilde{e} \neq e_1$ and $f_H(\tilde{e}) \cap f_H(e_1) = u_1$. Iterating this process one can again choose an edge e_2 with $f_H(e_2) \cap f_H(e_1) = u_1$, glue the graph corresponding to e_2 to the previously defined graph, choose an intersection u_2 in e_2 etc.

We have the following result concerning the classes of graph languages $L(H)$ and $L(IH)$.

Theorem 2.1. $L(H)$ and $L(IH)$ are incomparable but not disjoint.

Although hypergraph systems and intersection-based hypergraph systems were presented as grammatical (thus generative) devices there is a quite close analogy between those systems and finite automata one may view this quite naturally as a hypergraph system where all the edges are of cardinality two. Following edges in the transition graph of a finite automaton corresponds to following edges in the multigraph system and so it corresponds to an H system. On the other hand, following states (nodes) in the transition graph corresponds to following intersections in an IH system.

However, this analogy is not complete because in (the transition graph of) a finite automaton there are two additional components controlling the way it defines a language. Firstly, transitions are directed, and so if after a transition A a transition B follows it does not necessarily mean that A can follow B; in other words, transitions do not have to be "symmetric". Secondly, certain "places" (nodes) are distinguished as terminal places and a derivation following the transition graph is considered successful only if its last step corresponds to a terminal place in the graph. We will now consider these two additional "control features" within the framework of H systems and IH systems. In this way one can view H systems and IH systems as examples of the exhaustive approach to graph language definition : one takes into the language of a given system everything the system generates (each "intermediate" graph also belongs to the language). On the other hand the systems we will consider next may be viewed as an example of a "selective" approach to graph language definition : from the set of all graphs that a system generates one takes into the language of the system only those graphs that satisfy a certain "filtering condition".

Definition 2.3. A directed hypergraph system with final edges, abbreviated GFH system , is a system $G = (H,\Gamma,e_{in},E_{fin},C)$ where base $G = (H,\Gamma,e_{in})$ in an H system, E_{fin} is a subset of E_H and C is a subset of $E_H \times E_H$ such that $(e,\bar{e}) \in C$ implies that $f_H(e) \cap f_H(\bar{e}) \neq \emptyset$. □

A GFH system $G = (H,\Gamma,e_{in},E_{fin},C)$ defines a graph language L(G) in the same way as a H system, by gluing together graphs corresponding to edges $e_{in} = e_0,e_1,e_2,\ldots,e_r$ with $f_H(e_i) \cap f_H(e_{i+1}) \neq \emptyset$ for $0 \leq i \leq r-1$ but with the additional restrictions that $e_r \in E_{fin}$ and for $0 \leq i \leq r-1$, $(e_i, e_{i+1}) \in C$.

Definition 2.4. A directed intersection-based hypergraph system with final intersections (abbreviated GFIH system) is a system $G = (H,\Gamma,u_{in},I_{fin},C)$ where base $G = (H,\Gamma,u_{in})$ is an IH system, I_{fin} is a subset of int H and C is a subset of int H x int H such that $(u,\bar{u}) \in C$ implies that there exist e_1,e_2,e_3 in E_H with $f_H(e_1) \cap f_H(e_2) = u$ and $f_H(e_2) \cap f_H(e_3) = \bar{u}$. □

A GFIH system $G = (H,\Gamma,u_{in},I_{fin},C)$ defines a graph language L(G) in the same way as an IH system, by gluing together graphs corresponding to edges $H_0,\ldots,H_r$ using intersections $u_{in} = u_0, u_1, \ldots, u_r$ with $u_i = f_H(e_i) \cap f_H(e_{i+1})$ for $0 \leq i \leq r-1$, but with the additional restrictions that $u_r \in I_{fin}$ and that for $0 \leq i \leq r-1$, $(u_i, u_{i+1}) \in C$.

If in a GFH system $G = (H,\Gamma,e_{in},E_{fin},C)$, $E_{fin} = E_H$, then we omit E_{fin} from the specification of G. In this case G will be called a GH system. On the other hand if $G = (H,\Gamma,e_{in},E_{fin},C)$ and $C = \{(e,\tilde{e}) \mid e,\tilde{e} \in E_H$ and $f_H(e) \cap f_H(\tilde{e}) \neq \emptyset\}$ then we omit C from the specification of G and we say that G is a FH system. Analogously, we define GIH systems and FIH systems. A GIH system is a GFIH system $G = (H,\Gamma,u_{in},I_{fin},C)$ where $I_{fin} = \underline{int}(H)$; in this case, I_{fin} will be omitted from the specification of G. If on the other hand $G = \{(u,\tilde{u}) \mid u,\tilde{u} \in \underline{int}\ H)$ and there exist e_1, e_2, e_3 in E_H with $f_H(e_1) \cap f_H(e_2) = u$ and $f_H(e_2) \cap f_H(e_3) = \tilde{u}\}$ then we call G a FIH system and C is omitted from the specification of G.

We conclude this section with the following observation. Although the graph-language generating systems discussed in this section bear a certain similarity to finite automata defining string languages there are certain important differences between our systems and finite string-automata. Since our systems define graphs rather than strings, they are considerably more difficult to analyze. In particular, certain questions concerning the effectiveness of defining graph-languages by our systems turn out to be undecidable, while the corresponding questions for finite string-automata are "easily" decidable. Here is an example of such a situation.

Theorem 2.2. For two arbitrary given GFH systems G and $\overline{G}$, it is undecidable whether or not $L(G) \cap L(\overline{G})$ is empty.

The proof of the above result consists mainly of a special encoding of the post correspondence problem. A similar technique yields the analogous result for GFIH systems.

III. OVERLAPPING GRAPH SYSTEMS

In all the systems considered in the last section one notices the following phenomenon : even though two intersecting edges of a hypergraph (including their graph structure) may differ considerably, they always are identical within their intersection area. Since in a derivation step only intersecting edges may be used, this particular feature implies the following restriction : if X is an intermediate graph obtained in a derivation of a graph Y and X contains two nodes x_1, x_2 with no edge connecting them, then also in Y (the nodes corresponding to) x_1 and x_2 will have no edge connecting them. For this reason it seems natural to consider systems in which the basic building blocks will be graphs. Some of these graphs may have common nodes, however the structure of edges on the nodes common to two different graphs may be quite different. Such systems are considered in this section.

Definition 3.1. An overlapping graph system (abbreviated O system) is a pair $G = (\mathcal{H}, H_{in})$ where $\mathcal{H}$ is a finite nonempty set of graphs and $H_{in} \in \mathcal{H}$; H_{in} is called the initial graph of G. □

Observe that the sets of nodes of elements of $\mathcal{H}$ are not assumed to be disjoint.

An O system $G = (\mathcal{H}, H_{in})$ generates a graph language L(G) in the same way as a H system, except that the "building blocks" (i.e. the elements of $\mathcal{H}$) cannot be considered as subgraphs of one given graph (as in the case of H systems).

As in the case of H systems we will now define a counterpart of O systems based on intersections rather than on edges. (If M is a graph then by V_M we denote the set of nodes of M.)

Definition 3.2. Let $\mathcal{H}$ be a set of graphs. Then the set of intersections of $\mathcal{H}$, denoted by int $\mathcal{H}$, is the set $\{X \mid$ there exist distinct H, $\bar{H}$ in $\mathcal{H}$ with $X = V_H \cap V_{\bar{H}} \neq \emptyset\}$. □

Definition 3.3. An intersection based overlapping graph system (abbreviated IO system) is a system $(\mathcal{H}, u_{in})$ where $\mathcal{H}$ is a finite nonempty set of graphs and u_{in} is an element of int $\mathcal{H}$. □

An IO system G generates a graph-language L(G) in the same way as an IH system, except that the "building blocks" are the elements of H.

Analogously to previous definitions 2.1, 2.2, 2.3 and 2.4 one can introduce extra control features into the framework of O systems and IO systems; these control features correspond to the directed transitions and the final places of finite automata. The so obtained systems will be called <u>directed overlapping graph systems with final graphs</u> (abbreviated GFO <u>systems</u>) and <u>directed intersection-based overlapping graph systems with final intersections</u> (abbreviated GFIO <u>systems</u>) respectively.
This gives rise to GO systems, FO systems, GIO systems and FIO systems, analogously to GH, GH, GIH and FIH systems.

<u>Remark 3.1</u>. It is obvious that every XH system can be considered as a special form of a XO system where X stands for G, F, GF, GI, FI or GFI.

We have the following results comparing the graph-language generating power of the systems considered so far.

<u>Theorem 3.1</u>. L(GO) and L(FO) are incomparable but not disjoint.

<u>Theorem 3.2</u>. L(GH) and L(FH) are incomparable but not disjoint.

<u>Theorem 3.3</u>. $L(H) \subsetneq L(O)$.

<u>Theorem 3.4</u>. $L(GH) \subsetneq L(GO)$.

<u>Theorem 3.5</u>. The diagram of fig. 3.2. holds, where we denote $A \rightarrow B$ if $A \subseteq B$, $A \twoheadrightarrow B$ if $A \subsetneq B$ and $A \leftarrow\!\!\!\times B$ if A and B are incomparable but not disjoint. □

fig. 3.1. :

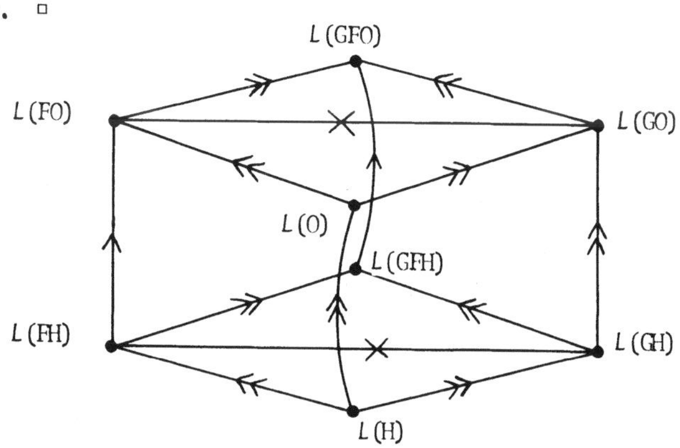

Next we consider systems based on intersections.

Theorem 3.6. L(FIO) and L(GIO) are incomparable but not disjoint.

Theorem 3.7. L(FIH) and L(GIH) are incomparable but not disjoint.

Theorem 3.8. $L(\mathrm{IH}) \subsetneq L(\mathrm{IO})$.

Theorem 3.9. $L(\mathrm{GIH}) \subsetneq L(\mathrm{GIO})$.

Theorem 3.10. The diagram of fig. 3.3. holds, where we denote A ⟶ B if $A \subseteq B$, A ⟶⟶ B if $A \subsetneq B$ and A ⟷ B if A and B are incomparable but not disjoint.□

fig. 3.2.

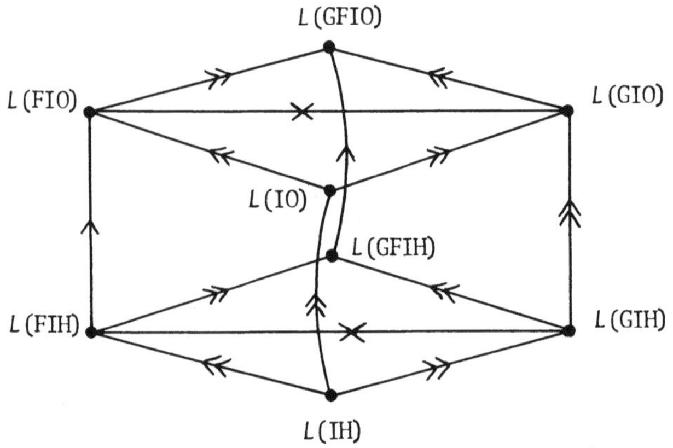

IV. FINITE GRAPH-AUTOMATA

The analogy between intersections in the systems we consider and states in finite automata cannot be pushed too far : whereas the states (nodes) in the transition graph of a finite automaton can be seen as intersections (of edges) consisting of one node only, in our systems we may have intersections of arbitrary cardinality. This implies that if an intersection involves m edges (graphs) then the graph structure of the pairswise intersections are not independent. To remove this obstacle we will equip our systems explicitly with states - a state being now an abstract entity remembering a specific information about the derivation performed so far. As usual, we will consider systems with a finite number of states only. Such systems are defined formally as follows.

Definition 4.1. A finite graph automaton, abbreviated FGA, is a system $A = (\Sigma,Q,q_0,F,\rho,\delta)$ where Σ is a set of graphs, such that no two elements of Σ are isomorphic. Σ is called the alphabet of A, Q is a finite nonempty set, called the set of states, q_0 is an element of Q, called the initial state, F is a subset of Q, called the set of final states, ρ is a function from Q into the set of all discrete graphs, and δ is a function from $Q \times \Sigma$ such that for each (q,H) in $Q \times \Sigma$, $\delta(q,H)$ is a finite set of elements of the form $(\overline{q},\gamma_{in},\gamma_{out})$ where $\overline{q} \in Q$ and γ_{in}, γ_{out} are injective homomorphisms from $\rho(q)$ into H and from $\rho(\overline{q})$ into H respectively. δ is called the transition function. □

Observe that in the above we do not require δ to be a total function.

A FGA $A = (\Sigma,Q,q_0,F,\rho,\delta)$ generates a language $L(A)$ in the following way : one starts in state q_{in}, with the graph $\rho(q_{in})$. In order to reach a new state q_1 one chooses a H_1 in Σ such that there exists a triple $(q_1, \gamma_{in}, \gamma_{out})$ in $\delta(q_{in},H_1)$ and one glues H_1 to $\rho(q_{in})$ along $\rho(q_{in})$ by the identity and $\gamma_{in} \cdot \gamma_{out}$ now specifies an "embedding" of $\rho(q_1)$ in the obtained graph. In the next step one chooses a H_2 such that $\delta(q_1,H_2)$ contains a triple $(q_2, \overline{\gamma}_{in}, \overline{\gamma}_{out})$ and one glues H_2 to the previous graph along $\rho(q_1)$ by $\overline{\gamma}_{in}$ and the embedding specified by γ_{out} ($\overline{\gamma}_{out}$ will be used in the following step). The language $L(A)$ is the set of graphs generated by iterating the above process and such that we end up in a state of F.

The following results show that finite graph-automata generalize both GFO systems and GFIO systems (the proofs can be found in the full version of the paper).

Theorem 4.1. For every GFO system G there exists an equivalent FGA A.

Theorem 4.2. For every GFIO system G there exists an equivalent FGA A.

V. DISCUSSION

Starting from the observation that the notion of a hypergraph generalizes the notion of a graph, we have shown that if one equips a hypergraph with an "ordinary" graph structure, then this hypergraph naturally defines a graph language. We have presented also a number of more general systems defining graph languages. The major objective of the paper was to introduce a formalism adequate to discuss these systems, to illustrate them by examples and to compare the classes of languages they generate.

As far as the comparison of the generative power of the systems is concerned, the basic missing results are the following : (i). for the edge-based approach we do not know whether the inclusion $L(GFH) \subseteq L(GFO)$ is strict, (ii). for the intersection-based approach we do not know whether the inclusion $L(GFIH) \subseteq L(GFIO)$ is strict, and (III). we do not know whether the inclusions $L(GFO) \subseteq L(FGA)$ and $L(GFIO) \subseteq L(FGA)$ are strict.

In our opinion four major relationships to be considered are the relationships between (i) $L(H)$ and $L(IH)$, (ii) $L(O)$ and $L(IO)$, (iii) $L(GFH)$ and $L(GFIH)$ and $L(GFO)$ and $L(GFIO)$. Theorem 2.1 settles (i). We are not able to settle (iii) and (iv) and (ii) is settled by the following result.

Theorem 5.1. $L(O)$ and $L(IO)$ are incomparable but not disjoint.

ACKNOWLEDGEMENT

The second author gratefully acknowledges the support of NSF grant number MCS 79-03838.

REFERENCES

[N2] M. Nagl, Graph-Grammatiken : Theorie, Anwendungen, Implementierung, Vieweg-Verlag, Wiesbaden, 1979.

[CER] V. Claus, H. Ehrig, G. Rozenberg, Graph Grammars and their applications to Computer Science and Biology, Springer-Verlag, Lecture Notes in Computer Science vol. 73, Berlin-Heidelberg-New York, 1979.

[E] H. Ehrig, Introduction to the algebraic theory of graph grammars, 1-69, in [CER].

[N1] M. Nagl, A tutorial and bibliographical survey on graph grammars, 70-126, in [CER].

LOWER BOUNDS FOR PROBLEMS DEFINED BY POLYNOMIAL INEQUALITIES

Jerzy W. Jaromczyk
Institute of Informatics
Warsaw University, PKiN VIII p.
00-901 Warsaw, Poland

1. Introduction

The decision tree model is one of the most powerful models for obtaining the lower bound time complexity of a given computational (or decision) problem. In this model the computation process involves selecting a path down a (ternary) tree, depending on the outcome of certain test functions applied along the way. When a leaf is reached the algorithm must be able to give an answer.

The methods commonly used in purpose to derive the lower bounds may be divided on two major groups; (A) ones based on the information-theoretic bounds of a given problem, (B) ones based upon the specific properties of the tests performed in the decision tree algorithm. In the latter the tests are mostly restricted to the class of linear polynomials, except the elegant paper of Yao [10] where the polynomials of degree 2 were allowed. The object of the present paper is to offer a new uniform method which enables us to deal with the wider class of the decision trees. Namely, the only restriction we put on the model is that the allowed tests must be polynomials with a bounded number of the irreducible factors (but they can be of any degree).

Given computational (or decision) task we consider a corresponding array of the polynomial inequalities called the (constructive) description. From the particular properties regarding a size and a so-called M-redundancy of this array the lower bound of the problem can be infered. Thus, the proposed method consists of studying the algorithm cost by reducing this to the combination of combinatorial, analytic and algebraic problems.

It is worth mentioning that our research was inspired by the fundamental paper of Rabin [6].

In Section 2 some needed definitions are presented. The main theorem is formulated in the Section 3. An application of the offered methods is given in Section 4.

2. Preliminaries and Definitions

In the sequel we will need some rudimentary knowledge of algebraic geometry. In order to make the paper understandable some elementary definitions are presented in this section.

Let R be a field of reals and $R[x_1,x_2,\dots,x_d]$ be a ring of polynomials in d indeterminants $x_1,x_2,\dots,x_d$ with coefitients in R. Any set $\{p_i\}_I$ of polynomials defines a real variety $V = V(\{p_i\}) = \{x \in R^d$: each $p_i(x)=0\}$.

A variety V is irreducible if $V = V_1 \cup V_2$ (V_1, V_2 varieties) implies $V = V_1$ or $V = V_2$. The dimension dim V of a variety V is the maximal integer n such that there exist distinct irreducible varieties $V_0, V_1, \dots, V_n$ such that $V_0 \subset V_1 \subset \dots \subset V_n \subset V$.

In the special case when $V = V(p)$ is a set of zeros of one polynomial p and dim $V = d-1$ the variety V is called hypersurface. If additionally p is a linear polynomial V is called hyperplane.

Let us introduce the following definition which is necessary because R is not algebraically closed:

We say that an irreducible hypersurface $V \subset R^d$ has the identity property if for any irreducible variety V_1 and any open set $U \subset R^d$ the equality $V \cap U = V_1 \cap U \neq \emptyset$ implies $V = V_1$.

All irreducible varieties $V \subset C^d$, C- the field of complex, have the identity property. Remark also that all hyperplanes have the identity property. Further facts considering algebraic geometry one can found e.g. in Kendig [5], van der Waerden [9].

Throughout the paper all functions denoted by p(x), d(x) with indices are polynomials in $R[x_1,x_2,\dots,x_d]$.

A system IN_i of inequalities $d_{i1}(x) \geq 0, \dots, d_{ik}(x) \geq 0$ is called simultaneously non-negative for $x \in D \subset R^d$ (shortly $SP(IN_i,x)$) if the conjunction $d_{i1}(x) \geq 0$ & ... & $d_{ik}(x) \geq 0$ is true.

Given an m x k array IN of inequalities

$$\begin{matrix} d_{11}(x) \geq 0 & \cdots & d_{1k}(x) \geq 0 \\ \vdots & & \vdots \\ d_{m1}(x) \geq 0 & \cdots & d_{mk}(x) \geq 0 \end{matrix} \quad .$$

IN describes the subset of points in R^d, it will be called alternatively a <u>description</u> .

An r x s array P of polynomials

$$\begin{matrix} p_{11}(x) & \cdots\cdots & p_{1s}(x) \\ \vdots & & \vdots \\ p_{r1}(x) & \cdots\cdots & p_{rs}(x) \end{matrix}$$

is a <u>complete proof for non-negativity</u> of IN in the set $D \subset R^d$ (shortly a complete proof for SP(IN,D)) if and only if for $x \in D$

(i) $SP(IN_i,x) \Rightarrow \exists j \ SP(P_j,x)$ for $1 \leq i \leq m$

(ii) $SP(P_j,x) \Rightarrow \exists i \ SP(IN_i,x)$ for $1 \leq j \leq r$,

where IN_i, P_i denotes the i-th row of arrays IN,P respectively.

In the case (i) of above definition we say that P_j verifies the $SP(IN_i,x)$. SP(IN,x) denotes that for certain i $SP(IN_i,x)$ is true. Additionally, Width(P) = s, Paths(P) = r.

The notion of the complete proof is due to Rabin [6] for the case m=1 and the linear polynomials in the array IN.

Some auxiliary definitions are needed.

The polynomials $d_1(x),\dots,d_k(x)$ are <u>sign independent</u> in $D \subset R^d$ if for any $(e_1,e_2,\dots,e_k) \in \{-1,0,+1\}^k$ there exists $x \in D$ such that sign $d_i(x) = e_i$ for $1 \leq i \leq k$ (Rabin [6]).

The open set $D \subset R^n$ is said to be <u>negatively dense</u> with respect to the array IN if and only if for any $x \in D$ such that for certain i,j, $1 \leq i \leq m$, $1 \leq j \leq k$ we have $d_{i1}(x) > 0,\dots,d_{ij}(x) = 0,\dots,d_{ik}(x) > 0$ and for any $\varepsilon > 0$ there exists $y \in D$, $\|y-x\| < \varepsilon$ such that non SP(IN,y) (i,e. the rows of the array IN are not simultaneously non-negative in y).

D is said to be <u>positively dense</u> (w.r.t. IN) if for any $x \in D$ such that for certain $1 \leq i \leq m$ we have $d_{i1}(x) \geq 0,\dots,d_{ik}(x) \geq 0$ and for any $\varepsilon > 0$ there exists $y \in D$ such that $d_{i1}(y) > 0,\dots,d_{ik}(y) > 0$ and $\|y-x\| < \varepsilon$

The array IN is called <u>M-redundant</u> if M is the minimal integer such that there exists a sequence of polynomials $d_1(x),\dots,d_m(x)$, each $d_i(x)$ is picked up from IN_i, such that for any sequence of points $x_1,\dots,x_m$ satisfying $SP(IN_i,x_i)$, $1 \leq i \leq m$, the inequality $d_i(x) \geq 0$, $1 \leq i \leq m$, holds for at most M distinct points $x_{i_1},\dots,x_{i_M}$ chosen from $x_1,\dots,x_m$.

Remark that the above definitions describe simple analytic properties of IN in D. M-redundancy measures how strongly the distinct rows of the description IN are mutualy tied up. On the other hand the sign independency ensures that the rows of IN are possibly short. The above definitions are illustrated in Section 4.

3. The Main Theorem

In this section we present a theorem answering the question about the complete proof size measured in the terms of Paths(P).

Let $D \subset R^d$ be an open set, let IN be an array m x k of irreducible polynomials:

$$\begin{matrix} d_{11} & \cdots\cdots & d_{1k} \\ \vdots & & \vdots \\ d_{m1} & \cdots\cdots & d_{mk} \end{matrix}$$

such that the varieties $V(p_{ij})$, $1 \leq i \leq m$, $1 \leq j \leq k$ are irreducible hypersurfaces having the identity property. Let IN_i, $1 \leq i \leq m$, be sign independent in D and D be positively and negatively dense w.r.t. IN. With the above assumptions we have the following

Theorem 1. If the description IN is M-redundant and P is any complete proof of SP(IN,D) (where all polynomials in the proof P are irreducible) then $Paths(P) \geq m/M$.

Proof (outline): The proof of the theorem is based upon the follow̲ig lemmas :

Lemma 1. For any $d_{ij}(x)$ from the row IN_i, $1 \leq i \leq m$, there exists such a row P(i,j) in the complete proof P in which the test polynomial $c \cdot d_{ij}(x)$ (c is a positive constant) must appear.

Lemma 2. There exists $x_{ij} \in D$, $1 \leq i \leq m$, $1 \leq j \leq m$, such that $SP(IN_i, x_{ij})$ and this is verified by the row P(i,j).

Let us collect the polynomials $d_1(x),\dots,d_m(x)$ (selected from distinct rows of IN) for which the description IN is M-redundant. From Lemma 1 and Lemma 2 there exist $x_1, x_2, \dots, x_m$ such that

(i) $SP(IN_i, x_i)$

(ii) the (i) is verified by P(i) i.e. the row including the test polynomial $c \cdot d_i(x)$.

Because of the M-redundancy of IN each P(i) can verify at most M points from $x_1,\dots,x_m$. On the other hand each x_i, $1 \leq i \leq m$, must be verified by a certain row of P. Thus, there are at least m/M rows in P. Hence $Paths(P) \geq m/M$. ▯

Remark that most of the assumptions pertaining the description IN and set D along with the assumption that the polynomials of P are irreducible are employed in order to prove the Lemmas 1 and 2. Their proofs may be found in Jaromczyk [4].

4. Application to the Decision Trees

We can use Theorem 1 to derive worst case lower bound time complexity in the model of decision trees for certain class of problems. In this model algorithms are the ternary trees T. In each internal node a test of the form p(x) : 0 is evaluated (with potentially infinite precision) and then the control passes to the left, middle or right son depending on the test result (<, = , > respectively). The outcome to the algorithm is given in the leaves either in the form of "yes" or "no" answer (for decision problems) or as a value of certain function from the specified class.

Cost(T) is defined as a length of the longest path of T, i.e. the maximal number of tests performed until the leaf is reached.

Let us consider a decision problem D-PR of the form "whether y , $y \in D \subset R^d$, has the property PR". Observe that if the problem has size n, i.e. the input consists of n items $x_1,\ldots,x_n$, we can treat $(x_1,..,x_n)$ as a point in R^n.

We say that an m x k array IN of the irreducible polynomial inequalities is a (constructive) description of D-PR if

$$y \in D \subset R^d \text{ has the property PR if and only if } SP(IN,y).$$

Usually the description size depends on n, i.e. m = m(n).

Now we are in a position to formulate the next theorem. Let the description IN of D-PR problem satisfies the assumptions to the Theorem 1.

Theorem 2. If IN is M-redundant and T is any decision tree algorithm (with tests being the irreducible polynomials) solving D-PR problem then $Cost(T) \geq \log_3(m/M)$.

Proof : Consider all paths leading to the leaves with the answer "yes". After certain (if necessary) sign changing the tests on these paths form the complete proof for SP(IN,D). □

Note that the Cost(T) is greater or equal to the maximum of two quantities : $\log_3$(number of paths in T) and the length of the longest path in T. In the present paper only the former is investigated, the latter is object of Rabin [6], Spira [8] and Jaromczyk [3].

On purpose to illustrate our methods let us consider the Convex hull problem in the plane. Unconstuctively, the convex hull of a given set of points $z_1, z_2, \dots, z_n$, $z_i \in R^2$, is the smallest convex set including all $z_1, \dots, z_n$. In fact, the convex hull is a convex polygon with vertices from the set $\{z_1, \dots, z_n\}$. We are interested in the following CH problem :

"Given n points $z_1, \dots, z_n \in R^2$ find the indices of the points forming the convex hull of the given set."

Let our algorithms be decision trees with tests restricted to the irreducible polynomials of a given input $(z_1, \dots, z_n) \in R^{2n}$. Then the following theorem holds :

Theorem 3. If T is any algorithm solving CH problem then Cost(T) is not less than $\log_3((n/2)!)$.

Remark : Theorem 3 was firstly proved, with separate techniques, by Yao [10] for the narrower class of algorithms, i.e. allowed tests were polynomials of degree $\leqslant 2$.

Proof : Let $z_i = (x_i, y_i) \in R^2$. Define $\det(z_i, z_k, z_j) = x_k(y_i - y_j) + y_k(x_j - x_i) + y_j x_i - y_i x_j$. Polynomial det can be treated as a function from R^{2n} to R depending only on the coordinates $x_i, y_i, x_j, y_j, x_k, y_k$. The sign of det has a simple geometrical interpretation, i.e. point z_k is to the right (on, to the left) of the directed line $z_i z_j$ if and only if $\det(z_i, z_k, z_j) < 0$ (= , >0 respectively). It can be shown that V(det) is irreducible hypersurface with the identity property.

Consider a decision version D-CH of Convex hull problem : "whether $z_1, z_2, \dots, z_n$ form the convex polygon". D-CH has the following description (see Fig. 1) :

$$IN_\pi : \det(z_1, z_{\pi(2)}, z_{\pi(3)}) \geqslant 0, \dots, \det(z_{\pi(n-1)}, z_{\pi(n)}, z_1) \geqslant 0 \text{ and}$$
$$\det(z_{\pi(n)}, z_1, z_{\pi(2)}) \geqslant 0,$$

where π is a permutation of $(2, \dots, n)$. We have $m = (n-1)!$ and $k = n$. Observe that M-redundancy of IN is equal to $(n-3)!$ (e.g. $\det(z_1, z_2, z_3)$ is greater or equal to zero for $(n-3)!$ permutations of $z_1, \dots, z_n$ forming the convex polygon). Thus, using Theorem 2 directly we do not obtain the expected lower bound.

Nevertheless our methods are powerful enough to end the proof. Assume that n is even. Consider

$$IN^+_\mu : \det(z_1, z_{\mu(2)}, z_3) = 0, \ \det(z_{\mu(2)}, z_3, z_{\mu(4)}) > 0, \dots,$$
$$\det(z_{2i-1}, z_{\mu(2i)}, z_{2i+1}) = 0, \ \det(z_{\mu(2i)}, z_{2i+1}, z_{\mu(2i+2)}) > 0, \dots$$

... $\det(z_{\mu(n)}, z_1, z_{\mu(2)}) > 0$,
where μ is a permutation of $\{2,4,\ldots,n\}$.
Adopting Lemmas 1,2 we can find for each permutation μ a point z_μ $z_\mu = (z_1, z_{\mu(2)}, z_3, \ldots, z_{n-1}, z_{\mu(n)})$ such that the control of the algorithm for the input z_μ passes down a path just including nodes with tests $c_i \cdot \det(z_{2i-1}, z_{\mu(2i)}, z_{(2i+1) \bmod n}) : 0$, $i = 1,\ldots,n/2$, c_i - positive constant, branching at these nodes to the middle sons, and z_μ satisfies IN^+_μ. It remains to observe that for $\mu \neq \mu'$ there exists i such that $\mu(2i) \neq \mu'(2i)$. Hence $\det(z_{2i-1}, z_{\mu(2i)}, z_{(2i+1) \bmod n}) \neq 0$ for $z_{\mu'}$, (see Fig. 2). In consequence $z_{\mu'}$ cannot pass down the same path the point z_μ passes. Therefore T has at least $(n/2)!$ distinct paths and $\mathrm{Cost}(T) \geq \log_3 (n/2)!$. □

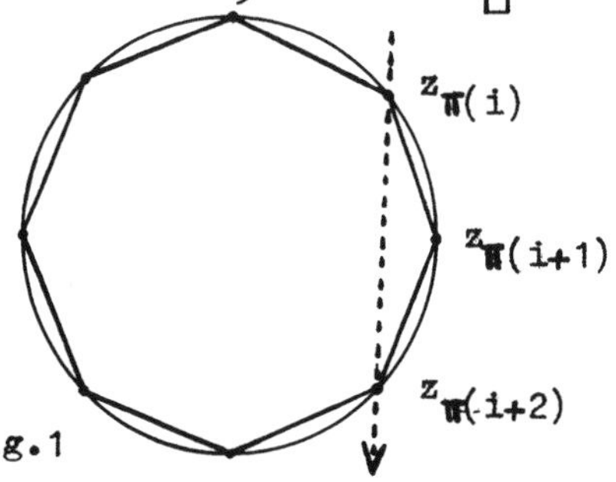

Fig.1

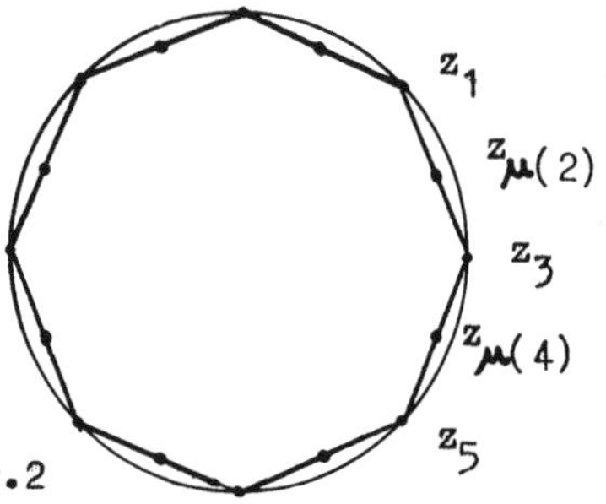

Fig.2

To complete this section let us note that the algorithms solving CH problem have to use det functions as primitives. Hence, the linear decision trees are too weak for this problem what was already mentioned in Yao 10 and independently proved in Avis 1 , Snir 7 and Jaromczyk 2 .

Let us also remark that in some cases, when the answer about the complexity in -notation is sufficient, the polynomials with bounded number of irreducible factors as test functions can be allowed. It follows from a simple observation that $p_1(x) \cdot p_2(x) \cdot \ldots \cdot p_l(x)$ 0 if and only if $e_1 p_1(x)$ 0, ..., $e_l p_l(x)$ 0 where $(e_1, \ldots, e_l)$ -1,+1 l and $e_1 \cdot \ldots \cdot e_l = +1$. Moreover, the above immediately yields that if the number of factors is unbounded with respect to description size then a large description can have even one element decision tree.

Conclusion

We have offered the method to study the lower bounds in a reasonable wide class of decision trees.
On the other hand we have shown that for certain problems wider classes decision trees are sometimes profitless. We belive that demonstrated ap-

proach can be helpful while considering problems in Computational Geometry.

At the end it is worth noticing that a case in which all the polynomials of the description are linear seems to be of the special interest.

References

1. Avis, D., Comments on a lower bound for convex hull determination. Inf. Proc. Let., 11 (1980), 126.

2. Jaromczyk, J., W., Linear decision trees are too weak for convex hull problem. to appear in Inf. Proc. Let.

3. Jaromczyk, J., W., A note on Rabin's complete proof notion (preliminary version). IInf UW Reports, 102 (1981).

4. Jaromczyk, J., W., to appear.

5. Kendig, K., Elementary Algebraic Geometry. Springer Verlag, New York, 1977.

6. Rabin, M., O., Proving simultaneous positivity of linear forms. J. Comp. Sys. Sci., 6 (1972), 639-650.

7. Snir, M., communicated to me by Peter van Emde Boas.

8. Spira, P., M., Complete linear proofs of systems of linear inequalities. J. Comp. Sys. Sci., 6 (1972), 205-216.

9. van der Waerden, B., L., Einführung in die algebraische Geometrie. Springer Verlag, Berlin, 1973.

10. Yao, A., C., A lower bound to finding convex hulls. Report STAN--CS-79-733, April 1979.

WHAT IS COMPUTABLE FOR ABSTRACT DATA TYPES ?

H. Kaphengst
VEB Robotron ZFT
DDR-8012 Dresden PSF 330

In recent discussions about programming methods abstract data types play a decisive role because it has been realized more and more that clear structures are necessary not only for algorithmic processes but even more for the data to be processed. There is not yet a generally accepted notion of data type. But it is clear that the question is of certain systems of value sets and operations for these values resembling a many-sorted algebra. We take (as most authors) the so-called initial approach what means that a data type is an initial element of a class of suitably generalized algebras defined by equations or conditional equations. Here we can build upon research in foundations of programming done in the nationally-owned firma Kombinat Robotron for many years. The operations of a data type are assumed to be effectively executable. One can ask whether further operations with values of the data type are also computable, i.e. executable algorithmically, on behalf of the executability of the base operations. It seems to be widely assumed that these are the operations implementable by programs which are built up from "calls" of these operations by the usual program connectives and which process values for variables of the various value sorts.

Let's have a look in this respect at the data type of formally infinite sequences $(n_0, n_1, n_2, \ldots)$ of natural numbers (i.e., only finitely many members n_i are different from 0) with the carrier sets Nat (set of natural numbers), Seq (set of the mentioned sequences) and the six operations 0, successor formation of natural numbers, Void = (0,0,0,...), the operation Put putting a natural number in front of a sequence, Get yielding the first member of a sequence and Pop omitting the first member of a sequence. We certainly can't compute in the above mentioned sense whether a given sequence is equal to Void because we can check the sequence only by Get. But if permanent application of Get and Pop yields always 0 then we can't know whether the continuation of the check will once yield an element different from 0 or not. On the other side the desired operation Ch is recursively defined by the equations

$$\mathrm{Ch}(\mathrm{Void}) = 1$$

$$\mathrm{Ch}(\mathrm{Put}(0,x)) = \mathrm{Ch}(x)$$

$$\mathrm{Ch}(\mathrm{Put}(n^+,x)) = 0 \qquad (n^+ \text{ is the successor of } n)$$

what should allow for the relative computability of Ch .
So, the necessity of an exact definition of the notion should be clear.
The foundations are contained in K-69, KR'71, HKR'80.

1. Partial many-sorted algebras

The generalization of the notion of an algebra used in this paper needs many sorts, partial operations and a certain non-standard agreement what it means that an equation holds.
A partial many-sorted algebra is a R-algebra for an operator domain R
Here an operator domain R is given by

(1) a set Sort(R) the elements of which are called the sorts of R
(2) a set Opr(R) the elements of which are called the operators of R
(3) a finite sequence $\underline{s} = (s_i /\ i \in n)$ of sorts s_i of R for every operator r of R called the arity of r (in R)
(4) a sort s for every operator r of R , the sort of r .

We write $r : (s_i /\ i \in n) \longrightarrow s$ (in R) if the operator r has in R the arity $(s_i /\ i \in n)$ and the sort s . Let $\underline{s}(i) = s_i$ for all $i \in n$

A R-algebra A is now given by a family $(A_s /\ s \in \mathrm{Sort}(R))$ of carrier sets A_s and a family $(r_A /\ r \in \mathrm{Opr}(R))$ of operations $r_A : \mathrm{Def}(r_A) \longrightarrow A_s$ with $\mathrm{Def}(r_A) \subseteq A_{\underline{s}} = \mathsf{X}\{A_{\underline{s}(i)} /\ i \in n\}$ for $r : (s_i /\ i \in n) \longrightarrow s$ (in R) .

Let's consider the direct products of sets $A_{\underline{s}}$ as the set of all sequences $(a_i /\ i \in n)$ with $a_i \in A_{\underline{s}(i)}$ for all $i \in n$. (By assuming for the natural numbers $n = \{0,1,2,\ldots,n-1\}$ we circumvent the notation $(a_o, a_1, \ldots, a_{n-1})$). Furthermore we assume that operators are always characters.
A homomorphism $h : A \longrightarrow B$ for R-algebras A,B is a family $(h_s /\ s \in \mathrm{Sort}(R))$ of maps $h_s : A_s \longrightarrow B_s$ (together with A,B) with the property that for every operator $r : (s_i /\ i \in n) \longrightarrow s$ of R the following is valid:
If $r_A(a_i /\ i \in n) = a$ then $r_B(h_{\underline{s}(i)}(a_i) /\ i \in n) = h_s(a)$.
This includes:
If $(a_i /\ i \in n) \in \mathrm{Def}(r_A)$ then $(h_{\underline{s}(i)}(a_i) /\ i \in n) \in \mathrm{Def}(r_B)$.

We define inductively for a variable domain X , i. e. a family $(X_s /\ s \in \mathrm{Sort}(R))$ of disjunct sets of characters - the variables of X must be different from the operators - and for every sort $s \in \mathrm{Sort}(R)$ R-terms over S of sort s as certain character strings in the following way:

(1) For every sort s the variable $x \in X_s$ is a R-term over X of sort s.
(2) For every operator $r : (s_i / i \in n) \longrightarrow s$ of R and R-terms t_i over X of sort s_i the character string

$$rt_o t_1 \ldots t_{n-1}$$

is a R-term over X of sort s.
(3) An object is R-term on behalf of (1),(2) only.

We imagine for every sort s a character $\underset{s}{=}$ (frequently abbreviated as "$\doteq$" or even as "=") and consider character strings of the form

$$t \underset{s}{=} t'$$

where t, t' are R-terms of sort s over X. We call these character strings R-equations over X. Now we define inductively R-implications over X with premises and end-equations:

(1) Every R-equation $t \underset{s}{=} t'$ over X is a R-implication over X with the empty set of premises. It is its end-equation itself.
(2) If p is a R-implication over X and $t = t'$ is a R-equation over X then the character string

$$t \underset{s}{=} t' \Longrightarrow p$$

is a R-implication over X. The premises are $t \underset{s}{=} t'$ and the premises of p, its end-equation is that of p.
(3) An object is a R-implication on behalf of (1),(2) only.

More precisly, a term, equation or implication can be considered as a character string only if a variable domain is fixed. In general they have to be considered as pairs of a variable domain and such strings.

We consider valuations $\underline{a} = (a_x / x \in X_s, s \in \mathrm{Sort}(R))$ of X in A for a R-algebra A and a variable domain X assigning every variable x in X of sort s an element $a_x \in A_s$ of the corresponding carrier set.

Every R-term t over X of sort s defines a composite operation t_A over A assigning to certain valuations $\underline{a}$ of X in A a value $t_A(\underline{a}) \in A_s$. To be more precise for all $x \in X_s$ and all valuations $\underline{a}$ of X in A there is $x_A(\underline{a}) = a_x$. For every operator $r : (s_i / i \in n) \longrightarrow s$ of R and all R-terms t_i over X of sort $s_i (i \in n)$ there is

$$(rt_o t_1 \ldots t_{n-1})_A = r_A \circ ((t_i)_A / i \in n) .$$

Here the superposition $r_A \circ (f_i / i \in n) : D \longrightarrow A_s$ for $f_i : D_i \longrightarrow A_{\underline{s}(i)}$ for all $i \in n$ is defined by the agreement that D is the set of the $d \in \bigcap \{D_i / i \in n\}$ for which $(f_i(d)/ i \in n) \in Def(r_A)$ is valid and that for $d \in D$

$$(r_A \circ (f_i / i \in n))(d) = r_A(f_i(d) / i \in n)$$

holds.

A valuation $\underline{a}$ of X in A is called a <u>solution</u> in A of the R-equation $t \underset{s}{=} t'$ if $\underline{a} \in Def(t_A)$, $\underline{a} \in Def(t'_A)$ and $t_A(\underline{a}) = t'_A(\underline{a})$. A R-implication p over a variable domain X <u>holds</u> in a R-algebra A if every valuation of X in A being a solution of all premises of p is also a solution of the end-equation of p. The insistence of X may be important because an empty carrier set may prevent the existence of valuations.

The combination of an operator domain R with a set of R-implications over finite variable domains is called a <u>signature</u> $\underline{S}$. An <u>$\underline{S}$-algebra</u> is a R-algebra in which all implications belonging to $\underline{S}$ hold. We call the R-homomorphisms, R-terms, ... $\underline{S}$-homomorphisms, $\underline{S}$-terms, ..., too. In putting down signatures we also use different notations for terms (e.g., infix operators) if those are closer to common usage.

An $\underline{S}$-algebra is called an <u>initial</u> $\underline{S}$-algebra if there is exactly one homomorphism $h : A \longrightarrow B$ to any $\underline{S}$-algebra B. We understand by a data type an initial $\underline{S}$-algebra for a finite signature $\underline{S}$.

The algebra of formally infinite sequences sketched in the introduction is initial for a signature $\underline{S}$ described by the following specifications:

$$Sort(\underline{S}) = \{nat, seq\}$$

$$\begin{aligned} Opr(\underline{S}) = \{ & 0 : \longrightarrow nat; & \_^{+} : nat \longrightarrow nat; \\ & void : \longrightarrow seq; & put : nat, seq \longrightarrow seq; \\ & get : seq \longrightarrow nat; & pop : seq \longrightarrow seq \} \end{aligned}$$

$$\begin{aligned} Imp(\underline{S}) = \{ (& n : nat;\ x : seq) \\ & put(0, void) \doteq void; \\ & get(put(n,x)) \doteq n; \\ & pop(put(n,x)) \doteq x; \\ & n^{+} \doteq n^{+} \quad \} \end{aligned}$$

In this example all implications are equations for a common variable domain X with $X_{nat} = \{n\}$, $X_{seq} = \{x\}$. The equation $n^{+} \doteq n^{+}$ ensures that the successor operation is a total one (i.e., defined everywhere).

2. Derivation of equations and free extensions

For our further considerations it is very important that the semantical consequences of a given set of equations may be obtained by formal derivation rules.

For a signature $\underline{S}$ we call the implications used in the formation of $\underline{S}$ and the following implications (there is a finite number of them for a finite operator domain) $\underline{S}$-axioms.

For every $s \in Sort(\underline{S})$ the following axioms belong to them:

$$x \underset{s}{=} x$$
$$x \underset{s}{=} y \Longrightarrow y \underset{s}{=} x$$
$$x \underset{s}{=} y \Longrightarrow y \underset{s}{=} z \Longrightarrow x \underset{s}{=} z$$

over X where $X_s = \{x, y, z\}$ and $X_{s'} = 0$ for all other sorts s'.

For every $r : (s_i / i \in n) \longrightarrow s$ of $\underline{S}$ and every $j \in n$ the following axioms belong to them:

$$x_j \underset{s}{=} y_j \Longrightarrow rx_0x_1\ldots x_{n-1} \underset{s}{=} z \Longrightarrow rx_0\ldots x_{j-1}y_jx_{j+1}\ldots x_{n-1} \underset{s}{=} z$$

over X where X is minimal with $x_i \in X_{\underline{s}(i)}$ for $i \in n$, $y_j \in X_{\underline{s}(j)}$ and $z \in X_s$.

Let $\underline{S}$ be a signature, X a variable domain and G a set of equations over X. We say that an $\underline{S}$-equation $t \underset{s}{=} t'$ over X is $\underline{S}$-derivable (or derivable with respect to $\underline{S}$) from the set G if this is true on behalf of the following rules (1) and (2):

(1) Every $t \underset{s}{=} t'$ G and every equation $x \underset{s}{=} x$ for $x \in X_s$, $s \in Sort(\underline{S})$ is $\underline{S}$-derivable from G.

(2) If q is an $\underline{S}$-axiom over Y and t_y is for every $y \in Y_s$, $s \in Sort(\underline{S})$ an $\underline{S}$-term over X of sort s which has been delivered as a subterm of an equation $\underline{S}$-derivable from G then the following applies: If p arises from q by substitution of the y with t_y (for all $y \in Y_s$, $s \in Sort(\underline{S})$) and if all premises of p are $\underline{S}$-derivable from G then the end-equation of p is $\underline{S}$-derivable from G, too.

Theorem 1. Let $\underline{S}$ be a signature, X a variable domain and G a set of $\underline{S}$-equations over X. Then an $\underline{S}$-equation $t \underset{s}{=} t'$ from G over X is $\underline{S}$-derivable iff for every valuation $\underline{a}$ of X in any $\underline{S}$-algebra A the following holds: If $\underline{a}$ is a solution of every equation in G then $\underline{a}$ is a solution of $t \underset{s}{=} t'$, too. An implication p follows from (the implications of) $\underline{S}$, i.e., is valid in all $\underline{S}$-algebras iff the endequation of p is $\underline{S}$-derivable from the premises of p.

The main idea in proving the theorem is the construction of an $\underline{S}$-algebra $D = D(X,G)$ with a valuation $\underline{d}$ of X in D the elements of which are certain equivalence classes of $\underline{S}$-terms such that $\underline{d}$ is a solution of all equations of G in a universal manner.

For a signature enlargement $\underline{R} \supseteq \underline{S}$ (containment of the corresponding sets of sorts, operators and implications) we can build the $\underline{S}$-part $A{\downarrow}\underline{S}$ of every $\underline{R}$-algebra A by forgetting all carrier sets and operations of A which don't refer to S ($(A{\downarrow}\underline{S})_s = A_s$ and $r_{A{\downarrow}\underline{S}} = r_A$ for all sorts s and operators r of $\underline{S}$).

The $\underline{R}$-algebra F is called a free $\underline{R}$-extension of the $\underline{S}$-algebra A if $A = F{\downarrow}\underline{S}$ and every homomorphism $g : A \longrightarrow B{\downarrow}\underline{S}$ for any $\underline{R}$-algebra B can uniquely be extended to a homomorphism $h : F \longrightarrow B$ (i.e., $h_s = g_s$ for all $s \in \mathrm{Sort}(\underline{S})$).

The elements $f \in F_{s'}$ ($s' \in \mathrm{Sort}(\underline{R})$) can be represented for a free $\underline{R}$-extension F of the $\underline{S}$-algebra A in the following way. We choose a variable domain X and a valuation $\underline{a}$ of X in A such that every $a \in A_s$ can be represented in the form $a = a_x$. Then there is an $\underline{R}$-term t over X and a finite set G of $\underline{S}$-equations over X for which $\underline{a}$ is a solution in A such that $t \underset{s'}{=} t$ is $\underline{R}$-derivable from G and $f = t_F(\underline{a})$ is valid. If f is representable in the same manner with t' and G' then $t \underset{s'}{=} t'$ is $\underline{R}$-derivable from $G \cup G'$.

For $\underline{S} = 0$ we deal with the special case of the initial $\underline{R}$-algebra. The signature with one sort nat and the operators $0 : \longrightarrow \mathrm{nat}$ and $\_^{+} : \mathrm{nat} \longrightarrow \mathrm{nat}$ and the implications $(\ )\ 0 \doteq 0$ and $(n : \mathrm{nat})\ n^{+} \doteq n^{+}$ shall be called $\underline{\mathrm{Nat}}$. The set $\mathrm{Nat} = \{0,1,2,\ldots\}$ of natural numbers with Null and Sucessor formation constitutes an initial $\underline{\mathrm{Nat}}$-algebra An . This is only the algebraic formulation of Dedekind's justification theorem for definitions by total induction. Furthermore there is a close connection between recursivity and free extensions.

> Theorem 2. A partial function $f : D \longrightarrow \mathrm{Nat}$, $D \subseteq \mathrm{Nat}^m$ is partial recursive iff there is a finite signature enlargement $\underline{R} \supseteq \underline{\mathrm{Nat}}$ for which the algebra An of natural numbers has a free $\underline{R}$-extension F such that for an operator $r : (\mathrm{nat}/\ i \in m) \longrightarrow \mathrm{nat}$ of $\underline{R}$ a representation $f = r_F$ exists.

A quite analogous theorem holds for the algebras of words over finite alphabets.

3. Languages for algebras

Let $\underline{S}$ be a signature.

A name domain is a family $N = (N_s / s \in Sort(\underline{S}))$ of disjunct sets of words $N_s \subseteq Z^*$ over a finite alphabet Z. A language l of an $\underline{S}$-algebra A is a map assigning to every name $w \in N_s$ of a given name domain N a meaning $l(w) \in A_s$ such that every element $a \in A_s$ is the meaning of a certain name u (i.e., $a = l(u)$). N is then called the name domain of l. If the name sets are recursively enumerable then l is often called formal. Let's call l strictly formal if even the sets of pairs $\{(u,v)/ l(u) = l(v);\ u,v \in N_s\}$ for all $s \in Sort(\underline{S})$ are recursively enumerable.

An operation $f : D \longrightarrow A_s$, $D \subseteq A_s$ is called - with reference to the language l - recursively implementable if there is a partial recursive word function g such that for all $(u_i / i \in n) \in N_s$ the following is valid: $f(l(u_i)/ i \in n)$ is defined iff $g(u_i / i \in n)$ is defined and if need be $f(l(u_i)/ i \in n) = l(g(u_i)/ i \in n))$. In other words: Computing starts with any names for the arguments and the meaning of a result is the desired function value.

With $\underline{S}$-algebras A we are, of course, only interested in those languages l with respect to which the operations r_A of A are recursively implementable.

It is possible to compare different languages with respect to their translatability into each other: We call the language k with the name domain K poorer than l if there is a partial recursive function f such that for every name $u \in K_s$, $s \in Sort(\underline{S})$ the following is valid:

$$l(f(u)) = k(u)$$

(in particular, $f(u) \in N_s$ is to be valid).

Languages k, l are called translation-equivalent if k is poorer than l and l is simultaneously poorer than k.

It is easy to see that with respect to trnaslation-equivalent languages the same operations are recursively implementable.

If $\underline{S}$ is a finite signature and A an initial $\underline{S}$-algebra (a data type) then we construct the canonical language $k = k_A$ of A in the following way: The attached name domain K should consist of all $\underline{S}$-terms t over the empty variable domain $X = (0/ s \in Sort(\underline{S}))$ for which the empty valuation $\underline{0}$ (i.e., the only valuation of X in A) belongs to the domain of t_A . Then let $k(t) = t_A(\underline{0})$.

<u>Theorem 3.</u> The canonical language k_A of the initial $\underline{S}$-algebra A for the finite signatur $\underline{S}$ makes every operation r_A of A recursively implementable and is poorer than every language l of A with this property; it is strictly formal. Every strictly formal language l of A making the operations of A recursively implementable is translation-equivalent to the canonical language k_A .

4. Characterization of relative computability

Let $\underline{s} = (s_i / \ i \in n)$.
A set $D \subseteq A_{\underline{s}}$ for an $\underline{S}$-algebra A is called <u>decidable</u> with respect to a language l of A if there is a partial recursive word function g defined for all n-tuples of names $(u_i / \ i \in n) \in N_{\underline{s}}$ such that $g(u_i / \ i \in n)$ is the empty word exactly for those $(\overline{u}_i / \ i \in n)$ for which $(l(u_i) / \ i \in n) \in D$.

One can easily specify a language l for the algebra An of natural numbers which makes the base operations of An recursively implementable but not all partial recursive functions. If, however, l makes at least the set $\{0\} \subseteq \mathrm{Nat}$ decidable then l is trnslation-equivalent to a strictly formal language, namely the canonical language k_{An} with

$$k_{An}(0\underbrace{+\cdots+}_{n}) = n .$$

Therefore we will restrict ourself to strictly formal languages.

For an $\underline{S}$-algebra A an operation $f : D \longrightarrow A_s$, $D \subseteq A_{\underline{s}}$ is <u>relatively computable</u> with respect to the operations of A if they are recursively implementable with reference to every strictly formal language l for A for which the operations of A are recursively implementable.

This makes precise the intuitive idea of relative computability for partial many-sorted algebras if one is willing to accept the following assumption: Computing means the transformation of character strings on behalf of formal rules and - by a language - a meaning is assigned to the argument and result character strings but in general not to the intermediate results (!). The result can depend on the chosen object names but not its meaning.

For data types, i.e. initial $\underline{S}$-algebras for finite $\underline{S}$, we now get easily the relatively computable operations. They are the operations recursively implementable with respect to a canonical language.
We can give a characterization generalizing theorem 2:

Theorem 4. Let A be an initial $\underline{S}$-algebra for a finite signature $\underline{S}$. Then an operation $f : D \longrightarrow A_s$, $D \subseteq A_{\underline{s}}$ is computable relatively to the operations of A iff there is a finite signature enlargement $\underline{R} \supseteq \underline{S}$ for which A has a free $\underline{R}$-extension F such that there is the representation $f = r_F$ for an operator $r : \underline{s} \longrightarrow s$ of $\underline{R}$.

With that, for example, the decision operation mentioned in the introduction is relatively computable although it cannot be constructed with the usual program connectives.

References

K'69 H. Kaphengst; Malzew-Räume, ein allgemeiner Begriff der rekursiven Abbildung, Z. f. math. Logik 15(1969), 63-76.

KR'71 H. Kaphengst, H. Reichel; Algebraische Algorithmentheorie. WIB Nr. 1, VEB Kombinat Robotron, Dresden 1971.

HKR'80 U. L. Hupbach, H. Kaphengst, H. Reichel; Initial Algebraic Specification of Data Types, Parametrized Data Types and Algorithms. WIB Nr. 15, VEB Robotron ZFT, Dresden 1980.

ON STRONGLY CUBE-FREE ω-WORDS GENERATED BY BINARY MORPHISMS

Juhani Karhumäki
Department of Mathematics
University of Turku
Turku, Finland

Abstract

An ω-word is called strongly cube-free if it does not contain a subword of the form vvfirst(v), with $v \neq \lambda$. We show that it is decidable whether a given morphism over a binary alphabet defines, when applied iteratively, a strongly cube-free ω-word. Moreover, an explicit and reasonably small upper bound for the number of iterations needed to be checked is given.

1. Introduction

Repetitions in words, i.e., the existence of occurrences of v^i, with $v \neq \lambda$ and $i \geq 2$, as subwords, was first studied by Thue in [T1] and [T2]. He proved, among other things, that there exists an infinite strongly cube-free word (cf. section 2) over a binary alphabet. Such a word is obtained by iterating the morphism $h(a) = ab$, $h(b) = ba$ when started at a. Few first words obtained are as follows: $a \to ab \to abba \to abbabaab \to abbabaabbaababba$.

Later on this sequence and its interesting properties have been rediscovered several times in different connections. This kinds of nonrepetative sequences have applications in many areas of discrete mathematics, for example in connection with unending games and in group theory to mention only few, cf. [MH].

We call a morphism $h: \Delta^* \to \Delta^*$ prefix-preserving if $h(a) = az$ for some a in Δ and $z \neq \lambda$. As is easily seen such a morphism defines, when applied iteratively starting at a, a unique ω-word. In this paper we are interested in under which conditions such a morphism over a binary alphabet generates a strongly cube-free ω-word and, in particular, whether this can be effectively decided. We shall show that such a morphism must not only be biprefix but also such that h(a) and h(b) must both start and end with a different letter. We call such morphisms strong biprefixes.

The answer to our decision problem is shown to be positive. Furthermore, we are able to give a relatively small upper bound for the number of iterations needed to guarantee an occurrence of a subword of the form vvfirst(v), with $v \neq \lambda$, if such will ever occur in the sequence.

The similar problems are considered in [B] and [K]. In [B] it is shown that the above decision problem for square-free ω-words in a three-letter alphabet is decidable, and in [K] that the same holds true for cube-free ω-words over a binary alphabet.

2. Preliminaries

We use only very basic notions of the formal language theory, see e.g. [H]. For clarity we want to specify the following.

The length of a word v is denoted by $|v|$. For two words u and v, $u^{-1}v$ (resp. vu^{-1}) means the left (resp. right) difference of v by u. Further we write u pref v (resp. u p-pref v) if u is a prefix (resp. a proper prefix) of v, i.e. $v = uw$ holds true for some word w (resp. for some word $w \neq \lambda, v$). For a word u the notation pref(u) denotes the set of all prefixes of u, while the notation $\mathrm{pref}_n(u)$ is used to specify the prefix of u of the length n. By definition, $\mathrm{pref}_n(u) = u$ if $|u| < n$. The corresponding notions for suffixes are obtained by replacing pref by sub. A word u is a subword of a word v if there exist words u' and u'' such that $v = u'uu''$. By saying that u is a subword in a language L we, of course, mean that u is a subword of some word in L.

By an ω-word we mean an infinite word (from left to right). A word or an ω-word is called strongly cube-free (resp. square-free, cube-free, or fourth power-free) if it does not contain as a subword any word of the form $vv\,\mathrm{first}(v)$ (resp. v^2, v^3 or v^4) with $v \neq \lambda$. Here first(v) denotes the first symbol of v, while last(v) is used to denote the last symbol of v. Of course, by saying that a language L is e.g. strongly cube-free we mean that all of its words are such.

Our basic notion is that of a morphism of a finitely generated free monoid Δ^*. Let $h: \Delta^* \to \Delta^*$ be a λ-free morphism, i.e., nonerasing. We say that h is prefix-preserving if

$$(*) \qquad h(a) = az \qquad \text{for some } a \text{ in } \Delta \text{ and } z \neq \lambda$$

If this is the case, then $h^2(a) = h(az) = h(a)\,h(z)$ and, in general,

$$h^i(a) = h^{i-1}(a)\,h^{i-1}(z) \qquad \text{for } i \geq 1.$$

Consequently, $h^{i-1}(a)$ is a proper prefix of $h^i(a)$ for each i, which means that the iterative application of h starting at a defines as a limit an ω-word. Infinite words thus obtained are called ω-words generated by morphisms. Morphisms satisfying (*) are called prefix-preserving morphisms or pp-morphisms for short.

As usual we call a morphism $h: \{a,b\}^* \to \{a,b\}^*$ a biprefix if each of $h(a)$ and $h(b)$ is neither a prefix nor a suffix of the other. By a strong biprefix we mean a morphism h over $\{a,b\}$ such that $\mathrm{first}(h(a)) \neq \mathrm{first}(h(b))$ and $\mathrm{last}(h(a)) \neq \mathrm{last}(h(b))$.

3. Simple Properties

In this section we present a necessary condition for a morphism to generate a strongly cube-free ω-word. We show that such a morphism must be a strong biprefix. The proof goes along the lines presented in [K], where it is shown that only biprefixes may generate cube-free ω-words.

The following lemma is established in [K] simply by generating, step by step, all words which are cube-free and does not contain aa as a subword.

Lemma 1 Every cube-free (and hence also strongly cube-free) word over a binary alphabet {a,b} and of the length at least 18 contains aa and bb as subwords.

Corollary 1 Every cube-free (and hence also strongly cube-free) ω-word over a binary alphabet {a,b} contains aa and bb as subwords.

From the corollary it is easy to conclude

Theorem 1 Every prefix-preserving morphism generating a strongly cube-free ω-word is a strong biprefix.

Combining the above with results in [K] we conclude this section with the following remarks concerning ω-words generated by morphisms over a binary alphabet. As is well-known square-free ω-words over {a,b} do not exist at all, and hence such cannot be generated by morphisms, either. As shown by Thue, strongly cube-free ω-words over {a,b} can be generated by morphisms, but, because of Theorem 1, only by strong biprefixes. If, in turn, we want to generate cube-free ω-words over {a,b}, then, as shown in [K], only biprefixes are suitable candidates. Finally, fourth power-free ω-words over {a,b} can be generated by nonbiprefixes, as shown in [K] by using the Fibonacci morphism: $h(a) = ab$, $h(b) = a$.

4. Main Result

In this section we prove our main result.

Theorem 2 It is decidable whether a given prefix-preserving morphism over a binary alphabet generates a strongly cube-free ω-word.

Proof Let $h: \{a,b\}^* \to \{a,b\}^*$ be a pp-morphism, say $h(a) = az$ with $z \neq \lambda$. We denote $h(a) = \alpha$, $h(b) = \beta$ and $L = \{h^n(a) \mid n \geq 0\}$. By Theorem 1, we may assume that h is a strong biprefix, i.e., $\mathrm{first}(\alpha) \neq \mathrm{first}(\beta)$ and $\mathrm{last}(\alpha) \neq \mathrm{last}(\beta)$.

The basic idea behind the proof is to show that if L contains long enough subwords of the form

(1) $$vvfirst(v) \quad \text{with } v \neq \lambda,$$

then it contains shorter, too.

We start with

Claim I Assume that $wwfirst(w)$ is a subword in L satisfying

(2) $$w = w_1\gamma w_2 = w_1'\gamma w_2'$$

for some w_1, w_2, w_1' and w_2' with $0 < ||w_1| - |w_1'|| < |\gamma|$ and $\gamma \in \{\alpha,\beta\}$. Then $wwfirst(w)$ is not the shortest subword of L of the form (1).

The claim is proved as follows. Without loss of generality let $|w_1| > |w_1'|$. Then (2) can be illustrated in the following way:

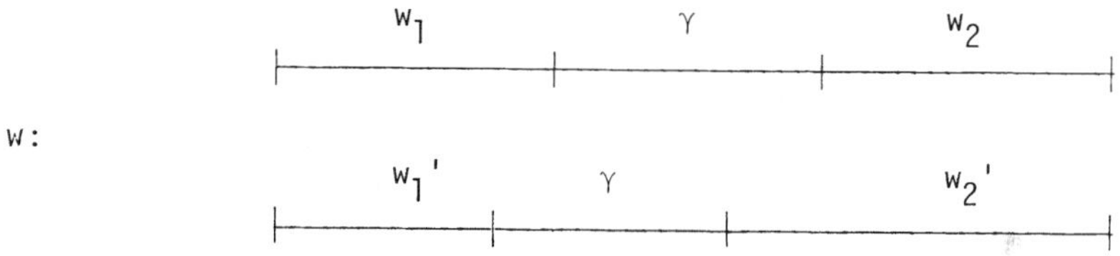

Consequently, the word w, and hence also L, has a subword $\gamma'\gamma'$ first (γ'), where $\gamma' = pref_{|w_1|-|w_1'|}(\gamma)$.

Claim II The shortest subword of L of the form (1) (if there are any) is of the length at most $4|\alpha\beta|$.

To prove claim II we proceed as follows. Let $xuufirst(u)y$ be a word in L such that $uu\,first(u)$ is a minimal subword (with respect to the length) of L of the form (1). We derive from the assumption

(3) $$|uufirst(u)| > 4|\alpha\beta|$$

a contradiction.

Let u_1 be a word satisfying

(4) $$x\,u_1 \in h(\Delta^*) \quad \text{with} \quad u_1 \notin \Delta^*\{\alpha,\beta\},$$
$$u_1\,\delta \text{ pref } u \quad \text{for some } \delta \text{ in } \{\alpha,\beta\}.$$

Without loss of generality we set $\delta = \alpha$.

Since xuufirst(u)y is in $h(\Delta^*)$ and h is a biprefix there exists a word u_2 such that

(5)
$$u_2 \in h(\Delta^*),$$
$$u_1\ \alpha\ u_2 \text{ pref } u\ u_1 \text{ p-pref } u_1\ \alpha\ u_2\ \psi \text{ pref } uu$$

for some ψ in $\{\alpha,\beta\}$. We show that $|\alpha u_2| = |u|$, which implies, since h is a biprefix, that $\psi = \alpha$, and therefore writing $u_2 = u_2'u_1$ we obtain

(6)
$$xuu = xu_1\alpha u_2'u_1\alpha u_2' \quad \text{with} \quad xu_1,\ xu_1\alpha u_2'u_1 \in h(\Delta^*).$$

To prove (6) assume that $|\alpha u_2| \neq |u|$ which means that $|\alpha u_2| < |u|$. If now $\psi = \alpha$, then, by (4) and (5), the assumptions of Claim I are satisfied with $w_1 = u_1$, $w_1' = u^{-1}u_1\alpha u_2$, $\gamma = \alpha$, and so we derive a contradiction with the minimality of uufirst(u). Consequently, it remains the case $\psi = \beta$. In this case we conclude from (4) and (5) that

(7)
$$uu_1\alpha,\ u_1\alpha u_2\beta \in \text{pref}(uu).$$

Moreover, since xu_1, $xu_1\alpha u_2$ and xuufirst(u)y $\in h(\Delta^*)$ and h is a biprefix, there exist words A and B in $\{\alpha,\beta\}^*$ such that uu pref $uu_1\alpha A$ and uu pref $u_1\alpha u_2\beta$ B. We show that any choice of A and B leads to a contradiction.

Assume first that $u_1\alpha u_2\beta$ p-pref $uu_1\alpha$. Please, observe that the equality $uu_1\alpha = u_1\alpha u_2\beta$ is excluded since h is a biprefix. Now B must start with β, otherwise we obtain, by claim I, a contradiction to the minimality of uufirst(u). It also follows that u_2 must end with α, otherwise we would contain three consequtive occurrences of β, again a contradiction to the minimality of uufirst(u). Consequently, by (3), $u_1\alpha u_2\beta\beta$ is a proper prefix of uu. If it is also a proper prefix of $uu_1\alpha$ we are done: there are no ways to continue $u_1\alpha u_2\beta\beta$ without a contradiction, since α-continuation is excluded by claim I and β-continuation since three consequtive β's in a word of L contradicts with (3). So it follows that $uu_1\alpha$ pref $u_1\alpha u_2\beta\beta$. Again the equality is excluded since h is a biprefix. Hence, by claim I, A must start with α. This means that last (u_1) must be different from last (α), otherwise we have a contradiction, and, consequently, (3) implies that $|u_1\alpha\alpha| < |u|$. So there is still a way to continue the proper prefixes $u_1\alpha u_2\beta\beta$ and $uu_1\alpha\alpha$ of uu. However, any continuation leads to a contradiction as in the case of the word $u_1\alpha u_2\beta\beta$ above.

Since the possibility $uu_1\alpha$ p-pref $u_1\alpha u_2\beta$ can be handled with the very same manner we have proved the identity (6). Now we consider the word uufirst(u) in the form

$$u_1\alpha u_2'\ u_1\alpha u_2' \text{ first } (u).$$

Let $h(p) = u_2'u_1$. If $u_1 = \lambda$, then first (u) = first (α) and, consequently, since h is a strong biprefix, $xuu\text{first}(u)y$ can be rewritten as $x\alpha u_2'\alpha u_2'\alpha y'$ for some word y'. So L contains a word $h^{-1}(\alpha u_2'\alpha u_2'\alpha)$ = apapa as a subword. This contradicts with the minimality of $uu\text{first}(u)$. Observe here that also the case when instead of setting $\gamma = \alpha$ we set $\gamma = \beta$ leads to a contradiction. If, in turn, $u_1 \neq \lambda$ then we may rewrite $xuu\text{first}(u)y$ as $x'\rho\alpha u_2'\, u_1\, \alpha u_2' u_1 y''$ for some words x' and y'' and some ρ in $\{\alpha,\beta\}$ such that ρ is a suffix of $u_2'u_1$. Consequently, L contains a subword last (p) a p a p which is again a contradiction with the minimality of $uu\text{first}(u)$.

So our proof for claim II is complete. Now, the theorem follows from claim II and from the following easily provable lemma (see [K], cf. also section 5).

Lemma 2 Given a morphism $h: \Delta^* \to \Delta^*$ and words w, $\omega \in \Delta^*$. It is decidable whether the language $\{h^n(\omega) \mid n \geq 0\}$ contains w as a subword.

5. An Effective Upper Bound

In this section we strengthen Theorem 2 by establishing an upper bound for the number of applications of h to guarantee the existence of a word of the form $vv\text{first}(v)$ as a subword in $\{h^n(a) \mid n \geq 0\}$ if such will ever occur. More precisely, we define for each $i \geq 1$ a number $\delta(i)$ as follows. Let h be a pp-morphism over $\{a,b\}$ such that $|h(a)| \geq i$, $|h(b)| \geq i$ and $h(a) = az$ for some $z \neq \lambda$. Then $\delta(i)$ is defined to be the smallest integer satisfying for every h of the above form:

$$\{h^n(a) \mid n \geq 0\} \quad \text{is strongly cube-free,}$$

if and only if,

$$\{h^n(a) \mid n \leq \delta(i)\} \quad \text{is strongly cube-free.}$$

We continue with three simple lemmas, the proofs of which can be easily derived from the arguments in section 6 of [K].

Lemma 3 If a prefix-preserving morphism over $\{a,b\}$ is not a strong biprefix, then it generates a word of the form $vv\text{first}(v)$, with $v \neq \lambda$, in not more than 7 steps.

Lemma 4 Any prefix-preserving strong biprefix h over $\{a,b\}$ satisfying $|h(a)| = 1$ or $|h(b)| = 1$ generates a word of the form vv first (v), with $v \neq \lambda$, in two steps.

Lemma 5 Any prefix-preserving strong biprefix h over $\{a,b\}$ satisfying $|h(a)| = |h(b)| = 2$ generates a word of the form vvfirst(v), with $v \neq \lambda$, in two steps, if at all.

Finally, we state our basic lemma of this section. For its proof we again refer to [K].

Lemma 6 Let h be a strong biprefix over $\{a,b\}$ such that $\min\{|h(a)|, |h(b)|\} \geq 2$, $\max\{|h(a)|, |h(b)|\} \geq 3$ and $h(a) = az$ for some $z \neq \lambda$. If a word x with the length at most $4|h(a)\ h(b)|$ occurs as a subword in the language $\{h^n(a) \mid n \geq 0\}$, it occurs also in $\{h^n(a) \mid n \leq 7\}$.

Now from Lemmas 3-5 and from claim II in the proof of Theorem 2, we conclude

Theorem 3 $\delta(1) \leq 7$ and hence also $\delta(2) \leq 7$.

In special cases Lemmas 3 and 5 can be strengthened to yield

Theorem 4 $\delta(3) \leq 6$ and $\delta(i) \leq 5$ for $i \geq 4$.

We are not claiming that our upper bounds for the values of $\delta(i)$ are the best possible. On the other hand, they are quite small.

6. Discussion

We have shown that it is decidable whether a given prefix-preserving morphism over a binary alphabet generates a strongly cube-free ω-word. Moreover, we proved that the problem can be settled by checking only relatively few iteration steps, and consequently it might be possible to use a computer in searching such morphisms.

The techniques we have used is similar to that used in [K], when we have shown that it is decidable whether a given pp-morphism over a binary alphabet generates a cube-free ω-word. However, our considerations are now shorter and, moreover, upper bounds for the number of iterations needed to be checked are smaller. So the approach seems to be very suitable for the problem solved here, i.e. for the strongly cube-freeness problem over a binary alphabet. We want to finish this section by mentioning that the same techniques can be used to establish the Berstel's result, cf. [B], which states that it is decidable whether a given prefix-preserving morphism over a three-letter alphabet generates a square-free ω-word. Moreover, our approach would give a constant upper bound independent of the given morphism for the number of applications of the morphism to be checked.

Acknowledgements

The author is grateful to Finnish Academy for the excellent working conditions under which this research was done. Completion of this paper was supported by the Natural Sciences and Engineering Council Canada, under grant A7403.

References

[B] J. Berstel, Sur les mots sans carré définis par un morphisme. Springer Lecture Notes in Computer Science v. 71, 1979, 16-25.

[H] M. Harrison, Introduction to Formal Language Theory, Addison-Wesley, Reading, 1978.

[K] J. Karhumäki, On cube-free ω-words generated by morphisms, manuscript.

[MH] M. Morse and G.A. Hedlund, Unending chess, symbolic dynamics and a problem in semigroups, Duke Math. J. 11, 1944, 1-7.

[T1] A. Thue, Über unendliche Zeichenreihen, Norsk. Vid. Selsk. Skr. I, Mat. - Nat. Kl. Nr. 7, 1906, 1-22.

[T2] A. Thue, Über die gegenseitige Lage gleicher Teile gewisser Zeichenreihen, Norsk. Vid. Selsk. Skr. I, Mat.-Nat. Kl. Nr. 1 (1912), 1-67.

ON THE ROLE OF SELECTORS IN SELECTIVE SUBSTITUTION GRAMMARS

H.C.M. Kleijn
G. Rozenberg
Institute of Applied Mathematics and Computer Science
Wassenaarseweg 80
2333 AL Leiden
The Netherlands

INTRODUCTION

Selective substitution grammars were introduced by Rozenberg, [2], to study a considerable number of seemingly different rewriting systems in a uniform way. A more "concrete" framework was studied by Rozenberg and Wood, [4] . This paper contains results of a continuation of this study.
One can consider context-free grammars (with the possibility of rewriting also terminals) and EOL systems as consisting of a set of elementary rewriting instructions (the productions) and a prescription (the selector) how to use them. Hence, a context-free grammar (with the possibility of rewriting also terminals) and an EOL system differ only in the way their productions are to be applied, i.e. how their selector is defined. The selector of a context-free grammar can be described as $\Sigma^*\bar{\Sigma}\Sigma^*$ (rewrite in each derivation step one occurrence of one symbol), whereas the selector of an EOL system can be described as $\bar{\Sigma}^+$ (rewrite in each derivation step all occurrences of all symbols). Since (it is well known that) the family of context-free languages is strictly contained in the family of EOL languages, it is a natural question to ask what makes the second type of selector more "powerful" than the first type.
To answer this and similar questions one has to find and formalize features of selectors, responsible for their language generating power. Intuitively the language generating power of a selector stems from the possibility it has to use information about the context during the rewriting process and to "block" derivations if they "go wrong". In this paper we impose various restrictions on selectors, each formalizing aspects of the above two features, and then investigate their effect on the language generating power. Some of those conditions yield characterizations of the family of context-free languages, for other conditions lower and upper bounds on the language generating power can be given. Also a natural notion of a class of "simple" rewriting

systems (pattern grammars) is introduced which turn out to possess surprisingly high language generating power.

PRELIMINARIES

We assume the reader to be familiar with formal language theory as, e.g. in the scope of Rozenberg and Salomaa, [3] , and Salomaa, [5]. Some notations, definitions and results need perhaps an additional explanation.

For a word w, $|w|$ denotes its length. Λ denotes the empty word. We consider two languages L_1 and L_2 equal if $L_1 \cup \{\Lambda\} = L_2 \cup \{\Lambda\}$. Two rewriting systems are equivalent if the languages they generate, are equal. The length set of a language L, denoted by LS(L), is defined by $LS(L) = \{|w| : w \in L\}$. A weak identity on an alphabet Σ is a mapping φ such that for all $a \in \Sigma$, either $\varphi(a) = a$ or $\varphi(a) = \Lambda$.

A context-free grammar is specified in the form $G = (\Sigma,h,S,\Delta)$, where Σ is its (total) alphabet, Δ its terminal alphabet, S its axiom and h the finite substitution on $\Sigma\backslash\Delta$ defining "the set of productions of G". (We only consider context-free grammars without erasing productions). It is often convenient (and essential in the general theory of rewriting systems) to extend the finite substitution of a context-free grammar to the whole alphabet (that is to provide productions also for the terminal symbols). The so obtained construct is referred to as an EPOS system . It is easy to see that EPOS systems generate precisely the class of context-free languages. Nevertheless the difference between EPOS systems and context-free grammars plays a role in the frame-work of selective substitution grammars, as considered in this paper.

An EOL system is specified in the form (Σ,h,S,Δ), where Σ, h, S and Δ are as in the EPOS case.

The classes of context-free, context-sensitive and EOL languages are denoted by $L(CF)$, $L(CS)$ and $L(EOL)$ respectively.

The following notions and definitions are typically for this paper. Barred versions of symbols are used with a special meaning: the original symbol is activated. If Σ is an alphabet, then the homomorphism iden from $(\bar{\Sigma} \cup \Sigma)^*$ into Σ^* is defined by iden $\bar{a} = a$ and iden $a = a$, for all $a \in \Sigma$.

An EPOS based s-grammar (abbreviated s-grammar) is a construct $H = (\Sigma,h,S,\Delta,K)$, where base $H = (\Sigma,h,S,\Delta)$ is an EPOS system and K, the selector of H, is a language over $\Sigma \cup \bar{\Sigma}$.

A context-free based s-grammar (abbreviated cf-s-grammar) is a construct $H = (\Sigma,h,S,\Delta,K)$ where base $H = (\Sigma,h,S,\Delta)$ is a context-free grammar and K, the selector of H, is a language over $\Sigma \cup \overline{\Sigma\backslash\Delta}$.

If $H = (\Sigma,h,S,\Delta,K)$ is an s-grammar or a cf-s-grammar, then the set of active symbols of H, denoted by A, is defined by $A = \Sigma$ if H is an s-grammar and by $A = \Sigma\backslash\Delta$ if H is a cf-s-grammar.

If v, w $\in \Sigma^*$, then $v \underset{H}{\Rightarrow} w$ if there exists a word $u \in K$, such that $u \neq v$, <u>iden</u> $u = v$ and $\tilde{h}(u) = w$; $\tilde{h}$ is the finite substitution from $(\bar{A} \cup \Sigma)^*$ into Σ^* defined by $\tilde{h}(\bar{a}) = h(a)$ for all $\bar{a} \in \bar{A}$ and $\tilde{h}(a) = a$ for all $a \in \Sigma$. As usual the language of H, denoted by L(H) is defined by $L(H) = \{w \in \Delta^* : S \underset{H}{\overset{*}{\Rightarrow}} w\}$, where $\underset{H}{\overset{*}{\Rightarrow}}$ is the transitive and reflexive closure of $\underset{H}{\Rightarrow}$.

The selector of an s-grammar (cf-s-grammar) forms the "programming part" of the grammar in the sense that actual application of productions depends on words from the selector. The restrictions as defined below aim at restricting the programming power of the selector:

(1) <u>bar-freeness</u> does not allow to program choices of particular places in a string to be rewritten;

(2) <u>symbol-freeness</u> does not allow to distinguish between symbols that should appear or should not appear at particular places in a word;

(3) <u>interspersion</u> forbids testing an "immediate" neighbourhood of letters;

(4) <u>universality</u> requires that every word can be rewritten, hence it forbids to program the rewriting in such a way that if something "goes wrong", then one gets a string that cannot be rewritten anymore;

(5) <u>occurrence universalness</u> is even stronger: no string contains an occurrence of an active letter that cannot be rewritten.

In every string of a selector occurrences of barred symbols correspond to occurrences of letters that have to be rewritten, all other occurrences of letters play the role of context. Taking this into account the restrictions concerning bar-freeness, symbol-freeness and interspersion are also considered separately for the case of activated symbols and for the case of context-symbols.

Let $H = (\Sigma, h, S, \Delta, K)$ be an s-grammar or a cf-s-grammar.

(1.i) H is <u>active bar-free</u> (abf) if for every $w_1, w_2 \in (\Sigma \cup \bar{A})^*$ and $a \in A$, whenever $w_1\bar{a}w_2 \in K$, then $w_1aw_2 \in K$.

(1.ii) H is <u>context bar-free</u> (cbf) if for every $w_1, w_2 \in (\Sigma \cup \bar{A})^*$ and $a \in A$, whenever $w_1aw_2 \in K$, then $w_1\bar{a}w_2 \in K$.

(1.iii) H is <u>bar-free</u> (bf) if H is both abf and cbf.

(2.i) H is <u>active symbol-free</u> (asf) if for every $w_1, w_2 \in (\Sigma \cup \bar{A})^*$ and $a \in A$, whenever $w_1\bar{a}w_2 \in K$, then $w_1\bar{A}w_2 \subseteq K$.

(2.ii) H is <u>context symbol-free</u> (csf) if for every $w_1, w_2 \in (\Sigma \cup \bar{A})^*$ and $a \in \Sigma$, whenever $w_1aw_2 \in K$, then $w_1\Sigma w_2 \subseteq K$.

(2.iii) H is <u>symbol-free</u> (sf) if H is both asf and csf.

(3.i) H is active interspersed (ai) if for every $w_1, w_2 \in (\Sigma \cup \bar{A})^*$ and $a \in A$, whenever $w_1\bar{a}w_2 \in K$ then $w_1\Sigma^*\bar{a}\Sigma^*w_2 \subseteq K$.

(3.ii). H is context interspersed (ci) if for every $w_1, w_2 \in (\Sigma \cup \bar{A})^*$ and $a \in \Sigma$, whenever $w_1aw_2 \in K$, then $w_1\Sigma^*a\Sigma^*w_2 \subseteq K$.

(3.iii). H is interspersed (i) if H is both ai and ci.
(4). H is universal (u) if for every $w \in \Sigma^*A\Sigma^*$ there exists a word $v \in K$ such that $v \neq w$ and iden $v = w$.
(5). H is occurrence universal (ou) if for every $w_1, w_2 \in \Sigma^*$, $a \in A$ there exists $v_1, v_2 \in (\Sigma \cup \bar{A})^*$ such that iden $v_1 = w_1$, iden $v_2 = w_2$ and $v_1\bar{a}v_2 \in K$.

THE EFFECT OF THE RESTRICTIONS

In this section we show how (the combinations of) the restrictions defined in the previous section affect or do not affect the language generating power of s-grammars and cf-s-grammars. The intention of defining of each of the restrictions on the selectors was to catch basic features of selectors responsible for various aspects of the language generating power of s-grammars and cf-s-grammars. In some sense each of those restrictions aims at forbidding a particular "context-sensitive" feature of an s-grammar or cf-s-grammar.
First of all we notice that $L(CF)$ constitutes a lower bound on the restrictions we consider.

Theorem 1. Let $L \in L(CF)$
(i). There exists an s-grammar H, such that $L(H) = L$ and H is bf, sf, i and ou.
(ii). There exists a cf-s-grammar H, such that $L(H) = L$ and H is bf, sf, i and ou.

Thus a restriction turns out to be "strong" if the s-grammar (cf-s-grammar) subject to it generates a context-free language.
It is not too difficult to see that the restrictions are "independent": if an s-grammar or cf-s-grammar H satisfies one of the conditions, then this does not imply that H must satisfy any other condition too, with the exception that, obviously, bf implies abf and cbf, sf implies asf and csf, i implies ai and ci, and ou implies u. However, it turns out that some combinations of restrictions imply other (combinations of) restrictions. We already mentioned that ou implies u. The next theorem shows that the combination of cbf and u implies ou.

Theorem 2. Let H be an s-grammar or an cf-s-grammar. If H is cbf and u, then H is ou.

For s-grammars we have the following equivalences of combinations of restrictions.

Theorem 3. (i). An s-grammar H is bf and csf if and only if H is bf and asf.
(ii). An s-grammar H is bf and ci if and only if H is bf and ai.

For cf-s-grammars, however, the following equivalences can be established.

Theorem 4. (i). A cf-s-grammar H is bf and csf if and only if H is bf and sf. (ii). A cf-s-grammar H is bf and ci if and only if H is bf and i.

The only difference between an s-grammar and a cf-s-grammar is the fact that in cf-s-grammars terminal symbols cannot be activated. That this can lead to differences when we impose restrictions on the selector, is shown by the following result (which was proved in Rozenberg and Wood, [4]).

Theorem 5. If H is a cf-s-grammar, such that H is u, then L(H) $\in$ L(CF).

(As can be seen from Table 1 s-grammars with even the ou restriction imposed on them can still generate arbitrary languages).
Our next result describes those cases, when cf-s-grammars can generate (at least) those languages that s-grammars generate.

Theorem 6. Let R = {abf, cbf, asf, csf, ai, ci} . Let H be an s-grammar, satisfying a subset V of R. There exists a cf-s-grammar equivalent with H, which also satisfies V.

Dually, the possibilities of a reduction from the case of cf-s-grammars to the case of s-grammars are described in the following result.

Theorem 7. Let R = {abf, cbf, asf, csf, ai, ci}. Let H be a cf-s-grammar. There exists an equivalent s-grammar H', such that if H satisfies a subset V of R, then H' also satisfies V, under the condition that if V contains both cbf and asf, then it must also contain csf.

We will not list in extension the results of the investigation of the effect all combinations of restrictions imposed on selectors have on the generative capacity of s-grammars and cf-s-grammars. (Results and proofs can be found in [1]). Instead we provide by means of Table 1 an (almost) complete survey of the situation for s-grammars.
Five different "bounds" on the language generating power of s-grammars subject to various combinations of restrictions are given. These bounds give rise to a division of the columns of the table in five "column sets" I through V. To the left of I through V the restrictions are listed.
Each column C of the table defines a set of restrictions R(C) containing exactly those restrictions that are marked in that column.
The interpretation of a column C is as follows.
If C belongs to I (L(CF)), then an s-grammar subject to R(C) generates a context-free

language. Moreover all s-grammars satisfying an equivalent (in the sense of Theorems 2,3 and 4) of R(C) or a combination of restrictions containing (an equivalent of) R(C) generate a context-free language.
If C belongs to II (inside L(CS)), then an s-grammar subject to (an equivalent of) R(C) or a combination of restrictions containing (an equivalent of) R(C) generates a context-sensistive language. Moreover if C is not the first column of II (meaning R(C) $\neq$ {abf, asf, ci}) then there exists an s-grammar subject to R(C) which generates a non context-free language. (It is an open problem whether or not L(CF) is strictly contained in the class of languages generated by s-grammars which are abf, asf and ci).
If C belongs to III (arbitrary LS), then there exists an s-grammar, subject to R(C), generating a language with an arbitrary complicated (even not recursively enumerable) length set.
If C belongs to IV (all + w.i.), then there exist for every arbitrary language an s-grammar H subject to R(C) and a weak identity φ, such that φ(L(H)) = L.
If C belongs to V (all), then there exists for every arbitrary language L an s-grammar, subject to R(C), generating L.
Hence, if C belongs to III, IV or V, then for all combinations of restrictions equivalent with R(C) or (equivalent with) a subset of R(C) no "reasonable upperbound" on the language generating power of s-grammars satisfying such combinations can be given.

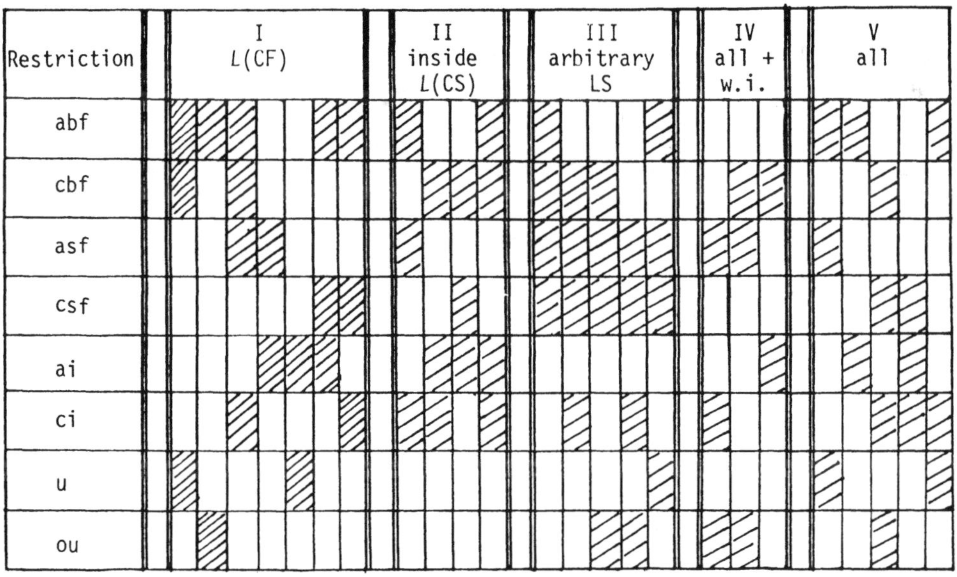

Restriction
I
L(CF)
II
inside
L(CS)
III
arbitrary
LS
IV
all +
w.i.
V
all
abf
cbf
asf
csf
ai
ci
u
ou

Table 1

For the case of context-free based s-grammars one should notice the following. Theorem 5, Theorem 6 and Theorem 7 imply that sets of restrictions that involve u or

or ou, or both cbf and asf but not csf may have different effects in the case of cf-s-grammars than in the case of s-grammars. However, it can be proved that Table 1 after some slight modifications also holds for cf-s-grammars.
If one removes the markings of u and ou from those columns which belong to III, IV or V and a column C with $R(C) = \{u\}$ is added to I (cf. Theorem 5), then the thus obtained table reflects the situation for cf-s-grammars.

Table 1 is complete in the sense that for any (except for three) combination of restrictions one can read from it, whether this combination imposed on an s-grammar (or a cf-s-grammar) leads to a context-free language, a context-sensitive language or that no "upperbound" for the generated language can be given. At the time of writing this paper we were still not able to resolve the following three combinations of restrictions: we miss upperbound results for {abf, i} (for s-grammars and cf-s-grammars), for {cbf, sf, ci, u} and {cbf, asf, ci, u} (for s-grammars).
Also sharpening the upperbounds as presented in Table 1 would contribute to our understanding of the role various restrictions on selectors have in determining the language generating power of selective substitution grammars.

PATTERN GRAMMARS

Two of the restrictions defined in this paper (asf and csf) are aimed to forbid that a symbol in an s-grammar carries any information relevant to the rewriting process. When combined these restrictions imply that the only way that the selector controls the rewriting in the so restricted s-grammar is that it imposes certain rewriting patterns, each of which is a word over an alphabet of two symbols: 1, standing for "rewrite", and 0, standing for "do not rewrite", independent of the actual symbols.
In this section sf context-free based s-grammars are investigated. (That this gives no loss of generality follows from Theorem 8). In particular we investigate the case when the selector is a regular language over the binary alphabet {0,1}.

A pattern grammar is a construct $H = (\Sigma,h,S,\Delta,K)$, where base $H = (\Sigma,h,S,\Delta)$ is a context-free grammar and K, the selector of H, is a language over {0,1}; if K is a context-free or a regular language then we say that H is a context-free pattern grammar or a regular pattern grammar, respectively.
If $v, w \in \Sigma^*$ then $v \underset{H}{\Rightarrow} w$ if $v = a_1 \dots a_n$, $w = \alpha_1 \dots \alpha_n$, $n \geq 1$, $a_1,\dots,a_n \in \Sigma$, $\alpha_1,\dots,\alpha_n \in \Sigma^+$, K contains a word $u \in \{0,1\}^* 1\{0,1\}^*$, $u = b_1 \dots b_n$, $b_1,\dots,b_n \in \{0,1\}$, such that for $1 \leq i \leq n$, either $a_i = \alpha_i$ and $b_i = 0$ or $\alpha_i \in h(a_i)$ and $b_i = 1$.
The language of H, denoted by L(H), is defined by $L(H) = \{w \in \Delta^* : S \underset{H}{\overset{*}{\Rightarrow}} w\}$, where $\underset{H}{\overset{*}{\Rightarrow}}$ is the transitive and reflexive closure of $\underset{H}{\Rightarrow}$.

Theorem 8. Let L be a language. The following statements are equivalent.
(1). L is generated by a (context-free or regular, respectively pattern grammar.
(2). L is generated by a sf cf-s-grammar (the selector of which is respectively a context-free or a regular language).
(3). L is generated by a sf s-grammar (the selector of which is respectively a context-free or a regular language).

Theorem 9. Let H be a context-free pattern grammar. If base H is a right-linear grammar, then L(H) is regular.

In contrast with the above theorem, in which conditions are imposed on the form of the productions, we now restrict the possible selectors.

Theorem 10. L(EOL) is strictly contained in the class of languages, generated by regular pattern grammars.

This theorem provides a lower bound for the generating power of regular pattern grammars. As a matter of fact it turns out that regular pattern grammars have a surprisingly strong language generating power, as shown by the next result .

Theorem 11. For every recursively enumerable language L there exists a regular pattern grammar H and a weak identity φ such that $\varphi(L(H)) = L$.

The above result indicates the following natural research problem. What subclass of regular languages should one consider such that the resulting class of regular pattern grammars has a "reasonable" language generating power (for example such that it yields a class of languages between L(EOL) and L(CS)) ? Some results in this direction will be presented in a forthcoming paper.

ACKNOWLEDGEMENTS

The second author gratefully acknowledges the support of NSF grant number MCS 79-03838. Both authors are indebted to A. Ehrenfeucht for useful discussions concerning the topic of this paper and to R. Verraedt for useful comments on the first draft of this paper.

REFERENCES

1. Kleijn, H.C.M. and Rozenberg, G., Context-freelike restrictions on selective rewriting, Inst. Appl. Math. and C.S., University of Leiden, Techn.Rep.No. 80-19.

2. Rozenberg, G., Selective substitution grammars (towards a framework for rewriting systems). Part I: definitions and examples. EIK, 13 /1977/, 455-463.
3. Rozenberg, G. and Salomaa, A., The Mathematical Theory of L Systems. Academic Press, New York, 1980.
4. Rozenberg, G. and Wood, D., Context-free grammars with selective rewriting. Acta Informatica, 13/1980/, 257-268.
5. Salomaa, A., Formal Languages. Academic Press, New York, 1973.

Classes Of Functions Over Binary Trees

Hans Kleine Büning
Institut für mathematische Logik
und Grundlagenforschung

D-44oo Münster/Germany

We will investigate classes of functions over binary trees defined by tree-counter-machines [2] and by operations such as bounded recursion, bounded iteration and weak bounded recursion.

1. Tree-counter-machines

The class Bm of binary trees is defined inductively as follows: $* \in Bm$ and if $\alpha, \beta \in Bm$ then $(\alpha\beta) \in Bm$. The length of a tree is defined as the number of stars $l(*)=1$ and $l((\alpha\beta)) = l(\alpha)+l(\beta)$ and the depth of a tree as $t(*)=0$ and $t((\beta\alpha))=\max\{t(\beta),t(\alpha)\}+1$.
Furthermore we require two elementary functions $(\alpha)_o, (\alpha)_1$ with $(*)_o=*$ and $(\alpha\beta)_o=\alpha$, $(*)_1=*$ and $(\alpha\beta)_1=\beta$.
A k-tree-counter-machine M consists of k counter and a finite sequence of instructions of the following form:

1) $I_q = q\, L_i\, q'$, i.e.

$$(q;\alpha_1,\ldots,\alpha_k) \overset{1}{\underset{M}{\rightarrow}} (q';\alpha_1,\ldots,(\alpha_i)_o,\ldots,\alpha_k)$$

2) $I_q = q\, R_i\, q'$, i.e.

$$(q;\alpha_1,\ldots,\alpha_k) \overset{1}{\underset{M}{\rightarrow}} (q';\alpha_1,\ldots,(\alpha_i)_1,\ldots,\alpha_k)$$

3) $I_q = q\, A_{ij}\, q'$, i.e.

$$(q;\alpha_1,\ldots,\alpha_k) \overset{1}{\underset{M}{\rightarrow}} (q';\alpha_1,\ldots,\alpha_{i-1},(\alpha_i\alpha_j),\alpha_{i+1},\ldots,\alpha_k)$$

4) $I_q = q\, P_i\, q^{=}\, q^{\neq}$, i.e.

$$(q_i\alpha_1,\ldots,\alpha_k) \overset{1}{\underset{M}{\rightarrow}} \begin{cases} (q^{=};\alpha_1,\ldots,\alpha_k), & \text{if } \alpha_i=* \\ (q^{\neq};\alpha_1,\ldots,\alpha_k), & \text{if } \alpha_i\neq * \end{cases}$$

5) I_q=stop

Let f be a function over Bm, then f is computable by M iff
$\forall\alpha: (1;\alpha,*,\ldots,*) \underset{M}{\rightarrow} (\text{stop};\alpha,f(\alpha),*,\ldots,*)$
Furthermore $\forall\alpha: t_{M,f}(\alpha)$ is the number of steps of M starting with α and

$$s^l_{M,f}(\alpha) = \max\{\sum_{1\le i\le k} l(\alpha_i) \mid C=(q;\alpha_1,\ldots,\alpha_k) \text{ configuration of M starting with } \alpha\} \text{ and}$$

$$s^t_{M,f}(\alpha) = \max\{\sum_{1\le i\le k} t(\alpha_i) \mid C=(q;\alpha_1,\ldots,\alpha_k) \text{ configuration of M starting with } \alpha\}$$

Let A,B be classes of functions over N, then (for q=l or q=t)

$R_q(A,B) := \{f \mid \exists k\text{-tree-counter-machine } M: \exists g \in A: \exists h \in B: M \text{ computes } f \text{ and } \forall \alpha: (t_{M,f}(\alpha) \leq g(q(\alpha)) \text{ and } S^q_{M,f}(\alpha) \leq h(q(\alpha)))\}$

and

$R^*_q(A,B) = \{Q \text{ predicate } \mid \chi_Q \in R_q(A,B)\}$

$$\text{where } \chi_Q(\alpha) = \begin{cases} (**), & \text{if } Q(\alpha) \\ *, & \text{otherwise} \end{cases}$$

Now let be $\psi_n(x) = 2^{2^{\cdot^{\cdot^{2^x}}}} \}$ n-times for any $x \in N, n \in N$, and $\forall n \in N$:

$nE \quad := \{f \mid \forall x \in N: f(x) \leq \psi_n(O(x))\}$

$nEP \quad := \{f \mid \exists \text{ polynomial } Q: \forall x \in N: f(x) \leq \psi_n(Q(x))\}$

$nELG := \{f \mid \exists \text{ polynomial } Q: \forall x \in N: f(x) \leq \psi_{n+1}(Q(\log(x)))\}$

$\Omega \quad = \bigcup_{n \in N} nE$ and $\theta = \bigcup_{n \in N} \{nE, nEP, nELG\}$

We abbreviate OE=L, OEP = P and OELG=LG.

Then we obtain as a trivial consequence of the above definitions

<u>Lemma 1:</u> $\forall A \in \theta$:

a) $R_1(\Omega,A) = R_1(EA,A)$

b) $R_1(A,\Omega) = R_1(A,EA)$

(EA=n+1E, if A=nE; EA=n+1EP if A=nEP and EA=n+1ELG if A=nELG)

The next theorem shows that classes of predicates can be identical while the associated classes of functions differ.

<u>Theorem 2:</u> $A,B \in \theta$:

a) $L \subsetneqq A \subset B \rightarrow R^*_1(A,B) = R^*_1(A,A)$

b) $A \subsetneqq B \rightarrow R_1(A,B) \supsetneqq R_1(A,A)$

By well known methods - constructing an universal tree-counter-machine - we obtain an enumeration theorem.

<u>Theorem 3:</u>

Let be $A_i, B_i \in \theta\ (i=1,2)$ and $A_1 \subsetneqq A_2, B_1 \subsetneqq B_2$ then

$\exists g \in R_1(A_2,B_2)\ \forall f^{(1)} \in R_1(A_1,B_1)\ \exists \beta \in Bm\ \forall \alpha \in Bm$:

$$f(\alpha) = g(\beta,\alpha).$$

As an immediate consequence of theorem 2 and lemma 1 we see:

<u>Theorem 4:</u> $\forall A \in \theta: R^*_1(\Omega,A) = R^*_1(A,\Omega)$

In some cases we can reduce inclusion or equality problems to well known problems for Turingmachines. For T(A,B) the set of languages accepted in time bounded by a function from A and space bounded by a function from B we can show the following

Theorem 5:

Let be $A_i, B_i \in \theta, A_i \supset B_i, A_i \supset P$ $(i=1,2)$, then $R_1^*(A_1,B_1) \supset R_1^*(A_2,B_2))$ iff $T(A_1,B_1) \subset T(A_2,B_2))$.

2. Machine independent description

For classes of functions and predicates given by bounded Turingmachines or counter-machines, in some cases a machine-independent classification has been proved, for example in [1,3,4].

At first we deal with bounded recursion for binary trees "BR_1", introduced in [2].

Definition 1: A class k of functions is closed under BR_1 iff for $g,h,g_1 \in k$ and

$$\forall \alpha, \beta, \sigma \in Bm: \quad f(*,\alpha) = g(\alpha)$$
$$f((\beta\sigma),\alpha) = h(\beta,\sigma,\alpha,f(\beta,\alpha),f(\sigma,\alpha))$$
$$f(\beta,\alpha) \leq_1 g_1(\beta,\alpha)$$

holds $f \in k$.

Definition 2: Let K be a class of functions; $A_{BR_1}(K)$ is the smallest class, such that

1) $K \subset A_{BR_1}(K)$

2) for $U_n^i(\alpha_1,\dots,\alpha_n) = \alpha_i, C_n^\alpha(\alpha_1,\dots,\alpha_n) = \alpha$ holds $U_n^i, C_n^\alpha \in A_{BR_1}(K)$

3) closed under substitution and BR_1.

Since for functions which increase exponentially substitution generates superexponential functions we introduce a restricted substitution

Definition 3: Let R,K be classes of functions, $A_R(K)$ is the smallest class of functions, such that

1) $R, K \subset A_R(K)$

2) If $g_i \in A_R(K)$ $(1 \leq i \leq r)$, $h \in R$ then $f = h \circ (g_1,\dots,g_r) \in A_R(K)$

3) If $g_i \in R$ $(1 \leq i \leq r)$, $h \in A_R(K)$ then $f = h \circ (g_1,\dots,g_r) \in A_R(K)$

In the following we make use of some special functions:

$\forall \alpha, \beta \in Bm$:

$$1)\ {}_{=}(\alpha,\beta) = \begin{cases} (**), & \text{if } \alpha \equiv \beta \\ *, & \text{otherwise} \end{cases}$$

$$2)\ \leq_1(\alpha,\beta) = \begin{cases} (**), & \text{if } l(\alpha) \leq l(\beta) \\ *, & \text{otherwise} \end{cases}$$

$$3)\ sg(\alpha) = \begin{cases} (**), & \text{if } \alpha = * \\ *, & \text{otherwise} \end{cases}, \quad \overline{sg}(\alpha) = \begin{cases} *, & \text{if } \alpha = * \\ (**), & \text{otherwise} \end{cases}$$

4) for $n \in N-\{0\}$: $\underset{\sim}{1} = *$, $\underset{\sim}{n+1} = (\underset{\sim}{n}*)$

5) for $n \in N$: $\underline{0} = *$, $\underline{n+1} = (\underline{n}\,\underline{n})$

6) $\varphi_1(\alpha) = \underline{1(\alpha)}$, $n \geq 1$: $\varphi_{n+1}(\alpha) = \varphi_1(\varphi_n(\alpha))$

7) $m(\alpha,\beta) = l(\alpha) \cdot l(\beta)$

Now we are able to define a hierarchy.

<u>Definition 4:</u> F_1-hierarchy

$F_1(L) = A_{BR_1}(\{(\alpha\beta), =, \leq_1, \underset{\sim}{1}, sg, \overline{sg}, (\alpha)_o, (\alpha)_1\})$

$F_1(P) = A_{BR_1}(\{(\alpha\beta), =, \leq_1, \underset{\sim}{1}, sg, \overline{sg}, (\alpha)_o, (\alpha)_1, m\})$

$F_1(LG) = A_{BR_1}(\{(\alpha\beta), =, {}_1, \underset{\sim}{1}, sg, \overline{sg}, (\alpha)_o, (\alpha)_1, m, \underset{\sim}{2^{\log^2 \circ 1}}\})$

$\forall n \geq 1$:

$F_1(nE) = A_{F_1(L)}(\{\varphi_n\})$

$F_1(nEP) = A_{F_1(P)}(\{\varphi_n\})$

$F_1(nELG) = A_{F_1(LG)}(\{\varphi_n\})$.

For these classes $F_1(A)$ we can find corresponding classes defined by tree-counter-machines.

<u>Theorem 6:</u> $\forall A \in \theta - \{L\}$: $F_1(A) = R_1(A,A)$

In order to accelerate the value of a function $f((\beta\sigma),\alpha)$ defined by means of BR_1, we need the left and right part of the tree $(\beta\sigma)$ and values $f(\beta,\alpha)$ and $f(\sigma,\alpha)$. For this reason we introduce a weak bounded recursion called WBR_1.

A class K is closed under WBR_1, iff for $g, h, g_1 \in k$ and

$$\forall \alpha, \beta, \sigma \in Bm: \quad f(*, \alpha) = g(\alpha)$$
$$f((\beta\sigma), \alpha) = h(\beta, \alpha, f(\beta, \alpha))$$
$$f(\beta, \alpha) \leq_1 g_1(\beta, \alpha)$$

holds $f \in K$.

Now substitute the operation BR_1 by WBR_1 in the hierarchy $F_1(A)$. For this new hierarchy - called $FW_1(A)$ - then holds:

<u>Theorem 7:</u> $\forall A \in \theta - \{L\}: FW_1(A) = R_1(A, A) = F_1(A, A)$

Our next step is to find a description of classes where the time restriction is greater than the space restriction.

We introduce a bounded iteration "BI_1" as follows:

K is closed under BI_1 iff for $f, g \in K$ and

a) $\forall \alpha \in Bm \; \exists n \in N: \; f^{n+1}(\alpha) = f^n(\alpha)$

b) $\forall \alpha \in Bm \; \exists n \in N: \; f^n(\alpha) \leq_1 g(\alpha)$

c) $\forall \alpha \in Bm: \; f'(\alpha) = \lim\limits_{n \in N} f^n(\alpha)$

holds $f' \in K$.

Now add to the classes in the FW_1-hierarchy the operation BI_1 (closure). For this new hierarchy - called FI_1 - we see

<u>Theorem 8:</u> $\forall A \in \theta: \; FI_1(A) = R_1(EA, A)$

If we ask whether bounded recursion BR_1 resp. WBR_1 has the same power as BI_1, we obtain for example for P:

$F_1(P) = FW_1(P) \overset{?}{=} FI_1(P)$ iff PTIME $\overset{?}{=}$ PTAPE

Thus we see that for Bm-functions the operation BI_1 corresponds to bounded recursion over N and BR_1 resp. WBR_1 to syntactical bounded recursion over N. The syntactical bounded recursion corresponds to PTIME [4].

If we consider classes $R_t(A,B)$ (bounded by depth) we get similar results but we require at least exponential time in order to find the depth of a tree.

<u>Theorem 9:</u> $\forall A \in \theta$: a) $R_t^*(EA, A) = R_t^*(EA, EA)$

b) $R_t(EA, A) \subsetneq R_t(EA, EA)$

For operations BR_t, WB_t, BI_t (bounded by depth not by length) and a slight modification we can prove only ($nE = A$, $n \geq 1$)

$(FW_t(A) = R_t(A,A) \underset{(1)}{\subset} F_t(A) \underset{(2)}{\subset} F_t(EA,A) \underset{(3)}{\subset} R_t(EEA,A) = FI_t(A)$, where for (1) and (2) one of them is a proper inclusion and for (3) the problem corresponds to wellknown problems for Turingmachines. Thus it is open whether BR_t and WBR_t are equivalent up to certain intial functions.

References

1) Cobham,A.: The intrinsic computational complexity of functions, Proc. 1964 Intern. Congr. on Logic, 24-3o

2) Cohors-Fresenborg,E.: Subrekursive Funktionenklassen über binären Bäumen. Dissertation, Münster 1971

3) Thompson,D.B.: Subrecursiveness:Machine-independent Notions of Computability in Restricted Time and Storage,MSTG(1972), 3-15

4) Wagner,K.: Bounded recursion and complexity classes.MFCS 1979, Lecture Notes 74

2. Rozenberg, G., Selective substitution grammars (towards a framework for rewriting systems). Part I: definitions and examples. EIK, 13 /1977/, 455-463.
3. Rozenberg, G. and Salomaa, A., The Mathematical Theory of L Systems. Academic Press, New York, 1980.
4. Rozenberg, G. and Wood, D., Context-free grammars with selective rewriting. Acta Informatica, 13/1980/, 257-268.
5. Salomaa, A., Formal Languages. Academic Press, New York, 1973.

5. A. Ehrenfeucht and G. Rozenberg, On the subword complexity of DOL languages with a constant distribution, Institute of Applied Mathematics and Computer Science, University of Leiden, The Netherlands, Technical Report No. 81-21, 1981.
6. K.P. Lee, Subwords of developmental languages, Ph.D. thesis, State University of New York at Buffalo, 1975.
7. G. Rozenberg and A. Salomaa, The mathematical theory of L Systems, Academic Press, London, New York, 1980.
8. A. Thue, Über unendliche Zeichenreihen, Norsk. Vid. Selsk. Skr. I Mat.-Nat.Kl. 7 (1906), 1-22.

MATHEMATICAL STRUCTURES UNDERLYING GREEDY ALGORITHMS

B. Korte, Inst. f. Ökonometrie und Operations Research,
Universitat Bonn, D-5300 Bonn, GFR

L. Lovász, Bolyai Institute, József Attila University,
H-6720 Szeged, Hungary

0. Introduction The principle of "greediness" plays a fundamental role both in the desigu of "continuous" algorithms (where it is called the steepest descent method) and of discrete algorithms. The discrete structure most closely related to greediness is a matroid; in fact, matroids may be characterized among independence structures by the optimality of the "greedy" solution to certain optimization problems. This is one of the very few examples when mathematical structures are defined by their algorithmic behavior (another example is a euclidean ring). This situation may be explained by the fact that algorithmic and procedural aspects have so far played a secondary role to propositional aspects throughout mathematics.

In this lecture we introduce a structure which is more general than matroids, but for which the notion of a "greedy algorithm" can be defined and again optimality of the "greedy" solution can be proved. This more general setting will include greedy algorithms like breadth first search and certain scheduling methods, which do not fit into the matroid framework. Proofs of the results will appear elsewhere.

1. Matroids and greedoids. The pair (E,J) is called an independence structure if E is a finite set, J is a set of subsets of E and the following two axioms are fulfilled:

(I1) $J \neq \emptyset$,

(I2) if $X \in J$ and $Y \subsetneq X$ then $Y \in J$.

(independence structures are often called hereditory setsystems or simplicial complexes). In algorithmic combinatorics one type of independence structures is of special impostance. The independence structure (E,J) is called a matroid if the following "exchange axiom" is fulfilled:

(M) if $X,Y \in J$ and $|X|>|Y|$ then there exists an $x \in X - Y$ such that $Y \cup \{x\} \in J$.

We cannot discuss the rich variety of nice properties and applications of matroids; see e.g. Welsh (1976).

Let E be a finite set and L a set of ordered subsets of E (in other words, the elements of L are words over E without repetition; we shall call them feasible words). The pair (E, L) is called a hereditary language if the following two axioms are fulfilled:

(H1) $L \neq \emptyset$,

(H2) if $\alpha \in L$ and β is a beginning section of α then $\beta \in L$.

We define a greedoid as a hereditary language which satisfies the following "exchange axiom":

(G) if $\alpha, \beta \in L$ and $|\alpha| > |\beta|$ then there exists an $x \in \alpha$ such that $\beta x \in L$.

The unordered subsets of E underlying the members of L will be called feasible, and we shall denote the set of all feasible subsets by J. It is immediate from the definition that J satisfies (I1) and (M). A certain converse of this statement is also true:

Theorem 1. Let E be a finite set and $J \subseteq 2^E$ such that (I1) and (M) are satisfied. Then there exists a unique greedoid on E whose feasible sets are the sets in J.

The maximal feasible words of a greedoid will be called basic words. It is clear that the set of all basic words determines the greedoid uniquely and that all basic words have the same length, which will be called the rank. The underlying sets of basic words will be called bases. We shall see examples which show that bases do not determine the greedoid uniquely.

Although in this paper we have no space to go into the structural aspects of greedoids, we ramark that if (E, L) is a greedoid and k is any natural number less than its rank, then the words in L of length at most k form another greedoid. This greedoid will be called the k-truncation of (E, L).

2. Examples of greedoids.

Example 2.1. Let $(E, \leq)$ be a partially ordered set and let the basic words of a greedoid be all linear extensions of this partial order. A set $X \subseteq E$ is then feasible if it is an ideal in the poset (i.e. $x \in X$, $y \leq x$ implies $y \in X$), and the feasible words are all lower ideals ordered compatibly with the given partial order. This greedoid plays an important role in the study of one-machine scheduling under precedence constraints: the basic words are just the feasible schedules. This greedoid will be called the schedule greedoid of the poset $(E, \leq)$.

Example 2.2. Theorem 1 above implies that the independent sets of a matroid form the feasible subsets of a (uniquely determined) greedoid. In this sense greedoids may be viewed as generalizations of matroids.

Example 2.3. Let G be a directed graph with a "root" $r \in V(G)$. Let $E = E(G)$ and let J be the set of arc-sets of all arborescences in G rooted at r. It is not difficult to show that (E,J) satisfies (I1) and (M) and so it defines a unique greedoid. Every basic word of this greedoid corresponds to a "search" of the graph. This greedoid shall be called the search greedoid of the rooted digraph G.

Example 2.4. Let T be a tree, $E = V(T)$, and let J consist of those subsets X V(T) for which $T - X$ is connected. Then again J is the set of feasible subsets of a greedoid on E, which will be called the shelling greedoid of T.

There are many ways to generalize the above examples, but we shall not go into this here.

3. The greedy algorithm. Let E be a finite set and L a hereditary language over E. Let

$$W : L \longrightarrow \mathbb{R}$$

be any objective function. We consider the optimization problem

(*) maximize $\{ w(\alpha) \mid \alpha \text{ is a basic word in } L \}$.

We define the greedy solution of (*) as follows. Choose, for $i = 1,2,\ldots,$ an element $x_i \in E$ such that $x_1 \ldots x_i \in L$ and

$$w(x_1 \ldots x_i) = \max \{ w(x_1 \ldots x_{i-1} y) \mid x_1 \ldots x_{i-1} y \in L \}.$$

Stop if no such x_i exists.

In general, the greedy solution will not be optimal even if (E,L) is a greedoid: one obviously needs some connection between the structures of L and w. We shall assume that the following axiom is fulfilled:

(W) If $\alpha \in L$, $\alpha x \in L$ and for all y, $w(\alpha x) \geq w(\alpha y)$ holds, then for all $y,z \in E$ and $\beta \in E^*$, we have $w(\alpha x \beta) \geq w(\alpha y \beta)$ and $w(\alpha x \beta y) \geq w(\alpha y \beta x) \geq w(\alpha y \beta z)$.

(If one of the words occuring here is not feasible then we consider the corresponding inequality vacuously satisfied.)

In quite a few applications the value of the objective function w(x) will not depend on the ordering of the elements of α (as long as the ordering is feasible). Let us call such an objective function stable. Thus a stable objective function may be viewed as a mapping $x : J \to \mathbb{R}$, where J is the set of feasible subsets of E. It is easy to see that for stable objective functions (W) is equivalent to the following, much simpler axiom:

(W') If $A,B \in J$, $A \subset B$ and $x \in E - B$ such that $A \cup \{x\} \in J$ and $B \cup \{x\} \in J$, and $w(A \cup \{x\}) \geq w(A \cup \{y\})$ for every y for which $A \cup \{y\} \in J$, then also $w(B \cup \{x\}) \geq w(B \cup \{y\})$ for every y for which $B \cup \{y\}$ is feasible.

We give some examples of objective functions satisfying these axioms.

Example 3.1. Let (E,L) be any greedoid and let $f : E \to \mathbb{R}$. Define a stable objective function, called the bottleneck objective function, by

$$w(X) = \min (f(x) \mid x \in X) .$$

This example generalizes to a non-stable objective function, which we shall call a generalized bottleneck objective function. Let $f : E \times \mathbb{N} \to \mathbb{R}$ be such that $f(x,k+1) \geq f(x,k)$ for every $x \in E$ and $k \in \mathbb{N}$. Define

$$w(x_1 \ldots x_k) = \min (f(x_1,1),\ldots,f(x_k,k)) .$$

Example 3.2. Let $(E,\mathcal{J})$ be a matroid, $f : E \to \mathbb{R}$ and define the (stable) linear objective function

$$w(X) = \sum (f(x) \mid x \in X) .$$

Example 3.3. Let (E,L) be the search greedoid of the digraph G rooted at r. Let, for every arborescence rooted at r, $-w(X)$ be the sum of lengths of paths in X, starting from r.

We now state the main result of this paper.

Theorem 2. Let (E,L) be a greedoid and let $w : L \to \mathbb{R}$ be an objective function satisfying (W). Then the greedy solution of (*) is optimum.

A certain converse of this theorem is also true.

Theorem 3. Let (E,L) be a hereditary language such that for every generalized bottleneck objective function the greedy solution of (*) is optimum. Then (E,L) is a greedoid.

Let us mention some applications of Theorem 2. If (E,L) is a schedule greedoid and we consider the generalized bottleneck objective function described in example 3.1, we obtain a theorem of Lawler (1973). If (E,L) is associated with a matroid and we consider the objective function in example 3.2, we of course obtain the result of Edmonds mentioned in the introduction. If (E,L) is a search greedoid and we consider the objective function in example 3.3, then the greedy algorithm yields a breadth first search arborescence, and theorem 2 implies that breadth first search yields a spanning arborescence in which the sum of distances of points from the root is minimum. Knowing that there is a spanning arborescence which simultaneously minimizes all distances, it follows that breadth first search finds the shortest path to every point from the root (Dijkstra 1959).

If we consider the negation of w in this last example, then the greedy algorithm executes a depth first search. We cannot apply Theorem 2, however. In fact, to find a spanning arborescence which maximizes the sum of distances from the root is an NP-hard problem (it includes the problem of finding a Hamilton path).

4. Problems. It seems to us that the examples and applications above justify the study of greedoids. The first task is to investigate to what extent do the results on the structure of matroids carry over to greedoids. (Some examinations in this direction are already needed to prove Theorems 1, 2 and 3.)

It is natural to ask whether the most important linear objective functions (example 3.1) can be maximized over the basic words of a general greedoid in polynomial time. The greedy algorithm will not do this in general, but there are other, more sophisticated algorithms known in other special cases. For example, the case of search greedoids is the "optimum branching" problem, solved by Edmonds (1967). Since the schedule and she shelling greedoids have only one basis, this problem is uninteresting for them, although it is interesting for appropriate truncations of them. A description of the convex hull of incidence vectors of feasible sets would also be interesting.

References.

E.W. Dijkstra, A note on two problems in connexion with graphs, Numer. Math. 1 (1959) 269-271.

J. Edmonds, Submodular functions, matroids, and certain polyhedra, in: Combinatorial Structures and their Applications, Gordon and Breach, 1970.

J. Edmonds, Optimum brauchings, J. Res. Nat. Bur. Stand. B71 (1967) 233-240.

E. Lawler, Optimal sequencing of a single machine subject to precendence constraints, Management Sci. 19 (1973) 544-546.

D. Welsh, Matroid Theory, Academic Press, 1976.

SOME PROPERTIES OF LANGUAGE FAMILIES GENERATED BY COMMUTATIVE LANGUAGES

Juha Kortelainen

Department of Mathematics
Faculty of Science
University of Oulu
Linnanmaa

SF-90570 Oulu 57 FINLAND

INTRODUCTION

In this paper the relationships between the language families generated by commutative languages are studied. Closely related to commutative languages are bounded languages. The properties of bounded trios and AFLs (= trios (AFLs resp.) generated by a family of languages containing only bounded languages) were first studied by Goldstine [3], [4] and Rovan [8]. Latteux in [5], [6] and [7] studies commutative trios and AFLs (= trios (AFLs resp.) generated by a family of languages containing only commutative languages).

In the first section some definitions and notations are given.

In the second section we prove that every commutative trio closed under union with {e} (e is the empty word) is closed under linear erasing. The result is then extended to commutative AFLs and to trios generated by commutative languages and closed under intersection. Since, for each k in $N_+ = \{1,2,\dots\}$, the smallest trio containing all k-bounded languages and closed under intersection $\mathcal{M}_\cap(\mathcal{B}_k)$ coincides with the smallest trio containing all k-commutative languages and closed under intersection $\mathcal{M}_\cap(\mathcal{COM}_k)$, the language family $\mathcal{M}_\cap(\mathcal{B}_k)$ is closed under linear erasing.

In [4] Goldstine shows that $\hat{\mathcal{F}}_\cap(\mathcal{B}_1)$ is the family of all languages. He also studies the inclusions $\mathcal{F}_\cap(\mathcal{B}_1) \subseteq \mathcal{F}_\cap(\mathcal{B}_2) \subseteq \cdots \subseteq \mathcal{F}_\cap(\mathcal{B}_k) \subseteq \cdots \subseteq \mathcal{F}_\cap(\mathcal{B})$ ($\mathcal{F}_\cap(\mathcal{B})$ is the smallest AFL containing all bounded languages and closed under intersection). It is not known whether the inclusions above are proper or not. Another problem considered in [4] is whether the language $DUP = \{a^n b^n \mid n \text{ in } N\}$ is in $\mathcal{F}_\cap(\mathcal{B}_1)$. In the third section we shall prove that DUP is even in $\mathcal{M}_\cap(\mathcal{B}_1)$. It is also shown that, for any two polynomials P(n) and Q(n) of n with coefficients in $N = \{0,1,2,\dots\}$, the language $L = \{a^{P(n)} b^{Q(n)} \mid n \text{ in } N\}$ is in $\mathcal{M}_\cap(\mathcal{B}_1)$.

I DEFINITIONS AND NOTATIONS

Some concepts used in the text are now briefly reviewed. We refer to [1] for undefined concepts and notations.

Let $\mathcal{L}$ be a family of languages. Then

- $\mathcal{M}(\mathcal{L})$ ($\hat{\mathcal{M}}(\mathcal{L})$) is the smallest (full) trio containing $\mathcal{L}$;
- $\mathcal{F}(\mathcal{L})$ ($\hat{\mathcal{F}}(\mathcal{L})$) is the smallest (full) AFL containing $\mathcal{L}$;
- $\mathcal{M}_\cap(\mathcal{L})$ ($\hat{\mathcal{M}}_\cap(\mathcal{L})$) is the smallest (full) trio containing $\mathcal{L}$ and closed under intersection;
- $\mathcal{F}_\cap(\mathcal{L})$ ($\hat{\mathcal{F}}_\cap(\mathcal{L})$) is the smallest (full) AFL containing $\mathcal{L}$ and closed under intersection;
- $\mathcal{M}^{lin}(\mathcal{L})$ ($\mathcal{F}^{lin}(\mathcal{L})$) is the smallest trio (AFL) containing $\mathcal{L}$ and closed under linear erasing.

A homomorphism h is linearly bounded on a language L if there exists $k>0$ such that for each x in L, $|x| \leq k\cdot\max\{1,|h(x)|\}$. A family $\mathcal{L}$ of languages is closed under linear erasing if h(L) is in $\mathcal{L}$ for each L in $\mathcal{L}$ and each homomorphism h linearly bounded on L .

Let Σ_1 be an alphabet. For each word x in Σ_1^* , let $|x|_a$ be the number of occurences of the symbol a in x .

For each x in Σ_1^* and each language $L\subseteq\Sigma_1^*$, define $c(x) = \{y \mid |y|_a = |x|_a$ for each a in $\Sigma_1\}$ and $c(L) = \bigcup_{x \text{ in } L} c(x)$. c(L) is called the commutative closure of L .

A language L is k-commutative , k in N_+, if $L = c(L)$ and the number of symbols in Σ_L is less than or equal to k . A language L is commutative if it is k-commutative for some k in N_+. Let $\mathcal{COM}_k$ be the family of all k-commutative languages and $\mathcal{COM}$ the family of all commutative languages.

A language family $\mathcal{L}$ is closed under commutation if c(L) is in $\mathcal{L}$ for each L in $\mathcal{L}$.

A language L is k-bounded , k in N_+, if there exists words $x_1,\ldots,x_k$ such that $L\subseteq x_1^*\ldots x_k^*$. A language L is bounded if it is k-bounded for some k in N_+. Let $\mathcal{B}_k$ denote the family of all k-bounded languages and $\mathcal{B}$ the family of all bounded languages.

Let L_1 and L_2 be languages. Define $\mathrm{shuf}(L_1,L_2) = \{x_1y_1\ldots x_ny_n \mid n\geq 1,\ x_1\ldots x_n \text{ in } L_1,\ y_1\ldots y_n \text{ in } L_2\}$.

Let $\mathcal{R}$ denote the family of all regular languages.

II LINEAR ERASING ON COMMUTATIVE TRIOS AND AFLS

It is now shown that if some assumptions are fullfilled, a trio (or AFL) generated by a language family which contains only commutative languages is always closed under linear erasing.

A family $\mathcal{L}$ of languages is a commutative trio if $\mathcal{L} = \mathcal{M}(\mathcal{L}')$ for some language family $\mathcal{L}' \subseteq \mathcal{COM}$.

LEMMA 1. Let $\mathcal{L}$ be a commutative trio. Then $h_1(L_1)-\{e\}$ is in $\mathcal{L}$ for each language L_1 in $\mathcal{L}$ and for each homomorphism h_1 linearly bounded on L_1 .

PROOF. Since $\mathcal{L}$ is a commutative trio, $\mathcal{L} = \mathcal{M}(\mathcal{L}_1)$ for some family $\mathcal{L}_1 \subseteq \mathcal{COM}$ of languages. Let L_1 be in $\mathcal{L}$ and h_1 a homomorphism linearly bounded on L_1. It can be shown that $h_1(L_1) = h(g^{-1}(L) \cap R)$ for some language L in $\mathcal{L}_1$, R in $\mathcal{R}$ and homomorphism h linearly bounded on $g^{-1}(L) \cap R$. Let k_1 be such that for all x in $g^{-1}(L) \cap R$, $|x| \leq k_1 \cdot \max\{1, |h(x)|\}$. Let $T = (K_1, \Sigma_R, \delta, q_0, F_1)$ be a deterministic finite state acceptor such that $R = T(A)$. Let r be the number of states in T , $W = \{a \text{ in } \Sigma_R \mid h(a) \neq e\}$ and $E = \Sigma_R - W$. Let $\mathcal{J}$ be the set of all $J \subseteq K_1$ such that q_0 is in J and $J \cap F_1 \neq \emptyset$. For each J in $\mathcal{J}$ we can define a finite state acceptor T_J such that x is in $T_J(A) = R_J$ if and only if

1. x is in R ; and
2. if y is a subword of x such that $h(y) = e$, then $|y| \leq r-1$; and
3. T uses exactly the states of J when reading x .

For each q in K_1, let $S(q) = \{x \text{ in } E^* \mid \delta(q,x)=q\}$ and for each $K' \subseteq K_1$ define $S_{K'} = (\cup\{S(q) \mid q \text{ in } K'\})^*$. Let $R'_J = \mathrm{shuf}(R_J, S_J)$ for each J in $\mathcal{J}$. Clearly R'_J is in $\mathcal{R}$. Now $h(g^{-1}(L) \cap R) = h(g^{-1}(L) \cap (\cup\{R'_J \mid J \text{ in } \mathcal{J}\})) =$ $= h(g^{-1}(L) \cap (\cup\{R'_J \mid J \text{ in } \mathcal{J}\}) \cap R_1)$ where $R_1 = \{e, E, \ldots, E^k\}(W^+\{e, E, \ldots, E^k\})^*$, $k = \max\{r-1, k_1\}$. Clearly h is e-limited on R_1, so (by a result in [1]) $h_1(L_1)-\{e\} = h(g^{-1}(L) \cap R)-\{e\}$ is in $\mathcal{L}$. □

The following theorem is an immidiate consequence from Lemma 1 .

THEOREM 1. Let $\mathcal{L}$ be a commutative trio closed under union with {e} . Then $\mathcal{L}$ is closed under linear erasing.

It should be clear that $\mathcal{M}(L)$ is closed under linear erasing for each L in $\mathcal{COM}$ such that e is in L .

Since $\mathcal{F}^{lin}(\mathcal{L}) = \hat{Sub}(\mathcal{R}, \mathcal{M}^{lin}(\mathcal{L}))$, we have

THEOREM 2. Assume $\mathcal{L} \subseteq \mathcal{COM}$ contains a language with e in it. Then $\mathcal{F}(\mathcal{L})$ is closed under linear erasing.

Consider the family $M_\cap(\mathcal{L})$, $\mathcal{L} \subseteq COM$, of languages. Each L in $M_\cap(\mathcal{L})$ can be presented in the form $L = h(g_1^{-1}(L_1) \cap \ldots \cap g_n^{-1}(L_n) \cap R)$, $n \geq 1$, where L_1, $L_2, \ldots, L_n$ are in $\mathcal{L}$, $g_1, \ldots, g_n$ and h are homomorphisms, h e-free and R in $\mathcal{R}$, by a result in [1] . On the other hand, $g_1^{-1}(L_1) \cap \ldots \cap g_n^{-1}(L_n)$ is a commutative language [5] . By the facts above, it is not difficult to see that $h_1(L)-\{e\}$ is in $M_\cap(\mathcal{L})$ for each L in $M_\cap(\mathcal{L})$ and each homomorphism h_1 linearly bounded on L . Thus we have

THEOREM 3. Assume $\mathcal{L} \subseteq COM$ is closed under union with {e}. Then $M_\cap(\mathcal{L})$ is closed under linear erasing.

A natural question arises if $F_\cap(\mathcal{L})$ is closed under linear erasing for each family $\mathcal{L} \subseteq COM$ of languages such that $\mathcal{L}$ contains a language with e in it. The answer is no. Turakainen has shown that $F_\cap(\{a^n b^n | n \text{ in } N\})$ is not closed under linear erasing [9] . On the other hand it can be shown that $F_\cap(\{a^n b^n | n \text{ in } N\}) = F_\cap(c(\{a^n b^n | n \text{ in } N\}))$.

Since $M_\cap(\mathcal{B}_k) = M_\cap(COM_k)$, $k \geq 1$, $M_\cap(\mathcal{B}_k)$ is closed under linear erasing by Theorem 3 .

It should be noted that not all trios generated by bounded languages and closed under union with {e} are closed under linear erasing.

III ONE-BOUNDED LANGUAGES AND INTERSECTION

In [4] Goldstine studies the language families $F_{\cap}(B_k)$ (k in N_+). Clearly $F_{\cap}(B_1) \subseteq F_{\cap}(B_2) \subseteq \ldots \subseteq F_{\cap}(B_k) \subseteq \ldots \subseteq F_{\cap}(B)$. Goldstine asks whether DUP = $= \{a^n b^n \mid n \text{ in } N\}$ is in $F_{\cap}(B_1)$ or not. Next theorem shows that the answer is yes.

<u>THEOREM 1</u>. The language $DUP = \{a^n b^n \mid n \text{ in } N\}$ is in $M_{\cap}(B_1)$.

<u>PROOF</u>. The languages $T_1 = \{a^{2^n} \mid n \text{ in } N\}$, $T_2 = \{b^{2^n} \mid n \text{ in } N\}$ and $T_3 = \{c^{2^n} \mid n \text{ in } N\}$ are clearly in $M_{\cap}(B_1)$. Let h_1 be the homomorphism on $\{a,b\}^*$ defined as $h_1(a) = h_1(b) = c$. Then the language $L_1 = T_1T_2 \cap h_1^{-1}(T_3)$ is in $M_{\cap}(B_1)$ since every trio closed under intersection is closed under concatenation. Now

$L_1 = T_1T_2 \cap h_1^{-1}(T_3) = \{a^{2^n} b^{2^m} \mid n,m \text{ in } N,\ 2^n + 2^m = 2^r \text{ for some } r \text{ in } N\}$.

It should be clear that $2^n + 2^m = 2^r$ for some n,m and r in N if and only if $n = m = r-1$, and thus $L_1 = \{a^{2^n} b^{2^n} \mid n \text{ in } N\}$. Let $L_2 = \{b^{2^n} c^{2^n} \mid n \text{ in } N\}$ and h_2 the homomorphism on $\{a,b,c\}^*$ defined as $h_2(a) = a$, $h_2(b) = b$ and $h_2(c) = c^2$. Then the language $L_3 = h_2^{-1}(L_1c^* \cap a^*L_2)$ is in $M_{\cap}(B_1)$ and $L_3 = \{a^{2^n} b^{2^n} c^{2^{n-1}} \mid n \text{ in } N_+\}$. Let a_1, b_1 and c_1 be new symbols and g_1, g_2 and g_3 homomorphisms on $\{a,b,c,a_1,b_1,c_1\}^*$ defined as follows: $g_1(a)=a$, $g_1(a_1)= =b$, $g_2(b)=a$, $g_2(b_1)=b$, $g_3(c)=a$, $g_3(c_1)=b$, $g_1(b)=g_1(b_1)=g_1(c)=g_1(c_1)= =g_2(a)=g_2(a_1)=g_2(c)=g_2(c_1)=g_3(a)=g_3(a_1)=g_3(b)=g_3(b_1)=e$. Let $L_4 = L_3\{a_1, b_1, c_1\}^* \cap g_1^{-1}(L_1) \cap g_2^{-1}(L_1) \cap g_3^{-1}(L_1)$ and g the homomorphism on $\{a,b,c\}^*$ defined as $g(a)=a_1$, $g(b)=b_1$, $g(c)=c_1$. Then it is easily seen that $L_4 = \{xx' \mid x \text{ in } L_3,\ x' \text{ in } c(g(x))\}$. Clearly L_4 is in $M_{\cap}(B_1)$. Let h_3 be the homomorphism on $\{a,b,c,a_1,b_1,c_1\}^*$ defined as $h_3(a_1)=a$, $h_3(b_1)=b$, $h_3(c_1)=c$, $h_3(a)=h_3(b)=h_3(c)=e$. Now h_3 is linearly bounded on L_4, so by Theorem 3 of the previous section, the language $L_5 = h_3(L_4)$ is in the family $M_{\cap}(B_1)$. Let h_4 be the homomorphism on $\{a,b,a_1,b_1,c_1\}^*$ defined as $h_4(a)=h_4(a_1)=a$, $h_4(b)=h_4(b_1)=b$, $h_4(c_1)=c$. Then the language

$$L_6 = h_4^{-1}(L_5) \cap a^*b^*(a_1b_1c_1)^*c_1^* = \{a^{2^n-j}b^{2^n-j}(a_1b_1c_1)^j c_1^{2^{n-1}-j} \mid n \text{ in } N_+,\ 0 \leq j \leq 2^{n-1}\}$$

is in $M_{\cap}(B_1)$. Let h be the following homomorphism on $\{a,b,a_1,b_1,c_1\}^*$: $h(a)=a$, $h(b)=b$, $h(a_1)=h(b_1)=h(c_1)=e$. Then the language $L_7 = h(L_6)$ is in $M_{\cap}(B_1)$ since h is linearly bounded on L_6 and $M_{\cap}(B_1)$ is closed under linear erasing. Obviously $DUP = L_7 \cup \{e\}$ and thus DUP is in $M_{\cap}(B_1)$ since $M_{\cap}(B_1)$ is closed under union with $\{e\}$. □

In the next theorem we shall see that the result above can be generalized.

THEOREM 2. For each integer $k\geq 2$, the language $P_k = \{a^n b^{n^k} \mid n \text{ in } N\}$ is in $M_\cap(B_1)$.

PROOF. First note, that the languages $T_1 = \{a^{2^n} \mid n \text{ in } N\}$, $T_2 = \{b^{2^{nk}} \mid n \text{ in } N\}$ and $T_3 = \{c^{2^n+2^{nk}} \mid n \text{ in } N\}$ are in $M_\cap(B_1)$. Let h_1 be the homomorphism on $\{a,b\}$ defined as $h_1(a)=h_1(b)=c$. Then, by the previous theorem, the language

$$L_1 = T_1 T_2 \cap h_1^{-1}(T_3) \cap \mathrm{DUPb}^* = \{a^{2^n} b^{2^{mk}} \mid n,m \text{ in } N,\ 2^n \leq 2^{mk},\ 2^n+2^{mk}=2^r+2^{rk} \text{ for some } r \text{ in } N\}$$

is in $M_\cap(B_1)$. Now the equation $2^n+2^{mk} = 2^r+2^{rk}$ where n,m and r are in N and $2^n \leq 2^{mk}$ is true if and only if $n=m=r$ and thus the language L_1 is in $M_\cap(B_1)$.

The next step in the proof is to show that the language $L_2 = \{a^{2^n} b^{(2^n+j)^k} \mid n,j \text{ in } N,\ 0\leq j<2^n$ is in $M_\cap(B_1)$. Let $b_1,\dots,b_{2^k}$ be ne symbols and h_2 and h_3 two homomorphisms on $\{a,b_1,\dots,b_{2^k}\}^*$ defined as follows: $h_2(a)=a$, $h_2(b_i)=b$, $h_3(a)=a$, $h_3(b_i) = b^i$, $i=1,\dots,2^k$. Then the language

$$L_3 = h_3(h_2^{-1}(L_1) \cap \{b_1,\dots,b_{2^k}\}^*\{b_1,\dots,b_{2^k-1}\}\{b_1,\dots,b_{2^k}\}^*)$$
$$= \{a^{2^n} b^h \mid n \text{ in } N,\ 2^{nk} \leq h < 2^{nk+k}\}$$

is in $M_\cap(B_1)$. Clearly the language $T_4 = \{c^{2^n+(2^n+j)^k} \mid n,j \text{ in } N,\ 0\leq j<2^n\}$ is in $M_\cap(B_1)$. Then the language

$$L_2' = L_3 \cap h_1^{-1}(T_4) = \{a^{2^n} b^h \mid n,h \text{ in } N,\ 2^{nk} \leq h < 2^{nk+k},\ 2^n+h=2^m+(2^m+j)^k \text{ for some } m,j \text{ in } N,\ 0\leq j<2^m\}$$

is in $M_\cap(B_1)$. It is easily seen that the equation $2^n+h = 2^m+(2^m+j)^k$ with the conditions n,m,h,j in N, $2^{nk} \leq h < 2^{nk+k}$, $0\leq j<2^m$ is true if and only if $n=m$ and $h=(2^m+j)^k$. Then $L_2=L_2'$ is in $M_\cap(B_1)$.

Now we shall see that the language $L_4 = \{a^{2^n+j} b^{(2^n+j)^k} \mid n,j \text{ in } N, 0\leq j<2^n\}$ is in $M_\cap(B_1)$. Since L_2 is in $M_\cap(B_1)$, it is obvious that the language $L_5 = \{a^{2^n+h} b^{(2^n+j)^k} \mid n,h,j \text{ in } N,\ 0\leq h,j<2^n\}$ is in $M_\cap(B_1)$. Also the language $T_5 = \{c^{2^n+j+(2^n+j)^k} \mid n,j \text{ in } N,\ 0\leq j<2^n\}$ is in $M_\cap(B_1)$. Then the language

$$L_4' = L_5 \cap h_1^{-1}(T_5) = \{a^{2^n+h} b^{(2^n+j)^k} \mid n,h,j \text{ in } N,\ 0\leq h,j<2^n,\ 2^n+h+(2^n+j)^k = 2^m+r+(2^m+r)^k \text{ for some } m \text{ in } N,\ 0\leq r<2^m\}$$

is in $M_\cap(B_1)$. Now the equation $2^n+h+(2^n+j)^k = 2^m+r+(2^m+r)^k$ for some n,h,j,m,r in N , $0\leq h,j<2^n$, $0\leq r<2^m$ is true if and only if $n=m$ and $h=j=r$. Thus the language $L_4=L_4'$ is in $M_\cap(B_1)$.

Since $P_k = L_4 \cup \{e\}$ and $M_\cap(B_1)$ is closed under union with $\{e\}$, the language P_k is in $M_\cap(B_1)$ for each $k\geq 2$. This completes the proof. □

We have the following corollary:

COROLLARY 1. Let r and s be in N_+ and $P(n) = l_0 n^r + \ldots + l_{r-1} n + l_r$ and $Q(n) = t_0 n^s + \ldots + t_{s-1} n + t_s$ two polynomials of n such that l_i, t_j are in N, $i=0,\ldots,r$, $j=0,\ldots,s$. Then the language $L = \{a^{P(n)} b^{Q(n)} \mid n \text{ in } N\}$ is in $M_\cap(B_1)$.

PROOF. Let a, b, c_i, d_j, $i=0,\ldots,r$, $j=0,\ldots,s$, be different symbols. The languages

$$V_i = \{c_i^n a^{l_i n^{r-i}} \mid n \text{ in } N\}, \quad i=0,\ldots,r, \qquad S_j = \{d_j^n b^{t_j n^{s-j}} \mid n \text{ in } N\}, \quad j=0,\ldots,s$$

are in $M_\cap(B_1)$ (by the previous theorem). Let $\Sigma_1 = \{a,b,c_i,d_j \mid i=0,\ldots,r, j=o,\ldots,s\}$ and h_i and g_j homomorphisms on Σ_1 defined as $h_i(c_i) = c_i$, $h_i(a)=a$, $h_i(x)=e$ for each x in $\Sigma_1 - \{a,c_i\}$, $i=0,\ldots,r$, $g_j(d_j) = d_j$, $g_j(b)=b$, $g_j(y)=e$ for each y in $\Sigma_1 - \{b,d_j\}$, $j=0,\ldots,s$. Then the language

$$L_1 = h_0^{-1}(V_0) \cap \ldots \cap h_r^{-1}(V_r) \cap g_0^{-1}(S_0) \cap \ldots \cap g_s^{-1}(S_s) \cap (c_0 \ldots c_r d_0 \ldots d_s)^* a^* b^* =$$

$$= \{(c_0 \ldots c_r d_0 \ldots d_s)^n a^{l_0 n^r + \ldots + l_{r-1} n + l_r} b^{t_0 n^s + \ldots + t_{s-1} n + t_s} \mid n \text{ in } N\}$$

is in $M_\cap(B_1)$. Let h be the homomorphism on Σ_1^* defined as follows: $h(a) = a$, $h(b) = b$, $h(x) = e$ for each x in $\Sigma_1 - \{a,b\}$. Then $L = h(L_1)$ is in $M_\cap(B_1)$ since h is linearly bounded on L_1 and $M_\cap(B_1)$ is closed under linear erasing. □

It has been proved by Turakainen [9] that $\hat{M}_\cap(B_1 \cup \{\{a^n b^{n^2} \mid n \text{ in } N\}\}) = \hat{M}_\cap(B_2) = \hat{M}_\cap(B)$. On the other hand, Latteux [5] shows that $M(COM) = \hat{M}(COM) = M_\cap(COM) = \hat{M}_\cap(COM)$. Since $M_\cap(COM) = M_\cap(B)$, we have

COROLLARY 2. $\hat{M}_\cap(B_1) = \hat{M}_\cap(B_2) = \ldots = \hat{M}_\cap(B_k) = \ldots = \hat{M}_\cap(B) = M_\cap(B) = \hat{M}_\cap(COM_1) = \hat{M}_\cap(COM_2) = \ldots = \hat{M}_\cap(COM_k) = \ldots = \hat{M}_\cap(COM) = M_\cap(COM) = \hat{M}(COM) = M(COM)$.

REFERENCES

1. S.Ginsburg, Algebraic and Automata-Theoretic Properties of Formal Languages, North-Holland Publishing Company, 1975.
2. S.Ginsburg and J.Goldstine, Intersection-Closed Full AFL and the Recursively Enumerable Languages,Inf. Control 22 (1973), 201-231.
3. J.Goldstine, Substitution and Bounded Languages, J.Comput.Syst.Sci. 6, (1972), 9-29.
4. J.Goldstine, Bounded AFLs, J.Comput.Syst.Sci. 12 (1976), 399-419.
5. M.Latteux, Cônes rationnels commutativement clos, RAIRO,Inf.Theor. 11 (1977), 29-51.
6. M.Latteux, Langages commutatifs, Thesè Sciences Mathematiques, Universitè Lille, 1978.
7. M.Latteux, Cônes rationnels commutatifs, J.Comput.Syst.Sci. 18 (1979), 307-333.
8. B.Rovan, Proving Containment of Bounded Languages, J.Comput.Syst.Sci. 11 (1975), 1-55.
9. P.Turakainen, On Some Bounded SemiAFLs and AFLs (to appear in Inf. Sci.).

ISOMORPHISM COMPLETENESS FOR SOME ALGEBRAIC STRUCTURES

L.Kučera and V.Trnková
Charles University
Prague, Czechoslovakia

The complexity of the problem of finding isomorphism of two graphs is one of the most famous problems in the complexity theory [9]. It is known that it belongs to the class NP; in some special cases, a polynomial algorithm is knlwn (planar graphs [4] , graphs of bounded valency [5]), but the computational complexity of the general problem is not known.

Nevertheless, a lot of problems has already been proved to be equivalent to the graph isomorphism problem (in the same sense as in the case of equivalence of any NP-complete problem to the problem of satisfiability of logical formulas). Problems equivalent to the graph isomorphism problem are usually called isomorphism complete. A survey of known results with outlines of proofs is presented in an expository article [1] .

The present paper is an contribution to the isomorphism completeness problem. We discuss the isomorphism problem in varieties of algebras with two unary idempotent operations and the isomorphism problem for group actions and semigroup actions (parts I and II). In the last part III, a connection with full embeddings of categories and some consequences of it are investigated.

I. Unary algebras

Unary algebras are of special interest because they are, in essence, sequential automata (Medvedeff sequential automata).

The isomorphism problem for unary algebras is isomorphism complete, even the isomorphism problem for unary algebras with two operation only is isomorphism complete, see [2]. Here, we investigate the variety I of unary algebras with two unary idempotent operations. We show that I is also isomorphism complete. Moreover, we present

a full discussion of all subvarieties of I; we show that for any subvariety V of I, either there exists a polynomial-time bounded algorithm for the isomorphism problem in V or the isomophism problem in V is isomorphism complete.

Let us describe the above results more in detail. Let I be the variety of all unary algebras with two operations, say (X,f,g), where $f:X \to X$ and $g:X \to X$ are the unary operations, fulfilling

$$(x)f^2 = (x)f \ , \ (x)g^2 = (x)g \ . \qquad (1)$$

Derived operations can be expressed as words in f and g. By the equations (1), we can investigate only words of the form

$$f,g,fg,gf,fgf,gfg,fgfg,gfgf,\ldots \qquad (2)$$

The number of symbols f,g in a word w will be denoted by $L(w)$ and called the length of w.

A subvariety V of I is given by the system (S) of equations

$$\begin{array}{ll} & (x)f^2 = (x)f \ , \ (x)g^2 = (x)g \\ (I). & (x)w_i = (x)w_i' , \quad i \in I \\ (II). & (x)v_j = (x)v_j' \ , \ j \in J \ . \end{array} \qquad (S)$$

Every equation of the system (II) determines a definable constant. If V has no definable constant, i.e. if the system (II) is missing, then V is said to be of the first kind.

Let us call the system S <u>reduced</u> if every word w_i, w_i', v_j, v_j' is of the form (2) and if, for every i, the word w_i is distinct from the word w_i'. Clearly, every subvariety V of I can be described by a reduced system of equations.

<u>Theorem</u> 1. Let a subvariety V of I be given by a reduced system of equations (S).

If all the words w_i, w_i' $(i \in I)$, v_j, v_j' $(j \in J)$ have the length at least 2 or the system (II) is missing and, for every $i \in I$, the word w_i begins with the same letter as the word w_i'

then the isomorphism problem in V is isomorphism complete

else there exists a polynomial-time bounded algorithm for the isomorphism problem in V.

We present the proof of the theorem. It depends on the following four lemmas.

<u>Lemma</u> 1. The isomorphism problem in the variety $I_1 \subset I$ given by the equations

$$(x)f^2 = (x)f \ , \ (x)g^2 = (x)g \ , \ (x)fg = (x)gf$$

is isomorphism complete.

Proof: by a polynomial procedure, we replace an undirected graph $G=(V,E)$ by an algebra $a(G)$ as follows:

$a(G) = (\{o\} \cup V \cup E \times \{0,1,2\}, f, g)$,
where f and g are defined as follows
$(o)f = (o)g = o$, $(v)f = o$ for all $v \in V$, $(v)g = v$ for all $v \in V$,
if $e = \{v_0, v_1\} \in E$ then
$(e,0)g = v_0$, $(e,0)f = (e,2)f = (e,2) = (e,1)f$, $(e,1)g = v_1$, $(e,2)g = o$.

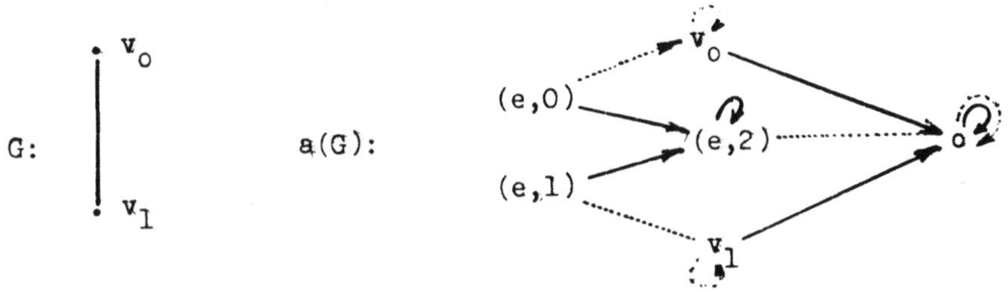

Then G is isomorphic to G´ iff a(G) is isomorphic to a(G´).

Lemma 2. The isomorphism problem in the variety I_2 I, given by the equations

$$(x)f^2 = (x)f, \quad (x)g^2 = (x)g, \quad (x)f = (x)fg, \quad (x)g = (x)gf$$

is isomorphism complete.

Proof: By a polynomial procedure, we replace a directed graph $G = (V, R)$ by an algebra a(G) as follows:

$a(G) = (V \cup R, f, g)$, where
$(v)f = (v)g = v$ for all $v \in V$,
if $r = (v_0, v_1)$ then $(r)f = v_0$, $(r)g = v_1$.

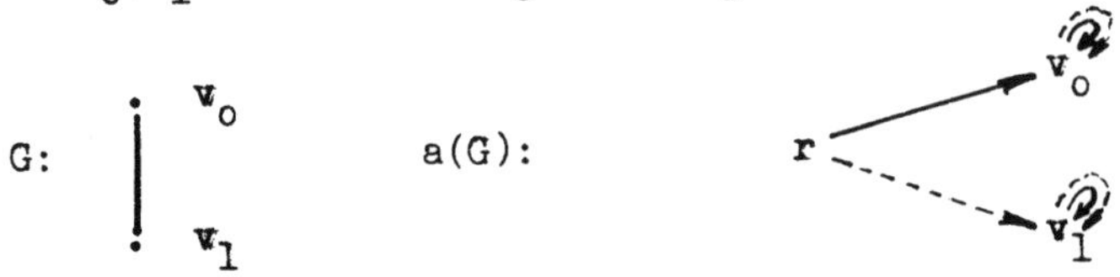

Then G is isomorphic to G´ iff a(G) is isomorphic to a(G´).

Lemma 3. There exists a polynomial-time bounded algorithm for the isomorphism problem in the class A(1) of all algebras with one unary operation and in the class A(1,0) of all algebras with one unary and one nullary operation.

Proof: An algebra (X,f) with one unary operation f can be viewed as a directed graph with the vertex set X and directed edges $(x,(x)f)$, $x \in X$. Components of such graphs are, roughly speaking, rooted trees (oriented toward the root), in which the root has been replaced by a cycle, see the picture on the next page.

A straigthforward generalization of the well known linear-time algorithm for the isomorphism of rooted trees gives the desired algorithm for the isomorphism problem in A(1). Using the more general algorithm for the isomorphism of rooted trees with weighted vertices, we obtain a solution of the problem in the class A(1,0).

Lemma 4. There exists a polynomial-time bounded algorithm for the isomorphism problem in the variety I(f), given by the equations

$$(x)f^2=(x)f\ ,\ (x)g^2=(x)g\ ,\ (x)f=(x)gf$$

and analogously in the variety I(g), given by the equations

$$(x)f^2=(x)f\ ,\ (x)g^2=(x)g\ ,\ (x)g=(x)fg\ .$$

Proof: If (X,f,g) is an algebra of the variety I(f) then the set $f^{-1}(x)$ is stable under g for every x in X.
Thus, we can solve the isomorphism problem in I(f) in the polynomial time using the algorithms searching an isomorphism with respected to f (see the proof of the lemma 3) combined with an intermediate use of the same algorithm searching an isomorphism of corresponding $f^{-1}(x)$'s with respect to the operation g.

Proof of the theorem: Let a subvariety V of I be given by a reduced system of equations (S).

a) If $L(w_i)\geq 2$, $L(w_i')\geq 2$, $L(v_j)\geq 2$, $L(v_j')\geq 2$ for all $i\in I$ and $j\in J$ then V contains I_1. For, any equation $(x)w_i=(x)w_i'$ and any equation $(x)v_j=(x)v_j'$ is fulfilled in I_1. Since $I_1\subset V$, V is isomorphism complete by the lemma 1.

b) Let the system (II) be missing in (S) and, for every $i\in I$, the word w_i begins with the same letter as the word w_i'. Then V contains I_2. For, any equation $(x)w_i=(x)w_i'$ is fulfilled in I_2. Since $I_2\subset V$, V is isomorphism complete by lemma 2.

c) Let us discuss the remaining cases: if neither a) nor b) is valid then necessarily one of the following cases occur:

1) there exists $j\in J$ such that either $L(v_j)=1$ or $L(v_j')=1$,
2) there exists $i\in I$ such that either $L(w_i)=1$ or $L(w_i')=1$ and the system (II) is not missing,
3) the system (II) is missing and there exists $i,i_1\in I$ such that the

words w_i and w_i' begins with distinct letters and either $L(w_{i_1})=1$ or $L(w_{i_1}')=1$.

We show that in all cases there exists a polynomial algorithm for the isomorphism problem in V. In 1), let us suppose that $v_j'=f$. Then $(x)f=(y)v_j=(y)f$ is fulfilled in V, hence the unary operation f can be replaced by a nullary operation and we can use the lemma 3. In 2), let us suppose that $w_i'=f$. If $w_i=g$ then both the operations coincide in V, and we use the lemma 3 again. If $w_i \neq g$ then necessarily $l(w_i) \geq 2$. Since (II) is not missing in (S), we have the following equations in (S):

$$(x)w_i = (x)f \quad , \quad (x)v_j = (y)v_j' \quad .$$

We can suppose that the words v_j and v_j' begins with distinct letters, otherwise we can consider the equations

$$(x)fv_j=(y)gv_j' \ ,$$

which is also fulfilled in V. Thus, we may suppose that one of the words v_j, v_j' begins with the same letter as w_i. Since $L(w_i) \geq 2$, we can find a natural number $k > 1$ such that the word $(w_i)^k$, rewritten in the word w of the form (2) fulfils

$$L(w) \geq L(v_j) \ .$$

Then there exists a word v such that $v_j v=w$. Thus we have

$$(x)f=(x)f^k=(x)w_i^k=(x)w=(x)v_j v=(y)v_j v=(y)f \ ,$$

hence the unary operation f can be replaced by the nullary operation and we can use the lemma 3 again. Finally, we discuss the case 3). Let us suppose that $w_{i_1}'=f$ and $w_i=ft$, $w_i'=gt'$ for some words t,t'. First, we show that an equation

$$(x)f = (x)gs$$

is fulfilled in V for some word s. If the word w_{i_1} begins with g then $w_{i_1}=gs$. Let us suppose that w_{i_1} begins with f. Let $k \geq 1$ be a natural number such that $(w_{i_1})^k$, rewritten in the word of the form (2) fulfils

$$L(w) \geq L(ft)$$

Then there exists a word v such that $ftv = w$. Then it is

$$(x)f=(x)f^k=(x)(w_{i_1})^k=(x)w=(x)ftv=(x)gt'v=(x)gs$$

for $s=t'v$. If $L(s) \geq 1$ then V is contained in thevariety I(f). For, $(x)f=(x)gs$ implies $(x)gf=(x)g^2s=(x)gs=(x)f$.

Then there exists a polynomial algorithm for the isomorphism problem in V by the lemma 4. If $L(s)=0$ then the operations f and g coincide in V and we use the lemma 3.

II. Group actions and semigroup actions.

Let us recall that a group acts on a set X if for every x in X and every element g of G a result $x.g \in X$ is given such that

$$(x.g).g' = x.(gg') \quad \text{for all } x \in X, \; g,g' \in G, \qquad (3)$$

i.e. a group action is a triple $\mathfrak{G} = (X,G,.)$, where X is a set (of states), G is a group (of actions) and $. : X \times G \to X$ is a map such that (3) is fulfilled.

Group actions are closely related to group automata. The cardinality of $\mathfrak{G} = (X,G,.)$ is the cardinality pf its state set X.

Denote by $\mathcal{G}$ the class of all group actions. Isomorphisms in $\mathcal{G}$ are as follows: a group action $\mathfrak{G} = (X,G,.)$ is isomorphie to $\mathfrak{G}' = (X',G',.')$ iff there exists a bijection f of X onto X' and an isomorphism h of G onto G' such that

$$(f(x.g) = f(x).h(g) \quad \text{for all } x \in X, \; g \in G.$$

Denote by $\mathcal{G}_k$ the class of all group actions $\mathfrak{G} = (X,G,.)$ with card $G \leq k$.

While the isomorphism problem in $\mathcal{G}$ is at least as difficult as the graph isomorphism problem, the following theorem is fulfilled:

Theorem 2. For every natural number k there exists a polynomial algorithm for the isomorphism problem in $\mathcal{G}_k$.

Proof: Without loss of generality we can restrict ourselves to the problem of connected group actions, i.e. actions $\mathfrak{G} = (X,G,.)$ such that for every $x,y \in X$ there exists $f \in G$ such that $f(x)=y$. Now, if k is fixed then the set of all connected group actions $(X,G,.)$ such that card $G \leq k$ is finite, since it must be card $X \leq$ card G.

Let us investigate a more general situation, where the graoup G is replaced by a semigroup. A semigroup action is a triple $\mathfrak{S} = (X,S,.)$, hwere X is a set, S is a semigroup and $. : X \times S \to X$ is a map such that

$$(x.s).s' = x.(ss') \quad \text{for all } x \in X, \; s,s' \in S.$$

The cardinality and the isomorphism of semigroup actions is defined in a similar way as in the case of group actions. Let us denote by $\mathcal{S}_k$ the class of all semigroup actions $\mathfrak{S} = (X,S,.)$ with card $S \leq k$. In a contrast to the above theorem, we have

Theorem 3. If it is k=1

then there exists a polynomial-time bounded algorithm for the isomorphism problem in $\mathcal{S}_k$

else the isomorphism problem in $\mathcal{S}_k$ is isomorphism complete.

Proof: If k=1 then $\mathcal{S}_k$ is contained in the class A(1) of unary algebras with one operations and hence a polynomial algorithm for the isomorphism exists in view of the lemma 3. If k is greater or equal to 2 then use the lemma 2.

III. Embeddings of categories.

Let us recall that a functor F from a category K into a category L is called a full embedding if, for any two objects a,b of the category K, F maps the set K(a,b) of all morphisms from a into b in K bijectively onto the set L(F(a),F(b)) af all morphisms from F(a) to F(b) in L.

A lot of constructions, used for tho polynomial reducibility are in fact full embeddings of suitable categories. Some of them were described in the literature in a language of full embeddings of categories before their rediscovery in the computer science.

Let us denote by Gra the category of all graphs and all their isomorphisms, by DiGra the category of all directed graphs and all their isomorphisms and by A(Δ) the category of all universal algebras of the type and all their isomorphisms. In [2] , [3] , [7] , full embeddings

Gra $\rightarrow$ A(Δ) whenever $\sum \Delta \geqq 2$,

A(Δ) $\rightarrow$ DiGra ,

Digra $\rightarrow$ Gra

are given by a polynomial procedure (even more complicated problems conserning homomorphisms are discussed in [2] , [3] , [7]). The immediate consequence: the isomorphism problem is of the same complexity for the structures Gra,DiGra and A(Δ) and the problem of the computing of the order of automorphism group has the same complexity in all these structures. A lot of other structures were constructed in many papers, most of the results about the problem of full embeddings of categories are summarised in the monography [8] .

Finally, let us mention that for any subvariety of the variety of algebras with 2 idempotent unary operations the problem of computing the order of automorphism group behaves in the same way as the isomorphism problem for V, i.e. it is either equivalent to the problem of computing the order of automorphism group of a graph or it is solvable in a polynomially bounded time, see theorem 1. For, it is sufficient to check that the constructions in the proofs of lemmas 1,2 gives full embeddings and to realize that the number of isomorphism of a rooted tree (ar an algebra with one unary operation) cam be computed in a polynomial time.

References.

1. Booth, K.S. and Colbourn, C.J., Problems polynomially equivalent to graph isomorphism, Tech.Rep. CS-77-04, Univ.of Waterloo, 1979.

2. Hedrlín, Z. and Pultr, A., On full embeddings of categories of algebras, Illinois J.Math., 10 (1966), 392-406.

3. Hedrlín,Z. and Pultr, A., O predstavlenii malych kategorij, Dokl.AN SSSR 160 (1965), 284-286.

4. Hopcroft, J.E and Wong., J.K., Linear time algorithm for isomorphism of planar graphs, Proc.6th Ann.ACM Symp. on Theory of Computing, 1974, 172-184.

5. Luks.,E.M., Isomorphism of bounded valency graphs can be tested in a polynomial time, Proc.21st Ann.Symp. on Foundations of Comp.Science, 1980, 42-49.

6. Miller, G.L., Graph isomorphism, general remarks, Proc. 9th Ann.ACM Symp.on Theory of Computing, 1977, 143-150.

7. Pultr,A., On full embeddings of conctete categories with respect to forgetful functors, Comment.Math.Univ.Carolinae, 9 (1968), 281-305.

8. Pultr,A. and Trnková, V., Combinatorial, Algebraic, adn Topological representation of Groups, Semigroups and Categories, Academia, Prague, 1980.

9. Read, R.C. and Corneil, D.G., The graph isomorphism disease, J.of Graph Theory 1(1977), 336-363.

REDUCING ALGEBRAIC TREE GRAMMARS

B. LEGUY

Equipe lilloise d'Informatique théorique et de
Programmation
Université de Lille I - FRANCE

I - INTRODUCTION

The usefulness of trees in computer science is well known now, and most of its recent aspects have been widely explained by Engelfriet [6].This paper is concerned with algebraic (i.e. context-free) tree grammars which are often viewed as non deterministic program schemes. The deterministic case has been studied by Nivat and the problems of reductions of the deterministic program schemes were solved by Guessarian [7].

In the non deterministic case, some reduction problems were studied, namely the $\emptyset$-reduction (i.e. carrying off symbols that generate empty sets) by Arnold, Boudol, Courcelle and Dauchet, and also the problem of useless arities by Boudol.

In this paper, tree grammars are viewed as a generalization of context free grammars defined in language theory. The main results are, first, a systematic investigation of the problem of reductions, secondly, we prove that erasing rules which generalize ε-rules in language theory, cannot always be carried off, and finally, we exhibit a new hierarchy in the class of algebraic forests. Definitions and results only are given. Proofs are to be found in [8].

We need an algebraic framework for studying precisely trees. The theory of the magmoïds developped by Arnold and Dauchet is a powerfull tool and part of its notations used in the paper, are defined in the section II.

The algebraic grammars and the generated forests are defined in part III. The different sorts of erasing rules are also defined here.

Part IV is concerned with definitions and results about reduced grammars.

Part V contains the counterexamples used in proving that erasing rules cannot always be avoided. The part V contains also the definition of the hierarchy.

II - PRELIMINAIRES

A *ranked alphabet* is a pair (Σ, d) where Σ is a finite set of symbols and d a function of Σ into $\mathbb{N}$. For any α in Σ we call $d(\alpha)$ the *degree* or *arity* of α. For any integer n we denote by Σ_n the set $d^{-1}(n)$. T_Σ is the *set of trees over* Σ. It denotes the smallest set such that $T_\Sigma = \Sigma_0 \cup \{\sigma(t_1,\ldots, t_n) \mid \sigma \in \Sigma_n$ and $t_1,\ldots, t_n \in T_\Sigma\}$. Let Σ be defined by $\Sigma_0 = \{\#\}$, $\Sigma_1 = \{a, \bar{a}\}$, $\Sigma_2 = \{b\}$, $\Sigma_4 = \beta$, then the tree $\bar{t} = b(\beta(a(\#), \#, \bar{a}(\#), b(\#, \#)), b(b(\#, \#), a(\#)))$ is in T_Σ.

Generally it is usefull to join a denumerable set of *variables* $\chi = \{x_1,\ldots, x_n,\ldots\}$ to Σ. The degree of such a variable is zero. For any integer n, the set $\{x_1,\ldots, x_n,\ldots\}$ is denoted by χ_n and $T(\Sigma)^1_n$ is the *set of trees over* Σ *indexed by* χ_n. It may be defined as $T_{\Sigma\cup\chi_n}$. So, the tree $\tau = b(\beta(x_1, \#, x_2, x_3),$ $b(x_3, x_1))$ is in $T(\Sigma)^1_3$ and $\tau_1 = a(\#)$, $\tau_2 = \bar{a}(\#)$ and $\tau_3 = b(\#, \#)$ are trees in $T(\Sigma)^1_0$. We say the vector $[\tau_1, \tau_2, \tau_3]$ is in $T(\Sigma)^3_0$ and more generally if $u_1 \in T(\Sigma)^1_p,\ldots, u_n \in T(\Sigma)^1_p$ then we say $[u_1, u_2,\ldots, u_n]$ is in $T(\Sigma)^n_p$. When we say that $\vec{u} \in T(\Sigma)^n_p$ we mean $\vec{u}$ is composed of n trees and any variable of χ_p may be in $\vec{u}$. So $x_2 \in T(\Sigma)^1_3$. It would be better to denote any vector $[t_1,\ldots, t_p]$ in $T(\Sigma)^p_q$ by the pair $\langle q ; [t_1,\ldots, t_p]\rangle$ because two vectors that are defined over different sets of variables are considered as different vectors by the two Magmoïd-operations. In this paper we will use only $[t_1,\ldots, t_p]$ because it is shorter.

The set $T(\Sigma) = \bigcup_{n,p\in\mathbb{N}} T(\Sigma)^n_p$ is called a *Magmoïd* when the two following operations are given. The *composition-product* of $\vec{u} = [u_1,\ldots, u_n] \in T(\Sigma)^n_p$ by $\vec{v} = [v_1,\ldots, v_p] \in T(\Sigma)^p_q$ is denoted by $\vec{u} \bullet \vec{v}$ and defined by

- $\vec{u} \bullet \vec{v} = [u_1 \bullet \vec{v}, u_2 \bullet \vec{v},\ldots, u_n \bullet \vec{v}]$, where $u_i \bullet \vec{v}$ is defined by :
- if $u_i = x_j \in \chi_p$ then $u_i \bullet \vec{v} = v_j$,
- if $u_i = \alpha \in \Sigma_0$ then $u_i \bullet \vec{v} = \alpha$,
- if $u_i = \sigma(t_1,\ldots, t_r)$ where σ is in Σ_r and $t_1,\ldots, t_r$ in $T(\Sigma)^1_p$ then $u_i \bullet \vec{v} = \sigma(t_1 \bullet \vec{v},\ldots, t_r \bullet \vec{v})$.

The second operation is the *tensorial-product*. If $\vec{u} = [u_1,\ldots, u_p] \in T(\Sigma)^p_q$ and $\vec{v} = [v_1,\ldots, v_{p'}] \in T(\Sigma)^{p'}_{q'}$, then their tensorial-product is $\vec{u} \otimes \vec{v} = [u_1, u_2,\ldots, u_p, v_1 \bullet [x_{q+1},\ldots, x_{q+q'}],\ldots, v_{p'} \bullet [x_{q+1},\ldots, x_{q+q'}]] \in T(\Sigma)^{p+q'}_{q+q'}$.
Both these operations are associative. The equality $\bar{t} = \tau \bullet (\tau_1 \otimes \tau_2 \otimes \tau_3)$ can be checked easily. Composition-product is for trees what concatenation is for words. Tensorial-product is used to put trees together and obtain vectors.

Any subset of $T(\Sigma)$ is called a *forest*. For any tree t in $T(\Sigma)^1$ like $\sigma(t_1,\ldots, t_n)$ where σ is in Σ_n, the *root* of t is σ. In any $\vec{u}$ in $T(\Sigma)$, each zero degree symbol is a *leaf*. The *yield* of $\vec{u}$ is the word over $\Sigma_0 \cup \chi$ that we obtain by putting the leafs together in the following way :

- For any $\alpha \in \Sigma_0 \cup X$, yield $(\alpha) = \alpha$,
- if $\sigma(t_1,\ldots, t_n) \in T(\Sigma)^1$ and $\sigma \in \Sigma$ then yield $(\sigma(t_1,\ldots, t_n))$ = yield (t_1) yield $(t_2),\ldots,$ yield (t_n),
- if $\vec{u} = [u_1,\ldots, u_p]$ then yield $(\vec{u})$ = yield (u_1) yield $(u_2),\ldots,$ yield (u_n).

The *depth* of $\vec{u}$ is an integer denoted $|\vec{u}|$ and defined by :

- For any $x_i \in X$, $|x_i| = 0$,
- for any $\alpha \in \Sigma_0$, $|\alpha| = 1$,
- if $\sigma(t_1,\ldots, t_n) \in T(\Sigma)^1$ and $\sigma \in \Sigma$ then $|\sigma(t_1,\ldots, t_n)| = 1 + MAX(|t_1|,\ldots, |t_n|)$,
- if $\vec{u} = [u_1,\ldots, u_p]$ then $|\vec{u}| = MAX(|u_1|,\ldots, |u_p|)$.

The set of variables that are actually in $\vec{u}$, is denoted by *var*$(\vec{u}) = \{x_i \mid \exists f_1, f_2 \in (\Sigma_0 \cup V)^*$ such that yield $(\vec{u}) = f_1\, x_i\, f_2\}$. A vector $\vec{v}$ in $T(\Sigma)_p^n$ is said *initial* in $T(\Sigma)_p^n$ if and only if yield $(\vec{v}) = u_0\, x_1\, u_1\, x_2\, u_2,\ldots, x_p\, u_p$ for $u_0, u_1,\ldots, u_p$ in $(\Sigma_0)^*$. We denote by $\tilde{T}(\Sigma)_p^n$ the set of initial vectors in $T(\Sigma)_p^n$ and in this paper any tree with the mark $\sim$ is initial. For any symbol σ in Σ_n it is convenient to denote by σ the initial tree $\sigma(x_1,\ldots, x_n)$. For any symbol a in Σ_1 we write $a^0 = x_1$, for any tree t, a t = a(t) and for any $i > 0$, $a^i = a\, a^{i-1}$. For any b in Σ_2 we write $\tilde{b}^0 = x_1$ and for any $i > 0$, $\tilde{b}^i = b(\tilde{b}^{i-1} \otimes \tilde{b}^{i-1})$. So $\tilde{b}^2 = b(b(x_1, x_2), b(x_3, x_4))$ is an initial balanced tree and its depth is 2.

Let t be a tree in $T(\Sigma)_p^1$ and let τ be a tree in $T(\Sigma)_n^1$. We say τ is a *subtree of* t or *occurs* in t if there exist $\vec{u} \in \tilde{T}(\Sigma)_q^1$, $\vec{v}_1 \in T(\Sigma)_{r_1}^{q_1}$, $\vec{v}_2 \in T(\Sigma)_{r_2}^{q_2}$, $\vec{w} \in T(\Sigma)_p^r$, where $q = q_1 + 1 + q_2$ and $r = r_1 + n + r_2$, such that $t = \vec{u} \cdot (\vec{v}_1 \otimes \tau \otimes \vec{v}_2) \cdot \vec{w}$. Note that $\vec{u}$ is initial in $\tilde{T}(\Sigma)_q^1$, then x_{q_1+1} is in yield $(\tilde{u})$, then we are sure τ occurs actually in the tree t. We say there is an occurrence of the symbol $\alpha \in \Sigma_n$ in the tree t if $\alpha(x_1,\ldots, x_n)$ is a subtree of t. An occurrence α of symbol is *above* an occurrence β in a tree t if there exists a subtree t' of t such that α is the root of t', β occurs in t' but is not its root. Then we say also β is *under* α.

III - GRAMMARS AND FORESTS

i) Definition of grammars

A (context-free or algebraic grammar is a generating system $G = \langle V, \Sigma, X_0, R\rangle$ where

- V is a ranked alphabet of symbols called *non-terminals*,
- Σ is a ranked alphabet of *terminal* symbols and $\Sigma \cap V = \emptyset$,
- X_0, called the *axiom*, is an element of V_0.
- R is a finite set of *rules* that are like $(X \to \tau)$ where X is in V and τ in $T(\Sigma \cup V)^1_{d(x)}$.

ii) Definition of derivations

- If $t \in T(\Sigma \cup V)^1$ contains an occurrence of a non-terminal X, (i.e. $t = \tilde{u} \cdot (\vec{v}_1 \otimes X \otimes \vec{v}_2) \cdot \vec{w}$ for some $\tilde{u}$, $\vec{v}_1$, $\vec{v}_2$, w), if there exists a rule like $r = (X \to \tau)$ and if $t' = \tilde{u} \cdot (\vec{v}_1, \tau, \vec{v}_2) \cdot \vec{w}$, we write $t \underset{G}{\Longrightarrow} t'$ and we say t is *derived* in t', by applying the rule r at the occurrence of X. As long as $\tilde{u}$ is initial, we can be sure a single occurrence of X has been changed.

Furthermore, if for every symbol above (resp. under) the derived occurrence of X in t is terminal, then the derivation is said descending or OI (resp. ascending or IO) and we write $t \overset{D}{\underset{G}{\Longrightarrow}} t'$ (resp. $t \overset{A}{\underset{G}{\Longrightarrow}} t'$). We write $t \overset{*}{\underset{G}{\Longrightarrow}} t'$ (resp. $t \; D\overset{*}{\underset{G}{\Longrightarrow}} t'$, $t \; A\overset{*}{\underset{G}{\Longrightarrow}} t'$) and say the *length* of the derivation is j if there exist $t_0, t_1, \ldots, t_j$ such that $t = t_0$, $t' = t_j$ and for each i in $[0, j-1]$, we have $t_i \underset{G}{\Longrightarrow} t_{i+1}$ (resp. $t_i \; D\underset{G}{\Longrightarrow} t_{i+1}$, $t_i \; A\underset{G}{\Longrightarrow} t_{i+1}$).

The descending derivation $X_0 \; D\Longrightarrow t_1 \; D\Longrightarrow t_2 \; D\Longrightarrow \ldots \; D\Longrightarrow t_{n-1} \; D\Longrightarrow t_n$ is said *initial* iff the roots of $t_1, t_2, \ldots, t_{n-1}$ are in V.

iii) Generated forest

Given the grammar G, for any τ in $T(\Sigma \cup V)^1$ we denote by $F(G, \tau)$ the forest $\{t \in T(\Sigma) \mid \tau \overset{*}{\underset{G}{\Longrightarrow}} t\}$ and similarly $F_{OI}(G, \tau) = \{t \in T(\Sigma) \mid \tau \; D\overset{*}{\underset{G}{\Longrightarrow}} t\}$, $F_{IO}(G, \tau) = \{t \in T(\Sigma) \mid \tau \; A\overset{*}{\underset{G}{\Longrightarrow}} t\}$. Then the forests generated by G are $F(G) = F(G, X_0)$, $F_{OI}(G) = F_{OI}(G, X_0)$ and $F_{IO}(G) = F_{IO}(G, X_0)$. It is well known that $F_{IO}(G) \subseteq F_{OI}(G) = F(G)$.

iv) Properties of rules

The rule $X \to \tau$ is :

- *strict* or ε-*free* iff $|\tau| > 0$, else it is an ε-*rule* ;
- *complete* iff $var(\tau) = var(X)$, else it is incomplete ;
- *linear* iff no variable x_i appears several times in yield (τ) ;
- *ordered* iff the indices of the variables are in a not decreasing order from left to-right in yield (τ) ;
- a *monadic* ε-*rule* when it is an ε-rule an $d(X) = 1$; then the rule is $X(x_1) \to x_1$ and it is complete.

Whenever every rule in G verifies a property P, we say G verifies P. If all the ε-rules in G are monadic, we say G is monadicly unstrict.

v) Example of grammar

The grammar G_1 is defined by $\Sigma_0 = \{\#\}$; $\Sigma_1 = \{a\}$; $\Sigma_2 = \{b, \delta, \gamma\}$; $V_0 = \{X_0\}$; $V_1 = \{Z\}$; $V_2 = \{X, Y\}$; $R = \{r_1 = X_0 \to X(\#, \delta(\#, \#))$; $r_2 = X(x_1, x_2) \to X(a(x_1), Y(a(x_1), x_2))$; $r_3 = X(x_1, x_2) \to Z(\gamma(x_1, x_2))$; $r_4 = Z(x_1) \to Z(b(x_1, x_1))$; $r_5 = Z(x_1) \to b(x_1, x_1)$; $r_6 = Y(x_1, x_2) \to \delta(x_1, x_1)$; $r_7 = Y(x_1, x_2) \to x_2\}$.

The only ε-rule is r_7 and it is not monadic. The two incomplete rules are r_6 and r_7. All the rules are ordered and r_1, r_3, r_6, r_7 are linear.

IV - REDUCED GRAMMARS

The Grammar $G = \langle V, \Sigma, X_0, R\rangle$ is OI-reduced (resp. IO-reduced) iff :

- $\forall X \in V$, $\exists t \in T(\Sigma \cup V)$ such that $X_0 \overset{*}{\underset{G}{\Longrightarrow}} t$ and X occurs in t ;
- $\forall X \in V$, $F_{OI}(G, X) \neq \emptyset$ (resp. $F_{IO}(G, X) \neq \emptyset$) ;
- $\forall X \in V$, $\forall x_i \in var(X)$, $\exists t \in F_{OI}(G, X)$ (resp. $F_{IO}(G, X)$) such that $x_i \in var(t)$.

The IO-reduction is a stronger property than the OI-reduction because $F_{IO}(G, X) \subseteq F_{OI}(G, X)$. G_1 is IO-reduced.

Theorem 1 For any grammar G such that $F_{IO}(G) \neq \emptyset$, an IO-reduced, complete and strict grammar G' can be found such that $F_{IO}(G) = F_{OI}(G')$.

Theorem 2 For any linear grammar G such that $F_{OI}(G) \neq \emptyset$ (resp. $F_{IO}(G) \neq \emptyset$), an IO-reduced, linear, complete, strict, ordered grammar G' (resp. G") can be found such that $F_{OI}(G') = F_{OI}(G)$ (resp. $F_{IO}(G'') = F_{IO}(G)$).

Theorem 3 For any grammar G such that $F_{OI}(G) \neq \emptyset$ an OI-reduced grammar G' can be found such that $F_{OI}(G') = F_{OI}(G)$ and every initial derivation in G' uses only complete and strict rules. Furthermore if G is complete or strict, so is G'.

Theorem 4 For any monadicly unstrict grammar G, a strict grammar G' can be found such that $F_{OI}(G') = F_{OI}(G)$.

Constructions and proofs for these results are given in [8, 9].

V - FORESTS WITH WIDE BALANCED TREES

Theorem 1 gives us a strong result about grammars and IO-generated forests. Unfortunately, in the OI case, we cannot get much more than theorem 3, particularly when the forest contain arbitrarily wide trees. So we study the wide balanced trees because the balance property makes the proofs easier.

For any ranked alphabet Σ that contains a symbol b of degree 2 and for any set $\mathcal{F}$ of forests in $T(\Sigma\backslash\{b\})^1_0$, we define the new forest :
$B\mathcal{F} = \{\tilde{b}^q \cdot [t_1,\ldots, t_m] \mid m = 2^q$ and $\exists F \in \mathcal{F}$ such that $t_1,\ldots, t_m \in F\}$.

Let be $G = \langle V, \Sigma, X_0, R\rangle$ a grammar that generates $B\mathcal{F}$.

For any τ in $T(\Sigma \cup V)^1_0$, we say the b-*level* of τ is defined with the value q iff every t in $F(G, \tau)$ is like $\tilde{b}^q \cdot [t_1,\ldots, t_m]$ where $t_1,\ldots, t_m$ are in $T(\Sigma\backslash\{b\})$.

Lemma 5 *For* G *we can find two constants* h *and* k *such that for each* t *in* $B\mathcal{F}$ *there exists a derivation* $X_0 \overset{*}{\underset{G}{\Rightarrow}} u \cdot \vec{v} \overset{*}{\underset{G}{\Rightarrow}} t$ *where* $u \in T(\{b\})^1_p$, $\vec{v} \in T(\Sigma \cup V)^p_0$, $p \leq k$ *and all the* p *components of* $\vec{v}$ *have the same* b-*level* $q \leq h$.

This lemma is like an iteration lemma about the width of trees, and we use it to prove the following results.

Let us write for each integer n, $F_{1,n} = \{\gamma(\delta(a^i\ \#, a^i\ \#), a^n\ \#) \mid i \leq n\}$ and $\mathcal{F}_1 = \{F_{1,n} \mid n \in \mathbb{N}\}$. Then $B\mathcal{F}_1$ is generated by the grammar G1 given above [1].

Proposition 6 $B\mathcal{F}_1$ *cannot be generated by an* ε-*free grammar.*

Let be $F_{2,n} = \{\alpha^i\ \gamma\ a^j\ \# \mid i + j = n\}$ for any integer n and $\mathcal{F}_2 = \{F_{2,n} \mid n \in \mathbb{N}\}$. The forest $B\mathcal{F}_2$ is generated by $X_0 \to X(\gamma\ \#, \#)$; $X_0 \to Z\ \gamma\ \#$; $X(x_1, x_2) \to Z\ Y(x_1, a\ x_2)$; $X(x_1, x_2) \to X(Y(X_1, a\ x_2), a\ x_2)$; $Z(x_1) \to Z\ b(x_1, x_1)$; $Z(x_1) \to b(x_1, x_1)$; $Y(x_1, x_2) \to \gamma\ x_2$; $Y(x_1, x_2) \to \alpha\ x_1$, which is an ε-free grammar.

Proposition 7 $B\mathcal{F}_2$ *cannot be generated by a complete grammar.*

Let be $F_{3,n} = \{\delta(a^j\ \#, a^n\ \#) \mid j \leq n\}$ for any integer n and $\mathcal{F}_3 = \{F_{3,n} \mid n \in \mathbb{N}\}$. Let be the set of rules $R = \{X_0 \to X(\#, \#)$; $X(x_1, x_2) \to Z\ \delta(x_1, x_2)$; $X(x_1, x_2) \to X(Y\ x_1, a\ x_2)$; $Z(x_1) \to b(x_1, x_1)$; $Z(x_1) \to Z\ b(x_1, x_1)$; $Y(x_1) \to a(x_1)\}$. The grammar, the rules of which are in $R \cup \{Y(x_1) \to x_1\}$ is (monadicly) unstrict but is complete and generates $B\mathcal{F}_3$. Furthemore if we take the set $R \cup \{Y(x_1) \to \#\}$, the new grammar is not complete but it is ε-free and generates $B\mathcal{F}_3$.

Proposition 8 $B\mathcal{F}_3$ *cannot be generated by a complete and* ε-*free grammar.*

Let us denote by OI (resp. S, C, CS, Lin) the class of forests generated by grammars the rules of which are of any sort (resp. strict, complete, complete and strict, linear).

Theorem 9 Lin $\subsetneq$ CS $\subsetneq$ C $\subsetneq$ S $\subsetneq$ OI.

Sketch of proof.

The definitions give CS $\subseteq$ C, CS $\subseteq$ S and S $\subseteq$ OI. Theorem 2 says Lin $\subseteq$ CS and it is easy to find forests in CS but not in Lin, then Lin $\subsetneq$ CS. By proposition 8 we have CS $\subsetneq$ C. Each complete and unstrict grammar is monadicly unstrict. Then theorem 4 gives C $\subseteq$ S and proposition 7 C $\subsetneq$ S. Finally proposition 6 says S $\subsetneq$ OI.

□

REFERENCES

[1] A. ARNOLD, M. DAUCHET, "*Forêts algébriques et homomorphismes inverses*", Information and Control. 37 (1978), pp. 182-196.

[2] A. ARNOLD, M. DAUCHET, "*Théorie des magmoïdes*", RAIRO inf. th. 12 (1978), n° 3, pp. 235-257 et 13 (1979), n° 2, pp. 135-154.

[3] G. BOUDOL, "*Langages polyadiques algébriques. Théorie des schémas de programme : Sémantique de l'appel par valeur*", Thèse de 3ème cycle, Paris VII (1975).

[4] B. COURCELLE, "*A representation of trees by languages*", Th. Comput. Sci. 6 (1978), pp. 255-279 and 7 (1978) pp. 25-55.

[5] J. ENGELFRIET and E.M. SCHMIDT, "*IO and OI*", J. Comput. System Sci. 15 (1977), pp. 328-353 and 16 pp. 67-99.

[6] J. ENGELFRIET, "*Some open questions and recent results on tree tranducers and tree languages*", Formal Language Theory edited by R.V. BOOK. Academic Press 1980.

[7] I. GUESSARIAN, "*Program transformation and algebraic semantics*", Publication interne LITP 78/21 (1978), A paraître dans TCS.

[8] B. LEGUY, "*Reductions, transformations et classification des grammaires algébriques d'arbres*", Thèse de 3ème cycle, Lille (1980).

[9] B. LEGUY, "*Grammars without erasing rules. The OI case*". C.A.A.P. 1981, Genova (Italy).

[10] M. NIVAT, "*On the interpretation of recursive program schemes*", Rapport IRIA, n° 84 (1974).

[11] W.C. ROUNDS, "*Mappings and grammars on trees*", Math. Systems theory. 4 (1968), pp. 257-287.

RATIONAL CONE AND SUBSTITUTION

by Jeannine LEGUY

UNIVERSITE DE LILLE I
UER IEEA - INFORMATIQUE
59655 VILLENEUVE D'ASCQ Cédex
(France)

ABSTRACT

Relations between $C(L)$, the rational cone (full trio) generated by L and subsets of $C(L)$ are given which allow to characterize the languages L which verify the relation :

$$\forall L', \; C(L) \square C(L') = C(L \uparrow L')$$

Languages satisfying this relation are said substitution complete languages.

INTRODUCTION

The syntatic substitution, denoted by $\uparrow$, has facilitated the study of the substitution operator and its consequences concerning the families of languages built from this operator on one hand and the rational cones on the other hand. S. Greibach has shown this important result :

$$\forall L \subset X^{*}, \; \forall L' \subset X'^{*} \text{ with } X \cap X' = \emptyset, \; \forall c \notin X \cup X'$$
$$C(L) \square C(L') = C(L \uparrow (L'c)^{*}).$$

Of course, $C(L \uparrow L') \subseteq C(L) \square \;(L')$ and if L' generates a full AFL (i.e $C(L') = C((L'c)^{*})$ we obtain :

$$\forall L : C(L) \square C(L') = C(L \uparrow L').$$

Our purpose is to characterize the substitution complete languages, namely the languages L such that :

$\forall L', \mathcal{C}(L) \square \mathcal{C}(L') = \mathcal{C}(L \uparrow L')$.

We already know [2] that the relation is true for Sym (The symetric language over two letters) and for $D_1'^{*}$, the Dyck-languages over one letter.

To show this result we need new notions, for instance, decreasing rational cone, saturated language ... L is a saturated language if for any language L', reached from L by rational transduction, we can find a decreasing rational transduction τ such that $L' = \tau(L)$. This means the language L is strong enough to make up for the loss of power of the decreasing rational transduction. A characterization of decreasing rational transduction has been proved in [10], so, the study of rational cone (full trio) generated by a saturated language is facilitated. In the first part of this paper we give the definitions we need later.

In the second part, we show properties concerning the saturated languages and we give a characterization of such languages.

Some very usefull relations between the rational cone $\mathcal{C}(L)$ and some particular subsets of $\mathcal{C}(L)$, like $\mathcal{C}^d(L)$ the decreasing rational cone or $\mathcal{C}^{bf}(L)$, the bifaithful rational cone, are reminded in the next part.

From these results we can conclude : L is complete under substitution if and only if it is saturated.

I - DEFINITIONS AND NOTATIONS

We assume known classical definitions of formal languages theory [2].

Definition 1 : let τ a rational transduction from X^* into Y^*

τ is a faithful transduction $\Leftrightarrow \forall w \in Y^* \quad \tau^{-1}(w)$ is a finite set

τ is a finite image transduction $\Leftrightarrow \forall w \in X^* \quad \tau(w)$ is a finite set

τ is a bifaithful transduction $\Leftrightarrow \tau$ is a faithful transduction and a finite image transduction

τ is a decreasing transduction $\Leftrightarrow \forall w \in X^*, \forall w' \in \tau(w)$ $|w'| \leq |w|$ (where $|w|$ is the length of w).

$\mathcal{C}(L)$ denotes the rational cone (full trio) generated by L.

Définition 2 : let L be a language :

$C^f(L) = \{L' \ / \ \exists\tau$ a faithful rational transduction such that $L' = \tau(L)\}$

$C^{if}(L) = \{L' \ / \ \exists\tau$ a finite image rational transduction such that $L' = \tau(L)\}$

$C^d(L) = \{L' \ / \ \exists\tau$ a decreasing rational transduction such that $L' = \tau(L)\}$

$C^{bf}(L) = \{L' \ / \ \exists\tau$ a bifaithful rational transduction such that $L' = \tau(L)\}$

$C^{df}(L) = \{L' \ / \ \exists\tau$ a decreasing faithful rational transduction such that $L' = \tau(L)\}$.

Définition 3 :

L is an erasable language $\iff C^f(L) = C(L)$ [6]

L is a saturated language $\iff C^{if}(L) = C(L)$

L is a bifaithful language $\iff C^{bf}(L) = C(L)$.

Définition 4 : let $\mathcal{L}$ and $\mathcal{L}'$ be families of languages. $\mathcal{L} \,\square\, \mathcal{L}'$ is a family of languages $s(L)$, where $L \in \mathcal{L}$ and s is a $\mathcal{L}'$-substitution ε free.

Définition 5 [5] : let L and L' be two languages,

$$L \uparrow L' = \{x_1 w_1 x_2 w_2 \ldots x_n w_n \ / \ x_1 x_2 \ldots x_n \in L,$$

$w_i \in L'$ and x_i is a letter, $\forall\, i \in [1,n]\}$.

Définition 6 : let L be a language. L is complete under substitution if and only if for any language L, defined over a disjoint alphabet,

$$C(L) \,\square\, C(L') = C(L \uparrow L').$$

Définition 7 : a morphism h from X^* into Y^* is ε-limited on $R \subseteq X^*$ if there exists $k \geq 1$ such that for all w in R with $|w| > k$ then $h(w) \neq \varepsilon$. h is ε-free if $h(X) \subseteq Y$

Theorem 8 [3] : a rational transduction τ from X^* into Y^* is a finite image (resp. bifaithful) transduction if and only if there exist an alphabet Z, two alphabetic morphisms h and g from Z^* into X^* and Y^* and a rational language $R \subseteq Z^*$ such that

$$\forall w \in X^*, \ \tau(w) = g(h^{-1}(w) \cap R)$$

and h is ε-limited on R (resp. h and g are ε-limited on R).

Theorem 9 [10] : a rational transduction τ from X^* into Y^* is decreasing if and only if there exist an alphabet Z, two alphabetic morphisms h and g from Z^* into X^* and Y^* h ε-free and a rational language $R \subseteq Z^*$ such that

$$\forall w \in X^*,\ \tau(w) = g(h^{-1}(w) \cap R).$$

II - SOME RESULTS ABOUT SATURATED LANGUAGES

The two following theorems have been shown by M. Latteux, the demonstrations can been found in [8, 9].

Theorem 10: any context-free generator and any Ocl-generator is a bifaithful language.

but

Theorem 11 : Any linear-generator is a saturated erasable language but it is not a bifaithful language.

Theorem 12 [11] : $L \subseteq a^*b^*$ is a saturated language if and only if L is an infinite rational language.

It follows that $C_1 = \{a^n b^n \ /\ n \geq 0\}$ is not a saturated language.

Before showing an interesting result about satured (resp. satured and erasable) languages we need the two following lemma :

Lemma 13.: for any language $L \subseteq X$ we have :

$$C(L) = C^d(L \uparrow c^*) = C^{if}(L \uparrow c^*)$$

$$C^f(L) = C^{uf}(L \uparrow c^*) = C^{bf}(L \uparrow c^*).$$

Lemma 14 : let L and L' be two languages,
if $L \uparrow c^*$ belongs to $C^{if}(L')$ (resp. $C^{bf}(L')$).
then $L \uparrow c^*$ belongs to $C^d(L')$ (resp. $C^{df}(L')$).

Sketch of the proof :

(The interested reader may find the complete proof in [11]).

If $L \uparrow c^*$ belongs to $C^{if}(L)$, from theorem 8, we can write $L \uparrow c^* = \tau(L') = g(h^{-1}(L') \cap R)$ where h is ε-limited. There exists k such that for allwords w with $|w| > k$, $h(w) \neq \varepsilon$. We are going to construct from τ, a transduction τ' which will erase some occurrences of the letter c, enough to have τ' decreasing, but not too much to preserve $L \uparrow c^* = \tau'(L') = g'(h'^{-1}(L') \cap R')$.

Definition of R'

We select in R all the words w such that $g(w) \in (X\, c^* c^k)^*$ (i.e. between two letters of X there are at least k occurrences of c). $R_1 = R \cap g^{-1}((X\, c^* c^k)^*)$. Now we "slice" each word of R_1 in k-lenth sub-words. To do so we define two new alphabets :

$$Z_1 = \underbrace{Z \times Z \times \ldots \times Z}_{k \text{ times}} \qquad Z_2 = \bigcup_{i=1}^{k-1} \underbrace{Z \times Z \times \ldots \times Z}_{i \text{ times}}$$

and the morphism :

$$t : (Z_1 \cup Z_2)^* \to Z^* : t(x_1,\ldots,x_n) = x_1 \ldots x_n,\ \forall\, (x_1,\ldots,x_n) \in Z'$$

where $Z' = Z_1 \cup Z_2$ then $R' = t^{-1}(R_1) \cap Z_1^*(Z_2 \cup \{\varepsilon\})$.

Définitions of h' and g'

$h' : Z'^* \to X'^*$

$g' : Z'^* \to X^*$

and $\forall\, z = (z_1,\ldots,z_n) \in Z'$

$$h'(z) = h(z_1,\ldots,z_n)$$

$$\begin{aligned} g'(z) &= x \text{ if } g(z_1,\ldots,z_n) \in c^* X c^* \\ &= c \text{ if } g(z_1,\ldots,z_n) \in c^+ \\ &= \varepsilon \text{ if } g(z_1,\ldots,z_n) = \varepsilon. \end{aligned}$$

Then $L \uparrow c^* = g'(h'^{-1}(L') \cap R')$.

The next theorem immediately follows from these two lemma :

<u>*Theorem 15*</u> [12] : given L, a language defined over X, these three properties are equivalent :

i) L is a saturated language

ii) $L \uparrow c^{\star} \in C^{if}(L)$

iii) $C(L) = C^{d}(L)$

and, if L is an erasable language, these three properties are equivalent :

i) L is an erasable saturated language

ii) $L \uparrow c^{\star} \in C^{bf}(L)$

iii) $C(L) = C^{df}(L)$.

III - RELATIONS BETWEEN $C^{df}(L)$, $C^{d}(L)$, $C^{f}(L)$ AND $C(L)$

Let # be a new letter, we denote $\#L = \{\#w \;/\; w \in L\}$, then :

<u>*Proposition 16*</u> : let $L_1 \subseteq X_1^{\star}$ and $L_2 \subseteq X_2^{\star}$ be languages with $X_1 \cap X_2 = \emptyset$, then

$$C^{d}(L_1) \square C^{d}(\# L_2) = C^{d}(L_1 \uparrow L_2)$$

$$C^{df}(L_1) \square C^{df}(\# L_2) = C^{df}(L_1 \uparrow L_2)$$

<u>*Corollary 17*</u> : for any language L we have :

$$C^{if}(L) = C^{d}(L) \square \mathrm{Fin}$$

$$C^{bf}(L) = C^{df}(L) \square \mathrm{Fin}$$

$$C(L) = C^{d}(L) \square \mathrm{Rat}$$

$$C^{f}(L) = C^{df}(L) \square \mathrm{Rat}$$

we can draw :

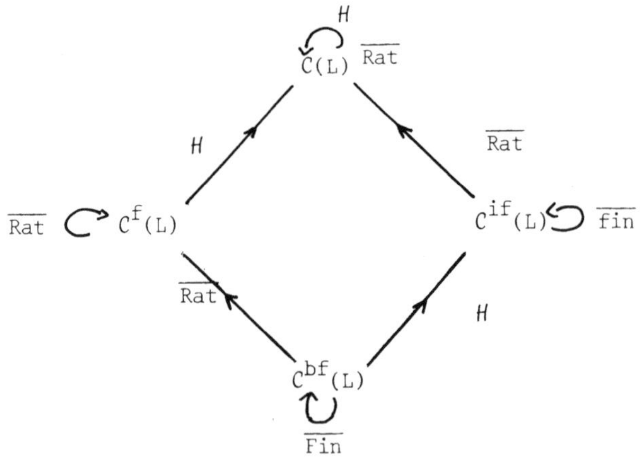

where $L \overset{H}{\to} L'$ means $L' = \{h(L) \ / \ L \in L$ and h is a morphism$\}$

$L \overset{\bar{M}}{\to} L'$ means $L' = L \,\square\, M$.

Corollary 19 : for all languages L_1 and L_2 defined over disjoint alphabets we have :

$$C(L_1 \uparrow L_2) = C^d(L_1) \,\square\, C(L_2)$$

$$C^f(L_1 \uparrow L_2) = C^{df}(L_1) \,\square\, C^f(L_2).$$

The complete proofs of the results of this part are done in [11].

IV - A CARACTERIZATION OF LANGUAGES COMPLETE UNDER SUBSTITUTION

Let us remind the M. Latteux's version of the syntactic lemma of substitution (due to S. Greibach [5]).

Theorem 19 [9] : let L_1 be a bifaithful rational cone, L_2 be a rational cone, $L_1 \subseteq X_1^*$, $L_2 \subseteq X_2^*$ be languages with $X_1 \cap X_2 = \emptyset$, then $L_1 \uparrow L_2 \in L_1 \,\square\, L_2$ involves $L_1 \in L_1$ or $(L_2 c)^* \in L_2$.

Theorem 20 : L is complete under substitution if and only if L is a saturated language.

Proof :

If L is complete under substitution, $C(L \uparrow (L'c)^*) = C(L) \square C(L') = C(L \uparrow L')$, the language $L \uparrow (L'c)^*$ belongs to $C(L \uparrow L')$ for all L'. Let us choose for L' the symetric language Sym, then we can write $L \uparrow (\mathrm{Sym}\ c)^* \in C(L \uparrow \mathrm{Sym})$. Since Sym is not a generator of a full AFL, we conclude, from the syntactic lemma and from theorem 15, that L is a saturated language.

Conversely, if L is a saturated language, then the theorem 15 involves
$C(L) = C^d(L)$.
$C(L \uparrow L') = C^d(L) \square C(L') = C(L) \square C(L')$ $\square$

BIBLIOGRAPHY

[1] J. BEAUQUIER, - A remark about a substitution property (1978). To appear in Math. Syst. Theory.

[2] J. BERSTEL, -"Transductions and context-free languages", Teubner Verlag, 1978.

[3] L. BOASSON and M. NIVAT, - Sur diverses familles de langages fermées par transductions rationnelles. Acta Informatica, 2 (1973) 180-188.

[4] C.C. ELGOT et J.F. MEZEI, - On relations defined by generalized finite automata. I.B.M. J. Res. Dev. vol 9, (1965) 47-68.

[5] S.A. GREIBACH, - Syntaxtic operators on full semi-AFLS. J. Comp. Syst. Sc. 6 (1972), 30-76.

[6] S.A. GREIBACH, - Erasing in context-free languages. Information and Control, 29, (1975), 301-326.

[7] M. LATTEUX, - Sur les générateurs Algébriques et Linéaires. Acta Informatica 13 (1980), 347-363.

[8] M. LATTEUX, - Quelques propriétés des langages à un compteur, 5 th GI, conference. Lecture Notes in Computer Sciences 104, springer (1981) 52-63.

[9] M. LATTEUX, - A propos du lemme de substitution, Theoretical Computer Science, 14 (1981), 119-123.

[10] J. LEGUY, - Transductions Rationnelles Décroissantes (1979), R.A.I.R.O. Informatique Théorique, Vol. 5, n° 2, (1981).

[11] J. LEGUY, - Transductions rationnelles décroissantes et substitution. Thèse 3e cycle, Université de Lille I, Lille 1980.

ON THE REGULARITY PROBLEM OF SF-LANGUAGES GENERATED BY MINIMAL LINEAR GRAMMARS

Matti Linna
Department of Mathematics
University of Turku
SF-20500 Turku 50, Finland

1. Introduction

Various decidability problems of SF-languages (sets of sentential forms generated by Chomsky type grammars) were investigated by Harju and Penttonen in [1]. Among other things, they studied the regularity problem for different families of SF-languages. They showed that it is undecidable whether the SF-language of a context-sensitive grammar is regular. On the other hand, the regularity problem is still open for the class of context-free grammars and even for the class of minimal linear grammars. In [3] some comparative problems between the Chomsky families and the Lindenmayer families were considered. In particular, it was shown that the regularity and context-freeness problems for the family of D0L languages are decidable. On the other hand, the corresponding problems for the family of 0L languages are still open and very little is known about them.

In this paper we consider the regularity problem of SF-languages generated by minimal linear grammars. We solve this problem in several special cases. Some of our results are essentially based on the finite power property theorem for regular languages (see [2] and [4]). It is easy to show that the family of SF-languages of context-free grammars is properly included in the family of 0L languages. So, this paper is also an attempt to approach the above mentioned regularity problem for 0L languages.

2. Preliminaries

For a context-free grammar we use the notation $G = (V_N, V_T, P, S)$, where V_N and V_T are the alphabets of nonterminals and terminals respectively, $S \in V_N$ is the start letter and P is the set of productions. The notation $x \Rightarrow^* y$ means that y is obtained from x by zero or more applications of productions. Words y in $(V_N \cup V_T)^*$ that can be derived from the start symbol S, i.e. $S \Rightarrow^* y$, are called <u>sentential forms</u>. The SF-<u>language</u> SF(G) of G is the set of all sentential forms of G.

The notation $w\backslash L$, where w is a word and L is a language, denotes the set $\{u \mid wu \in L\}$ and is called the left derivative of L with respect to w. Moreover, $|w|$ is the length of w.

In this paper we shall consider only minimal linear grammars, i.e. grammars $G = (V_N, V_T, P, S)$, where $V_N = \{S\}$ and P contains productions of the form

$$S \to wSv, \quad S \to x,$$

where $w,v,x \in V_T^*$. The problem is the following: Is it decidable whether the SF-language of a given minimal linear grammar is regular ? We shall answer this question in several special cases. The general case, however, remains open.

First we note that all productions of the form $S \to x$ can be ignored. This is seen in the following way. Let

$$SF(G) = L_1 \cup L_2,$$

where $L_1 = SF(G) \cap V_T^*$ and $L_2 = SF(G) - L_1$. So, in deriving a word in L_1 one must once apply a production of the form $S \to x$. If now L_2 is regular then also L_1 is regular, since L_1 is obtained from L_2 by a finite substitution. Hence, SF(G) is regular, too. Conversely, assume that L_2 is not regular. Then SF(G) cannot be regular, because otherwise the language $SF(G) \cap V_T^* S V_T^* = L_2$ would be regular. So SF(G) is regular if and only if L_2 is regular. From now on we shall assume that P contains only productions of the form $S \to wSv$, where $w,v \in V_T^*$.

The following lemma is easily verified.

<u>Lemma 1</u>. If P does not contain any production of the form $S \to wSv$, where $w,v \in V_T^+$, then SF(G) is regular.

<u>Lemma 2</u>. If P contains at least one production of the form $S \to wSv$, where $w,v \in V_T^+$, and no productions of the form $S \to uS$, where $u \in V_T^+$ (or $S \to St$, where $t \in V_T^+$), then SF(G) is not regular.

<u>Proof</u>. Assume that P contains a production $S \to wSv$, where $w,v \in V_T^+$, and no productions of the form $S \to uS$, where $u \in V_T^+$. Then for all positive integers n, $w^nS\backslash SF(G) \neq \phi$ and

$$\min\{|x| \mid x \in w^nS\backslash SF(G)\} \geqq n.$$

This implies that L has infinitely many different left derivatives.

3. Results

In the whole section we shall assume that P contains the productions

(1) $S \to w_1Sv_1|w_2Sv_2|\ldots|w_kSv_k,\ w_i,v_i \in V_T^+$ for $i = 1,\ldots,k,$

(2) $S \to u_1S|u_2S|\ldots|u_rS,\ u_i \in V_T^+$ for $i = 1,\ldots,r,$

(3) $S \to St_1|St_2|\ldots|St_s,\ t_i \in V_T^+$ for $i = 1,\ldots,s.$

Denote further

$$W = \{w_1,\ldots,w_k\},$$
$$V = \{v_1,\ldots,v_k\},$$
$$U = \{u_1,\ldots,u_r\},$$
$$T = \{t_1,\ldots,t_s\}$$

and

$$L_1 = U^*WU^*,\quad L_2 = T^*VT^*.$$

A language L is said to possess the finite power property if the set $\{L^i | i = 0,1,\ldots\}$ is finite, or in other words, if $L^{i+1} = L^i$ for some integer i.

Lemma 3. If SF(G) is regular then the regular languages $L_1 \cup \{\lambda\}$ and $L_2 \cup \{\lambda\}$ possess the finite power property (which is a decidable property by [2] and [4]).

Proof. Assume, e.g., that $L_1 \cup \{\lambda\}$ does not possess the finite power property. We shall show that SF(G) is not regular. By the assumption, for each $i \geq 1$, there exists a word x_i such that $x_i \in (L_1 \cup \{\lambda\})^i$ but $x_i \notin (L_1 \cup \{\lambda\})^{i-1}$. This means that when generating x_iSy for some y one must apply productions of the form (1) at least i times. Thus,

$$\min\{|y| \,|\, y \in x_iS\backslash SF(G)\} \geq i$$

for each i and so SF(G) has infinitely many different left derivatives.

Next we shall show that if P contains only one production of the form (1), i.e. k = 1, then also the converse of Lemma 3 holds true.

Theorem 1. If P contains only one production of the form (1) then SF(G) is regular if and only if the languages $L_1 \cup \{\lambda\}$ and $L_2 \cup \{\lambda\}$ possess the finite power property. Thus, the regularity problem is decidable in this case.

Proof. If SF(G) is regular then $L_1 \cup \{\lambda\}$ and $L_2 \cup \{\lambda\}$ possess the finite power property by Lemma 3.

Assume, conversely, that the languages $L_1 \cup \{\lambda\}$ and $L_2 \cup \{\lambda\}$ possess the finite power property. Then there are integers m and n such that

(4) $$L_1^m = \bigcup_{i=1}^{m-1} L_1^i \quad \text{and} \quad L_2^n = \bigcup_{i=1}^{n-1} L_2^i.$$

Denote $M = \max(m,n)$. We claim that

$$SF(G) = \bigcup_{i=1}^{M^3} L_1^i S L_2^i \quad \text{(a regular language).}$$

Obviously, the right hand side is included in the left hand side. To prove the converse, assume that $xSy \in SF(G)$. Then there exists an integer p such that $xSy \in L_1^p S L_2^p$. If $p \leq M^3$ we are done. So, assume that $p > M^3$. We shall show that there is an integer q, $q < p$, such that $xSy \in L_1^q S L_2^q$. Write

$$x = x_1 \ldots x_f x_{f+1},\ x_i \in L_1^m\ (i=1,\ldots,f),\ x_{f+1} \in L_1^h,$$
$$y = y_1 \ldots y_g y_{g+1},\ y_i \in L_2^n\ (i=1,\ldots,g),\ x_{g+1} \in L_2^{h'},$$

where $0 \leq h < m$, $0 \leq h' < n$, $fm+h = p$ and $gn+h' = p$. By (4), for $i = 1,\ldots,f$ and $j = 1,\ldots,g$ there exist integers m_i and n_j such that $0 < m_i < m$, $0 < n_j < n$ and $x_i \in L_1^{m_i}$, $y_j \in L_2^{n_j}$. By the choice of the bound M^3 and the inequality $p > M^3$, at least M integers of m_i´s are equal and also at least M integers of n_i´s are equal. So we may assume that

$$m_{i_1} = m_{i_2} = \ldots = m_{i_M} = m',\quad 0 < i_1 < i_2 < \ldots < i_M \leq f$$

and

$$n_{j_1} = n_{j_2} = \ldots = n_{j_M} = n',\quad 0 < j_1 < j_2 < \ldots < j_M \leq g.$$

If now $m-m' = n-n'$, we may choose $q = p-M(m-m')$. If $m' > n'$ or $m' < n'$, choose $q = p - (m-m')(n-n')$. In all cases $xSy \in L_1^q S L_2^q$. If now $q > M^3$, the above procedure can be repeated until $q \leq M^3$. So

$$SF(G) \subseteq \bigcup_{i=1}^{M^3} L_1^i S L_2^i.$$

This completes the proof of Theorem 1.

The following example, where $k = 3$, shows that Theorem 1 does not hold for all minimal linear grammars.

Example. Let P contain the productions

(5) $S \rightarrow aabaaSaabaa \mid baSbaba \mid babaSba$,

(6) $S \to aS|baaaS|baabaS|babaS,$

(7) $S \to Sa|Sbaaa|Sbaaba|Sbaba.$

We have $W = \{aabaa,ba,baba\} = V$ and $U = \{a,baaa,baaba,baba\} = T$. Denote $L_1 = U^*WU^*$ and $L_2 = T^*VT^*$. It can be shown that, for each $n \geq 1$, every derivation of the word $(aabaa)^nS(aabaa)^n$ contains at least n applications of productions of type (5). This implies that SF(G) is not regular. However, it can also be shown that both $L_1 \cup \{\lambda\}$ and $L_2 \cup \{\lambda\}$ possess the finite power property.

Next we shall consider some further special cases.

Lemma 4. Assume that $w_i, u_j \in a^*$ for $i = 1,\dots,k$, $j = 1,\dots,r$ and $v_i, t_j \in b^*$ for $i = 1,\dots,k$, $j = 1,\dots,s$. Then SF(G) is regular. In particular, SF(G) is regular if V_T consists of only one letter.

Proof. Assume that P contains the productions

$$S \to a^{m_1}Sb^{n_1}|a^{m_2}Sb^{n_2}|\dots|a^{m_k}Sb^{n_k} \quad (m_i,n_i \geq 1),$$

$$S \to a^{p_1}S|a^{p_2}S|\dots|a^{p_r}S, \quad 1 \leq p_1 < p_2 < \dots < p_r,$$

$$S \to Sb^{q_1}|Sb^{q_2}|\dots|Sb^{q_s}, \quad 1 \leq q_1 < q_2 < \dots < q_s.$$

Let

$$M = k(p_1q_1+1)\text{l.c.m.}(m_1,\dots,m_k,n_1,\dots,n_k).$$

Denote

$$L_1 = (a^{p_1} \cup \dots \cup a^{p_r})^*(a^{m_1} \cup \dots \cup a^{m_k})(a^{p_1} \cup \dots \cup a^{p_r})^*,$$
$$L_2 = (b^{q_1} \cup \dots \cup b^{q_s})^*(b^{n_1} \cup \dots \cup b^{n_k})(b^{q_1} \cup \dots \cup b^{q_s})^*.$$

It can be shown that

$$SF(G) = \bigcup_{i=1}^{M} L_1^iSL_2^i$$

which is a regular language.

For a word w, define its <u>primitive root</u> $\rho(w)$ to be the shortest word u such that $w = u^n$ for some $n \geq 1$. As a corollary of Lemma 4 we obtain

Lemma 5. SF(G) is regular if $\rho(w_1) = \dots = \rho(w_k) = \rho(u_1) = \dots = \rho(u_r)$ and $\rho(v_1) = \dots = \rho(v_k) = \rho(t_1) = \dots = \rho(t_s)$.

Lemma 5 is useful in the following particular case.

Theorem 2. Assume that P contains only one production of the form (2) and (3), i.e. $r = s = 1$. Then SF(G) is regular if and only if

(8) $$\rho(w_1) = \dots = \rho(w_k) = \rho(u_1)$$

and

(9) $$\rho(v_1) = \dots = \rho(v_k) = \rho(t_1).$$

Proof (outline). By Lemma 5, the conditions (8) and (9) imply that SF(G) is regular. Conversely, assume that SF(G) is regular. Let $x \in \{w_1,\dots,w_k,u_1\}^+$ and $|x| \geqq |u_1|$. The family $\{x^i S \backslash SF(G) \mid i=1,2,\dots\}$ of languages is finite because SF(G) is regular. Thus, we may assume that

$$x^{i_1} S \backslash SF(G) = x^{i_2} S \backslash SF(G) = \dots \quad \text{for } 1 \leqq i_1 < i_2 < \dots$$

Let z be some word in $x^{i_1} S \backslash SF(G)$. Corresponding to z, there exist integers m and n such that $x^m = u' u_1^n u''$, where $u''u' = u_1$ or $u''u' = \lambda$. In particular, it can be shown that there exist words u', u'' and integers m_i, n_i, $i = 1,\dots,k$ such that

$$w_i^{m_i} = u' u_1^{n_i} u'' \text{ for } i = 1,\dots,k$$

and

$$u_1 = u'u'' = u''u'.$$

These equations imply that $\rho(w_1) = \rho(w_2) = \dots = \rho(w_k) = \rho(u_1)$. The condition (9) is proved similarly.

References

1. Harju, T. and Penttonen, M., Some decidability problems of sentential forms, Intern. J. Computer Math. 7 (1979) 95-107.
2. Hashiguchi, K., A decision procedure for the order of regular events, Theoretical Computer Science 8 (1979) 69-72.
3. Salomaa, A., Comparative decision problems between sequential and parallell rewriting, Proc. Intern. Symp. Uniformly Structured Automata and Logic, Tokyo (1975).
4. Simon, I., Locally finite semigroups and limited subsets of a free monoid, a manuscript (1978).

CO-ALGEBRAS AS MACHINES FOR THE INTERPRETATIONS OF FLOW DIAGRAMS

W. Merzenich
Universität Dortmund
Abteilung Informatik
Postfach 50 05 00
D-4600 Dortmund 50

0. Introduction

Mathematical machines are used in theoretic computer science as a tool that maps programs into partial functions and thus are themselves special representations of functions from programs to partial functions [CC76]. The way how to get partial functions from programs using machines is a problem of operative semantics. In this paper an algebraic approach to this question is suggested. Programs will be regarded as flow-diagrams in the sense of Elgot [E77] and the machines that interprete these programs will be co-algebras which are dual to algebras. The concept of operatornetworks which we introduced earlier [DM80] is used to formalize programs, and results about solutions of algebraic equations [M79] enable us to treat the problem in a purely algebraic way. The subject is closely related to the approach using Cpo's and continuous algebras [ADJ77] and some remarks to this relationship will be made in section 4.

1. Algebras and co-algebras

The most general concept of an algebra seems to be that of an F-algebra [AT78] or F-dynamics [AM74] where $F : K \to K$ is an endofunctor on a category K and F-algebra is a pair (A,δ) with $\delta : AF \to A$. We are not going to treat the subject in that general way but regard the well known case where algebras are determined by operations on a set.

For a function $s : \Omega \to \mathbb{N}$ the pair (Ω,s) is called a ranked alphabet. If K is a category with finite products and (countable coproducts then we define the functor $\Omega_K : K \to K$ by:

$$A\Omega_K = \coprod_{\omega\in\Omega} A^{\omega s} \quad \text{for} \quad A \in K$$

A^k denotes the k-fold product of A.

An <u>Ω_K-algebra</u> is a pair (A,δ) with $\delta : A\Omega_K \to A$ a morphism in K and if (B,γ) is another Ω_K-algebra then a morphism $\phi : A \to B$ in K is called a <u>morphism of Ω_K-algebras</u> if (1.1) commutes.

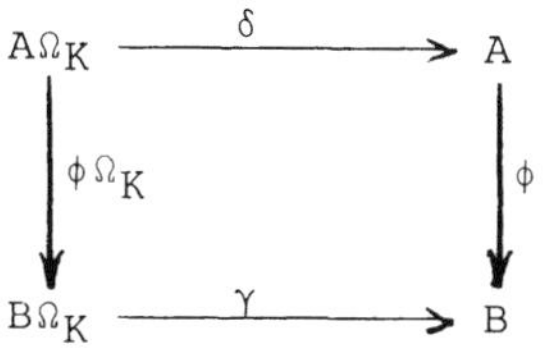

(1.1)

Let Ω_K-alg denote the category of Ω_K-algebras. As $A\Omega_K$ is a coproduct with injections $in_\omega : A^{\omega s} \to A\Omega_K$ δ is the same thing as a family of operations $(\delta_\omega : A^{\omega s} \to A \mid \omega \in \Omega)$ on A, defined by (1.2).

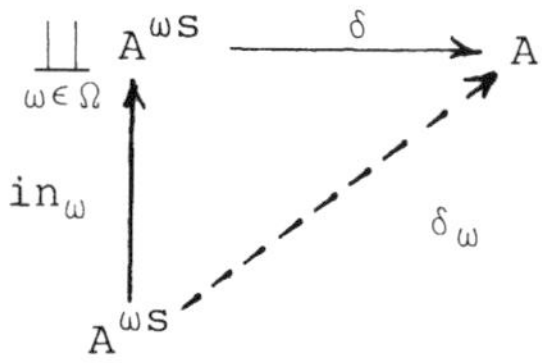

(1.2)

If K^{op} denotes the dual categroy of K than an $\Omega_{K^{op}}$-algebra is also called an <u>Ω_K-algebra</u> and if the base category K is clear from the context we only say Ω-algebra and Ω-co-algebra. In this paper we will only regard the case where K = <u>set</u> the category of sets and functions. In this case a k-ary operation $\delta_\omega : A^k \to A$ has its usual meaning. For Ω-co-algebras we have $\delta : A\Omega_{set^{op}} \to A$ in <u>set^{op}</u> which in set is a function $\delta : A \to \prod_{\omega\in\Omega} {}^{\omega s}A$ where ${}^kA$ denotes k-fold coproduct of A .

We also write $A\tilde{\Omega}$ for the set $A\Omega_{Set^{op}}$. In this case we have projections $pr_\omega : A\tilde{\Omega} \to {}^{\omega s}A$ and again δ is the same thing as a family of functions $(\delta_\omega : A \to {}^{\omega s}A \mid \omega \in \Omega)$ by (1.3).

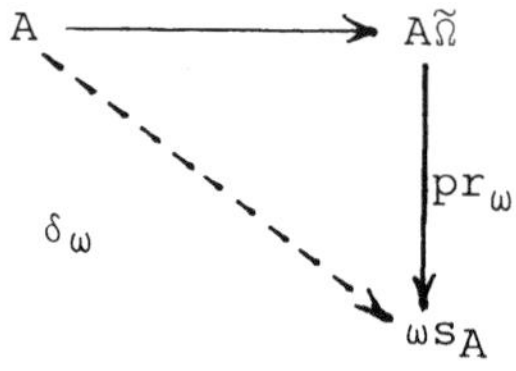

(1.3)

This is just the dual of (1.2). We call a function $\delta_\omega : A \to {}^kA$

a k-ary <u>co-operator</u> and also represent $^kA = A \times \underline{k}$ where $k = \{1,\ldots,k\}$. The intuitive meaning of a cooperation $\delta_\omega : A \to A \times \underline{k}$ is given by a box

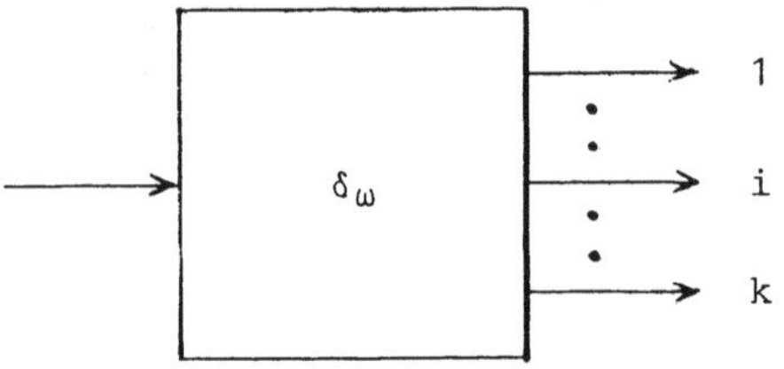

with one input line and k output lines. If we put $a \in A$ into the box then the element a' leaves on line i if $a\delta_\omega = (a',i)$. This is the fundamental component of a flow-diagram to be defined later.

We now regard some properties of free algebras. If A is a set, then a free Ω-algebra over A (generated by A) is an Ω-algebra $(X,\xi A)$ together with a function $\eta A : A \to X$ (inclusion of generators) such that for any Ω-algebra (B,γ) and any function $g : A \to B$ there exists a unique morphism $g^* : X \to B$ of Ω-algebras. g^* is the algebraic extension of g (1.4):

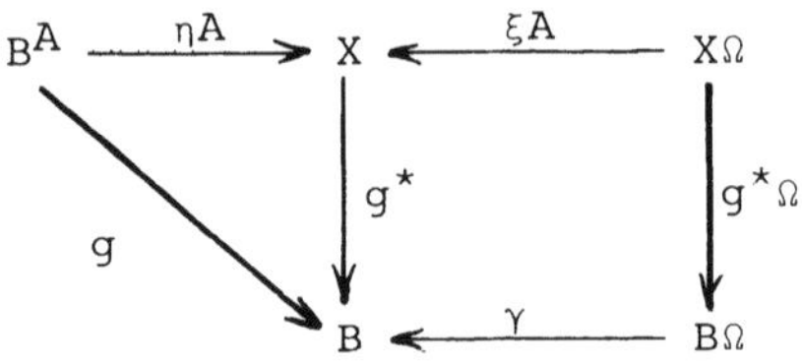

(1.4)

It is well known, that in <u>Set</u> free Ω-algebras always exist and that the free Ω-algebra generated by A is just given (up to isomorphism) by the set of all Ω-terms (expressions) AT_Ω in the generators $a \in A$. T_Ω : <u>Set</u> $\to$ <u>Set</u> is again a functor and we have two natural transformations η : <u>Set</u> $\to T_\Omega$ and $\xi : T_\Omega\Omega \to T_\Omega$ of which 1.4 shows one component. The algebra structure on AT_Ω is just forming new expessions out of expressions and an operatorsymbol.

For any Ω-algebra (B,γ) the elements $t \in AT_\Omega$ define operators $\gamma_t : B^A \to B$ of arity A simply by making the set $[B^A,B]$ of functions from B^A to B into an Ω-algebra:

Let $AQ_B := [B^A,A]$, $\omega \in \Omega$ with $\omega s = k$ and $f_1,\ldots,f_k \in AQ_B$ then define $(f_1,\ldots,f_k)\gamma^*_\omega$ by (1.5):

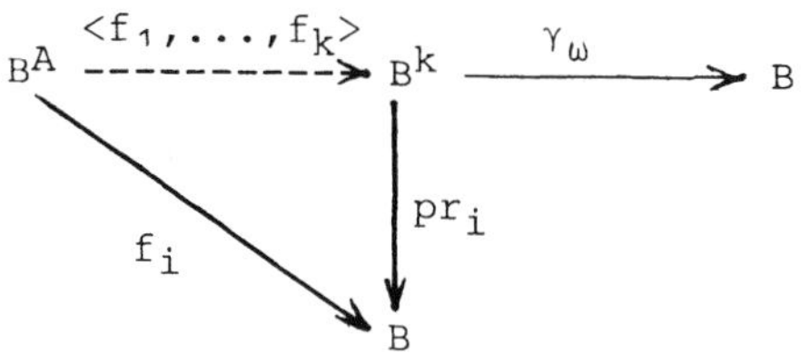

(1.5)

where $<f_1,\dots,f_k>$ is defined by the product property of B^k and we set $(f_1,\dots,f_k)\gamma^*_\omega = <f_1,\dots,f_k>\gamma_\omega$. This is just composition of functions. We only need to interprete elements $a \in A$ as operations $\hat{a} : B^A \to B$ and we chose the projections $pr_a : B^A \to B$. Let this function be denoted as $A\pi : A \to AQ_B$. Then (1.6) defines a unique morphism of Ω-algebras

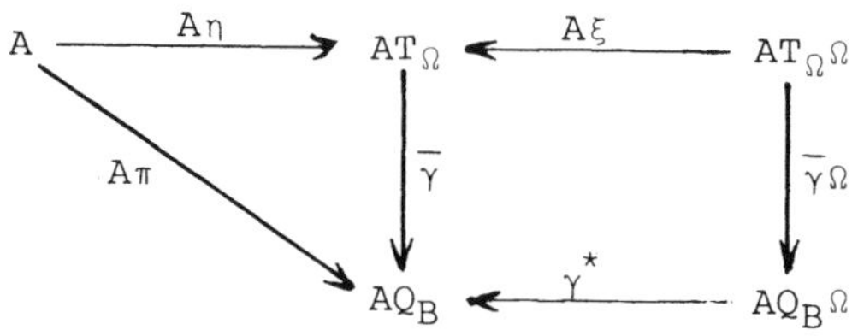

(1.6)

The interesting observation for this paper is the fact that if (B,γ) is an Ω-co-algebra then the set of co-operations $[B, B \times A]$ of arity A on B also admits a canonical Ω-algebra structure, and this is still true for the bigger set $Par[B, B \times A]$ of partial co-operators.

For sets A,B let AP_B denote the set $Par[B, B \times A]$ of partial co-operators $B \to B \times A$ of arity A . As $B \times A$ is the A-fold co-product of B in the category <u>Par</u> of sets and partial functions we can construct the composition of co-operators dual to (1.5). Let (B,δ) be a Ω-co-algebra and $\omega \in \Omega$ with $\omega s = k$ and $f_1,\dots,f_k \in AP_B$ then we define $(f_1,\dots,f_k)\hat{\gamma}_\omega$ by (1.7)

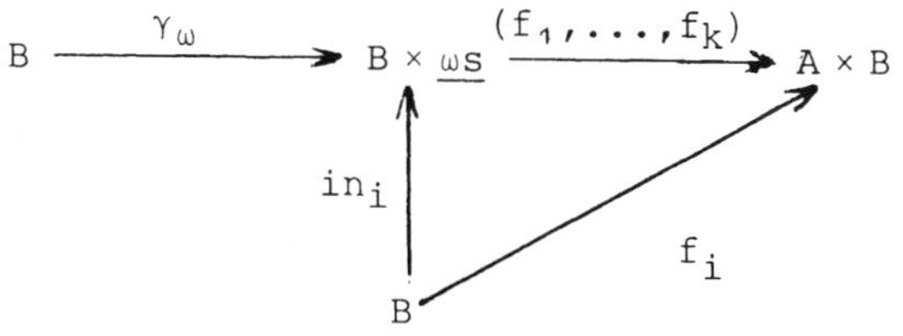

(1.7)

where $in_i : B \to B \times \underline{\omega s}$ are the co-product injections and $(f_1,\dots f_k)$

is defined by the co-product property. So $(f_1,\ldots,f_k)\hat{\gamma}_\omega := \gamma_\omega(f_1,\ldots,f_k)$ making $(AP_B,\hat{\gamma})$ into an Ω-algebra. Clearly $(f_1,\ldots,f_k)$ is a total function if $f_1,\ldots,f_k$ are all total and thus the total co-operators form a subalgebra. Again if we choose a function $A \to AP_B$ then there exists a unique morphism of Ω-algebras $\tilde{\gamma} : AT_\Omega \to AP_B$ which interprets any Ω-term over A as a (partial) co-operator on B of arity A. We choose the coproduct inclusions $in_a : B \to B \times A$ and thus get the unique $\tilde{\gamma} : AT_\Omega \to AP_B$. As the generators $a \in A$ are mapped into the total co-operators $in_a : B \to B \times A$ the image of AT_Ω only consists of total co-operations.

2. Operatornetworks and flow-diagrams

In this section we want to define Ω-nets and regard them as flow-diagrams which are then interpreted by Ω-co-algebras which are the machines.

By the constructions of the last section it was shown how such a machine interprets very restricted flow-diagrams (namely finite trees which are the elements of any AT_Ω). So we have to answer the question how can the machine (Ω-co-algebra) interprete arbitrary flow-diagrams.

Def.: Let (Ω,s) be a ranked alphabet, X,Y sets $(X \cap Y = \emptyset)$ and $F : X \to (X+Y)\Omega$. Then the triple $N = (X,Y,f)$ is called an Ω-net.

f is the "local" description of a directed graph and if $xf = (z_1,\ldots,z_k)\omega \in (X+Y)\Omega$ then we represent this graphically by:

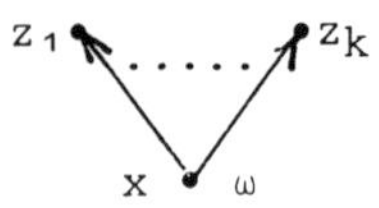

where $X + Y$ are the nodes, devided into "interior" nodes (X) and "boundary" nodes (Y) or exits in the language of flow-diagrams. Interior nodes are labelled with operator symbols such that the number of edges leaving x coincides with the arity of the label of x. Ω-nets thus represent all graphs of this kind which we also may regard as flow-diagrams if we assume that any interior node $x \in X$ with label $\omega \in \Omega$ represents a co-operator of arity ωs.

The interpretation of such a net is intuitively described as follows: Take an arbitrary Ω-co-algebra (A,δ) as a "machine", and the net $f : X \to (X+Y)\Omega$ as a "program". A is the "state space" of the machine and $X + Y$ the "control space" of the program. Choosing an arbitrary point $x \in X$ (entry point) in control space and an arbitrary state $a \in A$, then the program and the machine determine a new state a' and a new point z' of the control space $X + Y$ if $xf = (z_1,\ldots z_k)$ and $a\delta = (a',i)$ and $z' = z_i$.

If $z \in Y$ (exit) the machine "stops", else it continues. The total effect is that we may assign a partial co-operator $x\alpha : A \to A \times Y$ to any interior node $x \in X$, which tells us that we exit at point $y \in Y$ in state a' if we start at x with state a, or never reach an exit if $(a)x\alpha$ is not defined.

3. Equational interpretation of flow diagrams by co-algebras

We will use the terms flow diagram and Ω-net synonymously. If $f : X \to (X+Y)\Omega$ is a flow diagram we may choose any $x \in X$ as an entry point of the program. We are interested in the partial co-operation $x\alpha : A \to A \times Y$ which may be assigned to x, describing the meaning of the program started at x.

In [M79] an Ω-net is regarded as a formal system of equations. Given an Ω-algebra (B,γ) then this formal system can be interpreted as a genuine system which may or may not have solutions. In [M79] criteria for the existence of solutions were given and universal solutions constructed. We now want to use this idea to define the semantics of a flow-diagram as a solution of an equation defined by the interpreting co-algebra.

If $\alpha : X \to B$ and $\beta : Y \to B$ are functions we denote by $(\alpha,\beta) : X+Y \to B$ the unique function defined by the coproduct property of $X + Y$.

<u>Def.:</u> Let $N = (X,Y,f)$ be an Ω-net, (B,γ) an Ω-algebra and $\beta : Y \to B$ a function then diagram (3.1) defines a transformation $f_\gamma(\beta) : B^X \to B^X$:

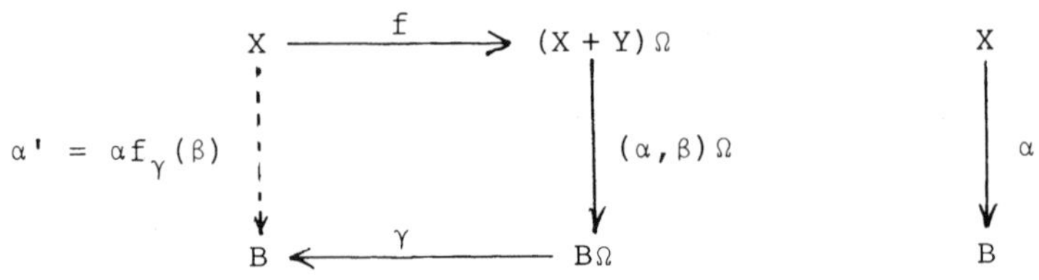

(3.1)

<u>Def.:</u> Referring to (3.1) $\alpha : X \to B$ is called a <u>solution</u> of f in (B, β) <u>under parameter</u> $\beta : Y \to B$ iff $\alpha f_{\gamma}(\beta) = \alpha$

It can be shown that for this situation there always exists a designated Ω-algebra in which f has a solution such that any other solution in any Ω-algebra is obtained from this (universal) solution by a unique morphism of Ω-algebras.

<u>Lemma:</u> If there exists an Ω-algebra (B', γ') and a solution $\alpha' : X \to B'$ under parameter $\beta' : Y \to B'$ and a morphism $\phi : B' \to B$ of Ω-algebras with $\beta = \beta'\phi$ then $\alpha'\phi$ is a solution of f under β.

So to find a solution in (B, γ) it is sufficient to construct an Ω-algebra (B', γ') with a solution α' and then define a suitable morphism $\phi : B' \to B$.

<u>Def.:</u> Let $N = (X, Y, f)$ be an Ω-net and (A, δ) an Ω-co-algebra. An <u>equational interpretation</u> of N in (A, δ) is a solution of f in the Ω-algebra (YP_A, δ^*) under parameter $Y\tau : Y \to YP_A$ (3.2).

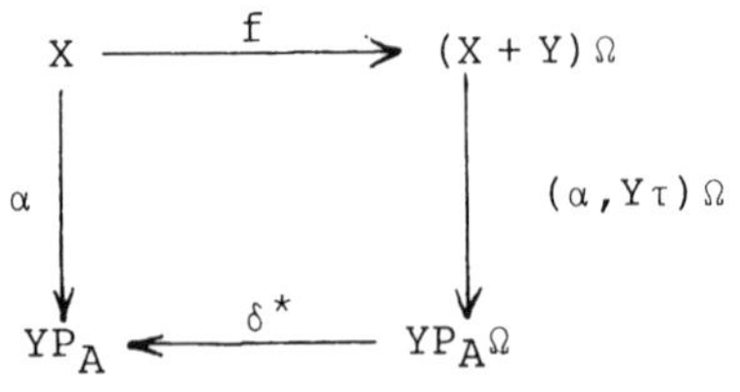

(3.2)

This is a precise definition of what an interpretation is but it does not tell whether solutions exist and if so whether they are unique.

We now outline the algebraic construction of solutions. Let $\mathbb{N}_+ := \mathbb{N} - \{0\}$ and for any set A define the function $|\,| : A\Omega \to \mathbb{N}$ by $|(a_1, \ldots, a_k)\omega| = k = \omega s$. Let A^* denote the free monoid over A with $\varepsilon \in A^*$ the empty word.

<u>Def.:</u> An <u>Ω-tree</u> is an Ω-net $M = (T,S,g)$ with $T,S \subseteq \mathbb{N}_+^*$ and

i) $\varepsilon \in T + S$

ii) $ui \in T + S$ iff $u \in T$ and $i \leq |ug|$

An Ω-tree has the graphical structure of a tree and ε is its root. We consider Ω-trees more detailed in [DM80]. Let Y be a set, $M = (T,S,g)$ an Ω-tree and $\lambda : S \to Y$ a function. Then the pair (M,λ) is called an Ω-tree over Y and YT_Ω^* is the set of all Ω-trees over Y (clearly $T_\Omega^* : \underline{Set} \to \underline{Set}$ defines a functor).

The Ω-trees over Y carry a canonical Ω-algebra structure $Y\xi^* : YT_\Omega^*\Omega \to YT_\Omega^*$, which intuitively is given by forming a new tree out of k trees $(t_1,\ldots,t_k)$ and an operatorsymbol ω by introducing a new root ε label it with ω and connect it to the roots of the given trees. We omit the formal construction in this paper. There is a natural inclusion $Y\eta^* : Y \to YT_\Omega^*$ which assigns the one point tree $S = \{\varepsilon\}$, $T = \emptyset$ and $\varepsilon\lambda = y$ to $y \in Y$.

<u>Theorem:</u> Let $N = (X,Y,f)$ be an Ω-net then it has a solution in $(YT_\Omega^*, Y\xi^*)$.

The solution is given by assigning to $x \in X$ its total unfoldment of f in the sense of [DM80].

<u>Theorem:</u> For any set Y and any Ω-co-algebra (A,δ) there exists a morphism of Ω-algebras $Y\bar{\delta} : YT_\Omega^* \to YP_A$ such that $Y\eta^* Y\bar{\delta} = Y\tau$ (3.3).

$$\begin{array}{ccccc} Y & \xrightarrow{Y\eta^*} & YT_\Omega & \xleftarrow{Y\xi^*} & YT_\Omega^*\Omega \\ & \searrow^{Y\tau} & \downarrow{Y\bar{\delta}} & & \downarrow{Y\bar{\delta}\Omega} \\ & & YP_A & \xleftarrow{\delta^*} & YP_A\Omega \end{array} \tag{3.3}$$

Together with the lemma above this gives the result:

<u>Theorem:</u> Any Ω-net $N = (X,Y,f)$ has an equational interpretation in any Ω-co-algebra (A,δ).

Such a solution is explicitely constructed by the above results.

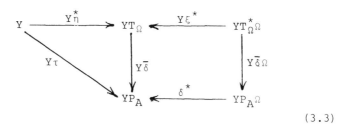

4. Connections with continous algebra interpretations

Though the results of the last section together with earlier results about unfoldments and coverings [DM80] were achieved mainly by constructions using categorical limits or colimits, the connection with continuous algebras [ADJ77] is quite close. It is not difficult to show that the Ω-algebra (YP_A, δ^*) induced by the Ω-co-algebra (A, δ) is indeed an ω-continuous Ω-algebra [ADJ77]. Further the transformation $f_\delta(\beta)$ defined by the diagram (3.1) is ω-continuous in the product space YP_A^X. A solution of f in the co-algebra (A, δ) is the same thing by definition as a fixpoint of $f_\delta(\beta)$ and thus the minimal fixpoint can be constructed in the usual way as the supremum of the sequence $(\perp f_\delta^i(\beta) \mid i \in \mathbb{N})$. Regarding (3.4) it should be mentioned that YT_Ω^* is the free continuous Ω-algebra generated by the set Y and as YP_A is also a continuous Ω-algebra $Y\bar{\delta} : YT_\Omega^* \to YP_A$ is then uniquely determined by $Y\tau : Y \to YP_A$ in the category of continuous Ω-algebras.

References

ADJ77 Goguen, J.A.; Thatcher, J.W.; Wagner, E.G.; Wright, J.B.: Initial Algebra Semantics and Continuous Algebras, J. Assoc. Comp. Mach. 24, 1977 (68 - 95)

AM74 Arbib, W.A.; Manes, E.G.: Machines in a category - an expository introduction, SIAM Review 16, 1974 (163 - 192)

AT78 Adámek, J.; Trnková, V.: Varietors and machines, Tech. Report COINS 78-6, University of Massachusetts, Amherst, 1978

CC76 Clark, K.L.; Cowell, D.F.: Programs, Machines and computations, McGraw-Hill, London 1976

DM80 Dittrich, G; Merzenich, W.: Unfoldments and coverings of operator-works (to appear)

E77 Elgot, C.C.: Some "geometrical" categories associated with flow-chart schemes, Lecture Notes in Comp. Sci. 56, Springer Heidelberg 1977 (256 - 259)

M79 Merzenich, W.: Allgemeine Operatornetze als Fixpunktgleichungen, Habilitationsschrift, Universität Dortmund, 1979

RANDOM ACCESS MACHINES AND STRAIGHT-LINE PROGRAMS

by

Friedhelm Meyer auf der Heide
University of Bielefeld
Faculty of Mathematics

4800 Bielefeld 1

W.-Germany

and

Anton Rollik
University of Dortmund
Faculty of Mathematics

4600 Dortmund

W.-Germany

Abstract: A method to simulate random access machines by decision trees with unbounded degree is used to prove that every random access machine with operation set {+,-,*} and indirect addressing which evaluates a polynomial can be simulated by a straight-line program without any loss of time.

Introduction: In (i) and (ii) a method is presented to simulate a random access machine (RAM) by a binary decision tree for nearly all inputs. This method was used to generalize lower time-bounds for decision trees to RAM's. In this paper we present a simulation of RAM's by trees whose degree is bounded by its depth. This simulation works for all inputs.

First we describe RAM's and straigth-line programs (SLP's). A RAM M is specified by

- a set of registers addressed by $Z = ..-1,0,1..$. The 0-th register is called the accumulator. Each register can store an element n of Z. Then n is the contents of the i'th register. Notation: $<i> = n$.

- an <u>instruction counter b</u>.
- a <u>program,</u> i.e. a finite sequence of numbered instructions, each from the set I specified below.

The actual contents of b determines the instruction to be executed next.

<u>The instruction set I:</u>

instruction	meaning	
halt	M stops	
c-load (k)	$<0> \leftarrow k$	$b \leftarrow b+1$
load (k)	$<0> \leftarrow <k>$	$b \leftarrow b+1$
i-load (k)	$<0> \leftarrow <<k>>$	$b \leftarrow b+1$
store (k)	$<k> \leftarrow <0>$	$b \leftarrow b+1$
i-store (k)	$<<k>> \leftarrow <0>$	$b \leftarrow b+1$
if $<0> > 0$ goto a	if $<0> > 0$ then	$b \leftarrow a$
	else	$b \leftarrow b+1$
add (k)	$<0> \leftarrow <0> + <k>$	$b \leftarrow b+1$
sub (k)	$<0> \leftarrow <0> - <k>$	$b \leftarrow b+1$
mult (k)	$<0> \leftarrow <0> * <k>$	$b \leftarrow b+1$
goto a		$b \leftarrow a$

We say <u>M started with $(x_1,\dots,x_n) \in Z^n$ computes $(y_1 \dots y_m) \in Z^m$,</u> if (1) and (2) implie (3):

1. Initially $<0> = n$, $<i> = x_i$ for $i = 1..n$, $<i> = 0$ for all other i and $b = 1$.
2. M successively executes the instructions given by b.
3. M stops with $<i> = y_i$ for $i = 1..m$.

<u>M computes the function $f: Z^n \to Z^m$</u> if M started with $\bar{x} \in Z^n$ computes $f(\bar{x})$.

The number of instructions M started with $\bar{x}$ executes is called <u>$t_M(x)$.</u>

Let M compute $f: Z^n \to Z^m$. If $\max(t_m(\bar{x}), \bar{x} \in Z^n)$ exists, we call it <u>T(M).</u> In this case, M is <u>a bounded RAM.</u>

A SLP S is a sequence of functions $A_1..A_\ell : R^n \to R$ (*) such that $A_i(x_1..x_n) = x_i$, $i = 1..n$ and for $j > n$ $A_j(x) = A_r(x) \circ A_s(x)$ with some $r,s < j$. "$\circ$" denotes an operation from $\{+,-,*\}$. The complexity measure is given by $\underline{T(S)} = \ell - n$.

S computes $f: R^n \to R^m$, if there are $i_1..i_m$ such that A_{i_j} is the j-th component of f.

Remark: SLP's can only compute polynomials.

For a function $f: R^n \to R^m$, let $f/_{Z^n}$ be the restriction of f on Z^n. In this paper we prove the following

Theorem: Let $p: R^n \to R^m$ be a polynomial such that $p/_{Z^n}$ only has value in Z^m. If a bounded RAM M computes $p/_{Z^n}$, then there is a SLP S which computes p with $T(S) \le T(M)$.

This theorem allowes to generalize all lower bounds for SLP's (compare for example (iii)) to RAM's.

1. "Unrolling" a RAM

In this chapter we present a method to attach a decision tree to a RAM.

Let M be a RAM which computes p as defined in the theorem. We represent a configuration of M by the value of b and by the addresses and contents of those registers which are already visited.

(*) R denotes the set of real numbers.

a configuration

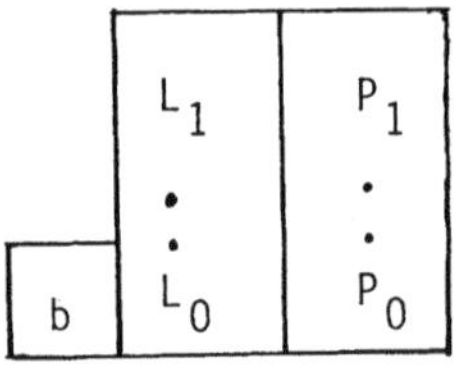

b is the instruction counter, L_i are the addresses of the registers with contents P_i .

Construction of a tree T for M

Each vertex of T is a configuration C of M . Some edges contain restrictions.

Let $k_1 .. k_r$ be those addresses which are refered to in M by "add(k_i)" etc. and which are larger than n .

Then the root of the tree is the following configuration:

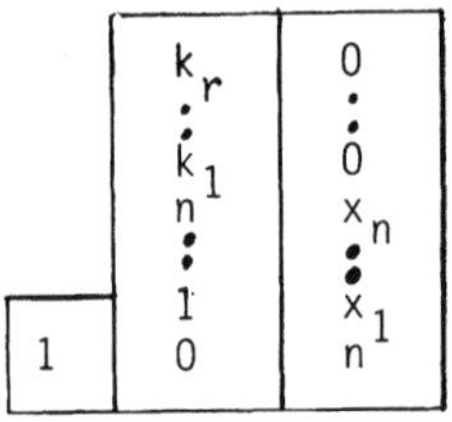

Let C be a vertex of T .

C =

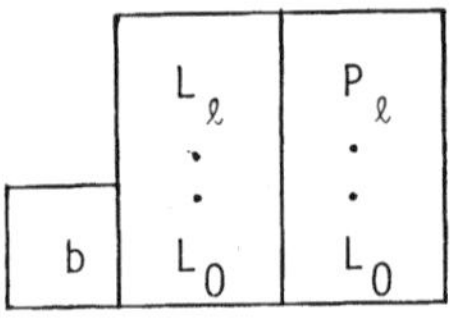

Case 1: b = "add(k)" .

Then C gets one son C' which differs from C only in P_0 and b . $P_0 \leftarrow P_0 + P_k$, $b \leftarrow b + 1$.

The construction is now obvious for all instructions except "if <0>>C goto a", "i-store(k)", and "i-load(k)".

Case 2: b = "if <0>>0 goto a" .
Then C gets two sons C_1 and C_2. They only differ from C by the instruction counter. It is $b+1$ for C_1 and a for C_2. The edge (C,C_1) gets the restriction "$P_0 \le 0$" and (C,C_2) gets "$P_0 > 0$" .

Case 3: b = i-store(k) .
Then C gets $\ell+1$ sons $C_1..C_{\ell+1}$. For $i \le \ell$, C_i differs from C by "$P_i \longleftarrow P_0$" . The edge (C,C_i) gets the restriction "$P_k - L_i = 0$" .

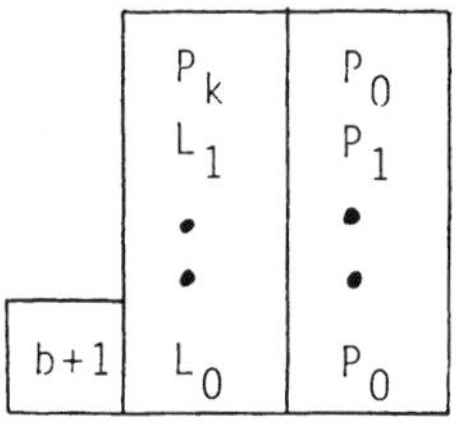

The edge $(C,C_{\ell+1})$ gets the restriction "$P_k - L_i \neq 0$ for all $i = 0..\ell$" .

The case b = "i-load(k)" is analogous.
C is a leaf, if b = "halt" .

For some leaf C of T let A(C) be the set of all input for M which fulfil all the restrictions on the path from the root to C .

We notice, that the sets A(C) for all leafs C of T are a partition of Z^n (i.e., the union of the A(C)'s is Z^n and they are pairwise disjoint).

By the variables and the operations on a path from the root to a leaf C , a SLP S(C) is defined in the obvious way. As M computes p , S(C) computes a function h_C with $h_C/_{A(C)} = p/_{A(C)}$. T(S(C)) is at most the length of the path which is at most T(M) .

Therefore the theorem is proved if the following proposition holds:

Proposition: T has a leaf C such that

$$h_C = p .$$

Let a set $B_1 x \ldots x B_n$, $B_i \subset R$ and $\# B_i \geq c$ for $i = 1..n$ be called a c-cube.

The proposition is a direct consequence of the following two lemmas:

Lemma 1: Let $p,q: R^n \to R^m$ be polynomials of degree d and B a $(d+1)$-cube, then $p/_B = q/_B$ implies $p = q$.

As the degree of h_C is bounded by $2^{T(M)}$, we now can conclude that, if we find a leaf C such that $A(C)$ contains a $(d+1)$-cube with $d = \max(\text{degree of } p, 2^{T(M)})$, the proposition is proved. The existence of such a leaf is guaranteed by

Lemma 2: Let $A_1..A_v$ be a partition of Z^n. Then for each c there exists an i such that A_i contains a c-cube.

Lemma 1 can be found in a similar form in (iv).

Lemma 2 can be proved with the help of methods from partition theory.

References

i) W.J. Paul, J. Simon: *Decision Trees and Random Access Machines.* Symp. über Logik und Algorithmik, Zürich, 1980

ii) P. Klein, F. Meyer auf der Heide: *Lower Time Bounds for the Knapsack problem on several Random Access Machines.* Proc. of the 10. GI-Jahrestagung, Saarbrücken, W.-Germany, 1980

iii) A. Borodin, I. Monroe: *The computational Complexity of algebraic and numeric Problems.* New York, Elsevier, 1975

iv) L. van der Waerden: *Algebra.* Band 1, Springer Verlag, Berlin, Heidelberg, New York, 1966

ON THE LBA PROBLEM

BURKHARD MONIEN

UNIVERSITÄT PADERBORN
PADERBORN, WEST - GERMANY

I. INTRODUCTION

The first definition of the linearly bounded automaton was given by John Myhill in 1960, (42), who generalized the notion of a two-way finite automaton, (43), and introduced an automaton which can read and write on the portion of the tape occupied by its input word. Noam Chomsky, (8), had defined in 1959 four classes of grammars and had pointed out the relationships between type 0 grammars and Turing machines and between type 3 grammars and finite state automata. The relationship between context-free grammars and pushdown automata was found in 1962 by N. Chomsky and M. Schützenberger, (9 , 47). In 1963 P.S. Landweber, (30), showed that every language acceptable by some deterministic linearly bounded automaton is a context-sensitive language and S.Y. Kuroda, (29), found in 1964 that the classes defined by context-sensitive grammars and by nondeterministic linearly bounded automata are identical.

Since this time it is an unanswered question whether DCSL, the class of languages acceptable by deterministic linearly bounded automata, is equal to CSL, the class of languages acceptable by nondeterministic linearly bounded automata. This problem is known now as LBA problem.

If DCSL = CSL holds then the complement of every context-sensitive language is also a context-sensitive language. Therefore the question whether CSL is closed under complementation is closely related to the LBA problem.

The LBA problem can be described also as a problem in complexity theory. This is true since a linearly bounded automaton can be considered to be a Turing machine accepting with a linear space bound. We want to give some definitions:

Let M be a (deterministic or nondeterministic) Turing machine. In order to be able to define also space complexity classes with sublinear space bounds, we assume that M has a two-way read-only input tape and a separate work tape. Let L(M) be the language accepted by M. For $x \in L(M)$, $SPACE_M(x)$ denotes the minimum number of work tape cells visited by M in a computation accepting x. For $x \notin L(M)$ we set $SPACE_M(x) = \infty$. For f a function from the nonnegative integers to the nonnegative integers, define the language recognized by M within space f, denoted by $LSPACE_f(M)$, to be the set $\{x \mid SPACE_M(x) \leq f(|x|)\}$. We shall say that M accepts within space f if $L(M) = LSPACE_f(M)$. Define

$DSPACE(f) = \{L \mid L = L(M) = LSPACE_f(M)$ for some deterministic Turing machine $M\}$

$NSPACE(f) = \{L \mid L = L(M) = LSPACE_f(M)$ for some nondeterministic Turing machine $M\}$

Some authors, (23, 24 , 50) define complexity classes by bounding the space of all computations or by using other models of Turing machines. In most cases this variation in the definition does not make a difference, however, if the bounding function f is "constructible". A function f is fully space-constructible if there is a deterministic Turing machine M such that $SPACE_M(x) = f(|x|)$ holds for all input strings x, (45,5o).

Using this notation CSL = NSPACE(n) and DCSL=DSPACE(n). Therefore the LBA problem is equivalent to the question whether DSPACE(n) = NSPACE(n) holds.

(We write also DSPACE(f(n)) instead of DSPACE(f), i.e. DSPACE(n) = DSPACE(f), where f is the function defined by f(n) = n for all n.)

In an analogous way let DTIME(f) and NTIME(f) be defined by using multitape Turing machines as machine model. We shall use the following abbreviations:

$\mathbb{P} = \bigcup_{k\geq 1} \mathrm{DTIME}(n^k)$, $\mathbb{NP} = \bigcup_{k\geq 1} \mathrm{NTIME}(n^k)$

$\mathbb{L} = \mathrm{DSPACE}(\log n)$, $\mathbb{NL} = \mathrm{NSPACE}(\log n)$.

In the last years there has been great interest in finding complete problems for various classes defined by Turing machines or formal grammars, respectively. This work is important in two aspects. First, to find a complete language for such a class is to show that a single problem represents the complexity of the whole class. Hence, the complexity of the class is better understood. Secondly, to identify a "natural" problem as being complete for a class is to classify the complexity of this problem. A great amount of this research activity has been concerned with problems which are complete for $\mathbb{NP}$, (12, 16, 28).

We shall now give some definitions:

Definition: Let Σ and Δ be alphabets and let $f: \Sigma^* \to \Delta^*$ be a function. f is log space computable if there is a deterministic Turing machine with a two-way read-only input tape, a one-way output tape, and a two-way read-write work tape, which when started with $x \in \Sigma^*$ on its input tape will halt having written $f(x) \in \Delta^*$ on its output tape and having visited at most $\log_2(|x|)$ distinct tape squares on its worktape.

Definition: Let F be a class of functions and let $A \subset \Sigma^*$, $B \subset \Delta^*$ be arbitrary sets. A is F-reducible to B, denoted by $A \leq_F B$, if there is a function $f: \Sigma^* \to \Delta^*$ such that $f \in F$ and

$$\forall x \in \Sigma^* \; [\, x \in A \Leftrightarrow f(x) \in B \,] .$$

(Note that $A \leq_F B$ implies $\bar{A} \leq_F \bar{B}$.)

We use the terms "log space reducible" ($\leq_{log}$) or "homomorphism reducible" ($\leq_{hom}$) or "polynomial time reducible" ($\leq_{pol}$) if F is the class of log space computable functions, or of homomorphisms, or of polynomial time computable functions, respectively. All of these three types of reductions are binary relations on languages. Considered as binary relations they are clearly reflexive and can be shown to be transitive (25). In general, let ≤ be a binary relation on languages and let Ω denote an arbitrary family of languages. Define $\mathrm{CLOSURE}_{\leq}(\Omega)$ to be the set $\{L \mid L \leq L' \text{ for some } L' \in \Omega\}$.

If $\leq$ is reflexive, as the given reducibility relations are, then clearly $\Omega \subseteq \text{CLOSURE}_{\leq}(\Omega)$. We shall say that Ω is closed under the relation $\leq$, if $\text{CLOSURE}_{\leq}(\Omega) \subseteq \Omega$. A language L is complete for Ω with respect to $\leq$, or simply complete for Ω when $\leq$ is understood, if $L \in \Omega$ and $\Omega \subseteq \text{CLOSURE}_{\leq}(\{L\})$. A language L is hard for Ω with respect to $\leq$, or simply hard for Ω when $\leq$ is understood, if $\Omega \subseteq \text{CLOSURE}_{\leq}(\{L\})$.

It is known that the complexity classes $\mathbb{P}$ and $\mathbb{NP}$ are closed under all three of the forms of reduction given. The complexity classes $\mathbb{L}$ and $\mathbb{NL}$ are closed under log space reductions and homomorphism reductions. Finally the classes DSPACE(n), NSPACE(n) and many formal language classes, such as the family of context-free languages (CFL) and the family of linear context-free languages (LINEAR-CFL) are closed under homomorphism reductions.

It is not difficult to construct a language L_o which is complete for NSPACE(n) with respect to $\leq_{hom}$. Consider the "universal" language, (22), $L_o = \{¢ M_i ¢ \text{ CODE}_i(a_1a_2...a_n) ¢ \mid a_1a_2...a_n \in \text{SPACE}_{id}(M_i)\}$, where $¢ M_i ¢$ denotes some simple encoding of the nondeterministic Turing machine M_i and $\text{CODE}_i(a_1a_2 ...a_n)$ is any symbol by symbol encoding of sequences over an arbitrary alphabet into a fixed alphabet $\{0,1,¢\}$ such that for every symbol a $|\text{CODE}_i(a)|$ is a least as large as the tape alphabet of M_i.

L_o belongs to NSPACE(n) since the encoding $\text{CODE}_i(a_1...a_n)$ provides enough space to simulate M_i and $L(M_i) \leq L_o$ for each nondeterministic Turing machine accepting within space n by means of the homomorphism mapping the left endmarker onto $¢ M_i ¢$, the right endmarker onto ¢ and each input symbol a onto $\text{CODE}_i(a)$. Therefore L_o is complete for NSPACE(n) with respect to $\leq_{hom}$. Since all the classes we have considered are closed under homomorphism reduction, this implies

$L_o \in \text{DSPACE}(n) \Leftrightarrow \text{NSPACE}(n) = \text{DSPACE}(n)$

$\bar{L}_o \in \text{NSPACE}(n) \Leftrightarrow \text{NSPACE}(n)$ is closed under complementation

$L_o \in \mathbb{P} \Leftrightarrow \text{NSPACE}(n) \subset \mathbb{P}$.

II. SAVITCH'S THEOREM AND COMPLETE PROBLEMS FOR $\mathbb{NL}$

Viewed as a problem in complexity theory the LBA problem is a problem on just one space bound. It is a more general question to ask how arbitrary nondeterministic space bounded computations can be simulated by deterministic Turing machines. Many of the techniques we shall describe can be supplied also to answer questions about the relationships between other complexity classes (relationship between deterministic and nondeterministic time bounded computations, relationship between space bounded and time bounded computations), (5 , 13 , 22 , 34 , 44 ,).

Relationships between complexity classes can be transferred in some sense from classes defined by small functions to classes defined by large functions. We will demonstrate this by considering the relationship between deterministic and nondeterministic space complexity classes, (45).

Let $L \subset \Sigma^*$ be some recursive language, let $g: \mathbb{N} \to \mathbb{N}$ be some function with $g(n) \geq n$ for all n and take some $a \notin \Sigma$. Set $T_g(L) = \{ wa^{g(|w|)-|w|} | w \in L\}$. We describe how the complexity of accepting L and $T_g(L)$ depend on each other.

An algorithm for deciding whether $v \in (\Sigma \cup \{a\})^*$ belongs to $T_g(L)$ is defined by

(1) Test whether $v = wa^{g(|w|)-|w|}$ for some $w \in \Sigma^*$

(2) Test whether $w \in L$.

If we assume that the function g is easily computable, i.e. that the function $\hat{g}: \{1\}^* \to \{0,1\}^*$, where $\hat{g}(1^n)$ is defined to be the binary decomposition of g(n), is fully tape constructible, then we can construct a Turing machine M which needs $\log |v|$ cells in order to perform (1). (This Turing machine marks off $\log|v|+1$ cells at the beginning. If during the computation of $\hat{g}(1^{|w|})$ this portion of the tape is left, then v is rejected. Otherwise $|\hat{g}(1^{|w|})| \leq \log|v|+1$ and M can check whether $|v|=g(|w|)$, holds.) In performing (2) our Turing machine M simulates a Turing machine accepting L and if this machine accepts within space fg(n) for some function f, then M needs $fg(|w|) = f(|v|)$ cells. Therefore we get

$L \in NSPACE(fg(n))$ implies $T_g(L) \in NSPACE(f(n))$

for any function f such that $f(n) \geq \log n$ for all n.

On the other hand if we know a Turing machine M accepting $T_g(L)$ within space f, then we can define a Turing machine $\hat{M}$ accepting L which simulates M step by step and which stores on its work tape the content of the work tape of M ($\hat{M}$ needs $fg(|w|)$ cells to do this) and the position of the input head of M in binary notation, again we assume that the function $\hat{g}$ is fully tape constructible and in this case $\hat{M}$ needs $\log g(|w|)$ cells to do the necessary computations. Therefore we get

$T_g(L) \in NSPACE(f(n))$ implies $L \in NSPACE(fg(n))$ for any function f such that $f(n) \geq \log n$ for all n.

Both results hold also for deterministic Turing machines and combining these results we get:

Theorem 1: Let α, f, h be arbitrary functions and suppose that f,h fullfill some smoothness conditions and that $\log n \leq f(n) \leq h(n)$ holds for all n.

Then $NSPACE(f) \subset DSPACE(\alpha f)$

implies $NSPACE(h) \subset DSPACE(\alpha h)$.

Proof: Let the smoothness conditions guarantee that $g = f^{-1}h$ is a welldefined function, that $\hat{g}$ is fully space constructible and that $f(n) \leq h(n)$ implies $g(n) \geq n$ for all n. Then the considerations above imply:

$L \in NSPACE(h) = NSPACE(fg)$

$\Rightarrow T_g(L) \in NSPACE(f) \subset DSPACE(\alpha f)$

$\Rightarrow L \in DSPACE(\alpha fg) = DSPACE(\alpha h)$ □

Example:

1. $NSPACE(\log n) \subset DSPACE((\log n)^{1+\delta})$
$\Rightarrow NSPACE(h(n)) \subset DSPACE((h(n))^{1+\delta})$ holds for every function h such that $h(n) \geq \log n$ for all n.

2. $\mathbb{NL} = \mathbb{L} \Rightarrow$ CSL = DCSL.

The above method holds also for $\{L \mid \bar{L} \in$ NSPACE(f)$\}$ and we get as an additional corollary

3. $\mathbb{NL}$ is closed under complementation

$\Rightarrow$ CSL is closed under complementation.

These considerations justify the study of particularly the nondeterministic log space bounded computations. On the other hand it is not known whether containment results can be transferred from classes defined by large functions to classes defined by small functions. Therefore, regarding our present knowledge, it could be true that $\mathbb{L} \neq \mathbb{NL}$ and DCSL = CSL. Walter Savitch, (45), found a relationship between nondeterministic and deterministic space bounded computations.

Theorem 2: NSPACE(log n) $\subset$ DSPACE($(\log n)^2$).

The proof can be found in (45) and in some text books (2 , 24). Savitch used an idea which had been applied previously by Lewis, Stearns and Hartmanis, (31), to show CFL $\subset$ DSPACE($(\log n)^2$). We will see later that in fact this earlier result is a generalization of Savitch's theorem.

Actually Savitch has proved a stronger theorem:

(c1) If L is accepted by some nondeterministic Turing machine within space f, $f(n) \geq \log n$, and time T, simultaneously, then L $\in$ DSPACE($f(n) \cdot \log T(n)$).

In the meantime there has been a lot of activity to improve Savitch's theorem and a number of related results have been proved, but we still don't know any space function f such that NSPACE(f) $\subseteq$ DSPACE($f(n)^{1+\delta}$) holds for some $\delta < 1$.

We list some related results:

(c2) CFL $\subseteq$ DSPACE($(\log n)^2$), (31).

(c3) If L is accepted by some nondeterministic auxiliary pushdown automaton (aux PDA) within space f($f(n) \geq \log n$) and time T, simultaneously, then L $\in$ DSPACE($f(n) \cdot \log T(n)$), (1o, 32). (An aux PDA has besides its working tape one additional pushdown tape and the cells used on the pushdown tape are not counted.)

(c4) Deterministic context free languages are acceptable by deterministic Turing machines within space $(\log n)^2$ and time $n^2/(\log n)^2$, simultaneously, (7,13).

(c5) If L is accepted by some probabilistic Turing machine within space f($f(n) \geq \log n$) and time T, simultaneously, then L $\in$ DSPACE($f(n) \cdot \log T(n)$), (27).

(c6) NSPACE(f) $\subseteq$ DSPACE($f(n) \cdot \log n$) for $\log \log n \leq f(n) \leq \log n$, (39).

We are now looking for a problem which is complete for $\mathbb{NL}$. Since we will consider mainly log space reductions we will understand always that log space reductions are used if it is not stated otherwise. Note that if L is complete for NSPACE(log n) then

$$L \in \text{DSPACE}((\log n)^{1+\delta})$$

$$\Leftrightarrow \mathbb{NL} \subset \text{DSPACE}((\log n)^{1+\delta}).$$

In (45) W. Savitch described a problem which he called problem of "threadable mazes" and which is now wellknown under the name "graph accessability problem" (GAP) and he showed that this problem has the same complexity as the whole class NSPACE(log n).

Let $G = (V,E)$ be a finite directed graph where $V = \{1,2,...k\}$ for some $k \geq 1$. Let code (G) denote some simple encoding of G given by its adjacency lists. GAP denotes the set of encodings code (G) of graphs $G = (V,E)$, $V = \{1,...,k\}$ for some k, which possess a path from node 1 to node k.

Since a nondeterministic Turing machine can guess the sequence of nodes on the path and since each node can be stored in binary notation the language GAP belongs to $\mathbb{NL}$. On the other hand let M be a nondeterministic Turing machine with $L(M) = L\,SPACE_{\log n}(M)$. Let $w = a_1a_2...a_n$ be an input word of length n.

Construct the graph $G_{M,w} = G = (V,E)$ which has the storage states (SST) of M on input w as nodes. (A configuration of M is given as a 5-tuple (s,w,h,u,j), where s denotes the state, w the input word, h the position of the input head, u the inscription on the work tape and j the position of the work tape head. A SST, when w is understood, is the remaining 4-tuple.) G has an edge from the SST x to the SST y if and only if y is one of the finite number of SSTs that can be entered in one step from x. Let k be the number of different storage states. Associate the number 1 to the initial SST of M on input w and the number k to the terminal SST. Then $G_{M,w} \in GAP \Leftrightarrow M$ accepts w. Thus we have proved:

<u>Theorem 3</u>: GAP is complete for $\mathbb{NL}$.

This result can be improved in the sense that also the graph accessability problem for acyclic graphs, which can be encoded as a context free language, is complete for NSPACE(log n). This can be derived from the fact that every language $L \in NSPACE(\log n)$ is reducible to some language $\hat{L} \in 1WAY\,N\,SPACE(\log n)$, where 1WAY N SPACE(log n) is the class of all languages acceptable by log space bounded Turing machines which move its input head to the right in every step. We will indicate the proof technique, this technique is widely applicated to classes of languages acceptable by automata operating within polynomial time, (13,34,53,55).

Let Σ be an alphabet and $0 \notin \Sigma$. Define a function $T: \Sigma^* \to (\Sigma \cup \{0\})^*$ by $T(a_1...a_n) = a_1\, 0^{2\lceil \log n\rceil}\, a_2\, 0^{2\lceil \log n\rceil} \dots a_n\, 0^{2\lceil \log n\rceil}$ and for $k \in \mathbb{N}$ set

$T_k(w) = T(w)^{|w|^k}$. Now let L be some language acceptable by some nondeterministic Turing machine within space log n and time n^k, simultaneously. We will define some nondeterministic one-way log space bounded Turing machine $\hat{M}$ accepting $T_k(w)$, $w \in \Sigma^*$, if and only if $w \in L$. $\hat{M}$ simulates M step by step and stores on its working tape additionally the position of the input head of M in binary notation. If M moves its input head to the right then so does $\hat{M}$. Simultaneously it performs the binary addition of +1 on its working tape. $\hat{M}$ needs at most $2\lceil \log n\rceil$ steps to do so and to move the work tape head to its original position again and in each of these steps it moves its input head one cell to the right (for this reason we introduced the string $0^{2\lceil \log n\rceil}$ in between two symbols of Σ) until the next symbol of Σ is reached. If the input head of M remains on the same cell or moves to the left then $\hat{M}$ moves its input head to the next copy of $T(w)$ in $T_k(w)$. $\hat{M}$ uses the head position of M stored on its work tape to reach the corresponding position inside the next copy of $T(w)$ again. The strings $0^{2\lceil \log n \rceil}$ are needed again so that $\hat{M}$ can perform the necessary binary operation and move its input head to the right in each step. Since M performs at most $|w|^k$ steps $\hat{M}$ can simulate the whole computation of M in this way. Thus we have proved

To every $L \in \mathbb{N}L$ there exists $\hat{L} \in$ 1WAY N SPACE(log n) such that $L \leq \hat{L}$.

Now let us consider the graph accessability problem for acyclic graphs. We assume that the set of nodes $V = \{1,\ldots,k\}$ is ordered in such a way that $(i,j) \in E$ implies $j > i$. In order to get a simply acceptable language we use a different encoding and denote by accode (G) the string

$$[U(\delta_1^{(1)}), U(\delta_2^{(1)}),\ldots,U(\delta_{d(1)}^{(1)})] \; [U(\delta_1^{(2)}), U(\delta_2^{(2)}),\ldots, U(\delta_{d(2)}^{(2)})]$$

$$\ldots [U(\delta_1^{(k-1)}), U(\delta_2^{(k-1)}),\ldots, U(\delta_{d(k-1)}^{(k-1)})],$$

where $U(j)$ denotes the string 1^j and

$\{j \mid (i,j) \in E\} = \{i + \delta_\ell^{(i)} \mid 1 \leq \ell \leq d(i)\} \cap \{1,\ldots,k\}$.

ACGAP denotes the set of encodings accode (G) of acyclic graphs $G = (V,E)$, $V = \{1,\ldots,k\}$, which possess a path from node 1 to node k.

The proof showing that GAP is complete for NSPACE(log n) can be changed into a proof for the completeness of ACGAP. We associate to SSTs numbers in such a way that for any two SSTs x and y the number associated to $x = (s_1,h_1,u_1,j_1)$ is smaller then the number associated to $y = (s_2,h_2,u_2,j_2)$ if $h_1 < h_2$. Thus to every one-way log space bounded machine M and to every input w there is associated a graph $G = (V,E)$ such that $(i,j) \in E$ implies $i < j$ and accode $(G) \in$ ACGAP $\Leftrightarrow w \in L$.

Theorem 4: ACGAP is complete for $\mathbb{N}L$.

Furthermore ACGAP is a context-free language, in fact ACGAP is accepted by some nondeterministic oneway counter automaton (consisting of a finite memory, a oneway read-only input tape and a counter). Let us denote by CL the class of languages acceptable by such counter automata.

A counter automaton accepting ACGAP is constructed in quite an obvious way. When it reaches block i (corresponding to node i) then it carries one of the distances $\delta_\ell^{(i)}$, $1 \leq \ell \leq d(i)$, over to its counter and uses this counter to reach block $i + \delta_\ell^{(i)}$. It accepts w if and only if it can reach the (empty) block k in this way.

Theorem 5: ACGAP $\in$ CL $\subset$ CFL.

Note that the above theorem implies that

NSPACE(log n) = CLOSURE$_\leq$ (NSPACE(log n)) $\subset$ CLOSURE$_\leq$ (CFL) and therefore the result of Lewis, Stearns, Hartmanis is a generalization of Savitchs theorem. Hal Sudborough, (55), has shown that CLOSURE$_\leq$ (CFL) is equal to the class of languages acceptable by aux PDA within space log n and polynomial time, simultaneously.

In formal language theory it is an appropriate method to discuss the closure properties of language families with respect to certain set operations such as

$\cup, \cap$, o (concatenation) and * (Kleene * operation). It can be easily seen, that $\mathbb{L}$ is closed under $\cap, \cup$, o and that $\mathbb{NL}$ is closed under $\cup, \cap$, o, *. Using the methods described above we shall prove, (33):

Theorem 6: $\mathbb{L} = \mathbb{NL}$

$\Leftrightarrow$ $\mathbb{L}$ is closed under the application of the Kleene * operation.

In order to prove this result let $\hat{L}$ be the set of strings of the following form

$$[U(\delta_1^{(1)}), U(\delta_2^{(1)}), \ldots, U(\delta_{d(1)}^{(1)})] \; [U(\delta_1^{(2)}), U(\delta_2^{(2)}), \ldots, U(\delta_{d(2)}^{(2)})] \ldots [U(\delta_1^{(p)}), U(\delta_2^{(p)}), \ldots, U(\delta_{d(p)}^{(p)})]$$

such that $p = \delta_j^{(1)} - 1$ for some j, $1 \leq j \leq d(1)$. Then $\hat{L} \in \mathbb{L}$ ($\hat{L}$ is accepted by some deterministic oneway counter automaton) and $(\hat{L})^* =$ ACGAP. Therefore, if $\mathbb{L}$ is closed under the application of * , then ACGAP $\in \mathbb{L}$ and because of theorem 4 this implies $\mathbb{L} = \mathbb{NL}$.

We have proved in theorem 4 that the one-way counter language ACGAP is complete for $\mathbb{NL}$. This result was slightly improved in (37) where it was shown that there exists also a language which is complete for $\mathbb{NL}$ and which is acceptable by some nondeterministic one-way partially-blind counter automaton (such an automaton is allowed only in its last step to test whether its counter stores the number zero, (19)). Furthermore, using the completeness of ACGAP, Hal Sudborough (53) showed that there exists also a linear context-free language which is complete for $\mathbb{NL}$. This language is described best by using the Greibach construction, (17). Let $L \subset \Sigma^*$ be any language. The language GRE(L) consists of all strings of the form

$$[w_1^{(1)}, w_2^{(1)}, \ldots, w_{n(1)}^{(1)}] \quad [w_1^{(2)}, w_2^{(2)}, \ldots, w_{n(2)}^{(2)}] \quad \ldots \quad [w_1^{(k)}, w_2^{(k)}, \ldots, w_{n(k)}^{(k)}] ,$$

where $w_j^{(i)} \in \Sigma^*$ for $1 \leq i \leq k$ and $1 \leq j \leq n(i)$, for which there exists a sequence of positive integers $i_1, i_2, \ldots, i_k$ such that the string $w_{i_1}^{(1)} \ldots w_{i_k}^{(k)}$ is in the language L. Sheila Greibach, (17), showed that the language GRE(D_2), where D_2 is the Dyck language generated by the context-free grammar $S \to SS$, $S \to aS\bar{a}$, $S \to bS\bar{b}$, $S \to \epsilon$, is complete for CFL with respect to homomorphism reductions, and Hal Sudborough, (52), proved that the language GRE(P), where $P = \{ww^R \mid w \in \{a,b\}^*\}$ is the set of palindroms, is a linear context-free language which is complete for $\mathbb{NL}$ with respect to log space reductions.

We have shown that the graph accessability problem GAP is complete for $\mathbb{NL}$ and that this remains to be true if only acyclic graphs are considered. Let UGAP denote the corresponding problem for undirected graphs. It is clear that the membership problem for UGAP is easier to solve (at least it is not more difficult) than the membership problem for GAP. It is not known whether UGAP belongs to $\mathbb{L}$ and there is some evidence that UGAP is not complete for $\mathbb{NL}$ because, (3), UGAP is accepted by a random Turing machine within space log n. We will see in the next section that some wellknown problems are equivalent to UGAP with respect to log space reductions.

III. RELATIONS TO OTHER COMPLEXITY PROBLEMS, MORE ABOUT REDUCTIONS INSIDE $\mathbb{N}$L, SUGGESTIONS FOR FURTHER RESEARCH

We have shown at the beginning of section II (theorem 1) that containment results for $\mathbb{N}L$ = NSPACE(log n) lead to containment results for CSL = NSPACE(n). This fact can be illustrated using an observation of Walter Savitch, (46). It is not difficult to see that for every language $L \in$ NSPACE(n), $L \subset \Sigma^*$, the language

$$\text{Unary}(L) = \{a^n \mid n = \text{CODE}_{|\Sigma|}(w),\ w \in L\}$$

belongs to NSPACE(log n), where $\text{CODE}_{|\Sigma|}$ is some simple bijective encoding from Σ^* onto $\mathbb{N}$. On the other hand if $L \in$ NSPACE(log n) and $L \subset \{a\}^*$ then the language

$$k\text{-nary}(L) = \{w \mid w = \text{CODE}_k^{-1}(n),\ a^n \in L\}$$

belongs to NSPACE(n) for every $k \geq 2$. The corresponding relationships hold also for deterministic space bounded and for time bounded computations and they can be generalized for other bounding functions f. As a corollary we get

NSPACE(n) $\subset$ DSPACE (α(n))

$\Leftrightarrow$ For every $L \subset \{a\}^*$:

$L \in$ NSPACE(log n) $\Rightarrow L \in$ DSPACE(α(log n))

and, of course, as a special case

CSL = DCSL

$\Leftrightarrow$ For every $L \subset \{a\}^*$: $L \in \mathbb{N}L \Rightarrow L \in \mathbb{L}$.

In (35) refinements of this result, (22), were improved and it was shown that every language $L \in \mathbb{N}L$, $L \subset \{a\}^*$, is reducible to some language $\hat{L} \subset \{a\}^*$ acceptable by some nondeterministic two-way one-counter automaton.

In section 2 we have described a technique how a language defined by some two-way automaton operating within polynomial time can be reduced to a language acceptable by some one-way automaton, and we have shown that there exists a language ACGAP which is complete for $\mathbb{N}L$ and which is accepted by some nondeterministic one-way one-counter automaton (and which therefore is a context-free language). This technique is not applicable if we consider only languages over a one-element alphabet and this is not surprising because otherwise we would have solved the LBA problem (remember that every context-free languages over a one-letter alphabet is a regular language).

Let us list these and some related results. For references see (15,34,35,53,54,55)

(e1) CSL $\subset$ DSPACE(α)

$\Leftrightarrow$ For every $L \subset \{a\}^*$, if L is accepted by some nondeterministic two-way one-counter automaton then $L \in$ DSPACE(α(log n)).

(e2) $\mathbb{N}L$ $\subset$ DSPACE(α(log n))

$\Leftrightarrow$ If L is accepted by some nondeterministic one-way one-counter automaton then $L \in$ DSPACE(α(log n)).

(e3) If L is accepted by some nondeterministic two-way k-head pushdown automaton within polynomial time, then $L \in$ DSPACE(α)

$\Leftrightarrow$ CFL $\subset$ DSPACE(α).

(e4) $\mathbb{P} \subset$ DSPACE(α)

$\Leftrightarrow$ If L is accepted by some nondeterministic two-way 1-head pushdown automaton, then L $\in$ DSPACE(α).

(e5) EXPTIME = $\bigcup_{d\geq 1}$ DTIME(d^n) $\subset$ DSPACE(α)

$\Leftrightarrow$ For every L $\subset$ {a}* , if L is accepted by some nondeterministic two-way 1-head pushdown automaton, then L $\in$ DSPACE(α(log n)).

We also state some further results related to our theorem 6, (6,18,33,36)

(e6) $\mathbb{L}$ = $\mathbb{NL}$

$\Leftrightarrow$ $\mathbb{L}$ is closed under the application of the Kleene * operation.

(e7) $\mathbb{P}$ = $\mathbb{L}$ (or $\mathbb{P}$ = $\mathbb{NL}$, respectively)

$\Leftrightarrow$ $\mathbb{L}$ (or $\mathbb{NL}$, respectively) is closed under the application of length preserving homomorphism.

(e8) $\mathbb{NP}$ = $\mathbb{P}$ (or $\mathbb{NP}$ = $\mathbb{NL}$, or $\mathbb{NP}$ = $\mathbb{L}$ respectively)

$\mathbb{P}$ (or $\mathbb{NL}$, or $\mathbb{L}$ respectively) is closed under the application of polynomial-erasing homomorphic replication.

At the end of section I we presented a language L_o which is complete for CSL with respect to homomorphsm reductions. We know quite a number of problems, (16), which are complete for $\mathbb{P}$SPACE=$\bigcup_{k\geq 1}$ DSPACE(n^k) with respect to log space reductions and it follows from the definition of completeness that if a language L belongs to CSL and if L is complete for PSPACE then L is complete also for CSL. On the other hand, for every language L$\in\mathbb{P}$SPACE, L$\in$DSPACE(n^p) for some p$\geq$0, the language T_g(L), where $g(n)=n^p$ for all n, belongs to DSPACE(n). Therefore CLOSURE$_{\leq log}$(DCSL)=CLOSURE$_{\leq log}$(CSL)=$\mathbb{P}$SPACE and this shows that log space reductions are not an appropriate type of reduction in order to study the relationship between DCSL and CSL. We have to look for a class F of funtions such that CSL is closed under $\leq_F$ and which is powerfull enough such that some "natural" problems can be shown to be complete for CSL with respect to $\leq_F$. We choose F to be the class of all log space computable functions f such that $|f(x)|\leq c\cdot|x|$ holds for some c>0 and for all x. For reductions defined by functions belonging to this class we use the term $\leq_{log\ lin}$. CSL is closed under $\leq_{log\ lin}$, (51), and it was shown in (22,51) that the "Inequivalence problem for regular expressions with $\cup$,o,*" and other problems concerning regular expressions are complete for CSL with respect to $\leq_{log\ lin}$.

Up to now we have described only one problem which is complete for $\mathbb{NL}$. In (25,26) there were presented a number of problems complete for $\mathbb{NL}$:

Unsatisfiability for formulas containing at most two literals per clause,

Unsatisfiability provable by unit resolution for formulas containing at most two literals per clause,

Generation of elements by an associative binary operation,

0-1 Integer Programming with at most two variables per inequality,

Strong connectivity of directed graphs ,

Inequivalence of deterministic generalized sequential machines with final states,

Non Simple Precedence Property for context-free grammars without ϵ-productions or nonproductive nonterminals,

Non L R(k) Property ($k \geq 1$ fixed) for context-free grammars without ϵ-productions or nonproductive nonterminals.

Furthermore in (26) there were given some problems which are equivalent to UGAP, where UGAP denotes the accessability problem for undirected graphs:

Nonbipartiteness for undirected graphs

Exact Cover where each element appears at most twice

Hitting Set where each set contains at most two elements

Uncoverability by two cliques for undirected graphs .

B. Monien and I.H. Sudborough, (4o , 41) , considered graphs of bounded bandwidth and showed that the complexity of many $\mathbb{NP}$ - complete graph problems is reduced by bounding the bandwidth of the graphs. We say that a graph $G = (V,E)$, $V = \{1,...,k\}$ has bandwidth m if and only if $|i - j| \leq m$ for all $(i,j) \in E$. This definition can be extended to some more problems not containing a graph in their formulation.

It was shown that the following problems are complete for $\mathbb{NL}$ for graphs $G = (V,E)$ of bandwidth $\log(|V|)$:

3-Satisfiability, Maximum-2-Satisfiability, Independent Set, Vertex Cover, Simple Max Cut, Partition into Triangles, 3-DIM Matching, Dominating Set, Exact Cover by 3 sets, Graph Grundy Numbering, 3-Colorability

In fact the results of (4o ,41) are somewhat more general. They show that for any function f, $\log n \leq f(n) \leq n$, these problems are complete for NTISP(poly,f),if only graphs $G = (V,E)$ of bandwidth $f(|V|)$ are considered. NTISP(poly, f) denotes the class of all languages acceptable by nondeterministic Turing machines within polynomial time and space f, simultaneously.

Instead of bounding the bandwidth of a graph problem you can bound the largest number occuring in a "number" problem. This restriction leads, (38), also to a large number of problems belonging to $\mathbb{NL}$. Some of these problems are complete for $\mathbb{NL}$, (38), but the real complexity of most of them is not known.Consider for example the problems "Subset Sum" and "Scheduling on k Processors" with a linear bound on the largest number

$$SUB = \{(a_1,...,a_n,b) \mid b \leq n \text{ and } \exists\, I \subset \{1,...,n\}: \sum_{i \in I} a_i = b\}$$

$$MPS(k) = \{(a_1,...,a_n,D) \mid D \leq n \text{ and } \exists I_1,...,I_k \subset \{1,...,n\}:$$

$$\sum_{i \in I_j} a_i \leq D \text{ for all } 1 \leq j \leq k\} .$$

Furthermore let us consider $GRE(\{a^n b^n \mid n \in \mathbb{N}\})$, where GRE denotes again the Greibach construction. All these problems belong to $\mathbb{NL}$ and it is not difficult to see that

$$SUB \equiv MPS(2) \leq MPS(3) \leq \ldots \leq MPS(k) \leq \ldots \leq GRE(\{a^n b^n \mid n \in \mathbb{N}\}).$$

We do not know whether one of these problems is complete for $\mathbb{NL}$ or whether one of them belongs to $\mathbb{L}$. Each of these problems is reducible, (38), to some language acceptable by some nondeterministic one-way counter automaton which changes the direction on its counter only once during its computation. Let us denote the class of languages acceptable by such an automaton by RCL. Since RCL is a class of languages defined by a very restricted type of automaton it would be very interesting to get some answer to one of the following questions:

- does there exist a language L $\in$ RCL which is complete for NL?
- does there exist some $\delta < 1$ such that RCL $\subset$ DSPACE$((\log n)^{1+\delta})$?

Note that in the case of nondeterministic one-way pushdown automata a bound on the number of reversals of the pushdown tape leads from the class of context-free languages to the class of linear context-free languages.

In this paper we have presented numerous results concerning the LBA problem. Bringing this presentation to an end we want to draw your attention to some open questions related to the material described above whose solution would help to again a better understanding of the LBA problem.

(q1). W. Savitch has put a landmark by showing that the space bound has at most to be squared by going from nondeterministic space bounded computations to deterministic space bounded computations. Of course, an improvement of Savitch's result for all space bounds $f(n) \geq \log n$ would be an outstanding result also. It might be more realistic to proceed in the following way.

(1.1) Find some fully space constructible function f such that
NSPACE(f) $\subset$ DSPACE$((f(n))^{1+\delta})$ for some $\delta < 1$.

Note that the mapping "Unary" maps languages from NSPACE(n) [or from $\bigcup_{d\geq 1}$ NSPACE$(f(d^n))$ respectively] onto languages over a one-letter alphabet which belong to NL [or to CSL, respectively]. We know, (21), that there does not exist a language over a one-letter alphabet which is complete for CSL. Therefore one possibility to solve (1.1) could be:

(1.2) Show that L $\in$ CSL, L $\subset \{a\}^*$
implies L $\in$ DSPACE$(n^{1+\delta})$ for some $\delta < 1$.

One can also study the deterministic space complexity of subclasses of NSPACE(f) which are not defined itself by space bounded computations (see also q4). One of these questions is the following:

(1.3) Does there exist some $\delta < 1$ such that SUB $\subset$ DSPACE$((\log n)^{1+\delta})$?

Containment results for complexity classes can be transferred "upward". It is an open question whether they can be transferred also "downward".

(1.4) Does there exist some function f, $f(n) \gg \log n$, such that for every function α
NSPACE(f) $\subset$ DSPACE(αf) implies NSPACE(log n) $\subset$ DSPACE$(\alpha(\log n))$?

(q2) We have shown that the language ACGAP, which is complete for NL , is accepted by some nondeterministic one-way counter automaton. Let us denote by FA(k) [or NFA(k), respectively] the class of languages acceptable by deterministic [or nondeterministic, respectively] k-head two-way finite state automata. We know, (2o), L $= \bigcup_{k\geq 1}$FA(k) and NL $= \bigcup_{k\geq 1}$NFA(k). Then for the problem

(2.1) Look for the largest k you can find such that NFA(k) $\subset$ DSPACE$((\log n)^{1+\delta})$ holds for some $\delta < 1$

we already have given a partial answer, because NFA(1) = DSPACE(0) and ACGAP $\in$ CL $\subset$ NFA(2) (and therefore NFA(2) $\subset$ DSPACE($(\log n)^{1+\delta}$) implies $\mathbb{N}$L $\subset$ DSPACE($(\log n)^{1+\delta}$)). Because of the same reason any positive answer to the question

(2.2) Does there exist some $k \geq 1$ such that NFA(2) $\subset$ FA(k)?

would prove $\mathbb{L}$ = $\mathbb{N}$L. Therefore it promisses to be more interesting to ask this question the other way around, i.e. to ask whether one can show that nondeterminism cannot be replaced by some additional heads.

(2.3) Look for the largest k you can find such that NFA(2) $\not\subseteq$ DFA(k).

At the moment it is not even known whether NFA(2) $\neq$ DFA(2) holds.

(q3) Similar questions as in (q2) can be asked about the relationship between pushdown automata and finite state automata. Let PDA(k) [or NPDA(k), respectively] denote the class of languages acceptable by deterministic [or nondeterministic, respectively] k-head two-way pushdown automata. We know that, (1 , 11), $\mathbb{P}$ = $\bigcup_{k\geq 1}$PDA(k) = $\bigcup_{k\geq 1}$NPDA(k) and that there exists some language L $\in$ PDA(1) which is complete for $\mathbb{P}$. Therefore PDA(1) $\subseteq$ DSPACE($(\log n)^k$) would imply $\mathbb{P}$ $\subset$ DSPACE($(\log n)^k$) and PDA(1) $\subset$ NFA(k) [or PDA(1) $\subset$ FA(k), respectively] for some $k \geq 1$ would imply $\mathbb{P}$ = $\mathbb{N}$L [or $\mathbb{P}$ = $\mathbb{L}$, respectively] . Again it seems to be more interesting to ask whether a pushdown tape can be replaced by an additional number of heads or by adding nondeterminism.

(3.1) Look for the largest k you can find such that PDA(1) $\not\subseteq$ NFA(k) or PDA(1) $\not\subseteq$ FA(k), respectively.

We know, (54), that NFA(2k) $\subset$ NPDA(k) holds for $k \geq 1$ and because of the hierarchy result for nondeterministic multi-head two-way finite state automata, (34), this result implies NPDA(k) $\not\subseteq$ NFA(2k-1) for $k \geq 1$. This gives no new information in the case k = 1 because we already know that NPDA(1) contains non-regular sets. We do not know whether NPDA(1) $\not\subseteq$ NFA(2) or PDA(1) $\not\subseteq$ FA(2) holds. Recently, (14), it was shown that a deterministic one-head two-way pushdown automaton is more powerful than a deterministic two-way counter automaton.

(q4) We have described three subclasses of $\mathbb{N}$L whose complexity we cannot classify. These are the classes

Ω_1 = CLOSURE$_\leq$({L | L $\subset$ {a}* , L $\in$ $\mathbb{N}$L}),

Ω_2 = CLOSURE$_\leq$({L | L $\in$ RCL}),

Ω_3 = CLOSURE$_\leq$({UGAP}),

where UGAP denotes the graph accessability problem for undirected graphs and RCL the class of languages acceptable by nondeterministic reversal bounded one-way counter automata. It is clear that UGAP is complete for Ω_3 and we have already stated that there exists a nodeterministic two-way counter language which is complete for Ω_1.

(4.1) Does there exist some $\delta < 1$ such that $\Omega_i \subset$ DSPACE($(\log n)^{1+\delta}$) holds for some $\delta < 1$, i = 1,2,3?

(4.2) Find "simple" complete languages for Ω_i, i = 1,2,3.

(4.3) Study reductions between languages belonging to these classes.

Some proposals for subclasses between Ω_1 and $\mathbb{L}$ were given in (35) and (56). Reductions between some languages which are neither known to be complete for $\mathbb{NL}$ nor known to belong to $\mathbb{L}$ were studied in (38).

REFERENCES

(1) Aho, A.V., J.E. Hopcroft and J.D. Ullman, (1968). Time and tape complexity of pushdown automaton languages, Info. and Control 13, pp. 186 - 2o6.

(2) Aho, A.V., J.E. Hopcroft and J.D. Ullman,(1974) The design and analysis of computer algorithms, Addison-Wesley Publ. Comp.,Reading, Massachussetts .

(3) Alelinuas, R., R.M. Karp, R.J. Lipton, L. Lovasz and C. Rackhoff, (1979). Random walks, universal sequences and the complexity of maze problems, Proc. 2oth IEEE Symp. on Foundations of Computer Science.

(4) Book, R.V. (1972). On languages accepted in polynomial time, SIAM J. Computing 4, 281 - 287.

(5) Book, R.V.,(1976). Translational lemmas, polynomial time, and $(\log n)^j$-space, Theor. Comput. Sci. 1, pp 215 - 226.

(6) Book, R.V.,(1978). Simple representations of certain classes of languages, J.Ass. Comp. Mach. 25, 23 - 31.

(7) von Braunmühl, B. and R. Verbeek, (198o). A recognition algorithm for deterministic CFLs optimal in time and space, Proc. 21st Annual Symp. Foundations of Computer Science pp. 411 - 42o.

(8) Chomsky, N., (1959). On certain formal properties of grammars, Inform. and Control 2, 133 - 167.

(9) Chomsky, N., (1962). Context-free grammars and pushdown storage, Quart. Progr. Rept. No. 65, MIT, pp. 187 - 194.

(1o) Cook, S.A., (197o). Path Systems and language recognition,Proc. 2nd ACM Symp. Theory of Computing, 7o - 72.

(11) Cook, S.A., (1971). Characterizations of pushdown machines in terms of time-bounded computers, J. Assoc. Comput. Mach. 18, pp. 4 - 18.

(12) Cook, S.A.,(1971). The complexity of theorem proving procedures, Proc. 3rd Annual ACM Symp. on Theory of Computing, 151 - 158.

(13) Cook, S.A., (1979) Deterministic CFLs are accepted simultaneously in polynomial time and log squared space, Proc. 11th Annual ACM Symp. on Theory of Computing, 338 - 345.

(14) Duris, P. and Z. Galil, (198o). Fooling a two-way automaton or One pushdown store is better than one counter for two-way machines, submitted for publication.

(15) Galil, Z., (1977) Some open problems in the theory of computation as questions about two-way deterministic pushdown automata languages, Math. Systems Theory 1o, pp. 211 - 218.

(16) Garey, M.R. and D.S. Johnson, (1979). Computers and Intractability: A guide to the Theory of NP-Completeness, W.H. Freeman and Co., San Francisco.

(17) Greibach, S.A., (1973). The hardest context-free language, SIAM J. Computing 2, pp. 3o4 - 31o .

(18) Greibach, S.A., (1977) A note on NSPACE(log n) and substitution, RAIRO Informatique théorique 11, 127 - 132.

(19) Greibach, S.A., (1978) Remarks on blind and partially blind one-way multicounter machines, Theoretical Comp. Sci. 7, 311 - 324.

(2o) Hartmanis, J., (1972) On non-determinacy in simple computing devices, Acta Informatica, 334 - 336.

(21) Hartmanis, J., (1978). On log-tape isomorphisms of complete sets, Theoretical computer Science 7, 273 - 286.

(22) Hartmanis, J. and H.B. Hunt, (1973). The LBA problem and its importance in the theory of computing, Cornell University, Technical Report.

(23) Hartmanis, J. and R.E. Stearns, (1965). On the computational complexity of algorithms, Transactions of American Math. Society 117, 285 - 3o6.

(24) Hopcroft, J.E. and J.D. Ullman, (1969; new edition = 1979), Formal Languages and their Relation to Automata, Addison-Wesley, Reading. Mass., USA.

(25) Jones, N.D. (1975). Space bounded reducibility among combinatorial problems, J. Comput. System Sci. 11, 62 - 85.

(26) Jones, N.D., Y.E. Lien and W.T. Laaser. (1976). New problems complete for nondeterministic Log space, Math. Systems Theory 1o, 1 - 17.

(27) Jung, H., (1981), Relationships between probabilistic and deterministic tape complexity, submitted for publication.

(28) Karp, R.M., (1972). Reducibility among combinatorial problems, in Complexity of Computer Computation, (R. Miller and J. Thatcher, eds.) Plenum Publishing Co., New York, pp. 85 - 1o3.

(29) Kuroda, S.Y., (1964). Classes of languages and linear-bounded automata, Inform. and Control 7, 2o7 - 223.

(3o) Landweber, P.S. (1963). Three theorems on phrase structure grammars of type 1, Inform. and Control 6, 131 - 136.

(31) Lewis, P.M., R.E. Stearns, and J. Hartmanis, (1965). Memory bounds for the recognition of context-free and context-sensitive languages, Proc. 6th Annual IEEE Cinf. on Switching Circuit Theory and Logical Design, pp. 191 - 2o2.

(32) Monien, B., (1972).Relationships between pushdown automata and tape-bounded Turing machines, Proc. First Symp. on Automata Languages and Programming, North-Holland Publ. Comp. Amsterdam, pp. 575 - 583.

(33) Monien, B. (1975). About the deterministic simulation of nondeterministic (logn)-tape bounded Turing machines, Lecture Notes in Comp. Sci. 33, Springer Verlag, pp. 118 - 126.

(34) Monien, B., (1977). Transformational methods and their application to complexity problems, Acta Informatica 6, pp. 95-1o8, Corrigenda, Acta Informatica 8, pp. 383 - 384.

(35) Monien, B., (1977). The LBA-problem and the deterministic tape complexity of two-way one-counter languages over a one-letter alphabet, Acta Informatica 8, 371 - 382.

(36) Monien, B., (1977). About the derivation languages of grammars and machines, Lecture Notes in Comp. Sci. 52, Springer Verlag, pp. 337 - 351.

(37) Monien, B. (1979) Connections between the LBA problem and the knapsack problem, Proceedings Frege-Konferenz, University Jena, pp. 262 - 28o.

(38) Monien, B., (198o). On a subclass of pseudonomial problems, Lecture Notes in Comp. Sci. 88, Springer Verlag, pp. 414 - 425.

(39) Monien, B. and I.H. Sudborough, (1979). On eliminating nondeterminism from Turing machines which use less than logarithm worktape space, Lecture Notes in Comp. Sci. 71, Springer Verlag, pp. 431 - 445.

(4o) Monien, B. and I.H. Sudborough, (198o). Bounding the bandwidth of NP-complete problems, Lecture Notes in Comp. Sci. 1oo, Springer Verlag, pp. 279 - 292.

(41) Monien, B. and I.H. Sudborough, (1981). Bandwidth constrained NP-complete problems, Proc. 13th ACM Symp. Theory of Computing.

(42) Myhill, J., (196o). Linear bounded automata, Wright Air Development Division, Tech. Note No. 6o - 165, Cincinnati, USA.

(43) Rabin, M.O. and D. Scott, (1959). Finite automata and their decision problems, IBM.J.Res. 3:2, 115 - 125.

(44) Ruby, S. and P.C. Fischer, (1965). Translational methods and computational comlexity, Proc. 6th Annual IEEE Conf. on Switching Circuit Theory and Logical Design, pp. 173 - 178.

(45) Savitch, W.J., (197o). Relationships between nondeterministic and deterministic tape complexities, J. Comput. System Sci. 4, 177 - 192.

(46) Savitch, W.J.,(1973). A Note on multihead Automata and context-sensitive languages, Acta Informatica 2 (1973), pp. 249 - 252.

(47) Schützenberger, M., (1963). Context-free languages and pushdown automata, Inform. and Control 6, 246 - 264.

(48) Seiferas, J. I., (1977). Techniques for separating space complexity classes, J. Comput. System Sci. 14, pp. 73 - 99.

(49) Seiferas, J.I. , (1977). Relating refined space complexity classes, J. Comput. System Sci. 14, pp. 1oo - 129.

(5o) Stearns R.E., J. Hartmanis, and P.M. Lewis II, (1965). Hierarchies of memory limited computations, Proc. 6th Annual IEEE Conf. on Switching Circuit Theory and Logical Design, 179 - 19o.

(51) Stockmeyer, L.J. and A.R. Meyer, (1973). Word problems requiring exponential time, Proc. 5th Annual ACM Symp. Theory of Comput. pp. 1 - 9.

(52) Sudborough, I.H., (1975). A note on tape bounded complexity classes and linear context-free languages, J. Assoc. Comput. Mach. 22, pp. 5oo - 5o1.

(53) Sudborough, I.H., (1975). On tape bounded complexity classes and multihead finite automata, J. Comput. System Sci. 1o, pp. 62 - 76.

(54) Sudborough, I.H., (1977). Some remarks on multihead automata, R.A.I.R.O. Informatique thèorique/Theoretical Computer Science II, pp. 181 - 195.

(55) Sudborough, I.H., (1978). Relating open problems on the tape complexity of context-free languages and path system problems, Proc. Conf. on Info. Sciences and Systems, The John Hopkins University, Baltimore (USA).

(56) Voelkel, L., (1979). Language recognition by linear bounded and copy programs, Proc. Fundamentals of Computation Theory, Akademie-Verlag Berlin, pp. 491-495.

DYNAMIC ALGEBRAS OF PROGRAMS

I. Németi

Mathematical Institute of the Hungarian Academy of Sciences
Budapest, Reáltanoda u. 13-15, H-1053, Hungary

Dynamic algebras are two-sorted algebras. For the theory of many-sorted or heterogeneous algebras see §II.2 of [10] pp.137-171.

A many-sorted similarity type d is a function $d: \Sigma \longrightarrow S^{\oplus}$ where $S^{\oplus} \stackrel{d}{=} \bigcup\{ {}^nS : 0<n\in\omega \}$ is the set of nonempty finite strings over the alphabet S . We call the elements of Σ function symbols and S is called the set of sorts. If $d(f) = \langle s_0,\ldots,s_n\rangle$ then the operation corresponding to f goes from the universes of sorts $s_0,\ldots,s_{n-1}$ into the universe of sort s_n , in every algebra of type d.

Now we define the <u>similarity type</u> d <u>of dynamic algebras</u>.:
A two element set {actions, Boolean} of sorts is fixed.
The function d is defined as follows:
$\mathrm{Dom}\, d \stackrel{d}{=} \{ \vee , ; , {}^{*} , \cdot , - , \Diamond \}$ and
$d(\vee) = d(;) = \langle \text{actions},\text{actions},\text{actions}\rangle$,
$d({}^{*}) = \langle \text{actions},\text{actions}\rangle$,
$d(\cdot) = \langle \text{Boolean},\text{Boolean},\text{Boolean}\rangle$,
$d(-) = \langle \text{Boolean},\text{Boolean}\rangle$,
$d(\Diamond) = \langle \text{actions},\text{Boolean},\text{Boolean}\rangle$.

Alg(d) denotes the class of all (heterogeneous) algebras of similarity type d .

Next we define the class Ds of dynamic <u>set</u> algebras. [11] denoted the class Ds by K and called its members <u>Kripke structures</u>.
Let U be a set. Then Sb U denotes the set of all subsets of U .

<u>DEFINITION 1</u> ([9] , [11] p.9)

a) An algebra $D = \langle \underline{\underline{A}},\underline{\underline{B}},\Diamond\rangle$ of type d is defined to be a <u>dynamic set algebra</u> (in short, a Ds) if there is a set U (called the <u>base</u> of D) such that 1) - 3) below are satisfied.
   1) $\underline{\underline{B}}$ is a <u>Boolean set algebra</u> of some subsets of U (i.e. $B\subseteq \mathrm{Sb}\, U,\ldots$)
   2) $\underline{\underline{A}}$ is a <u>Kleene set algebra</u> with base U , i.e. $A\subseteq \mathrm{Sb}(U\times U)$ and the interpretations of the operations $\vee,;,{}^{*}$ in $\underline{\underline{A}}$ are <u>set</u>

<u>union</u> $\cup$, the usual <u>composition</u> $|$ of relations, and <u>transitive reflexive closure</u> $^+$ respectively where

$a^+ = \bigcap\{ b \in Sb(U\times U) :$ b is a transitive and reflexive relation on U and $a \le b \}$.

3) $\Diamond(a,p) = \{ u \in U : (\exists v \in p) \langle u,v\rangle \in a \}$, for any $a\in A$ and $p\in B$.

Ds denotes the class of all dynamic set algebras.

b) An algebra of type d is called <u>representable</u> if it is isomorphic to a dynamic set algebra. <u>END(Def.1)</u>

For any class $K \subseteq Alg(d)$ of algebras we use the notations **I**K, **H**K, **S**K, **P**K, **Up** K, **Uf** K as the standard textbook [8] introduces them. A deviation from the other standard textbook [7] of universal algebra is that here **IP** K = **P**K, **IUp** K = **Up** K, **IUf** K = **Uf** K. By II.2.6(10) of [10], the variety Mod Eq K generated by K is **HSP** K in many-sorted algebra theory too.

<u>DEFINITION 2</u> ([3], [4], [7] p.380 §63 and p.338 §57)
Let κ be a cardinal and consider the set $L^d_{\infty\omega}$ of infinitary formulas of type d . By an <u>infinitary quasiequation</u> of type d we understand a formula $((\bigwedge_{i<\varrho} e_i) \to e_\varrho) \in L^d_{\infty\omega}$ where $(\forall i \le \varrho)$ e_i is an atomic formula that is, in our present case, e_i is an equation of type d . The quasiequation $((\bigwedge_{i<\varrho} e_i) \to e_\varrho)$ is κ-ary iff $\varrho < \kappa$ (that is iff $((\bigwedge_{i<\varrho} e_i) \to e_\varrho) \in L^d_{\kappa\omega}$).

$Qeq_\kappa K \stackrel{d}{=} \{ \varphi \in L^d_{\kappa\omega} : K \models \varphi$ and φ is an infinitary quasiequation $\}$.

$Qeq_\infty K \stackrel{d}{=} \bigcup\{ Qeq_\kappa K : \kappa$ is a cardinal $\}$.

K is an <u>infinitary quasivariety</u> if K is axiomatizable by infinitary quasiequations that is if $K = Mod\ Qeq_\infty K$.

<u>THEOREM 1</u>

(i) **I** Ds is an infinitary quasivariety, that is **I** Ds = Mod Qeq_∞ Ds

(ii) **I** Ds = **SP** Ds . **SUp** Ds = **PSUp** Ds .

(iii) **I** Ds $\neq$ **Up** Ds .

(iv) Eq Ds is recursively enumerable.

(v) For every sets X,Y of generators and set $R \subseteq {}^2Fr_{X,Y,d}$ of defining relations, the algebra $\underline{\underline{Fr}}^{(R)}_{X,Y,d}$ Ds $\in$ **I** Ds Ds-freely generated by the sets X,Y (of sorts actions and Boolean) under R exists in Ds . See e.g. [8] p.146 for terminology.

(vi) Ds is an epireflective subcategory of Alg(d) .

PROBLEM 1: Is $\mathbf{I}\,Ds = Mod\, Qeq_{\kappa} Ds$ or $\mathbf{I}\,Ds = Mod\, L_{\kappa\omega} Ds$ for some cardinal κ ? Is $Eq\, Ds$ recursively axiomatizable ? (Finitely ?). Is $Qeq_{\omega} Ds$ recursively enumerable ?

The class $Cn \subseteq Alg(d)$ of *-continuous dynamic algebras was introduced by [9]. We shall define this class later.

THEOREM 2

(i) Cn is a finitely based ω_1-ary quasivariety.

(ii) There is an ω_1-ary quasiequation $ce \in L^{d}_{\omega_1\omega}$ and a finite set Ed of equations of type d such that $Cn = Mod(Ed \cup \{ce\})$.

(iii) $\mathbf{SP}\, Cn = Cn \neq \mathbf{Up}\, Cn$.

ABBREVIATIONS: $y_1+y_2 \overset{d}{=} -((-y_1)\cdot(-y_2))$, $0 \overset{d}{=} y\cdot(-y)$, $1 \overset{d}{=} -0$, $y_1 \leq y_2$ iff $y_1\cdot y_2 = y_1$, $x_1 \leq x_2$ iff $x_1 \vee x_2 = x_2$, $\Box(x,y) \overset{d}{=} -\Diamond(x,-y)$.

In the language $L^{d}_{\infty\omega}$ the variables x, x_i ($i \in Ord$) are of sort "actions" and y, y_i ($i \in Ord$) are variables of sort "Boolean". That is $\Diamond(x,y)$ is a term but $\Diamond(y,x)$ is not a term of type d .

$\Diamond(x^{o},y) \overset{d}{=} y$ and $(\forall n \in \omega)\ \Diamond(x^{n+1},y) \overset{d}{=} \Diamond(x,\Diamond(x^{n},y))$.

DEFINITION 3

(i) The set $Ed \subseteq L^{d}_{\omega\omega}$ is defined to consist of (1)-(5) below.

(1) The equations defining Boolean algebras.

(2a) $\Diamond(x,0) = 0$.

(2b) $\Diamond(x,y_1+y_2) = \Diamond(x,y_1)+\Diamond(x,y_2)$.

(3) $\Diamond(x_1 \vee x_2,y) = \Diamond(x_1,y)+\Diamond(x_2,y)$.

(4) $\Diamond(x_1;x_2,y) = \Diamond(x_1,\Diamond(x_2,y))$.

(5) $y+\Diamond(x,\Diamond(x^{*},y)) \leq \Diamond(x^{*},y) \leq y+\Diamond(x^{*},\Diamond(x,y)\cdot(-y))$.

(ii) $Da \overset{d}{=} Mod\, Ed$, hence $Da \subseteq Alg(d)$ is a variety.

(iii) $sep \in L^{d}_{\omega\omega}$ is defined to be the formula

$\forall x_1 \forall x_2 \exists y(\ \Diamond(x_1,y)=\Diamond(x_2,y) \rightarrow x_1=x_2\ )$ and

$Sda \overset{d}{=} Mod(\{sep\} \cup Ed)$, hence $Sda \subseteq Da$ and $\mathbf{P}^{r}\, Sda = Sda$.

(iv) $ce \in L^{d}_{\omega_1\omega}$ is defined to be the infinitary quasiequation

$(\ \bigwedge\{\ y_1 \geq \Diamond(x^{n},y)\ :\ n \in \omega\ \} \rightarrow y_1 \geq \Diamond(x^{*},y)\ )$.

(v) $Cn \overset{d}{=} Mod(\{ce\} \cup Ed)$.

(vi) $Qeq^{B}_{\kappa} \subseteq Qeq_{\kappa}$ is defined by restricting equations only for terms of sort Boolean that is

$\mathrm{Qeq}^{B}_{\kappa} \overset{d}{=} \{ ((\bigwedge_{i<\varrho} \tau_i = \sigma_i) \to \tau_\varrho = \sigma_\varrho) \in \mathrm{Qeq}_{\kappa} : (\forall i \le \varrho)(\tau_i, \sigma_i$ are terms of sort Boolean$) \}$.

Note that $\{ce\} \cup \mathrm{Ed} \subseteq \mathrm{Qeq}^{B}_{\omega_1}$.

Mod $\mathrm{Qeq}^{B}_{\kappa}$ K is defined the natural way. END(Def.3)

Pratt proved that **HSP** Sda = **HSP** Ds that is Eq Sda = Eq Ds . Kozen proved Cn $\neq$ **I** Ds hence we have Cn $\supsetneq$ **SP** Ds .

Let K $\subseteq$ Alg(d) be any class of algebras. Let κ be a regular cardinal. Then $\mathbf{P}^{r\kappa}$ K denotes the class of all reduced products of families of members of K modulo κ-complete filters. $\mathbf{P}^{r}$ K denotes $\mathbf{P}^{r\omega}$ K that is $\mathbf{P}^{r}$ K is the class of reduced products of members of K. **Ud** K is the class of directed unions of subsets of K .

THEOREM 3: Let $D = \langle \underline{\underline{A}}, \underline{\underline{B}}, \Diamond \rangle \in \mathbf{SP}^{r}$ Sda. Then * is that closure operator on $\underline{\underline{A}}$ which assigns to each $a \in A$ the smallest closure operator $b \in A$ on $\underline{\underline{B}}$. More precisely, let $a \in A$. Then

$a^{*} = \min\{ b \in A : \Diamond(b,-) \in {}^{B}B$ is a closure operator on $\underline{\underline{B}} \}$.

Note that since both $\underline{\underline{A}}$ and $\underline{\underline{B}}$ are partially ordered, the notions of a closure operator $^{*}: A \longrightarrow A$ on $\underline{\underline{A}}$ and a closure operator $\Diamond(a^{*},-): B \longrightarrow B$ on $\underline{\underline{B}}$ are well defined.

Kozen has found a Π_4 formula distinguishing Ds and Sda . Thm.4(i) below improves Kozen's result.

THEOREM 4

(i) There is a Π_2 formula $\pi \in L^{d}_{\omega\omega}$ such that Ds $\models \pi$ and Sda $\not\models \pi$.

(ii) **S** Sda = **SPUp** Sda = Mod Qeq_{ω} Sda .

(iii) The infinitary quasivariety Cn coincides with the class of *-continuous Da-s of [9]. That is, let $D \in$ Da . Then $D \models ce$ iff $D \models \forall x \forall y (\Diamond(x^{*},y) = \sup\{\Diamond(x^{n},y) : n \in \omega\})$.

(iv) Cn = Mod $\mathrm{Qeq}^{B}_{\omega_1}$ Ds .

(v) Cn $\notin$ **SPUp** Ds = $\mathbf{SP}^{r}$ Ds and Cn = $\mathbf{SP}^{r\omega_1}$ Cn $\notin$ Mod Qeq_{ω} Sda .

(vi) There exists Sda $\supsetneq$ K $\supsetneq$ (Ds $\cap$ Sda) such that K = **Ud Uf Up** K . Further Sda = **Ud Uf** $\mathbf{P}^{r}$ Sda .

(vii) **H** Ds $\notin$ **SPUp** Sda. Hence the quasivariety **SPUp** Sda seems to be more useful for the representation theory than the variety **HSP** Ds.

(viii) Qeq_ω Sda is recursively enumerable.

PROOF: <u>(i) and (vi)</u>: Let $\psi(x,y,y_1)$ be the equation $y\cdot\Diamond(x,y_1)=0$. Let $\pi \in L^d_{\omega\omega}$ be the Π_2 formula

$$\forall x\forall y\forall y_1\exists y_2([\psi(x,y,y_2)\to\psi(x,y,\Diamond(x,y_2))]\to$$
$$\to[\psi(x,y,y_1)\to\psi(x,y,\Diamond(x^*,y_1))])\quad .$$

Then $Ds \models \pi$ and $Sda \not\models \pi$.

<u>(v) and (vii)</u>: Let qe be the quasiequation $(\Diamond(x,1)=0 \to x\le x_1)$. Clearly $Sda \models qe$. Choosing $D = \langle \underline{\underline{A}}, \underline{\underline{B}}, \Diamond\rangle$ with $|B| = 1$, $|A|>1$, we obtain $Cn \not\models qe$. [4] completes the proof of (v) and (vii) since $D \in \mathbf{H}\, Ds$.

<u>QED</u>.

<u>PROBLEMS 2</u>

(i) Is $\mathbf{S}$ Sda = $\mathbf{SPUp}$ Ds ? Is Qeq_ω Sda = Qeq_ω Ds ?

(ii) Is there a Σ_2 formula or a Π_1 formula valid in Ds but not in Sda ?

(iii) What is the smallest prenex-complexity of $L^d_{\omega\omega}$ formulas needed to distinguish Sda and Sda $\cap$ Ds , is it e.g. Σ_1 or Δ_2 or Δ_1 ? We know that it is not larger than Π_2, but is Π_2 the smallest possible ? Is there a Π_2 Horn formula valid in Ds but not in Sda ?

(iv) Let Thda $\overset{d}{=} L_{\omega\omega}$ (Ds). That is Thda = $\{\varphi \in L^d_{\omega\omega} : Ds \models \varphi\}$. Is Thda recursively enumerable ? Is Thda recursively axiomatizable ? Is Thda axiomatizable by (not necessarily universal) Horn formulas ?

(v) Does there exist a cardinal κ such that Mod Qeq^B_∞ Ds = Mod Qeq^B_κ Ds ?
(Clearly ω_1 is such a cardinal for Cn .
Is Mod Qeq_∞ Cn = Mod Qeq_κ Cn for some κ ?)

(vi) Is Qeq_ω Sda or Qeq_ω Ds recursively (or finitely) axiomatizable ?

By recursive enumerability of Qeq_ω Sda and by $\mathbf{H}$ Ds $\not\subseteq$ Mod Qeq_ω Sda, the <u>finitary</u> quasivariety Mod Qeq_ω Sda seems to be more important than the variety Mod Eq Sda. We note that the theory of quasivarieties is just as well developed as that of varieties in universal algebra and nearly all nice properties carry over e.g. free algebras exist.

DEFINITION 4 ([11])

a) The similarity type t is defined as $t \stackrel{d}{=} d \cup \{\langle ?, \langle \text{Boolean}, \text{actions} \rangle\rangle\}$. $Alg(t)$ denotes the class of algebras of type t.

b) A dynamic test set algebra (in short, a Dts) is defined to be an algebra D of similarity type t such that the ?-free reduct of D is a Ds and the interpretation $?^D$ of the symbol ? in D is the following:

$?^D(p) = \{\langle u,u\rangle : u \in p\}$, for every element p of Boolean sort of D.

Dts denotes the class of all dynamic test set algebras.

c) The class Dta of dynmaic test algebras is defined as a variety of similarity type t by the equations (1),(2a),(2b),(3),(4),(5) (which define Da in Def.3) together with an additional equation

(6) $\Diamond(?(y_1), y_2) = y_1 \cdot y_2$.

Relativization is a central tool in classical logic, see e.g. [8]. (One of the logical uses of relativization is to interpret one theory (or logic) in another by using an extra predicate w.) In algebraic logic of classical propositional logic, relativization is always a homomorphism. We shall see that the situation in dynamic logic is different.

DEFINITION 5: Let $D = \langle \underline{\underline{A}}, \underline{\underline{B}}, \Diamond, ? \rangle$ be an algebra of type t. Let $w \in B$. We define the relativization function rl^D_w as a pair $rl^D_w \stackrel{d}{=} \langle rl^A_w, rl^B_w \rangle$ of functions, where

$rl^A_w : A \longrightarrow A$ and $rl^B_w : B \longrightarrow B$ are defined as follows:

$rl^A_w(a) \stackrel{d}{=} ?(w);(a;?(w))$ and $rl^B_w(p) \stackrel{d}{=} p \cdot w$, for every $a \in A$ and $p \in B$.

Note that if $D \in Dts$ then $rl^A_w(a) = a \cap w \times w$ and $rl^B_w(p) = p \cap w$ for every $a \in A$ and $p \in B$.

THEOREM 5 ([6]) : Let $D = \langle \underline{\underline{A}}, \underline{\underline{B}}, \Diamond, ? \rangle \in$ **HSP** Dts and let $w \in B$. Then (i) and (ii) below are equivalent.

(i) rl^D_w is a homomorphism on D.

(ii) $(\forall a \in A)\ w \leq \Box(a,w)$.

THEOREM 6 ([6]) : Let $D = \langle \underline{\underline{A}}, \underline{\underline{B}}, \Diamond, ? \rangle$ be a separable Dta. Let $w \in B$ be such that $(\forall a \in A)\ w \leq \Box(a,w)$.

Then rl^D_w is a homomorphism on D.

<u>THEOREM 7</u> ([6] , necessity of some conditions in Thm.6)
Let $D = \langle \underline{\underline{A}}, \underline{\underline{B}}, \Diamond, ? \rangle \in Alg(t)$ and let $w \in B$.
Consider statements (a), (a'), (b), (c) below.
(a) All the axioms of Dta are satisfied in D except the axiom(2a).
(a') All the axioms of Dta are satisfied in D except the axiom(2b).
(b) $(\forall a \in A)\ w \leq \Box(a,w)$.
(c) rl_w^D is <u>not</u> a homomorphism on D .

Now there are a separable D and a w such that statements (a),(b), (c) hold for them, and there are a separable D and a w such that statements (a'),(b),(c) hold for them.

<u>DEFINITION 6</u>: Let $D \in Alg(t)$.

a) Let R be a congruence on D . Then R is said to be a <u>direct factor congruence</u> on D iff there are D_0 , $D_1 \in Alg(t)$ and an isomorphism h between D and the direct product $D_0 \times D_1$ such that $R = ker(pj_1 \circ h)$ where pj_1 is the projection function $pj_1 : D_0 \times D_1 \longrightarrow D_1$.

b) Let R be a system $\langle R_i : i \in I \rangle$ of congruences on D . Then R is said to be a <u>direct decomposition</u> of D iff there is a system $\langle D_i : i \in I \rangle$ of algebras of type t and there is an isomorphism h between D and $P_{i \in I} D_i$ such that $(\forall i \in I)\ R_i = ker(pj_i \circ h)$ where $pj_i : P_{j \in I} D_j \longrightarrow D_i$ is the i-th projection of the direct product $P_{i \in I} D_i$.

<u>THEOREM 8</u> ([6]) : Let $D = \langle \underline{\underline{A}}, \underline{\underline{B}}, \Diamond, ? \rangle \in$ **HSP** Dts and let R be a congruence on D . Then (i) and (ii) below are equivalent.
(i) R is a direct factor congruence on D .
(ii) $(\exists w \in B)\left[\ R = ker(rl_w^D) \text{ and } (\forall a \in A)\ \Diamond(a,w) \leq w \leq \Box(a,w)\ \right]$.

Let $D = \langle \underline{\underline{A}}, \underline{\underline{B}}, \Diamond, ? \rangle \in Alg(t)$. We define the partial ordering $\leq^A$ on the set A of actions as follows: $(\forall a,b \in A)\left[\ a \leq^A b \iff a \vee b = b\ \right]$. If $X \subseteq A$ then $sup^A X$ denotes the supremum of X w.r.t. $\leq^A$.

<u>THEOREM 9</u> ([6]) :
Let $D = \langle \underline{\underline{A}}, \underline{\underline{B}}, \Diamond, ? \rangle \in$ **HSP** Dts. Then (i) and (ii) below hold.
(i) Let $w = \langle w_i : i \in I \rangle \in {}^I B$ be a system of elements of B .
Assume that w satisfies conditions 1) - 3) below.
1) $(\forall i,j \in I)\left[\ i \neq j \Rightarrow w_i \cdot w_j = 0\ \right]$.

2) $(\forall a,b \in A)\left[(\forall i \in I)(rl^{A}_{w_i}(a)=rl^{B}_{w_i}(b)) \Rightarrow a=b \right]$.

3) $(\forall a \in {}^{I}A)\left[\sup^{A}\{ ?(w_i);a_i : i \in I \} \text{ exists} \right]$.

Then $\langle \ker rl^{D}_{w_i} : i \in I \rangle$ is a direct decomposition of D .

Note that this implies that $rl^{D}_{w_i}$ is a homomorphism on D for every $i \in I$.

(ii) To every direct decomposition $\langle R_i : i \in I \rangle$ of D there is a $w \in {}^{I}B$ satisfying 1)-3) in (i) above such that $R_i = \ker rl^{D}_{w_i}$ for all $i \in I$.

THEOREM 10 ([6])

(i) Conditions 1)-3) in Thm.9 are independent of each other. I.e. for every infinite set I there are $D = \langle \underline{\underline{A}},\underline{\underline{B}},\Diamond,? \rangle \in Dts$ and $w \in {}^{I}B$ such that D and w satisfy two of conditions 1)-3) but D and w do not satisfy the third condition.

(ii) Consider conditions 4)-7) below.

4) $\sup^{A}\{ w_i : i \in I \} = 1^{B}$.

5) $(\forall p \in {}^{I}B) \sup^{A}\{ w_i \cdot p_i : i \in I \}$ exists.

6) $(\forall i \in I)(\forall a \in A)\ \Diamond(a,w_i) \leq w_i \leq \Box(a,w_i)$.

7) $(\forall a,b \in A)\left[(\forall i \in I)(?(w_i);a=?(w_i);b) \Rightarrow a=b \right]$.

Then statements a)-c) below hold.

a) Condition 2) cannot be replaced with the conjunction of conditions 4),5) and 6). Moreover, there are $D = \langle \underline{\underline{A}},\underline{\underline{B}},\Diamond,? \rangle \in Dts$ and $w \in {}^{I}B$ such that D and w satisfy conditions 1), 4)-6), 3) but D and w do not satisfy condition 7) (which is a weaker form of 2))

b) Condition 3) cannot be replaced with the conjunction of 4)-7), i.e. there are $D = \langle \underline{\underline{A}},\underline{\underline{B}},\Diamond,? \rangle \in Dts$ and $w \in {}^{I}B$ such that D and w satisfy 1), 2), 4)-7) but D and w do not satisfy 3).

c) Conditions 1)-7) imply each other in the way indicated on Figure 1 E.g. conditions 1) and 2) together imply condition 6).

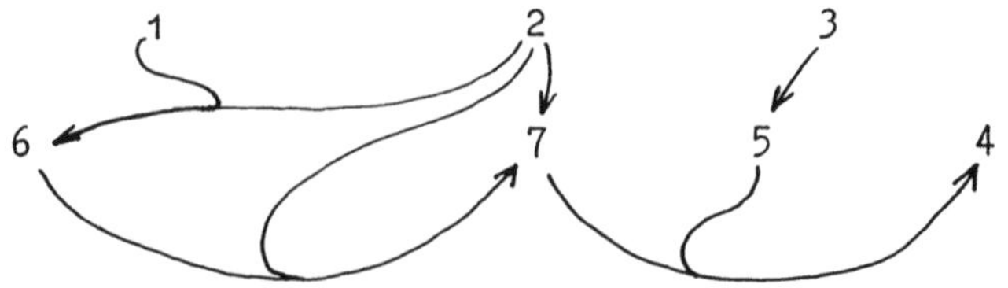

Figure 1

THEOREM 11 : **I** Dts = **SP** Dts .

I Dts is an infinitary quasivariety. **I** Dts = Mod Qeq_{∞} Dts .

DEFINITION 7: Sta $\stackrel{d}{=}$ $\{ D \in Dta : D \models sep \}$.

NOTATION: Let $K \subseteq Alg(t)$. Let X,Y be two sets. Then $Fr_{X,Y} K$ denotes the free **SP** K-algebra with free generators X of sort actions and Y of sort Boolean.

PROBLEM 3: Is **HSP** Dts = **HSP** Dta ? Let X,Y be any two sets. Is then $Fr_{X,Y}$ Sta $\in$ Sta ? We know that $Fr_{X,Y}$ Sda $\in$ **I** Ds but our proof does not generalize to dynamic algebras with tests.

PROBLEM 4: Are the epimorphisms surjective in Dts, **SPUp** Dts, **HSP** Dts, **SPUp** Sta, Sta, and Dta ? It was proved by Andréka-Németi [1], [2] that this is the algebraic logic form of Beth definability property of a logic (in the present case of PDL).

PROPOSITION 12: The epimorphisms are not surjective neither in Ds nor in **SPUp** Sda .

PROOF: **SPUp** Sda $\models \forall x_1 \forall x_2 (\Diamond(x_1,1)=0 \rightarrow x_1 \leq x_2)$. This is not the only reason. **SPUp** Sda $\models ([\Diamond(x_0,1)=0 \wedge x_1^*=x_2^*=x_0^* \wedge \Diamond(x_1,1)=y=\Diamond(x_2,1)] \rightarrow x_1=x_2)$.

Therefore f(y) determines the value of $f(x_1)$ if f is a homomorphism. That is there are $D_1 \subseteq D_2 \in Ds$, $a_1 \in A_2 \sim A_1$, $p \in B_1$, $a_0 \in A_1$ such that $(D_2,a_0,a_1,p) \models (\Diamond(a_0,1)=0 \wedge a_1^*=a_0^* \wedge \Diamond(a_1,1)=p)$.

Let $k,f \in Hom(D_2,D_3)$ be arbitrary with $D_3 \in$ **SPUp** Sda. Assume $k(p)=f(p)$. Then by the above we have $k(a_1)=f(a_1)$. Hence the identical embedding Id: $D_1 \rightarrowtail D_2$ is an epimorphism. QED.

Actually we proved that ?(y) is implicitely definable in **SPUp** Sda.

THEOREM 13: **H** Dts $\not\subseteq$ $\mathbf{SP}^r$ Sta .

PROOF: Let qe be the quasiequation $(x=x;x \wedge x_1^* \leq x) \rightarrow x=x^*$. Then Sta $\models$ qe by Lemmas 3.1, 3.3 of [11] but **H** Dts $\not\models$ qe , since there is $\langle \underline{\underline{A}},\underline{\underline{B}},\Diamond \rangle \in$ **H** Dts with $|B|=2$ and $A = \{?(0), ?(1), a, a^*\}$ and $?(0) < ?(1) < a < a^*$ and $a;a=a$. QED.

DEFINITION 8: Let $D = \langle\langle A,;,v,^* \rangle, \underline{\underline{B}} , \Diamond \rangle \in Alg(d)$. Then $D \in Nds$ is defined to hold iff (i)-(ii) below hold.

(i) $D' \stackrel{d}{=} \langle\langle A,v,; \rangle , \underline{\underline{B}} , \Diamond \rangle$ is a *-free dynamic set algebra that is D' is a subalgebra of a reduct of some Ds .

(ii) For all $a \in A$,

$$a^{*} = \min\{ b \in A \ : \ a \leq b \text{ and } b \text{ is transitive and reflexive} \}.$$

THEOREM 14: $Sda \subseteq \mathbf{I}\, Nds$.

REFERENCES

1. Andréka,H.-Gergely,T.-Németi,I., On universal algebraic construction of logics. Studia Logica 36, 1977, Nr.1-2, pp.9-47.
2. Andréka,H.-Németi,I., A simple, purely algebraic proof of the completeness of some first order logics. Algebra Universalis 5, 1975, pp.8-15.
3. Andréka,H.-Németi,I., Generalization of variety and quasivariety concept to partial algebras through category theory. Dissertationes Mathematicae (Rozprawy) No.204. To appear.
4. Andréka,H.-Németi,I., A general axiomatizability theorem formulated in terms of cone-injective subcategories. Contributions to Universal Algebra (Proc.Coll.Esztergom 1977) Colloq.Math.Soc.J. Bolyai Vol.29, North-Holland. In press.
5. Andréka,H.-Németi,I., Every free algebra in the variety generated by the representable dynamic algebras is separable and representable. Theoretical Computer Science. To appear.
6. Andréka,H.-Németi,I.-Sain,I., Relativization of dynamic algebras a tool for decomposition of dynamic theories. Preprint, 1980.
7. Grätzer,G., Universal algebra. (Second Edition) Springer Verlag, New York, 1979.
8. Henkin,L.-Monk,J.D.-Tarski,A., Cylindric algebras Part I. North-Holland, 1971.
9. Kozen,D., A representation theorem for models of *-free PDL. Report RC7864, IBM Research, Yorktown Heights, New York, 1979.
10. Lugowski,H., Grundzüge der Universellen Algebra. Teubner Verlag, Leipzig, 1976.
11. Pratt,V.R., Dynamic algebras: examples, constructions, applications. Report MIT/LCS/TM-138. July 1979.

THE EQUIVALENCE PROBLEM FOR LL- AND LR-REGULAR GRAMMARS(1)

(extended abstract)

Anton Nijholt
Koninginneweg 179
1075 CP Amsterdam The Netherlands

. INTRODUCTION

Questions whether or not two grammars belonging to a family of grammars gener-.te the same language have extensively been studied in the literature. These problems ıre called equivalence problems and if there exists an algorithm which for each pair ·f grammars of this family gives an answer to this question then the equivalence pro-·lem for this family of grammars is said to be decidable. Otherwise the problem is aid to be undecidable. For example, the equivalence problem for the family of regular ;rammars is decidable. On the other hand, the equivalence problem for the family of ontext-free grammars is known to be undecidable.

The equivalence problem is open for various classes of grammars which gener-.te deterministic languages. For simple deterministic and LL(k) grammars the problem ıas been solved. In this paper we study the equivalence problem for the class of LL-egular grammars and languages. The class of LL-regular grammars is obtained from the lass of LL(k) grammars by allowing regular look-ahead instead of finite look-ahead, f. Jarzabek and Krawczyk |8|, Nijholt |10,11,12| and Poplawski |16| for results on L-regular grammars and languages. The class of LL(k) grammars is properly included n the class of LL-regular grammars and the class of LL(k) languages is properly in-luded in the class of LL-regular languages. Contrary to the other families of lang-ıages which have been studied from the point of view of the equivalence problem, the lass of LL-regular languages contains languages which are not deterministic.

It will be shown that the equivalence problem for LL-regular grammars is deci-.able. Apart from extending the known result for LL(k) grammar equivalence to LL-egular grammar equivalence, we obtain an alternative proof of the decidability of

1) The preparation of this paper was partially supported by a Natural Sciences and Engineering Research Council of Canada Grant No. A-7700 during the author's stay at McMaster University, Hamilton, Ontario, Canada.

LL(k) equivalence. From |21| we understand that the equivalence problem for LL-regular grammars has been studied before, but not solved. Our proof that this equivalence problem is decidable is simple. However, this is mainly because we can reduce the problem to the equivalence problem for real-time strict deterministic grammars, which is decidable, see Oyamaguchi, Honda and Inagaki |15| and Ukkonen |18|.

In this extended abstract the proofs of the theorems are not included. A complete paper will appear elsewhere.

Preliminaries

We assume that the reader is familiar with Aho and Ullman |1| or Harrison |3|. For notational reasons we review some concepts.

A context-free grammar (CFG for short) is denoted by the quadruple $G = (N,\Sigma,P,S)$, where N consists of the nonterminal symbols, Σ consists of the terminal symbols, $N \cap \Sigma = \emptyset$ (the empty set); $N \cup \Sigma$ is denoted by V (elements of V will be denoted by X, Y and Z; elements of $V^\times$ will be denoted by α, β, γ, δ and ω). We use ε to denote the empty word. The elements of $\Sigma^\times$ will be denoted by x, y, z and w. The set P of productions is a subset of $N \times V^\times$ (notation $A \to \alpha$ if (A,α) is in P) and $S \in N$ is called the start symbol of the grammar.

We have the usual notation $\Rightarrow$, $\underset{L}{\Rightarrow}$ and $\underset{R}{\Rightarrow}$ for derivations, leftmost derivations and rightmost derivations, respectively. The superscripts + and × will be used to denote the transitive and the reflexive-transitive closures of these relations.

For any string $\alpha \in V^\times$ define

$$L(\alpha) = \{w \in \Sigma^\times \mid \alpha \overset{\times}{\Rightarrow} w\}.$$

The language L(G) of a CFG G is the set L(S). Two grammars G_1 and G_2 are said to be equivalent if $L(G_1) = L(G_2)$.

For any string $\alpha \in V^\times$ we use α^R to denote the reverse of α. If L is a set of strings, then $L^R = \{w^R \mid w \in L\}$. If $\alpha \in V^\times$ then $|\alpha|$ denotes the length of α. For any $\alpha \in V^\times$ and non-negative integer k we use $k : \alpha$ to denote the prefix of α with length k if $|\alpha| \geq k$ and otherwise $k : \alpha$ denotes α. A production $A \to \varepsilon$ is called an ε-production; a CFG without ε-productions is called an ε-free grammar.

A CFG $G = (N,\Sigma,P,S)$ is said to be right linear if each rule is of the form $A \to uB$ or $A \to u$, with $A,B \in N$ and $u \in \Sigma^\times$. A subset L of $\Sigma^\times$ is said to be regular if there exists a right linear grammar G such that $L(G) = L$.

For any set Q, a partition π of Q is a finite set of mutually disjoint subsets of Q such that each element of Q is in one of these subsets. The elements of a partition are called blocks or equivalence classes. If two elements x and y belong to the same block $B \in \pi$ then we write $x \equiv y \pmod{\pi}$.

DEFINITION 1.1. Let $\pi = \{B_1, B_2, \ldots, B_n\}$ denote a partition of $\Sigma^\times$, where Σ is a finite set, into n blocks. Partition π is said to be a regular partition of $\Sigma^\times$ if all the sets B_i are regular. Partition π is a left congruence (right congruence) if for any strings x, y and z in $\Sigma^\times$, $x \equiv y \pmod{\pi}$ implies $zx \equiv zy \pmod{\pi}$ ($xz \equiv yz \pmod{\pi}$).

A partition $\pi' = \{B_1', B_2', \ldots, B_m'\}$ is a refinement of a regular partition $\pi = \{B_1, B_2, \ldots, B_n\}$ of $\Sigma^\times$ if each B_i of π is the union of some of the blocks of π'. It is well-known that every regular partition of a set $\Sigma^\times$ has a refinement of finite index which is both a left and a right congruence (which we call a congruence for short) (see Hopcroft and Ullman |7|).

In the forthcoming sections it is assumed that the grammars under consideration are reduced. We recall the definitions of strict deterministic and real-time strict deterministic grammars (cf. Harrison and Havel |4,5|).

DEFINITION 1.2. Let $G = (N, \Sigma, P, S)$ be a CFG and let ψ be a partition of V. Partition ψ is called strict if

(i) $\Sigma \in \psi$, and

(ii) For any $A, A' \in N$ and $\alpha, \beta, \beta' \in V^\times$, if
$A \to \alpha\beta$ and $A' \to \alpha\beta'$ are in P and $A \equiv A' \pmod{\psi}$, then either:
(a) both $\beta, \beta' \neq \varepsilon$ and $1 : \beta \equiv 1 : \beta' \pmod{\psi}$,
or
(b) $\beta = \beta' = \varepsilon$ and $A = A'$.

Now a grammar $G = (N, \Sigma, P, S)$ is called strict deterministic if there exists a strict partition of V.

In general, a strict deterministic grammar can have more than one strict partition of V. Let ψ_1 and ψ_2 be two partitions of V with induced equivalence relations $\equiv_1$ and $\equiv_2$, respectively, then $\psi_1 \leq \psi_2$ if and only if $\equiv_1 \subseteq \equiv_2$. The partitions form a semi-lattice with this ordering and under the meet-operation. In Harrison and Havel |4| an algorithm is given which computes the minimal strict partition of a strict deterministic grammar.

A strict deterministic grammar $G = (N, \Sigma, P, S)$ with minimal strict partition ψ is called a real-time strict deterministic grammar if it is ε-free and for all $A, A', B, B' \in N$; $\alpha, \beta \in V^\times$, if $A \to \alpha B$ and $A' \to \alpha B'\beta$ are in P, then $A \equiv A' \pmod{\psi}$ implies $\beta = \varepsilon$.

2. THE EQUIVALENCE PROBLEM FOR GRAMMARS WITH LOOK-AHEAD

One way to generalize definitions of classes of deterministically parsable grammars is to let the decisions in the parsing process of these grammars be determined by look-ahead of the input string. This look-ahead may be finite or regular. Finite look-ahead is for instance used in the definition of LL(k) and LR(k) grammars. Regular look-ahead is used in the definitions of LL-regular and LR-regular grammars. In this section we will introduce regular look-ahead for strict deterministic and real-time strict deterministic grammars. Then it will be shown how the equivalence problems for these grammars with look-ahead can be reduced to the equivalence problems for strict deterministic and real-time strict deterministic grammars. In the following section we will study LL-regular grammars as a special case of the (real-time) strict deterministic grammars with regular look-ahead.

The generalization which we give here for (real-time) strict deterministic grammars conforms the generalizations in |13| for finite look-ahead. We use the following notation. Let $G = (N,\Sigma,P,S)$ be a CFG and let $\pi = \{B_1,B_2,\ldots,B_n\}$ be a regular partition of $\Sigma^\times$. For any $\alpha \in V^\times$,

$$\mathrm{BLOCK}(\alpha) = \{B_k \in \pi \mid L(\alpha) \cap B_k \neq \emptyset\}.$$

<u>DEFINITION 2.1.</u> A CFG $G = (N,\Sigma,P,S)$ is <u>strong</u> <u>SD(π)</u>, where π is a regular partition of $\Sigma^\times$, if there exists a partition ψ of $V = N \cup \Sigma$ such that

(i) $\Sigma \in \psi$

(ii) For any $w_1, w_2 \in \Sigma^\times$; $A, A' \in N$; $\alpha, \beta, \beta', \omega_1, \omega_2 \in V^\times$ with $A \equiv A' \pmod{\psi}$ and derivations

(a) $S \underset{L}{\overset{\times}{\Rightarrow}} w_1 A \omega_1 \underset{L}{\Rightarrow} w_1 \alpha\beta\omega_1$

(b) $S \underset{L}{\overset{\times}{\Rightarrow}} w_2 A' \omega_2 \underset{L}{\Rightarrow} w_2 \alpha\beta'\omega_2$

the condition

$$\mathrm{BLOCK}(\beta\omega_1) \cap \mathrm{BLOCK}(\beta'\omega_2) \neq \emptyset$$

always implies that either

(1) both $\beta, \beta' \neq \varepsilon$ and $1 : \beta \equiv 1 : \beta' \pmod{\psi}$, or

(2) $\beta = \beta' = \varepsilon$ and $A = A'$.

A strong SD(π) grammar $G = (N,\Sigma,P,S)$ with a minimal partition ψ is now called <u>strong</u> <u>real-time</u> <u>SD(π)</u> if G is ε-free and the following condition is satisfied:

For all $A, B, A', B' \in N$ and $\alpha, \beta \in V^\times$, if $A \to \alpha B$ and $A' \to \alpha B'\beta$ are in P with $A \equiv A' \pmod{\psi}$ then if

$$S \overset{\times}{\underset{L}{\Rightarrow}} w_1 A \omega_1 \underset{L}{\Rightarrow} w_1 \alpha B \omega_1$$

$$S \overset{\times}{\underset{L}{\Rightarrow}} w_2 A' \omega_2 \underset{L}{\Rightarrow} w_2 \alpha B' \beta \omega_2$$

and

$$\mathrm{BLOCK}(B\omega_1) \cap \mathrm{BLOCK}(B'\beta\omega_2) \neq \emptyset \qquad (\times)$$

then $\beta = \varepsilon$.

Clearly, the real-time strict deterministic grammars are a special case (no look-ahead) of this definition. Notice that because of ($\times$) $B \equiv B' \pmod{\psi}$.

We now show that the equivalence problem for strong real-time SD(π) grammars is decidable. We start with a strong real-time SD(π) grammar and convert it into a real-time strict deterministic grammar. The conversion will be done in such a way that two strong real-time SD(π) grammars are equivalent if and only if their associated real-time strict deterministic grammars are equivalent.

Let $G = (N,\Sigma,P,S)$ be any CFG without ε-productions and let $\pi = \{B_0, B_1, \ldots, B_n\}$ be a regular partition of $\Sigma^\times$. Without loss of generality we may assume that π is a left congruence and that $B_0 = \{\varepsilon\}$. It follows that $\pi^R = \{B_0^R, B_1^R, \ldots, B_n^R\}$ is a right congruence. Then π^R defines the states and the transitions of a (deterministic) finite automaton $M_\pi = (Q, \Sigma, \delta, q_0)$, where

Q is the set of states, $Q = \{q_0, q_1, \ldots, q_n\}$,
$q_0 \in Q$ is the initial state,
Σ is the input alphabet
$\delta : Q \times \Sigma \to Q$ is the transition function

and δ satisfies

$$B_i^R = \{w \mid \delta(q_0, w) = q_i\}$$

for $0 \leq i \leq n$.

Now let p_0 be a symbol not in Q and let $\perp$ be a special symbol not in Σ. Define a grammar $G_\pi = (N', \Sigma', P', S')$ as follows:

$$N' = \{S'\} \cup (Q \times N \times Q)$$

$\Sigma' = (Q \cup \{p_0\}) \times (\Sigma \cup \{\perp\}) \times Q$

and P' contains productions

(i) $S' \to \langle p_0 \perp p\rangle\langle pSq_0\rangle$ for all $p \in Q$

(ii) If $A \to X_1X_2 \ldots X_r$ is in P then $\langle pAq\rangle \to \langle pX_1p_1\rangle\langle p_1X_2p_2\rangle \ldots \langle p_{r-1}X_rq\rangle$ is in P', for any $p, q, p_1, \ldots, p_{r-1}$ in Q such that if $X_j \in \Sigma$, then $\delta(p_j, X_j) = p_{j-1}$, for $1 < j < r$; if $X_1 \in \Sigma$ then $\delta(p_1, X_1) = p$ and if $X_r \in \Sigma$, then $\delta(q, X_r) = p_{r-1}$.

We can reduce grammar G_π. Throughout this paper, whenever we use the subscript π then we refer to the grammar which is obtained with this construction.

Let G and G_π be as above. Define a homomorphism $\rho\colon V'^{\times} \to V^{\times}$ by

$\rho(\langle p_0 \perp p\rangle) = \varepsilon$ for every $p \in Q$

$\rho(\langle pXq\rangle) = X$ for each $p, q \in Q$ and $X \in V$

The proofs of the following claims are straightforward and therefore omitted.

CLAIM 2.1. For any $\langle rXs\rangle \in V'$ and $y \in \Sigma'^{\times}$, if $\langle rXs\rangle \overset{\times}{\Rightarrow} y$, then $\delta(s, \rho(y^R)) = r$.

Clearly, this claim can easily be extended to an arbitrary string $\alpha = \langle rX_1s_1\rangle\langle s_1X_2s_2\rangle \ldots \langle s_{n-1}X_ns_n\rangle$ in $V'^{\times}$. If $\alpha \overset{\times}{\Rightarrow} y$, where $y \in \Sigma'^{\times}$, then $\delta(s, \rho(y^R)) = r$.

CLAIM 2.2. For any $\langle pXq\rangle \in V'$, if $\langle pXq\rangle \overset{\times}{\Rightarrow} \alpha\langle rYs_1\rangle\langle s_2Zt\rangle$ for some string $\alpha\langle rYs_1\rangle\langle s_2Zt\rangle\beta$ in $V'^{\times}$, then $s_1 = s_2$.

CLAIM 2.3. For any $\langle pXq\rangle \in V'$ and $\omega' \in V'^{\times}$, if $\langle pXq\rangle \underset{L}{\overset{\times}{\Rightarrow}} \omega'$ in G_π then $X \underset{L}{\overset{\times}{\Rightarrow}} \rho(\omega')$ in G.

From Claim 2.3 it is immediately clear that $L(G) = \rho(L(G_\pi))$, where we have extended the definition of ρ to sets of strings.

CLAIM 2.4. For any $w, x \in \Sigma'^{\times}$, $\langle pXq\rangle \in V'$ and $\omega \in V'^{\times}$, if $S' \underset{L}{\overset{\times}{\Rightarrow}} w\langle pXq\rangle\omega \underset{L}{\overset{\times}{\Rightarrow}} wx$ in G_π, then $\rho(x) \in B_p$, where B_p is a block of partition $\pi = \{B_0, B_1, \ldots, B_n\}$.

With the help of these claims it is now straightforward to prove the following lemmas.

LEMMA 2.1. If G is an ε-free strong SD(π) grammar then G_π is an ε-free strict deterministic grammar.

LEMMA 2.2. If G is a strong real-time SD(π) grammar then G_π is a real-time strict deterministic grammar.

Now consider two ε-free grammars G_1 and G_2 which are strong (real-time) SD(π_1) and strong (real-time) SD(π_2), respectively. Here π_1 and π_2 are regular partitions of the same set $\Sigma^\times$. Then G_1 and G_2 are both strong (real-time) SD(π) with respect to the regular partition

$$\pi = \{B \mid B_i \cap B_j = B,\quad B \neq \emptyset,\ B_i \in \pi_1,\ B_j \in \pi_2)$$

For π we can construct the sequential machine M_π and the (real-time) strict deterministic grammars G_π^1 and G_π^2. Clearly, if $L(G_1) = L(G_2)$ then $L(G_\pi^1) = L(G_\pi^2)$ and if $L(G_1) \neq L(G_2)$ then $L(G_\pi^1) \neq L(G_\pi^2)$. It follows that we have reduced the equivalence problem for strong (real-time) SD-regular grammars to the problem for (real-time) strict deterministic grammars.

Any real-time strict deterministic grammar can be converted into an equivalent real-time deterministic pushdown automaton (cf. Harrison |3|) which accepts with empty stack. In Oyamaguchi, Honda and Inagaki|15| the decidability of the equivalence problem for these automata has been shown.

COROLLARY 2.1. The equivalence problem for strong real-time SD(π) grammars is decidable.

In the following section it will be shown that each strong LL-regular grammar is a strong real-time SD-regular grammar. It is wellknown that strong LL-regular grammars can generate non-deterministic languages. The language

$$L = \{a^n b^k a^n,\ a^k b^n c^n \mid n \geq 1,\ k \geq 1\}$$

is an example of a language which is not real-time strict deterministic but it is deterministic. Moreover, L is a strong real-time SD-regular language.

Culik and Cohen |2| use a slightly different method than is presented here to convert an LR-regular grammar into an LR(0) grammar. Clearly, the argument which we gave above holds for LR-regular grammars as well. That is, we have the following proposition:

PROPOSITION 2.1. The equivalence problem for LR-regular grammars is decidable if and only if the equivalence problem for LR(0) grammars is decidable.

3. THE EQUIVALENCE PROBLEM FOR LL-REGULAR GRAMMARS

We start this section with the definition of LL-regular grammars (Nijholt |11|, Poplawski |16|).

DEFINITION 3.1. Let $G = (N,\Sigma,P,S)$ be a CFG and let $\pi = \{B_0,B_1,\ldots,B_n\}$ be a regular partition of $\Sigma^\times$. Grammar G is an LL(π) grammar if, for each $w, x, y \in \Sigma^\times$; $\alpha, \gamma, \delta \in V^\times$ and $A \in N$, the conditions

(i) $S \overset{\times}{\underset{L}{\Rightarrow}} wA\alpha \underset{L}{\Rightarrow} w\gamma\alpha \overset{\times}{\underset{L}{\Rightarrow}} wx$

(ii) $S \overset{\times}{\underset{L}{\Rightarrow}} wA\alpha \underset{L}{\Rightarrow} w\delta\alpha \overset{\times}{\underset{L}{\Rightarrow}} wy$

(iii) $\mathrm{BLOCK}(\gamma\alpha) \cap \mathrm{BLOCK}(\delta\alpha) \neq \emptyset$

always imply that $\gamma = \delta$.

Notice that if $\mathrm{BLOCK}(\gamma\alpha) \cap \mathrm{BLOCK}(\delta\alpha) \neq \emptyset$ then there exist strings $x \in L(\gamma\alpha)$ and $y \in L(\delta\alpha)$ such that $x \equiv y \pmod{\pi}$. A CFG G is called LL-regular if it is LL(π) for some regular partition π of $\Sigma^\times$. Notice that a grammar G is LL(k) if G is LL(π_k) for the regular partition

$$\pi_k = \{\{u\} \mid u \in \Sigma^\times \text{ and } |u| < k\} \cup \{\{uw \mid w \in \Sigma^\times\} \mid u \in \Sigma^k\}$$

where Σ^k is the set of all words over Σ with length k.

As in the case of LL(k) grammars it is possible to define strong LL-regular grammars.

DEFINITION 3.2. Let $G = (N,\Sigma,P,S)$ be a CFG and let $\pi = \{B_1,B_2,\ldots,B_n\}$ be a regular partition of $\Sigma^\times$. Grammar G is a strong LL(π) grammar if, for each $w_1, w_2, x, y \in \Sigma^\times$; $\alpha_1, \alpha_2, \gamma, \delta \in V^\times$ and $A \in N$, the conditions

(i) $S \overset{\times}{\underset{L}{\Rightarrow}} w_1A\alpha_1 \underset{L}{\Rightarrow} w_1\gamma\alpha_1 \overset{\times}{\underset{L}{\Rightarrow}} w_1x$

(ii) $S \overset{\times}{\underset{L}{\Rightarrow}} w_2A\alpha_2 \underset{L}{\Rightarrow} w_2\delta\alpha_2 \overset{\times}{\underset{L}{\Rightarrow}} w_2y$

(iii) $x \equiv y \pmod{\pi}$

always imply that $\gamma = \delta$.

The class of LL-regular grammars properly includes the class of strong LL-regular grammars. However, the language families coincide. In Poplawski |16| a transformation can be found which converts any LL-regular grammar into an equivalent strong LL-regular grammar. Hence, without loss of generality we may assume that the LL-regular grammars which are considered are strong.

The language

$$L = \{aa^nba^{2n}b,\ ba^nba^{2n},\ aa^nba^na,\ ba^nba^nb \mid n \geq 0\}$$

is an example of a non-deterministic language which is LL-regular (cf. |11|). Language

$$L = \{a^nb^ka^n,\ a^kb^nc^n \mid n \geq 1,\ k \geq 1\}$$

is an example of an LL-regular language which is not real-time strict deterministic.

Let G be an LL-regular grammar. The method which is given in |1| for eliminating ε-productions from an LL(k) grammar can easily be modified in order to obtain the result that for every LL-regular grammar we can find an equivalent ε-free LL-regular grammar. As mentioned above, we may assume that the LL-regular grammars under consideration are strong. The proof of the following theorem is again in the complete paper.

<u>THEOREM 3.1.</u> If G is an ε-free strong LL-regular grammar, then G is a strong real-time SD-regular grammar.

From Corollary 2.1 and Theorem 3.1 we may now conclude:

<u>COROLLARY 3.1.</u> The equivalence problem for LL-regular grammars is decidable.

It is natural to ask whether it is possible to convert LL-regular grammar G to an LL(1) grammar G_π. The method which is given in Culik and Cohen |2| yields for each LR-regular grammar G an LR(0) grammar G_π. Therefore it is not necessary to develop a parsing method for LR-regular grammars since the method for LR(0) grammars can be used. Unfortunately, the conversion which we use here does not necessarily yield an LL(1) grammar. In |12| a method has been given which converts an LL(π) grammar G into an LL(1) grammar G' such that $L(G_\pi) \subseteq L(G')$. Here G_π is the grammar which is obtained from LL(π) grammar G with the method described above. If we were able to obtain from LL(π) grammar G an LL(1) grammar G', with $L(G') = L(G_\pi)$ then we should have reduced the equivalence problem for LL-regular grammars to the equivalence problem for LL(1) grammars.

REFERENCES

1. Aho, A. V. and Ullman, J. D. The Theory of Parsing, Translation, and Compiling, Vols. 1 and 2. Prentice Hall, Inc., Englewood Cliffs, N.J., 1972 and 1973.

2. Culik, K. and Cohen, R. LR-regular grammars -- an extension of LR(k) grammars. J. Comput. System Sci. 7 (1973), 66-96.

3. Harrison, M. A. Introduction to Formal Language Theory. Addison-Wesley, Reading, Mass., 1978.

4. Harrison, M. A. and Havel, I. M. Strict deterministic grammars. J. Comput. System Sci. 7 (1973), 237-277.

5. Harrison, M. A. and Havel, I. M. Real-time strict deterministic grammars. SIAM J. Comput. 1 (1972), 333-349.

6. Harrison, M. A. Havel, I. M. and Yehudai, A. An equivalence of grammars through transformation trees. Theoret. Comput. Sci. 9 (1979), 173-206.

7. Hopcroft, J. E. and Ullman, J. D. Formal Languages and their Relation to Automata. Addison-Wesley, Reading, Mass., 1969.

8. Jarzabek, S. and Krawczyk, T. LL-regular grammars. Information Processing Letters 4 (1975), 31-37.

9. Korenjak, A. J. and Hopcroft, J. E. Simple deterministic languages. Conf. Record of 7th Annual Symp. on Switching and Automata Theory 1966, 36-46.

10. Nijholt, A. On the parsing of LL-regular grammars. In: Math. Foundations of Computer Sci., A. Mazurkiewicz (ed.), LNCS 45, Springer, Berlin, 1976, 446-452.

11. Nijholt, A. LL-regular grammars. Int. J. of Computer Math. 8 (1980), 303-318.

12. Nijholt, A. From LL-regular to LL(1) grammars. Report IR-61, Amsterdam, May 1980.

13. Nijholt, A. A framework for classes of grammars between the LL(k) and LR(k) grammars. TR No. 80-CS-25, McMaster University, Hamilton, 1980.

14. Olshansky, T. and Pnueli, A. A direct algorithm for checking equivalence of LL(k) grammars. Theoret. Comput. Sci. 4 (1977), 321-349.

15. Oyamaguchi, M., Honda, N. and Inagaki, Y. The equivalence problem for real-time strict deterministic languages. Information and Control 45 (1980), 90-115.

16. Poplawski, D. A. On LL-regular grammars. J. Comput. System Sci. 18 (1979), 218-227.

17. Rosenkrantz D. J. and Stearns, R. E. Properties of deterministic top-down grammars. Information and Control 17 (1970), 226-255.

18. Ukkonen, E. A decision method for the equivalence of some non-real-time deterministic pushdown automata. 12th Ann. S. on Theory of Computing, 1980, 29-38.

19. Wood, D. Some remarks on the KH algorithm for s-grammars. BIT 13 (1973), 476-489.

20. Wood, D. Lecture notes on top-down syntax analysis. J. of the Computer Society of India 8 (1978), 1-22.

21. Zubenko, V. V. Simple pushdown storage automata and the equivalence problem in certain classes of LL(π) grammars (in Russian). Theory and Practice of systems programming, Inst. Kibernet., Akad. Nauk Ukrain. SSR, Kiev, 1976.

CONTEXT-FREE LANGUAGES OF INFINITE WORDS AS LEAST FIXPOINTS

Axel Poigné
Informatik II
Universität Dortmund
Postfach 50 05 00
D-4600 Dortmund 50, F.R.G.

). INTRODUCTION

basic theorem on non-determinism is Schützenberger's result [12], which says given context-free grammar the generated language may be characterized as the least fixoint of an appropiate function. Let for example $G = (X,T,x,P)$ with $X:=\{x\}$, $T:=\{a,b\}$ $:=\{x \to ax|b\}$ be a grammar. Then the generated language $L(G,x)$ is the least fixoint of the function $\hat{G}: P(T^*) \to P(T^*)$, $S \to \{a\}\cdot S \cup b$, the least fixpoint being deined by $Fix(\hat{G}) := \bigcup_n \hat{G}^n(\phi)$ ($P(T^*)$- power set of T^*, order is inclusion).

he theorem is restricted to words of finite length. There have been spent considerable fforts to achieve a similar fixpoint theorem for context-free languages of infinite ords. A first solution to this problem is due to Nivat [8]:
et T^∞ be the set of finite and infinite words over T with a suitably defined monoid ultiplication. Then given a context-free grammar G the generated language of finite nd infinite words is obtained as the greatest fixpoint of a function which in our xample is $\hat{G}: P(T^\infty) \to P(T^\infty)$, $S \to \{a\}\cdot S \cup b$. The greatest fixpoint is given by $\bigcap \hat{G}^n(T^\infty)$. The result holds if the grammar is in Greibach form.

e suggest a different approach which makes use of ω-complete partial orders and of he Knaster-Tarski theorem on least fixpoints:
o model the non-determinism inherent in grammars we use commutative idempotent semiroups appropiately enriched with ω-complete partial orders (compare [9]). Now certain onoids, we call left-strict, may be defined with respect to the tensorproducts of hese commutative idempotent semigroups. We shall show that free structures $CP^*(T)$ of this ind consist of "finitely generated" subsets of the set $T_\perp^\infty$ of finite and infinite words ver the flat domain $T_\perp$ (i.e. T plus a least element $\perp$) which are ordered by cofinaliy induced by the order on $T_\perp^\infty$. Then the generated language is obtained as the least ixpoint of a ω-continuous function $\hat{G}: CP^*(T)^X \to CP^*(T)^X$, the least fixpoint being as sual $\bigsqcup_n \hat{G}^n(\perp))$. This again holds if the grammar is in Greibach form.

n objective of our work is to obtain some insight into the interdependence of the noion of derivation and completion with respect to limits. We are guided by Scott's oint of view that an infinite amount of information should be approximated by finite ieces of information. This is somewhat opposed to Nivat's approach [8]. Moreover we feel that computing on a syntactic level should be reflected by the use of free structures. As a consequence we have to introduce and investigate several new structures which allow to transfer the proof techniques used in [10] to the case of context-free languages. Thus we establish a connection between formal computations of non-deterministic recursive schemes and derivations of context-free languages similar to that to be found in the works of Arnold and Nivat [2],[7].

Our paper is divided into three sections: In the first one we give the basic definitions on infinite words and infinite derivations, in the second one we discuss the algebraic framework and in the third one we sketch the proof of the fixpoint theorem.

Due to limitation of space we are only able to sketch the proofs. Full proofs are to be found in [11]. Our approach strongly depends on the ideas exposed in [7],[9]. For category theory we refer to [5], [6].

1. GENERATION OF INFINITE WORDS

Let X be a (finite) set. A partial function $w: \mathbb{N} \to X$ from the set of integers into X is called *word* over X iff $n+1 \in def(w)$ implies that $n \in def(w)$. A word is called *finite* iff $def(w)$ is finite otherwise *infinite*. The set of words is denoted by X^{∞}, that of finite resp. infinite words X^{*} resp. X^{ω}. The length $|w|$ of a word w is the cardinality of $def(w)$. With a multiplication

$$vw(n) := \begin{cases} v(n) & \text{if } n \in def(v) \\ w(n-|v|) & \text{else} \end{cases}$$

and with the totally undefined function λ as unit X^{∞} is a monoid, X^{*} with the restricted multiplication a free monoid.

Let $X := \{x_i \mid i \in \underline{n}\}$ $(\underline{n} := \{0,\dots,n-1\})$ be a finite set of *non-terminals* and T be a finite set of *terminals*. Then a *context-free grammar* G will be written as a system of equations

$$G = \{x_i = S_i \mid i \in \underline{n}\}$$

with S_i be finite, non-empty subsets of $(X+T)^{*}$ (Note: we assume the grammar to be reduced in the sense that there are no superfluous non-terminals).

For a grammar G we define a relation $\underset{G}{\to}$ on $(X+T)^{*}$ by

$v \underset{G}{\to} w$ iff $\exists v', v'' \in (X+T)^{*} \exists i \in \underline{n} \exists w' \in S_i : v = v'x_iv''$ and $w = v'w'v''$.

Let $\underset{G}{\overset{*}{\to}}$ be the reflexive and transitive closure of $\underset{G}{\to}$ (For convenience we shall omit the index G in the following).

The *language generated by a grammar* G *and a word* $v \in (X+T)^{*}$ is defined by

$$L(G,v) := \{w \in T^{*} \mid v \overset{*}{\to} w\} \quad .$$

We may consider *infinite derivations* $w_o \to w_1 \to w_2 \to \dots \to w_n \to \dots$ and ask for the generated word which should consist of terminals. In [8] the following approach is proposed:

On X^∞ a partial order $\le$ is defined by $v \le w$ iff $\forall\, n \in \mathrm{def}(v)\colon v(n) = w(n)$. One says that v is a *left factor* of w. Then any increasing ω-chain $w_o \le w_1 \le \dots \le w_n \le \dots$ has a least upper bound $\bigsqcup_n w_n$.

For all $w \in (X+T)^\infty$ let a(w) denote the *largest terminal left factor:*

$$a(w) := \bigsqcup\{v \in T^\infty \mid v \le w\}.$$

Then $v \to w$ implies $a(v) \le a(w)$ and for any infinite derivation $w_o \to w_1 \to \dots \to w_n \to \dots$ an increasing chain $a(w_o) \le a(w_1) \le \dots \le a(w_n) \le \dots$ which has a least upper bound $\bigsqcup_n a(w_n)$. An infinite derivation is *successful* iff $\bigsqcup_n a(w_n)$ is an element of T^ω. For successful derivations we say that $w = \bigsqcup_n a(w_n)$ is the *generated word* and use $w_o \overset{\omega}{\to} w$ as notation.

The *ω-language generated by a grammar* G *and a word* $v \in (X+T)^*$ then is

$$L^\omega(G,v) := \{w \in T^\omega \mid v \overset{\omega}{\to} w\}$$

We denote $L^\infty(G,v) := L(G,v) \cup L^\omega(G,v)$.

From a structural point of view this approach has a drawback; the monoid multiplication is not monotone with respect to the left factor order. For example for $X:=\{a,b,c,d\}$ $a \le ab$ and $c \le cd$ does not imply $ac \le abcd$. This is why we suggest a slightly different approach which is more appropiate to our purposes:

Let X be a (finite) set and $X_\perp := X+\{\perp\}$. We factorize the monoid $X_\perp^\infty$ by the least congruence relation containing the relation $\{(\perp w, \perp) \mid w \in X_\perp^\infty\}$ where $\perp\colon \mathbb{N} \to X$ is the constant function $\perp(n) = \perp$. The carrier of the resulting monoid is isomorphic to

$$C^*(X) := X^* \cup X^* \cdot \{\perp\} \cup X^\omega$$

the multiplication being defined by

$$v \cdot w := \begin{cases} vw & \text{if } v \in X^\infty \\ v & \text{if } v \in X^* \cdot \{\perp\} \end{cases} .$$

On $C^*(X)$ a partial order is defined by $v \le w$ iff $v = w$ or ($v = v'\perp$ and $w = v'v''$). The submonoid consisting of finite words is denoted by $F^*(X)$ (carrier is $X^* \cup X^* \cdot \{\perp\}$).

It is straightforward to ensure:

.1 <u>*Lemma:*</u> (i) The monoid multiplication of $F^*(X)$ is monotone.

(ii) $C^*(X)$ is ω-complete (i.e. every ω-chain has a least upper bound).

(iii) The monoid multiplication of $C^*(X)$ is ω-continuous (lub's are preserved).

We note that the order is almost the same as the left-factor order except that $\perp$ and not the empty word is the smallest left factor for any word.

An important feature is that $F^*(X)$ and $C^*(X)$ are free structures:
Let pos (ω-pos) denote the category of (ω-complete) posets and monotone (ω-continuous) maps [1]. A $\text{pos}_\perp$- *enriched* ($\omega\text{-pos}_\perp$-*enriched*) *monoid* is a monoid such that the carrier is a (ω-complete) poset with a least element $\perp$ and that the monoid multiplication is monotone (ω-continuous). With morphisms being strict [1] monotone (ω-continuous) monoid homomorphisms these data define categories $\text{mon-pos}_\perp$ ($\text{mon-}\omega\text{-pos}_\perp$).

A $\text{pos}_\perp$-enriched monoid is called *left-strict* iff it satisfies the equation $\perp\cdot x = \perp$.
Let $\text{lsmon-pos}_\perp$ and $\text{lsmon-}\omega\text{-pos}_\perp$ be the full subcategories of the above categories with objects being left-strict. Then we have

1.2 *Proposition:* $F^*(X)$ resp. $C^*(X)$ is a free left-strict $\text{pos}_\perp$- resp. $\omega\text{-pos}_\perp$-enriched monoid over a set X.

Now let $b: C^*(X+T) \to C^*(X+T)$ be the unique $\text{lsmon-}\omega\text{-pos}_\perp$ morphism induced by freeness from

$$X+T \to C^*(X+T), \quad z \to \begin{cases} \perp & \text{if } z \in X \\ t & \text{if } t \in T \end{cases} .$$

$b(w)$ is the largest terminal left factor of w except that $b(w)$ may end with a $\perp$. So ω-languages can be defined in the same way as above using $b(w)$ instead of $a(w)$.

2. NON-DETERMINISTIC MONOIDS

Context-free grammars are inherently non-deterministic; replacing a non-terminal one has to choose among a finite number of productions. We shall characterize this situation by algebraic methods:

A binary choice operator $|$, used for instance in the Backus-Naur notation for context-free grammars, naturally satisfies the axioms

$$x \mid (y \mid z) = (x \mid y) \mid z$$
$$x \mid y = y \mid x$$
$$x \mid x = x \quad .$$

This suggests that commutative idempotent semigroups may be the appropiate domain to model this kind of non-determinism.

Let cis be the category of *commutative idempotent semigroups* (i.e. a set X together with a binary operation such that the above axioms are satisfied) with *linear* mappings ($f: X \to X'$ such that $f(x+y) = f(x) + f(y)$) as morphisms. It is well known that the forgetful functor $\text{cis} \to \text{set}$ to the underlying sets has a canonical left adjoint P_f: $\text{set} \to \text{cis}$ defined by $P_f(X) := \{S \subseteq X \mid S \text{ finite, non empty}\}$, the structure map being the union. Thus the free structure immediately yields the natural interpretation of the choice construct, which is the union.

The adjunction lifts to the level of monoids: Given a monoid $(M,\cdot,e)$ $P_f(M)$ together with the multiplication $S\cdot T := \{x\cdot y \mid x \in S,\ y \in T\}$ and the unit $\{e\}$ is a monoid. It is even a free monoid but only with respect to tensorproducts in cis [4].

Let X,Y,Z be commutative idempotent semigroups. A mapping $f\colon X \times Y \to Z$ is called *bilinear* iff for all $x \in X$, $y \in Y$ the mappings $f(x,\_)\colon Y \to Z$, $y \to f(x,y)$ and $f(\_,y)\colon X \to Z$ $x \to f(x,y)$ are linear. Then a *tensorproduct* of X and Y (in cis) is an object $X \otimes Y$ together with a (universal) bilinear mapping $\gamma\colon X \times Y \to X \otimes Y$ such that for all bilinear mappings $f\colon X \times Y \to Z$ there exists an unique linear mapping $\bar{f}\colon X \otimes Y \to Z$ such that

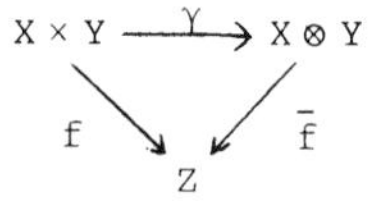

commutes. It is well known that cis has tensorproducts.

A *monoid with respect to the tensorproduct* in cis is a monoid with a carrier being a commutative idempotent semigroup and a multiplication being a bilinear mapping (or equivalently: a linear mapping out of the tensor product). With the obvious notion of homomorphism this defines a category $\text{mon}_\otimes\text{-cis}$.

2.1 *Proposition:* The forgetful functor $\text{mon}_\otimes\text{-cis} \to \text{set}$ has a left adjoint $P^\star\colon \text{set} \to \text{mon}_\otimes\text{-cis}$ defined by

$$P^\star(X) := \{S \subseteq X^\star \mid S \text{ finite, non empty}\}$$

$$S + T := S \cup T$$

$$S \cdot T := \{x\cdot y \mid x \in S,\ y \in T\}$$

$$e := \{\lambda\}$$

(compare [4]).

As we deal with infinite derivations we have to add some kind of limit process which according to the previous section should be ω-completeness. There are several possibilities to enrich commutative idempotent semigroups with order [9]. We choose the following one:

A commutative idempotent semigroup (X,+) may be understood as a poset such that any finite, non empty subset has a least upper bound (lub), the ordering is given by $x \le y$ iff $x+y = y$ for all $x,y \in X$. Linear maps then are monotone maps preserving these lubs. If the induced order is already ω-complete and if the structure mapping is already ω-continuous we call the commutative idempotent semigroups *ω-complete with respect to join ordering*. The appropiate notion of morphism then is linear mappings being ω-continuous. These data define a category ω-⊔-pos.

Again it is well known [9] that $\underline{\omega\text{-}\sqcup\text{-pos}}$ has tensorproducts. So it makes sense to define $\underline{\omega\text{-}\sqcup\text{-pos}}_{\perp}$-enriched monoids with respect to the tensorproduct. The carrier of such a monoid is a $\underline{\omega\text{-}\sqcup\text{-pos}}$ - object with a least element, the multiplication is a bilinear mapping (or linear out of the tensorproduct) being ω-continuous. With morphisms being strict linear ω-continuous monoid homomorphisms this defines a category $\underline{\text{mon}_{\otimes}\text{-}\omega\text{-}\sqcup\text{-pos}_{\perp}}$. The full subcategory of left-strict monoids is denoted by $\underline{\text{lsmon}_{\otimes}\text{-}\omega\text{-}\sqcup\text{-pos}_{\perp}}$.

Our approach now basically depends on the fact that we can charactrize free structures of this kind:

2.2 *Definition*: Let $S \subseteq C^{*}(X)$. An element $s \in S$ is called *maximal relative* S iff for all $s' \in S$ $s \leq s'$ implies $s = s'$.

For $S \subseteq C^{*}(X)$ let $\hat{S} := \{x \in F^{*}(X) \mid \exists\, s \in S: x \leq s\}$ the *ideal of finite words generated by* S and

$$\mathcal{C}(S) := \{\bigsqcup_n x_n \mid (x_n \mid n \in \mathbb{N}) \text{ is a } \omega\text{-chain in } \hat{S}\}$$

be the *chain-closure* of S.

A subset $S \subseteq C^{*}(X)$ is called *finitely generated* iff there exists a ω-chain $(S_n \mid n \in \mathbb{N})$ of finite, non empty subsets S_n of $F^{*}(X)$ with $S_n \sqsubseteq S_{n+1}$ for all $n \in \mathbb{N}$ such that

$$S = \mathcal{C}_m(\bigcup_n S_n) := \{x \in \mathcal{C}(\bigcup_n S_n) \mid x \text{ maximal relative } \mathcal{C}(\bigcup_n S_n)\}$$

(where $S \sqsubseteq S'$ iff $\forall\, s \in S\ \exists s' \in S': s \leq s'$).

2.3 *Theorem*: Let X be a set. Then a free $\underline{\text{lsmon}_{\otimes}\text{-}\omega\text{-}\sqcup\text{-pos}_{\perp}}$ object over a set X is given by

$$CP^{*}(X) := \{S \subseteq C^{*}(X) \mid S \text{ finitely generated}\}$$

$$S \leq S' \text{ iff } S \sqsubseteq S'$$

$$\perp := \{\perp\}$$

$$\bigsqcup_n S_n := \mathcal{C}_m(\bigcup_n S_n) \quad \text{for } \omega\text{-chains } (S_n \mid n \in \mathbb{N})$$

$$S + S' := \mathcal{C}_m(S \cup S')$$

$$S \cdot S' := \mathcal{C}_m(\{x \cdot y \mid x \in S,\ y \in S'\})$$

$$e := \{e\}$$

The proof is all along the structure. In fact we use the same techniques as used for ω-completion of posets and construction of free monoids with respect to tensorproducts, but stepwise and lifted to the level of our kind of structures.

The problem is to show that cofinality gives a partial order. This immediately follows from

2.4 *Lemma:* Let $(S_n \mid n \in \mathbb{N})$ be a chain of finite, non empty subsets S_n of $F^*(X)$ with $S_n \sqsubseteq S_{n+1}$ for all $n \in \mathbb{N}$. Then $C(\bigcup_n S_n) \sqsubseteq C_m(\bigcup_n S_n)$.

This is not obvious as there may be some ω-chain in $C(\bigcup_n S_n)$ which does not have some upper bound in $C(\bigcup_n S_n)$. The proof depends on the following: Let $(x_i \mid i \in \mathbb{N})$ be a ω-chain in $\widehat{\bigcup_n S_n}$. W.r.g. we assume that $x_n \in S_n$. Then the sets

$$K_n := \{\bigsqcup T \mid T \subseteq \hat{S}_n \text{ and } \{x_n\} \sqsubseteq T \text{ and } \forall j \leq n\ \exists s_j \in \widehat{\bigcup_n S_n}: T \cup \{x_j\} \sqsubseteq \{s_j\}\}$$

are finite and non empty and for all $n \in \mathbb{N}$ $K_n \sqsubseteq K_{n+1}$. Now there exists a ω-chain $(y_n \mid n \in \mathbb{N})$ such that $y_i \in MAX(K_n) := \{x \in K_n \mid x \text{ maximal relative } K_n\}$. But $\bigsqcup_n y_n$ is maximal relative $C(\bigcup_n S_n)$ and $\bigsqcup_n x_n \leq \bigsqcup_n y_n$.

2.5 *Remark:* There seems to be a close connection to structures used by Back [3] to model unbounded non-determinism. The monoid $C^*(X)$ is exactly what is called set of execution path' in [2], and $CP^*(X)$ seems to be the very same structure as the set $H(\Sigma_\omega)$.

3. THE FIXPOINT THEOREM

As all finite, non empty subsets of $X+T^*$ are elements of $CP^*(X+T)$ a grammar may be understoood as a mapping $G: X \to CP^*(X+T)$.

Now let $f: X \to CP^*(T)$ be a mapping and $[f,\eta]^{\#}: CP^*(X+T) \to CP^*(T)$ be the unique (smon$_\otimes$-ω-$\sqcup$-pos$_\perp$) homomorphic extension of $[f,\eta]: X+T \to CP^*(T)$ with $[f,\eta](x) := f(x)$ for $x \in X$ and $[f,\eta](t) := t$ for $t \in T$.

3.1 *Lemma:* The function $\hat{G}: CP^*(T)^X \to CP^*(T)^X$, $f \to [f,\eta]^{\#} \cdot G$ is ω-continuous (with respect to the order induced by $CP^*(T)$).

Our main theorem now is

3.2 *(Fixpoint-) Theorem:* Let $G: X \to CP^*(X+T)$ be a grammar in *Greibach form*, i.e. for all $x \in X$ $w \in G(x)$ implies that $w \in T\cdot(X+T)^*$. Then for all $x \in X$

$$L^\infty(G,x) = \bigsqcup_n \hat{G}^n(\perp)\ (x)$$

where $\perp: X \to CP^*(T)$ is the constant function.

Our proof heavily depends on the following lemma derived by using freeness of $CP^*(X+T)$:

3.3 *Lemma*: Let $\phi: X \to CP^{*}(X+T)$, $\psi: T \to CP^{*}(X+T)$ be the canonical embeddings. Then a grammar $G: X \to CP^{*}(X+T)$ induces an arrow $[G,\psi]^{\#}$ in

$$\begin{array}{ccc} X+T & \longrightarrow & CP^{*}(X+T) \\ & {\scriptstyle [G,\psi]} \searrow & \downarrow {\scriptstyle [G,\psi]^{\#}} \\ & & CP^{*}(X+T) \end{array}$$

$[G,\psi]$ being induced by coproduct properties. We abbreviate $G_n := [G,\psi]^{\# n}$. Then

(i) for all $S \in CP^{*}(X+T)$ $G_n(S)$ is finite and non empty if S is so.

(ii) for all $n \in \mathbb{N}$

$$\hat{G}^{n}(\bot) = b \cdot G_n \cdot \phi$$

with $b = [\bot, \eta]^{\#}$ (We note that for $w \in (X+T)^{*}$ $b(w)$ corresponds to the monoid homomorphism defined at the end of section 1).

(i) is obvious and (ii) says more or less that either words may be substituted for non-terminals in right hand sides of productions or that right hand sides of productions may be substituted in words.

We now need a series of simulation lemmas. By induction on length of terms we obtain:

3.4 *Lemma*: Let $S \subseteq (X+T)^{*}$. Then

(i) for all $v \in S$ there exists a $w \in G_n(S)$ s.t. $v \to^{*} w$.

(ii) for all $w \in G_n(S)$ there exists a $v \in S$ s.t. $v \to^{*} w$.

(iii) for all $n \in \mathbb{N}$ $b(G_n(S)) \sqsubseteq b(G_{n+1}(S))$.

(iv) there exists an infinite derivation $w_o \to^{*} w_1 \to^{*} \dots \to^{*} w_n \to^{*} \dots$ such that $w_n \in G_n(S)$.

3.5 *Lemma*: Let $w_o \to w_1 \to \dots \to w_n \to \dots$ be an infinite derivation. Then for all $n \in \mathbb{N}$ there exists a $w'_n \in G_n(\{w_o\})$ such that $w_n \to^{*} w'_n$. The proposition holds analoguously for finite derivations.

For 3.4(iv) we apply König's lemma (cf [7] for example). The proof of 3.5 is done by induction considering several cases. The main argument is that if $v \to w$ then for all $w' \in G_n(\{w\})$ there exists a $v' \in G_{n+1}(\{v\})$ such that $w' \to^{*} v'$.

Using all these lemmas we conclude

3.6 *Proposition*: For all $x \in X$ $L^{\infty}(G,x) \sqsubseteq C_m(\bigcup_n b(G_n(\{x\})))$.

The proof of the reverse direction is based on the fact that for a grammar G being Greibach any word $w \in G_n(S)$ starts with at least n terminals, because n substitutions are executed. This is used in the proof of

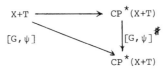

3.7 *Lemma:* Let G be Greibach and $S \subseteq (X+T)^*$. Then

(i) if $v \in G_n(S)$, $w \in G_{n+i}(S)$ with $v \to^* w$ then $v_n = w_n$ for all $n, i \in \mathbb{N}$ where v_n, w_n denote the words consisting of the first n letters of n resp. w.

(ii) Let $w_o \to^* w_1 \to^* \dots \to^* w_n \to^* \dots$ be a derivation such that $w_n \in G_{j(n)}(\{w_o\})$ and $j(n) \leq j(n+1)$ for all $n \in \mathbb{N}$. Then $\bigsqcup_n b(w_n) \in T^\omega$, i.e. the derivation sequence is successful.

3.8 *Proposition:* If G is Greibach then $C_m(\bigcup_n b(G_n(\{x\})) \sqsubseteq L^\infty(G,x)$.

The argument is as follows:

Let $(x_n | n \in \mathbb{N})$ be a ω-chain in $\widehat{\bigcup_n b(G_n(\{x\}))}$ such that the lengt' $|x_n|$ are increasing. Then for any $n \in \mathbb{N}$ there exists a $j(n)$ and a $w \in G_{j(n)}(\{x\})$ such that $x_n \leq b(w)$. Let $|x_n| \leq j(n)$ then by 3.4(ii) there exists a $v \in G_{|x_n|}(\{x\})$ such that $v \to^* w$. But as $v_{|x_n|} = w_{|x_n|}$ (3.7(i)) $x_n \leq b(w)$ implies $x_n \leq b(v)$. If $j(n) \leq |x_n|$ then by 3.4(iii) $\{x\}_n \sqsubseteq b(G_{|x_n|}(\{x\}))$.

We conclude that the sets

$$S_n := \{w \in G_{|x_n|}(\{x\}) \mid x_n \leq b(w)\}$$

are finite and non empty.

Now let $w' \in S_{n+1}$. Then there exists a $w \in G_{|x_n|}(\{x\})$ such that $w \to^* w'$. But $x_n \leq x_{n+1} \leq b(w')$ and $w_{|x_n|} = w'_{|x_n|}$ implies $x_n \leq b(w)$ resp. $w \in S_n$.

Applying König's lemma we obtain an infinite derivation $w_o \to^* w_1 \to^* \dots \to^* w_n \to^* \dots$

By 3.7(ii) the derivation is successful, thus $\bigsqcup_n b(w_n) \in L^\infty(G,x)$, but $\bigsqcup_n x_n \leq \bigsqcup_n b(w_n)$.

Similar arguments for the case that $(x_n | n \in \mathbb{N})$ is not increasing yields the result.

By 3.3, 3.6 and 3.8 follows the fixpoint theorem.

4. CONCLUDING REMARKS

Looking at the above results we doubt if the Greibach condition is as canonical for languages of infinite words as for languages of finite words. Let us give two arguments:

1) Given a grammar in Greibach form it is possible to recover the language of infinite words from the language of finite words.

   Proposition: If $G: X \to CP^*(X+T)$ is a grammar in Greibach form then for all $x \in X$

   $$L^\infty(G,x) = C_m(\ L(G,x)\ ).$$

   A similar result is stated for adherences in [7].

2) If we consider the grammar $G = (\{x\}, \{a,b\}, x, \{x \to xa | b\})$ then $L^\infty(G,x) = L(G,x)$. But there is no grammar G' in Greibach form such that $L^\infty(G',x) = L(G,x)$, as any

grammar being Greibach such that L(G',x) is infinite implies that $L^{\omega}(G',x)$ is not empty [7].

It seems that this asymmetry of the Greibach condition inherently corresponds to the definition of left-strict monoids (or to the definition of the monoid multiplication in [7]), as a consequence the fixpoint theorem is restricted to the Greibach case. We feel that from a computational point of view the theorem is not satifactory as according to (1) it does not tell too much about the computational power of grammars with respect to infinite computations. To obtain a more general fixpoint theorem it seems to be necessary to consider structures such as generalized words as in [13],[14], or to use non-deterministic domains, i.e. commutative idempotent semigroups which are enriched by order structure in the sense of [9].

Let us remark at last that the order approach has the advantage (in contrast to the metrical approach of [7]) that the structures used are monoidal closed, which implies the existence of higher types. So we hope to be able to extend our methods to non-deterministic schemes of higher types.

REFERENCES

[1] ADJ-group: Some fundamentals of order algebraic semantics. MFCS '76, Lect. Notes in Comp. Sci. 45, 1976

[[2] Arnold,A., Nivat,M.: Formal computations of non deterministic recursive program schemes. Math. Syst. Th. 13, 1980

[3] Back,J.: The semantics of unbounded non-determinism. ICALP '80, Lect. Notes in Comp. Sci. 85, 1980

[4] Huwig,H.,Poigné,A.: Continuous and non-deterministic completions of algebras. 3rd Hungarian Comp. Sci. Conf., 1981

[5] MacLane,S.: Kategorien. Berlin-Heidelberg-New-York 1972

[6] Manes, E.G.: Algebraic theories. Berlin-Heidelberg-New-York 1976

[7] Nivat,M.: Infinite words. Found. of Comp. Sci.III, ed: deBakker,J. ,van Leuwen,J., Amsterdam 1979

[8] Nivat,M.: Mots infinie engendré par une grammaire algébrique. R.A.I.R.O. informatique theorique Vol. 11, 1977

[9] Hennessy,M.C.B.,Plotkin,G.: Full abstraction of a simple programming language. MFCS '79, Lect. Notes in Comp. Sci. 74. 1979

[10] Poigné,A.: Using least fixed points to characterize formal computations of non-deterministic equations. Proc. Formal. of Progr. Concepts, Peniscola, Lect. Notes in Comp. Sci. 105, 1981

[11] Poigné,A.: A least fixpoint semantics of non-deterministic schemes. forthcoming

[12] Schützenberger,M.P.: Push-down automata and context-free languages. Inf. and Contr. Vol 6, 1963

[13] Courcelle,B.: Frontiers of infinite trees. RAIRO Inf.Theor. 12, 1978

[14] Heilbrunner,S.: An algorithm for the solution of fixed-point equations for infinite words. RAIRO Inf. Theor. Vol 14, 1980

REMARKS ON THE NOTION OF CONCURRENCY RELATION IN THE CASE OF SYSTEMS

Piotr Prószyński

Institute of Mathematics, Warsaw Technical University
Pl.Jedności Robotniczej 1, 00-661 Warsaw / Poland

Introduction.

The main aim of this paper is to present some remarks on the method of investigation of Petri nets properties on the basis of concurrency relation. We are going to consider wide class of nets, including cyclic and non-deterministic ones. The notion of K-density - a property of occurence nets (see [1,5,6]) - will be generalized for concurrent systems.

The method of the description of nets is derived from R.Janicki [3]. In his approach the special attention is paid to the class of so called "proper" nets, which may be decomposed into subnets representing sequential subsystems. In the subsequent considerations we will limit ourselves to that class of nets, however, the results concerning the concurrency relation are valid for all symmetric and irreflexive relations (called in the paper concurrency-like relations).

The present paper has been divided into two parts. The first of them contains basic notions and theoretical results , the second - interpretations of these notions based on the examples.

In principle, this paper is the continuation and the complement of [7] , however it can be read independently.

1. Theoretical approach.

1.Simple and proper nets.

Let X be a set and let $left: X \times X \to X$, $right: X \times X \to X$ be the functions:

$$(\forall (x,y) \in X \times X) \quad left((x,y)) = x , \quad right((x,y)) = y.$$

Df. By a simple net we mean any pair N=(T,P) , where:

T is a set of transitions,

$P \subseteq 2^T \times 2^T$ is a relaton (also interpreted as a set of places),

$(\forall a \in T)(\exists p,q \in P) \quad a \in left(p) \cap right(q)$.

The net is interpreted thus as a set of transitions and, a description of the way in which these transitions are interlinked.

We shall only consider finite simple nets.

To denote the fact $\{\{a_1,\dots,a_n\} , \{b_1,\dots,b_m\}\} \in P$ we shall write:

$[a_1,\dots,a_n : b_1,\dots,b_m] \in P$ and represent this fact graphically:

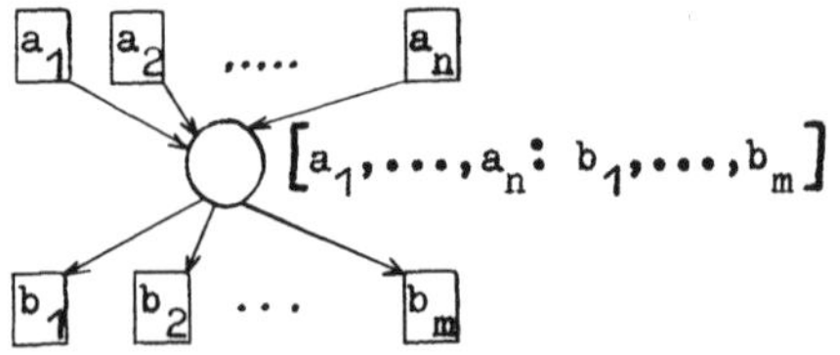

Let us define relation $F \subseteq T \times P \cup P \times T$:

$(x,y) \in F \iff x \in left(y) \cup y \in right(x)$.

We can note that triple (T,P,F) is the standard representation of Petri net (see [5,6]).

We will say that net $N_1=(T_1,P_1)$ is a subnet of $N=(T,P)$ and write $N_1 \subseteq N$ iff $P_1 \subseteq P$. It was proved that "$\subseteq$" is a partial order relation and that the set of all simple nets with this relation is a lattice. In this lattice we have:

$N_1 \cup N_2 = (T_1 \cup T_2,\ P_1 \cup P_2)$ (see also [3,4]).

Now we introduce a class of proper nets. To simplify the consideration we shall use the well-known notation:

1). $(\forall p \in P)$ $p^\bullet = right(p)$, $^\bullet p = left(p)$

2). $(\forall a \in T)$ $a^\bullet = \{p \in P \mid a \in left(p)\}$, $^\bullet a = \{p \in P \mid a \in right(p)\}$.

<u>Df.</u> A simple net N=(T,P) is called <u>elementary</u> iff:

1. $(\forall a \in T)$ $|^\bullet a| = |a^\bullet| = 1$

2. $(\forall x,y \in T \cup P)$ $(x,y) \in (F \cup F^{-1})^*$ (N is connected)

Note that an elementary net represents a sequential system, generally - a non-deterministic one.

<u>Df.</u> A simple net N is said to be <u>proper</u> iff it is a union of its elementary subnets.

In other words, a proper net represents a system, which can be decomposed into sequential subsystems.

Let us define the notion of marked net. This notion enables us to take into consideration the dynamic structure of system.

Let $N=(T,P)$ be a simple net and $R1 \subseteq 2^P \times 2^P$ be the following relation:

$(M_1,M_2) \in R1 \iff (\exists a \in T)\ M_1 - {}^\bullet a = M_2 - a^\bullet \ \&\ {}^\bullet a \subseteq M_1 \ \&\ a^\bullet \subseteq M_2$.

The relation $RN=(R1 \cup R1^{-1})^*$ is called the <u>reachability relation of a net N</u>. RN is an equivalence relation and for every $M \in 2^P$ the equivalence class of RN containing M will be denoted : $[M]_{RN}$.

<u>Df.</u> By a <u>marked simple net</u> we mean any triple MN=(T,P,Mar), where N=(T,P) is a simple net, $Mar \subseteq 2^P$ is a set of <u>markings</u> of MN, and:

$Mar = \bigcup \{[M]_{RN} \mid M \in Mar\}$.

We will say that :

A transition $a \in T$ is <u>fireable</u> iff $(\exists M_1,M_2)$ $^\bullet a \subseteq M_1$ & $a^\bullet \subseteq M_2$.

A marked net is <u>locally fireable</u> iff every transition of this net is fireable.

A marked net is compact iff $(\forall M \in Mar)\ Mar = [M]_{RN}$.

A marked net is safe iff $(\forall C \in 2^P)(\forall a \in T)$

$$({}^{\bullet}a \cap C = \emptyset \ \&\ (\exists M \in Mar)\ {}^{\bullet}a \cup C \subseteq M) \Leftrightarrow (a^{\bullet} \cap C = \emptyset \ \&\ (\exists M' \in Mar)\ a^{\bullet} \cup C \subseteq M').$$

2. Concurrency-like relations.

The relations we are going to use as a model of concurrency relations are symmetric and irreflexive (sir-relations - 2).

Let X be a set and let $id \subseteq X \times X$ be the identity relation. For every sir-relation $C \subseteq X \times X$ let kens(C) and $\overline{kens}(C)$ be the following families of subsets of X :

$$A \in kens(C) \Leftrightarrow \begin{array}{l} 1.(\forall a,b \in A)\ \ (a,b) \in C \cup id \\ 2.(\forall c \notin A)(\exists a \in A)\ \ (a,c) \in C \end{array}$$

$$A \in \overline{kens}(C) \Leftrightarrow \begin{array}{l} 1.(\forall a,b \in A)\ \ (a,b) \notin C \\ 2.(\forall c \notin A)(\exists a \in A)\ \ (a,c) \in C \end{array} .$$

It can be proved that :

1. kens(C) and $\overline{kens}(C)$ are coverings of X .
2. The relation $\overline{C} = X \times X-(C \cup id)$ is sir-relation and $kens(\overline{C})=\overline{kens}(C)$ and $\overline{kens}(\overline{C})=kens(C)$.

Df. Let cov be a covering of X. A sir-relation $sir(cov) \subseteq X \times X$ defined as follows

$$(a,b) \subseteq sir(cov) \Leftrightarrow \{A \mid A \in cov \ \&\ a \in A\} \cap \{A \mid A \in cov \ \&\ b \in A\} = \emptyset$$

is called the sir-relation defined by the covering cov.

Let N=(T,P) be a proper net, and let $E=\{N_1,\dots,N_m\}$ be a set of elementary subnets of N such that: $N=N_1 \cup N_2 \cup \dots \cup N_m$. Define $cov_E=\{P_1,\dots,P_m\}$, where P_i (i=1,...,m) is a set of places of net N_i. All places belonging to the same sequential component are dependent on each other, thus on the basis of this covering we can state which places can ''coexist'' .

Df. The relation $coex_E=sir(cov_E)$ is called the coexistancy relation defined by the covering E.

In particular, when E consists of all elementary subnets of N, this relation is called the coexistency defined by whole structure of net N and denoted by $coex_N$.

When N is the net of occurrences, then the relation $coex_N$ is the concurrency relation from Petri ([5]) restricted to places and minus identity.

The following theorem is valid ([3]):

For every proper net N=(T,P) the marked simple net $MN=(T,P,kens(coex_E))$ is safe and locally fireable.

3. Consistency of coexistency relation with the net structure.

The fact that the decomposition of any proper net into sequential components (described by its static structure) and the decomposition given by net's concurrency relation are consistent can be denoted as follows:

$$\overline{kens}(coex_E) = cov_E \ .$$

We shall write generally that the covering cov and sir-relation sir(cov) are consistent iff

$$\overline{kens}(sir(cov)) = cov \ .$$

However, there are nets, without this property (see part II, Example 1,2) .

Conditions for consitency of the covering with the sir-relation defined by this covering are desribed in [7] . But a little weaker property,namely :

$$cov \quad \overline{kens}(sir(cov))$$

is more useful. In this case we shall write that covering cov and the relation sir(cov) are semiconsistent.

Now we shall formulate such condition which would enable us to investigate semiconsistency without the necessity of building the relation. The reason why we are treating the semiconsistency as important will be shown more clearly in the following sections.

Let cov be a covering of X, where X is a finite set, and let S be an arbitrary subset of X.

Df. A family $R_S(cov) \subseteq 2^X$ is said to be hooked on S iff

$$R_S \subsetneq cov \quad \& \quad (\forall P \in R_S) \quad P \cap S \neq \emptyset \ .$$

Let $H_S(cov)$ denote the set of all families hooked on S and let $\overline{H}_S$ denote the set of all families hooked on S which do not contain S.

Df. A covering cov is said to be replete set covering (abbr. RS-covering) iff

$$(\forall S \in cov)(\forall R \in H_S(cov)) \quad S \subseteq \bigcup R \implies \bigcap R \subsetneq S \ .$$

Theorem 3.1.

The families $\overline{kens}(sir(cov))$ and kens(sir(cov)) are RS-coverings.

Theorem 3.2.

$cov \subsetneq \overline{kens}(sir(cov)) \iff$ cov is a RS-covering.

4. C-density.

In his ''Non-sequential processes'' ([5]) Petri has defined the notion of K-density of the form (in our notation) :

''A sir-relation C is K-dense iff

$$(\forall A \in kens(C))(\forall B \in \overline{kens}(C)) \quad A \cap B \neq \emptyset \ . ''$$

The K-density defined for concurrency relation of occurrence nets has the following interpretation: every sequential process (component) and every '' case '' have common element. By ''case '' we mean the global state of concurrent process, consisting of the sum of the local states at certain time (see [1]).

Now we are going to describe a notion similar to K-density. Our intention here is to save the above interpretation also in the case of systems. At this point we should make it clear that in further considerations we shall use the simple marked nets $MN=(T,P,kens(coex_N))$ (where $N=(T,P)$ is proper net) as the models of concurrents systems. We can not accept definition of K-density in the previously accepted form. The reason for it is the fact that not all sets from family

$\overline{\text{kens}}(\text{coex}_E)$ represent real sequential components (see Example 1,2) . Accordingly, we shall describe the following notion (from [4]):

Df. A sir-relation sir(cov) defined by the covering cov is said to be C-dense for covering cov iff

$$(\forall A \in \text{cov})(\forall B \in \text{kens}(\text{sir}(\text{cov}))) \quad A \cap B = \emptyset \text{ .}$$

From the above definition it follows that C-density is rather a property of the covering; however, in the case of the relation coex_E the covering is always understood. We want also to pay attention to the fact that interpretation of C-density is as follows: if the relation coex_E is C-dense then every sequential subsystem and every global state of the system have a common place, or, in other words: each global state of the system is described by the local states of all sequential subsystems. Permissible states of the system are, in this case, described by the family $\text{kens}(\text{coex}_E)$. Comparing the interpretation of K- and C-density it is easily seen that in the second one the phrase ''at certain time'' is omitted. Since our systems may be non-deterministic, a statement about states ''at time'' can lead to misunderstanding.

Let us now formulate the conditions which would enable us to investigate C-density.

Theorem 4.1.

If a covering is not the RS-covering, then the relation sir(cov) defined by thiscovering is not C-dense for this covering.

So, RS-coverability (equivalent to the semiconsistency) is the necessary condition for C-density; although this condition is not sufficient.

Theorem 4.2.

A relation sir(cov) defined by RS-covering cov is C-dense iff

$$(\forall A \in \text{cov})(\forall R \in \overline{H}_A(\text{cov}))\left[A \subset \bigcup R \Rightarrow \sim(\exists W \in X)(\forall Z \in R)(\forall Y \in \text{cov})\ |W \cap (Z-A) = 1 \ \& |W \cap Y \leqslant 1\right].$$

The condition contained in this theorem is very difficult to verify in practice, but on the basis of it we can formulate some necessary conditions of C-density (see [7]). One of these (which follows from Theorem 4.2 immediately) is of the form:

Corollary.

If a covering cov is minimal (e.g. $(\forall A \in \text{cov})$ cov- $\{A\}$ is not any covering), then the relation sir(cov) is C-dense for the covering cov.

5. Concurrency of actions.

Finally, we want to pay attention to the problem of concurrency of transitions (actions) , the most important one in all concurrent systems theory. In fact, thinking about ''concurrency'' we mean that some actions can be executed simultaneously. If we consider any concurrency of actions in intuitive terms then we do agree that:

1) execution of these actions in an arbitrary sequence, or

2) simultaneous execution of them

gives the same result. The second condition <u>can not</u> be omitted, since it states difference between concurrency and non-determinism (which will be pointed out in Example 3). From this point of view, concurrency-like relations, which have been considered earlier in this paper, describe only the coexistancy of the local states of the system. However, in this way we are able to describe all permissible global states of this system, and consequtively, describe the notion of transitions' concurrency.

<u>Df.</u> The transitions $a_1,\dots,a_n$ of the net $MN=(T,P,kens(coex_E))$ are said to be <u>concurrent</u> iff

$$(\forall a_i,a_j)\ i\neq j \Rightarrow {}^\bullet a_i \cap {}^\bullet a_j = a_i^\bullet \cap a_j^\bullet = {}^\bullet a_i \cap a_j^\bullet = a_i^\bullet \cap {}^\bullet a_j = \emptyset$$
$$\& \quad (\exists M \in kens(coex_E)) \quad \bigcup_{i=1}^{n} {}^\bullet a_i \subset M .$$

Note that the above definition is valid, if the marked net is safe; any analogous definition without this assumption is more complicated.

Note also that if the n transitons are concurrent, then every $2,3,\dots,n-1$ of them are concurrent too, but not vice versa.

II. Interpretations.

Now we shall consider a few examples, which should show the meaning of notions defined in this paper, and, in the same time, should improve the understanding of concurrency phenomena.

<u>Example 1.</u>

Let us consider the following net:

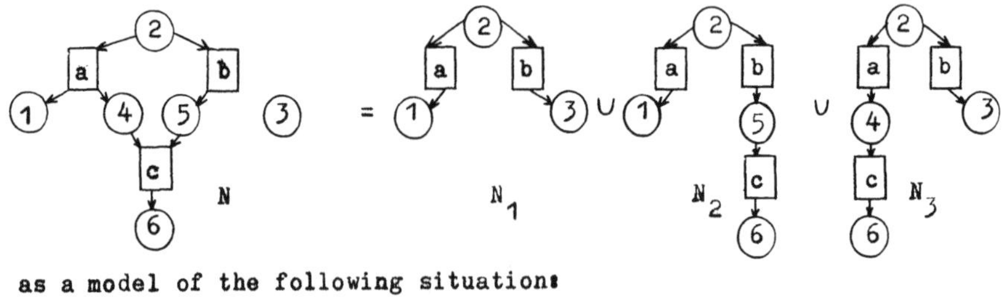

as a model of the following situation:

''Two tasks may be activated, but available storage is enough only for one of them. If they are both active, they can resolve the given problem.''

Of course, the request is self-contradictory and the net has bad properties (it does not represent a ''well-defined'' system) . Let us interpret places and transitions as follows:

(1) - storage has been allocated to task 1

(2) - storage allocation is possible

(3) - storage has been allocated to task 2

(4) - task 1 is active

⑤ - task 2 is active

⑥ - the problem has been resolved

[a] - storage is allocated to task 1 and task is activated

[b] - storage is allocated to task 2 and task is activated

[c] - the problem is resolved

We have thus a covering :

$cov_N = \{\{1,2,3\}, \{1,2,5,6\}, \{2,3,4,6\}\}$ - which is not the RS-covering, so the relation $coex_N = sir(cov_N)$ is not consistent with cov (with the net stucture), and $coex_N$ is not C-dense.

The families $kens(coex_N)$ and $\overline{kens}(coex_N)$ have the following form :

$$kens(coex_N) = \{\{2\}, \{6\}, \{1,4\}, \{3,5\}, \{4,5\}\}$$

$$\overline{kens}(coex_N) = \{\{1,2,3,6\}, \{1,2,5,6\}, \{2,3,4,6\}\} .$$

Let us observe that the set $\{4,5\}$ of the family $kens(coex_N)$ represents the state of system, in which a state of the subsystem represented by the net N_1 is ''undefined''. The net N has also bad properties then, if we consider the relation $coex_E$ defined by another covering :

$$cov_E = \{\{1,2,5,6\}, \{2,3,4,6\}\} .$$

This covering is minimal, so the relation $coex_E$ is C-dense (but not compact) . However, if we ascribe to the net the above interpretation then it will apear that we have lost from our sight a very important component, namely the process of storage allocation. Now we have :

$$\overline{kens}(coex_E) = cov_E$$

$$kens(coex_E) = \{\{2\}, \{6\}, \{1,3\}, \{1,4\}, \{4,5\}, \{3,5\}\} .$$

We can see now that the state $\{1,3\}$ is permissible. This fact means that the condition ''storage can be allocated to one task only'' - which is fulfilled by the net structure - is violated by the system's dynamic structure. It being so, only the net $(T,P,kens(coex_N))$ models our problem.

Example 2.

''Two tasks ask in the same moment for allocation of the same tape unit; the unit is allocated to one of them.''

The net which is the model of this situation is of the form:

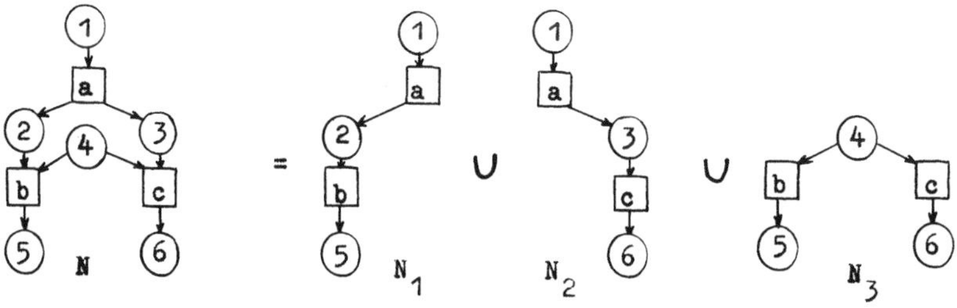

$$cov_N = \{\{1,2,5\}, \{1,3,6\}, \{4,5,6\}\}$$

The places and transitions have here the following meaning:

(1) - tasks are active

(2) - task 1 is waiting for allocation

(3) - task 2 is waiting for allocation

(4) - allocation is possible

(5) - the unit has been allocated to task 1

(6) - the unit has been allocated to task 2

[a] - the tasks are asking for allocation

[b] - the unit is allocated to task 1

[c] - the unit is allocated to task 2

The interpretation of subnets N_1, N_2, N_3 is clear: N_1 describes the allocation process for task 1, N_2 - allocation process for task 2, and N_3 - a choice of task (non-deterministic). The covering cov_N is minimal, which implies that the relation $coex_N$ is semiconsistent with the net structure and C-dense. The family

$$\overline{kens}(coex_N) = \{\{1,2,5\}, \{1,3,6\}, \{4,5,6\}, \{1,5,6\}\} ,$$

so cov_N $\overline{kens}(coex_N)$, however $cov_N \neq \overline{kens}(coex_N)$.
Since, $kens(coex_N) = \{\{1,4\}, \{2,6\}, \{3,5\}, \{2,3,4\}\}$ - we can easily show that net $(T,P,kens(coex_N))$ is compact.

Example 3.

Now we shall consider the next situation:

"Two tasks ask in the same moment for storage allocation; there is enough storage for many tasks." This situation may be represented by the net:

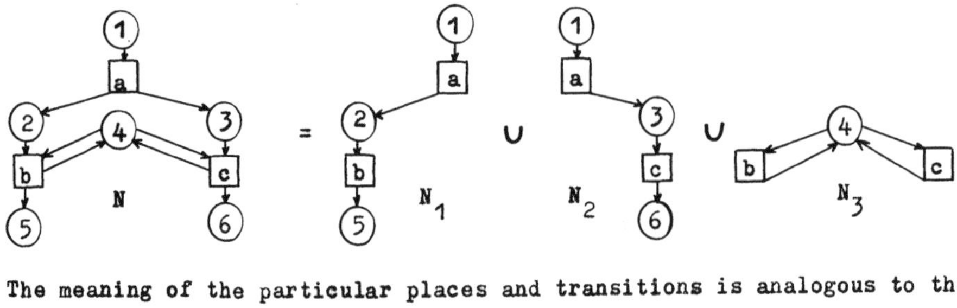

The meaning of the particular places and transitions is analogous to that seen in Example 2. We have here:

$$cov_N = \{\{1,2,5\}, \{1,3,6\}, \{4\}\} .$$

This covering is a minimal one, thus the relation $coex_N$ is C-dense.

$$\overline{kens}(coex_N) = cov_N$$

$$kens(coex_N) = \{\{1,4\}, \{2,3,4\}, \{2,4,6\}, \{3,4,5\}, \{4,5,6\}\}$$

We can state that this net is compact too.

It is argued that the firing of transitions b and c in arbitrary sequence gives the same result (and we are not able to say which one of them will be fired first). They <u>can not</u> be, however, fired simultaneously, so they <u>are not</u> concurrent ones. The definition of transitions' concurrency accepted in the present paper states this fact directly: it is enough to remind the reader that the transitions b and c have a common input place.

Example 4.

''There is no bridge across the river, and people have to use a boat. But there is only one boat. Two persons are coming independently to the river and are going to cross it, each from the other bank. To simplify the situation we assume that the boat is put ashore on one of the banks.''

Let us consider as a model of this situation the following net:

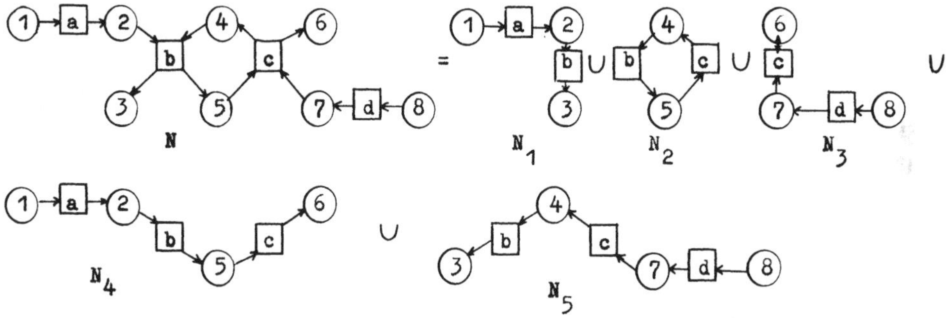

So, we have $cov_N = \{\{1,2,3\}, \{4,5\}, \{6,7,8\}, \{1,2,5,6\}, \{3,4,7,8\}\}$.

This covering is a RS-covering, however the relation $coex_N$ is not C-dense, since:

$\{4,5\} \subset \{1,2,5,6\} \cup \{3,4,7,8\}$ and exist the set $W=\{1,8\}$, such that has exactly one common element with the sets $\{1,2,5,6\}$ and $\{3,4,7,8\}$, and has at most one common element with every set of covering cov_N (see Theorem 4.2). The lack of C-density means that there are global states of system, in which states of some subsystems are undefined. Hence, comparing this fact to description of situation we should suppose that the dynamic structure of the net (with the coexistency relation $coex_N$) is quite diffrent than expected. The analysis of the set of markings ($kens(coex_N)$) endorses our supposition.

Let us describe in exact terms the meaning of places and transitions of this net:

- (1) - there is a person on left bank
- (2) - the person on the left bank is going to use the boat
- (3) - the person from the left bank has got across the river
- (8)(7)(6) - analogous description for second person
- (4) ((5)) - the boat is on the left (right) bank
- [a] - the person on the left bank is coming to the river
- [b] - the person is crossing the river from the left to the right bank
- [d] [c] - analogous description for second person.

Now, looking closer the subnets N_1-N_5 we can see that subnets N_4 and N_5 have wrong interpretation, since the fact that one of the persons has used the boat does not imply that the second person would also use it (the second person could use it earlier). That is the reason why we have to constuct the concurrency relation using another covering, namely :

$cov_E = \{\{1,2,3\}, \{4,5\}, \{6,7,8\}\}$.

This is the minimal covering, and the relation $coex_E$ defined by this covering is C-dense. However, this net is not compact. The main equivalence classes in this case are $[\{1,4,8\}]_{RN}$ and $[\{1,5,8\}]_{RN}$. Their markings have an interpretation consistent with the described situation. But there are also markings, for example $\{3,4,7\}$, which are not adequate to it; such marking represents the following state of the system: two persons on one bank of the river (one waiting for the boat, and the other which had just used the boat) but the boat is on another bank of the river (!). The equivalence class generated by this marking consist of two markings: $\{3,4,7\}$ and $\{3,4,8\}$ and is a covering of N_5 (the same is with the class generated by $\{2,5,6\}$ and the net N_4). In [3,4] R.Janicki suggests that these markings (''degenerated'' markings) which cover the omitted elementary subnets should be also omitted.

Finally we should observe that the transitions a and d, a and c, b and d are concurrent.

Conclusion.

This paper is an attempt to investigate certain properties of concurrent systems on the basis of concurrency-like relations. Some general properties of simmetric and irreflexive relations have been stated. However, the result obtained here are not satisfatory - only properties of so simple nets as those which were used in the examples could be investigated without difficulties.

The author supposes that only precise definition of the concurrency relation may improve our understanding of the concurrency phenomena and would enable us to investigate any properties of concurrent systems.

Acknowledgement.

The author would like to thank R.Janicki for many invaluable discusions, stimulating papers and for help in the formulation of many problems. The author whishes also to thank P.E.Lauer and P.R.Torrigiani for their critical remarks on the interpretation of C-density, and for their remarks on the nature of concurrency.

References.

[1] Best E., The Relative Strength of K-density, Lecture Notes in Comp. Sci., vol.84, Springer-Verlag, 1980, pp.261 - 276.

[2] Janicki R., A Characterization of Concurrency-like Relations, Lecture Notes in Comp. Sci., vol.70, Springer-Verlag, 1979, pp.109 - 122.

[3] Janicki R., On Atomic Nets and Concurrency Relations, Lecture Notes in Comp. Sci., vol.88, Springer-Verlag, 1980, pp.320 - 333.

[4] Janicki R., An Approach to the Phenomenon of Concurrency, ICS Reports, to appear.

[5] Petri C.A., Non-Sequential Processes, ISF Report 70-01, GMD, Bonn, 1977.

[6] Petri C.A., Concurrency as a Basis of System Thinking, ISF Report 78-06, GMD, Bonn, 1978.

[7] Prószyński P., Petri Nets and Concurrency-like Relations, Lecture Notes in Comp. Sci., vol.107, Springer-Verlag, 1981.

ON THE SIZE OF CONJUNCTIVE REPRESENTATIONS OF n-ARY RELATIONS

Aleš Pultr
Charles University, Prague

Let us have a finite system $\mathcal{A}$ of symbols. If we wish to encode a set with an n-ary relation, (X,R), by means of $\mathcal{A}$, we represent he elements $x \in X$ as strings $a_1(x)\dots a_{r(x)}(x)$ of symbols from $\mathcal{A}$ and describe the relation by a system $\underline{F}$ of predicates on $\mathcal{A}$, say as follows

$$(x_1,\dots,x_n) \in R \quad \text{iff} \quad \mathcal{F}(a_1(x_1),\dots,a_{r(x_1)}(x_1),\dots,a_{r(x_n)}(x_n)) \quad \text{for some } \mathcal{F} \in \underline{F} .$$

n a way, the simplest case is that where all the strings are equally long and we have predicates $\mathcal{F}_i$ such that

$$(x_1,\dots,x_n) \in R \quad \text{iff} \quad \forall i \ \mathcal{F}_i(a_i(x_1),\dots,a_i(x_n))$$

such an encoding will be referred to as a conjunctive representation).

In this note we are concerned with the necessary size of conjunctive representations of general relations (and also of some special ypes of them). We show that the length of strings increases in the rder of $|X|^{n-1}$ (for more precise formulations see §1).

Conventions : The sets will be finite, in the infinite cases we speak rather of classes or systems (such as, e.g., the system of all inite n-ary relations). The cardinality of X will be denoted by $|X|$. The system of all natural numbers is denoted by $\mathbb{N}$, the natural numbers themselves are considered, as usual, as the sets $n = \{0,1,\dots,n-1\}$. Further conventions (concerning categorial terminology) will be introduced at appropriate places.

§1. Formulation of the problem

1.1. Let us have a finite system $\mathcal{A}$ of symbols. We say that an n-ary relation R on a set X is r-<u>conjunctively represented</u> by means of $\mathcal{A}$ if the $x \in X$ are encoded as strings $(a_1(x),\dots,a_r(x)) \in \mathcal{A}^r$ and we have got predicates $\mathcal{F}_1,\dots,\mathcal{F}_r$ on $\mathcal{A}$ such that

$$(x_1,\dots,x_n) \in R \quad \text{iff} \quad \forall i\ \mathcal{F}_i(a_i(x_1),\dots,a_i(x_n)) .$$

<u>1.2.</u> We will be concerned with the increase of the necessary length r with increasing X. Let us introduce the following notation :

$d_{s,n}(X,R)$ is the minimum r such that there is an r-conjunctive representation of (X,R) by means of $\mathcal{A}$ with $|\mathcal{A}| = s$.

$D_{s,n}(k;\mathcal{C}) = \max\{d_{s,n}(X,R) \mid (X,R) \in \mathcal{C},\ |X| \leq k\}$, where $\mathcal{C}$ is a class of n-ary relations.

If $\mathcal{C}$ is the class of all n-ary relations, we write simply

$$D_{s,n}(k) .$$

<u>1.3.</u> We will show that for any s there is a positive α such that

$$D_{s,n}(k) > \alpha \cdot k^{n-1} \qquad \text{(see 3.5 below)}$$

and that for sufficiently large s

$$D_{s,n}(k) \leqslant k^{n-1} \qquad \text{(see 4.1 below)} .$$

Similar estimates will be obtained also for some further classes (e.g. for that of all symmetric relations).

§2. <u>Reformulation</u>

<u>2.1.</u> <u>Conventions and notation</u> : All the <u>categories</u> $\mathcal{C}$ in the sequel will be concrete categories, i.e., they will be assumed endowed with a fixed forgetful functor $U:\mathcal{C} \to \mathrm{Set}$ where Set is the category of sets and all their mappings. The symbol for the forgetful functor in the indications of categories will mostly be omitted; U will be supposed to preserve monomorphisms. In fact, the categories in question will always be concrete subcategories of

$$\mathrm{Rel}(n) ,$$

the category of finite sets with n-ary relations and their usual homomorphisms (endowed with the obvious forgetful functor sending (X,R) to X).

The product of a system $(C_i)_{i \in J}$ is denoted by $\bigtimes_{i \in J} C_i$.

The cardinality of an object is that of its underlying set (i.e., $|C| =$

$= |U(C)|$).

A <u>subobject</u> (this notion seems to having been used in this form first in [2]) in $(\mathcal{C},U)$ is a monomorphism $\mu:C\to D$ such that whenever $U(\mu)\circ f = U(\psi)$ for a $\psi:E\to D$, there is a morphism $\varphi:E\to C$ with $f = U(\varphi)$. Sometimes we use the expression "C is a subobject of D" to indicate the existence of a subobject $\mu:C\to D$.

A system $\mathcal{B}$ of objects of $\mathcal{C}$ is said to be a <u>system of generators</u> if each C from $\mathcal{C}$ is a subobject of a product $\times_{i=1}^{n} B_i$ with $B_i \in \mathcal{B}$.

From category theory only basics are assumed (such as may be found in the more elementary chapters of textbooks, e.g.[3]).

<u>2.2.</u> Consider a system $\mathcal{B}$ of generators of a category $\mathcal{K}$. For $K\in\mathcal{K}$ denote the minimum necessary r in a subobject

$$\mu:K\longrightarrow \times_{i=1}^{r} B_i \quad \text{with} \quad B_i\in\mathcal{B}$$

by

$$d(K;\mathcal{B},\mathcal{K}) .$$

For a subcategory $\mathcal{C}$ of $\mathcal{K}$ put

$$D(k;\mathcal{B},\mathcal{C},\mathcal{K}) = \max\{d(K;\mathcal{B},\mathcal{K}) \mid K\in\mathcal{C},\ |K|\leq k\} .$$

Thus, if we take $\mathcal{B}_s = \{(X,R)\in \mathrm{Rel}(n) \mid |X|\leq s\}$, we see easily that (recall 1.2)

$$d_{s,n}(X,R) = d((X,R);\mathcal{B}_s,\mathrm{Rel}(n)) ,$$
$$D_{s,n}(k;\mathcal{C}) = D(k;\mathcal{B}_s,\mathcal{C},\mathrm{Rel}(n)) .$$

§3. <u>Lower estimate</u>

<u>3.1.</u> An object S of $\mathcal{C}$ is said to be <u>subdirectly irreducible</u> in $\mathcal{C}$ if, for every subobject $\mu:S\to\times_{i=1}^{n} C_i$ such that for p_i the natural projections all the $p_i\cdot\mu$ are onto, some of the $p_i\cdot\mu$ is an isomorphism (see [5]; this is an obvious generalization of the notion from [1] - in categories of algebras, of course, all the monomorphisms are subobjects).

<u>3.2.</u> One sees immediately that every equalizer $\mu:A\to B$ is a subobject. In general, subobjects need not be equalizers, but in the category Rel(n) they always are.

3.3. <u>Proposition</u> : Let $\mathcal{C}$ be a coreflectve subcategory of $\mathcal{K}$, let in $\mathcal{K}$ the equalizers coincide with the subobjects. Let $\mathcal{S}$ be the system of all subdirectly irreducible objects of $\mathcal{C}$ and $\mathcal{B}$ be a finite system of generators of $\mathcal{K}$. Let, for a $\varphi: \mathbb{N} \to \mathbb{N}$,

$$D(k; \mathcal{S}, \mathcal{C}, \mathcal{C}) \geq \varphi(k) .$$

Then there is an $\alpha > 0$ such that

$$D(k; \mathcal{B}, \mathcal{C}, \mathcal{K}) \geq \alpha \cdot \varphi(k) .$$

Proof : Denote by ϱ the coreflection functor $\mathcal{K} \to \mathcal{C}$. For every $B \in \mathcal{B}$ choose a subobject

$$\varrho(B) \longrightarrow \bigtimes_{j=1}^{u(B)} S(B,j) \quad \text{with} \quad S(B,j) \in \mathcal{S} .$$

For each k choose an object $C_k \in \mathcal{C}$ such that

$$|C_k| = k \quad \text{and} \quad d(C_k; \mathcal{S}, \mathcal{C}) = D(k; \mathcal{S}, \mathcal{C}, \mathcal{C}) ,$$

and a subobject

$$C_k \longrightarrow \bigtimes_{i=1}^{r} B_i , \quad B_i \in \mathcal{B} .$$

The functor ϱ preserves equalizers and products. Thus, according to the assumption on subobjects in $\mathcal{K}$, we have a composite subobject

$$C_k = \varrho(C_k) \longrightarrow \bigtimes_{i=1}^{r} \varrho(B_i) \longrightarrow \bigtimes_{i=1}^{r} \bigtimes_{j=1}^{u(B_i)} S(B_i, j) .$$

Thus,

$$\varphi(k) \leq d(C_k; \mathcal{S}, \mathcal{C}) \leq \bigtimes_{i=1}^{r} u(B_i) .$$

Since $\mathcal{B}$ is finite, we have a $u_0 \geq u(B)$ for all $B \in \mathcal{B}$ and hence finally

$$r \geq u_0^{-1} \cdot \varphi(k) . \quad \square$$

3.4. Proposition : Denote by $\mathcal{C}_0$ the category of all symmetric reflexive n-ary relations, $n \geq 2$. Let $\mathcal{S}$ be the system of its subdirectly irreducible objects. Then

$$D(k; \mathcal{S}, \mathcal{C}_0, \mathcal{C}_0) \geq \alpha . k^{n-1} \quad \text{for some } \alpha > 0 .$$

Proof : Consider the functor $P_n^+ : \mathrm{Set} \to \mathrm{Set}$ defined by

$$P_n^+(X) = \{ Y \mid Y \subset X, |Y| \leq n \} , \quad P_n^+(f)(Y) = f(Y) .$$

$\mathcal{C}_0$ can be viewed as the category with objects (X,R) where $\{\{x\} \mid x \in X\} \subset R \subset P_n^+(X)$, and morphisms $f:(X,R) \to (Y,S)$ the mappings $f: X \to Y$ with $P_n^+(f)(R) \subset S$. We will consider two cases :

I. $n > 2$: By [5; Thm 3.3] (cf. also [4; 2.6, 2.8]) we see easily that a subdirectly irreducible object S in $\mathcal{C}_0$ is either of a form

(1) $\quad (X, P_n^+(X))$

or of a form

$$(2) \qquad (X, P_n^+(X) \setminus \{Y\}) \qquad (1 < |Y| \leq n\ ).$$

Consider the objects

$$C_k = (k, \{Y \mid |Y| = n \text{ or } 1\}).$$

Let $\mu : C_k \longrightarrow \prod_{i=1}^{r} S_i$ be a subobject, $S_i \in \mathcal{S}$; in case that S_i is of the form (2) we use the notation $S_i = (X_i, P_n^+(X_i) \setminus \{Y_i\})$. Let μ_i: $C_k \longrightarrow S_i$ be the compositions of μ with the projections.

If $U \subset k$, $|U| = n-1$, we have to have an $i = i(U)$ such that

$$\mu_i(U) = P_n^+(\mu_i)(U) = Y_i \quad .$$

Let $U \neq V$, $|U| = |V| = n-1$. We have a $v \in V \setminus U$ and $|U \cup \{v\}| = n$. If $i = i(U) = i(V)$, we obtain $\mu_i(U \cup \{v\}) = Y_i$ in contradiction with μ_i being a morphism. Thus, $i(U) \neq i(V)$ and we obtain

$$(3) \qquad r \geqslant \binom{k}{n-1} .$$

Obviously, $\binom{k}{p} > p^{-p}.k^p$ (since $\frac{k-x}{p-x} > \frac{k}{p}$ for $x > 0$). Thus, by (3),

$$d(C_k; \mathcal{S}, \mathcal{C}_o) > (n-1)^{n-1}.k^{n-1} \quad .$$

II. n=2 : By [5, 4.6], all the subdirectly irreducibles are subobjects of $D = (3, \{\{0\}, \{1\}, \{2\}, \{0,1\}, \{1,2\}\})$. Put

$$C_{2k} = (k \times 2,\ P_2^+(k \times 2) \setminus \{\{(x,0),(x,1)\} \mid x \in k\}) .$$

Consider a subobject $\mu : C_{2k} \longrightarrow D^r$, let, again, $\mu_i : C_{2k} \longrightarrow D$ be its compositions with the projections. For every $x \in k$ we have an $i = i(x)$ such that $\{\mu_i(x,0), \mu_i(x,1)\} = \{0,2\}$. For $x \neq y$ we have $i(x) \neq i(y)$ since otherwise we would have, for some j_1, j_2, $\mu_i(\{(x,j_1),(y,j_2)\}) = \{0,2\}$. Thus, $d(C_{2k}; \mathcal{S}, \mathcal{C}_o) \geqslant k$. □

<u>3.5.</u> One checks easily that Rel(n) and $\mathcal{C}_o$ satisfy the conditions of 3.3. Further, obviously $D(k; \mathcal{B}, \mathcal{C}, \mathcal{K}) \geqslant D(k; \mathcal{B}, \mathcal{D}, \mathcal{K})$ for $\mathcal{C} \supset \mathcal{D}$. Thus, 3.4 and 3.3 immediately yield

<u>Theorem</u> : Let $\mathcal{C}$ be an arbitrary system of n-ary relations containing $\mathcal{C}_o$. Then for each finite system of generators $\mathcal{B}$ there is a positive α such that

$$D(k; \mathcal{B}, \mathcal{C}, \mathrm{Rel}(n)) > \alpha.k^{n-1} \quad \text{for all } k .$$

In particular, for each s there is an $\alpha > 0$ such that

$$D_{s,n}(k; \mathcal{C}) > \alpha.k^{n-1} . \quad □$$

§4. Upper estimate

4.1. Theorem : For $s \geqslant 2^{n+1}$,

$$D_{s,n}(k) \leqslant k^{n-1}.$$

Moreover, there is a one-element generating system $\mathcal{B} = \{B\}$ with $|B| = s$ such that

$$D(k;\mathcal{B},\mathrm{Rel}(n),\mathrm{Rel}(n)) \leqslant k^{n-1};$$

thus, the predicates $\mathcal{F}_i$ in 1.1 can be requested to coincide.

Proof : Put $B = (\{A \mid A \subset n+1\}, R_0)$ where

$$(A_1,\ldots,A_n) \in R_0 \quad \text{iff} \quad (((j \leqslant n-1) \Rightarrow j \in A_j) \Rightarrow n \in A_n).$$

For an n-ary relation (X,R) consider $J = X^{n-1}$ and define a mapping

$$\mu : (X,R) \to B^J$$

by $p_\eta \circ \mu = \mu_\eta : (X,R) \to B$ where for $\eta = (x_1,\ldots,x_{n-1})$

$$\mu_\eta(x) = \begin{cases} \{i \mid x = x_i\} \cup \{n\} & \text{if } (x_1,\ldots,x_{n-1},x) \in R \\ \{i \mid x = x_i\} \cup \{0\} & \text{if } (x_1,\ldots,x_{n-1},x) \notin R. \end{cases}$$

We have

(1) for $\eta = (x_1,\ldots,x_{n-1}) \neq (y_1,\ldots,y_{n-1})$ always

$$(\mu_\eta(y_1),\ldots,\mu_\eta(y_n)) \in R_0$$

(since if $(j \leqslant n-1) \Rightarrow j \in \mu_\eta(y_j)$ then $j \in \{i \mid y_j = x_i\}$, i.e. $y_j = x_j$ for all $j \leqslant n-1$)

and

(2) for $\eta = (x_1,\ldots,x_{n-1})$,

$$(x_1,\ldots,x_n) \in R \quad \text{iff} \quad (\mu_\eta(x_1),\ldots,\mu_\eta(x_n)) \in R_0$$

(since here we have $i \in \mu_\eta(x_i)$ for $i \leqslant n-1$, and $(x_1,\ldots,x_n) \in R$ iff also $n \in \mu_\eta(x_n)$).

From (1) and (2) it follows easily that

$$(x_1,\ldots,x_n) \in R \quad \text{iff} \quad \forall \eta \; (\mu_\eta(x_1),\ldots,\mu_\eta(x_n)) \in R_0.$$

Thus, since μ is obviously one-one, μ is a subobject. □

4.2. Remarks : (1) If $\mathcal{C}$ is a coreflective subcategory of Rel(n) (e.g., the category of all symmetric relations, or that of symmetric reflexive ones, or arising from requesting only some particular symmetries as e.g. $(x_1,x_2,x_3) \in R \Rightarrow (x_2,x_3,x_1) \in R$, etc.) we see that one has again

$$D(k;\{B'\},\mathcal{C},\mathcal{C}) \leqslant k^{n-1}$$

for a suitable $B' \in \mathcal{C}$. It suffices to take $B' = \varrho(B)$ where ϱ is the appropriate coreflection (recall the argument in the proof of 3.3.)

(2) We have been concerned with at least binary relations. The case of the unary one is trivial, of course not with $k^{n-1} = k^{0} = 1$ but with $\log k$ at the places of k^{n-1} in both the upper and lower estimates.

R e f e r e n c e s :

[1] G.Birkhoff : Lattice Theory , AMS Colloq.Publ.Vol.25, Providence, R.I.(1967)

[2] J.F.Kennison : Reflective functors in general topology and elsewhere, Trans.AMS 118(1965), 303-315

[3] S.MacLane : Categories for the working mathematician, Graduate Texts in Math., Vol.5, Springer (1971)

[4] A.Pultr : On product dimensions in general and that of graphs in particular, Vorträge zu Grundlagen der Informatik, Heft 27/77, TU Dresden (1977), 66-79

[5] A.Pultr and J.Vinárek : Productive classes and subdirect irreducibility, in particular for graphs, Discr.Math. 20(1977), 159-176

ON SUBWORDS OF FORMAL LANGUAGES

G. Rozenberg
University of Leiden
The Netherlands

A way to understand the structure of a language is to investigate the set of all subwords that occur in the (words of the) language. A natural first step in such an investigation is simply to count the number of subwords of a given length in the language. Let for a language K, $\underline{sub}(K)$ denote the set of subwords of K, $\underline{sub}_n(K)$ denote the number of subwords of length n occurring in K and let $\pi_K(n)$ denote the cardinality of $\underline{sub}_n(K)$. Thus π_K is a function of positive integers assigning to each n the number of subwords of length n that occur in K; we refer to π_K as the *subword complexity* function of K. One can say that investigating the subword complexity of a language K forms a *numerical* approach to the investigation of the subwords of K.

In the first part of this paper we investigate the subword complexity of arbitrary languages. In particular we investigate to what extent a homomorphic mapping can influence the number of subwords.

Rather soon it becomes evident that to get a theory of subword complexity one has to consider languages that have "some structure" (as opposed to arbitrary languages). We choose to consider the class of languages generated by TOL systems and its subclasses. In the second part of this paper we demonstrate how the subword complexity (which is a *global* property in the sense that it is defined on a language independently of a system that generates it) of a TOL language is influenced by *local* restrictions (that is restrictions concerning the set of productions available) on a TOL system that generates it.

In the last part of this paper we consider global *structural* restrictions on the set of subwords of a given DOL language. For example we consider (following [8]) the restriction that no subword of a language is of the form xx where x is a non-empty word; such a language is called *square-free*. It turns out that the square-free condition on a DOL language restricts the number of possible subwords (of any length) quite considerably. In this way we see how a structural global restriction influences the global numerical measure.

This paper surveys results concerning subword complexity of formal languages obtained in the last few years. The proofs are not given, they can be found in the cited references.

PRELIMINARIES

We assume the reader to be familiar with the basic formal language theory. We use standard language theoretic notation and terminology. Perhaps the following points require an additional explanation. In this paper we consider finite alphabets only. On the other hand, since the problems considered become trivial otherwise, we consider infinite languages only (and consequently rewriting systems which generate infinite languages). For a finite set Z, #Z denotes its cardinality. For a word α, alph(α) denotes the set of all letters occurring in α and $|\alpha|$ denotes the length of α; Λ denotes the empty word. A word α is a subword of a word β if $\beta = \gamma\alpha\delta$ for some words γ,δ. For a language K, sub(K) denotes the set of all subwords (occurring in the words) of K and $\mathrm{sub}_n(K)$ denotes the set of subwords of K of length n.

The following is the central notion of this paper. For a language K its subword complexity, denoted π_K, is the function of positive integers such that $\pi_K(n) = \#\mathrm{sub}_n(K)$ for each positive integer n.

I. ARBITRARY LANGUAGES

In this section we investigate the subword complexity of arbitrary languages. First of all we establish the lower bound on the subword complexity of a language; we notice that there do not exist sublinear (but not constant) subword complexities.

Theorem 1. ([3]). Let K be a language. Either

(1). $\pi_K(n) \geq n+1$ for every positive integer n, or

(2). there exists a positive integer C such that $\pi_K(n) \leq C$ for every positive integer n. □

Then we turn to the investigation of the effect that a homomorphic mapping can have on a subword complexity. That is we investigate the relationship between $\pi_{h(K)}$ and π_K for a language K and a homomorphism h. It turns out that in general nothing meaningful can be said about this relationship.

Theorem 2. ([3]). For every positive integer e there exist alphabets Δ,Σ, a positive integer C, a language $K \subseteq \Delta^*$ and a homomorphism $h : \Delta^* \to \Sigma^*$ such that $\#\Sigma = e$ and $\pi_K(n) \leq Cn$, $\pi_{h(K)}(n) = e^n$ for every positive integer n. □

Even if we restrict ourselves to Λ-free homomorphisms the situation is "quite bad": no polynomial upper bound exists for the ratio $\dfrac{\pi_{h(K)}(n)}{\pi_K(n)}$.

Theorem 3. ([3]). There exists a language K and a Λ-free homomorphism h such that for no polynomial f, $\pi_{h(K)}(n) \le f(n)\pi_K(n)$ for all positive integers n. □

To get a reasonable upper bound one has to put some structure on a language K. A natural first step in this direction is to require that π_K is a nondecreasing function.

Theorem 4. ([3]). Let $K \subseteq \Delta^*$ be a language such that π_K is a nondecreasing function and let h be a Λ-free homomorphism on Δ^*. Then there exists a positive integer constant C such that, for every positive integer n, $\pi_{h(K)}(n) \le Cn\pi_K(n)$. □

We would like to remark here that the above result is not true for arbitrary (not necessarily Λ-free) homomorphisms.

II. LANGUAGES GENERATED BY GRAMMARS; THE EFFECT OF LOCAL RESTRICTIONS

In this section we investigate the subword complexity of languages generated by grammars; we have chosen to investigate languages generated by TOL systems (see, e.g., [7]).

A TOL system is a triple $G = (\Delta, H, \omega)$ where Σ is an alphabet, H is a nonempty finite set of finite substitutions (called tables) on Δ (into the subsets of Δ^*) and ω, the axiom, is an element of Σ^*. If for $h \in H$ and $a \in \Delta$, $\alpha \in h(a)$ then we say that $a \to \alpha$ is a production in G. The language of G, denoted L(G), is defined by $L(G) = \{x \in \Delta^* : x = \omega \text{ or } x \in h_1 \ldots h_m(\omega) \text{ where } k \ge 1 \text{ and } h_1, \ldots, h_m \in H\}$; L(G) is called a TOL language. We say that G is a deterministic TOL system, abbreviated DTOL system, if for every $h \in H$ and every $a \in \Delta$, $\#h(a) = 1$; accordingly L(G) is called a DTOL language.

Clearly, the set of all words over an alphabet Δ is a TOL language, so nothing specific can be said about the subword complexity of TOL languages in general. However, it turns out that the subword complexity (which is a global feature of a language) is sensitive to various local restrictions (that is restrictions on the sets of productions available in TOL systems). First of all it turns out that the (effect of the) deterministic restriction on TOL systems can be "detected by" looking at the subword complexity of generated languages.

Theorem 5. ([1]). Let Δ be a finite alphabet such that $\#\Delta = m \ge 2$. If K is a DTOL language, $K \subseteq \Delta^*$ then $\lim\limits_{n\to\infty} \frac{\pi_K(n)}{m^n} = 0$. □

If $G = (\Delta, H, \omega)$ is a DTOL system such that $\#H = 1$ then we say that G is a DOL system (and L(G) is a DOL language). In this case if $H = \{h\}$ then we specify G in the form (Δ, h, ω).

Again, the restriction of DTOL systems to systems with one table only has an effect on the subword complexity of generated languages.

Theorem 6. ([2], [6]). Let K be a DOL language. There exists a positive integer constant C such that $\pi_K(n) \leq Cn^2$ for every positive integer n. □

The above result yields the best upper bound because there exist DOL languages with a subword complexity of order n^2 ([2], [6]).

A natural local restriction on a DOL system is a restriction on the length of (the right-hand side of) productions. A DOL system $G = (\Delta, h, \omega)$ is called growing, abbreviated as a GDOL system, if $\alpha = h(a)$ for $a \in \Delta$ implies that $|\alpha| \geq 2$; L(G) is referred to as a GDOL language. A DOL system $G = (\Delta, h, \omega)$ is called uniformly growing, abbreviated as a UGDOL system, if there exists a $t \geq 2$ such that if $\alpha = h(a)$ for $a \in \Delta$ implies that $|\alpha| = t$; L(G) is referred to as a UGDOL language.

Theorem 7. ([2], [6]). Let K be a DOL language.

(i). If K is a GDOL language then there exists a positive integer C such that $\pi_K(n) \leq Cn \log_2 n$ for every positive integer n.

(ii). If K is a UGDOL language then there exists a positive integer C such that $\pi_K(n) \leq Cn$ for every positive integer n. □

Also the above results ((i) and (ii)) yield the best upper bounds for the subword complexity of GDOL and UGDOL languages ([2], [6]).

As far as the effect of homomorphisms on the subword complexity is concerned we have the following results.

Theorem 8. ([3]). Let $K \subseteq \Delta^*$ be a DOL language and let h be a homomorphism of Δ^*. There exists a positive integer constant C such that $\pi_{h(K)}(n) \leq Cn^2$ for every positive integer n. □

Theorem 9. ([3]). There exists a UGDOL language $K \subseteq \Delta^*$, a positive real C and a homomorphism h of Δ^* such that $\pi_{h(K)}(n) \geq Cn^2$ for every positive integer n. □

The situation looks quite different if one considers Λ-free homomorphisms.

Theorem 10. ([3]). Let $K \subseteq \Delta^*$ be a DOL language and let h be a Λ-free homomorphism of Δ^*.

(i). If K is a GDOL language then there exists a positive integer C such that $\pi_{h(K)}(n) \leq Cn \log_2 n$ for every positive integer n.

(ii). If K is a UGDOL language then there exists a positive integer C such that $\pi_{h(K)}(n) \leq Cn$ for every positive integer n. □

III. LANGUAGES GENERATED BY GRAMMARS; THE EFFECT OF GLOBAL RESTRICTIONS

In this section we consider "structural" restrictions on the distribution of subwords in a DOL language.

Following [8] we say that a word is square-free if it does not have a subword of the form xx where x is a nonempty word. A language is called square-free if it consists of square-free words only. We will consider square-free DOL languages now. Clearly, the square-free restriction is a global restriction (its formulation is independent on a DOL system that generates the DOL language under consideration). It is also a structural restriction in the sense that it talks about the structure of (the distribution of) subwords in words of a language. Again: also this restriction can be detected by the subword complexity function.

Theorem 11. ([4]). Let K be a square-free DOL language. There exists a positive integer C such that $\pi_K(n) \leq Cn \log_2 n$ for every positive integer n. □

Theorem 12. ([4]). There exist a square-free DOL language K and a positive integer constant D such that $\pi_K(n) \geq Dn \log_2 n$ for every positive integer n. □

To put the above results in a proper perspective we report also the following two results (the first of which is stated for arbitrary languages).

Theorem 13. ([4]). If K is a square-free language then $\pi_K(n) \geq n$ for every positive integer n. □

Theorem 14. ([4]). There exists a square-free DOL language K and a positive integer constant C such that $\pi_K(n) \leq Cn$ for every positive integer n. □

Another type of a global structural restriction is the following one. We say that a language K has a constant distribution if there exist an alphabet Δ and a positive integer constant C such that $\underline{alph}(\alpha) = \Delta$ for every word in $\underline{sub}_C(K)$. Also this structural global restriction is detectable by the subword complexity function.

Theorem 15. ([5]). Let K be a DOL language that has a constant distribution. There exists a positive integer constant C such that $\pi_K(n) \leq Cn$ for every positive integer n. □

ACKNOWLEDGEMENTS.

The author is deeply indebted to A. Ehrenfeucht for the continuous cooperation in the research reported. The author gratefully acknowledges the support by NSF grant MCS 79-03838

REFERENCES

1. A. Ehrenfeucht and G. Rozenberg, A limit for sets of subwords in deterministic TOL systems, Information Processing Letters, 2 (1973), 70-73.
2. A. Ehrenfeucht, K.P. Lee and G. Rozenberg, Subword complexities of various classes of deterministic developmental languages without interactions, Theoretical Computer Science, 1 (1975), 59-76.
3. A. Ehrenfeucht and G. Rozenberg, On subword complexities of homomorphic images of languages, Rev. Fr. Automat. Inform. Rech. Opér., Ser. Rouge, to appear.
4. A. Ehrenfeucht and G. Rozenberg, On the subword complexity of square-free DOL languages, Theoretical Computer Science, to appear.

5. A. Ehrenfeucht and G. Rozenberg, On the subword complexity of DOL languages with a constant distribution, Institute of Applied Mathematics and Computer Science, University of Leiden, The Netherlands, Technical Report No. 81-21, 1981.
6. K.P. Lee, Subwords of developmental languages, Ph.D. thesis, State University of New York at Buffalo, 1975.
7. G. Rozenberg and A. Salomaa, The mathematical theory of L Systems, Academic Press, London, New York, 1980.
8. A. Thue, Über unendliche Zeichenreihen, Norsk. Vid. Selsk. Skr. I Mat.-Nat.Kl. 7 (1906), 1-22.

FIRST ORDER DYNAMIC LOGIC WITH DECIDABLE PROOFS AND WORKABLE MODEL THEORY

I. Sain

Mathematical Institute of the Hungarian Academy of Sciences

Budapest, Reáltanoda u. 13-15, H-1053, Hungary

ω is the set of natural numbers.

By a similarity type d we understand a function $d : \Sigma \longrightarrow \omega \times \{F,R\}$ with $F \neq R$. We call Σ the signature or set of nonlogical symbols of d and if $d(f)=\langle n,F\rangle$ and $d(P)=\langle n,R\rangle$ then f and P are said to be n-ary function symbols and relation symbols respectively.

$X \stackrel{d}{=} \{x_i : i \in \omega\}$ and $Y \stackrel{d}{=} \{y_i : i \in \omega\}$ are used as sets of variables.

F_d is the set of all first order formulas (with equality) of type d with variables in X.

$Term_d$ is the set of all terms of type d with variables in X.

M_d is the class of all classical models of type d.

Underlined capitals like $\underline{\underline{D}}$ denote classical first order models and the corresponding capital D without underlining denotes the universe of the model $\underline{\underline{D}}$.

For any function f, the range $Rng\ f$ of f is defined to be $Rng\ f \stackrel{d}{=} \{ r : (\exists s) \langle s,r\rangle \in f \}$.

${}^{A}B$ is the set of all functions from A into B, that is $\forall f\ (f : A \longrightarrow B \text{ iff } f \in {}^{A}B)$.

$\underline{\underline{D}} \models \varphi[q]$ means that the valuation $q \in {}^{\omega}D$ satisfies $\varphi \in F_d$ in the model $\underline{\underline{D}} \in M_d$.

$H \upharpoonright f \stackrel{d}{=} \{\langle a,b\rangle \in f : a \in H\}$ is the function f restricted to domain H.

$A \sim B \stackrel{d}{=} \{ a \in A : a \notin B \}$.

DEFINITION 1

Let d be a similarity type. By a dynamic model M of type d we understand a tuple

$M \stackrel{d}{=} \langle T^M, first^M, next^M, \underline{\underline{D}}^M, I^M\rangle = \langle T, first, next, \underline{\underline{D}}, I\rangle$ where

$\underline{\underline{D}} \in M_d$, $I \subseteq {}^T D$, $first \in T$, and $next : T \longrightarrow T$.
DM_d denotes the class of all dynamic models of type d .

DEFINITION 2

The class P_d of programs of type d is defined to be the smallest class satisfying (i) - (ii) below.

(i) $() \in P_d$ and $(x_i \leftarrow \tau) \in P_d$ for all $i \in \omega$ and $\tau \in Term_d$.

(ii) Assume that $\chi \in F_d$ is quantifier free and let $p, p_1, p_2 \in P_d$. Then $(p_1; p_2) \in P_d$ and $\text{while } \chi \text{ do } [p] \in P_d$. END(Def.2)

We call () the empty program.

DEFINITION 3

Let $\underline{\underline{D}} \in M_d$ and $q \in {}^{\omega}D$.

(I) We define a function $nxt(\underline{\underline{D}}q) : P_d \longrightarrow P_d$ by postulating (i)-(iii) below for all $p, p_1, p_2 \in P_d$, $\chi \in F_d$, $\tau \in Term_d$ and $i \in \omega$.

(i) $nxt(\underline{\underline{D}}q)(p) \stackrel{d}{=} ()$ if $[p=() \text{ or } p=(x_i \leftarrow \tau)]$.

(ii) $$nxt(\underline{\underline{D}}q)(p_1;p_2) \stackrel{d}{=} \begin{cases} p_2 & \text{if } p_1=() \\ (nxt(\underline{\underline{D}}q)(p_1);p_2) & \text{otherwise} \end{cases} .$$

(iii) $$nxt(\underline{\underline{D}}q)(\text{while } \chi \text{ do } [p]) \stackrel{d}{=} \begin{cases} (p;\text{while } \chi \text{ do } [p]) & \text{if } \underline{\underline{D}} \models \chi[q] \\ () & \text{otherwise} \end{cases} .$$

(II) We define $Head : P_d \longrightarrow P_d$ as follows. Let $p, p_1, p_2, \chi, \tau, i$ be as in (I) above. Then

$Head(p) \stackrel{d}{=} p$ if $p \in \{(), (x_i \leftarrow \tau), \text{while } \chi \text{ do } [p_1]\}$, and

$Head(p_1;p_2) \stackrel{d}{=} Head(p_1)$. END(Def.3)

Next we define the meanings or denotations of programs in dynamic models i.e. we define the semantics of P_d in DM_d .

DEFINITION 4

Let $p \in P_d$ and $M = \langle T, first, next, \underline{\underline{D}}, I \rangle \in DM_d$.

(I) The number v_p of variables in p is defined to be

$$v_p \stackrel{d}{=} \min\{n \in \omega : (\forall i \in \omega)[\text{ if } x_i \text{ occurs in } p \text{ then } i<n]\} .$$

(II) For any sequence $\bar{s} \in {}^{n}I$ and $t \in T$ we define $\bar{s}(t) \stackrel{d}{=} \langle s_i(t) : i \in n \rangle$.

(III) By a <u>trace</u> $\bar{\bar{s}}$ of p in M we understand a pair $\bar{\bar{s}} \stackrel{d}{=} \langle \bar{s}, c \rangle$ such that (i) - (iv) below hold for $v \stackrel{d}{=} v_p$.

(i) $\bar{s} \in {}^{v}I$ that is $\bar{s} = \langle s_0, \ldots, s_{v-1} \rangle$ with $s_0, \ldots, s_{v-1} \in I$.

(ii) $c : T \longrightarrow P_d$ is such that $|\mathrm{Rng}\ c| < \omega$, $c(\mathrm{first})=p$, and $(\forall t \in T)\ c(\mathrm{next}(t)) = \mathrm{nxt}(\underline{\underline{D}}\ \bar{s}(t))(c(t))$. Here we note that by $\bar{s}(t) \in {}^{v}D$ and by $v \leq \omega$ this definition is meaningful.

(iii) Let $t \in T$, $i \in \omega$, $\tau \in \mathrm{Term}_d$, $\chi \in F_d$, $p_1 \in P_d$. Then (1)-(2) below hold.

(1) Assume $\mathrm{Head}(p) \in \{(\,),\ \text{while } \chi \text{ do } [p_1]\}$. Then $\bar{s}(\mathrm{next}(t))=\bar{s}(t)$.

(2) Assume $\mathrm{Head}(p) = (x_i \leftarrow \tau)$. Then $s_i(\mathrm{next}(t)) = \tau[\bar{s}(t)]$ computed in $\underline{\underline{D}}$, and $(\forall j < v)[\ j \neq i \Rightarrow s_j(\mathrm{next}(t)) = s_j(t)\ ]$.

(iv) $(\forall e,g \in \mathrm{Rng}\ c)(e \neq g \Rightarrow \exists f[f\colon \mathrm{Rng}\ c \to D,\ (f \circ c) \in I \text{ and } f(e) \neq f(g)])$.

(IV) Let $k,h \in {}^{\omega}D$ and $v \stackrel{d}{=} v_p$. Then h is defined to be an <u>output</u> of p with input k in M iff (*) below holds.

(*) There is a trace $\langle \bar{s}, c \rangle$ of p in M with $\langle k_i : i < v \rangle = \bar{s}(\mathrm{first})$ and $(\exists t \in T)[\ c(t)=(\,) \text{ and } \langle h_i : i < v \rangle = \bar{s}(t)\ ]$ and $(\omega \sim v) \upharpoonright k \subseteq h$.

The trace $\langle \bar{s}, c \rangle$ is said to <u>terminate</u> at time $t \in T$ iff $c(t)$ is the empty program $(\,)$. <u>END(Def.4)</u>

<u>DEFINITION 5</u>

Let d be a similarity type.

(I) Tm_d and Fm_d are defined to be the set of all terms and the set of all first order formulas of type d <u>with variables in</u> $X \cup Y$. Note that $\mathrm{Term}_d \subsetneq \mathrm{Tm}_d$ since $y_i \notin \mathrm{Term}_d$ and $y_i \in \mathrm{Tm}_d$. Hence $F_d \subsetneq \mathrm{Fm}_d$ too.

(II) DF_d is defined to be the smallest set satisfying (i)-(iii) below.

(i) $\mathrm{Fm}_d \subseteq \mathrm{DF}_d$.

(ii) $\neg\varphi,\ \exists x_i \varphi,\ \exists y_i \varphi,\ (\varphi \wedge \psi),\ \Box(p,\varphi) \in \mathrm{DF}_d$ for all $\varphi, \psi \in \mathrm{DF}_d$, $p \in P_d$ and $i \in \omega$.

(iii) $\mathrm{Alw}\ \varphi$, $\mathrm{First}\ \varphi$, $\mathrm{Next}\ \varphi \in \mathrm{DF}_d$ for all $\varphi \in \mathrm{DF}_d$.

Definition 6 below is in the standard style of Kripke-model theory, see p.3 of [1] or Defs 2.1-2.7 in [4].

DEFINITION 6

Let $M = \langle T, \text{first}, \text{next}, \underline{\underline{D}}, I\rangle \in TM_d$. Let $k \in {}^{\omega}D$, $q \in {}^{\omega}I$ and $t \in T$.
Let $i \in \omega$, $\langle f,\langle n,F\rangle\rangle,\langle P,\langle m,R\rangle\rangle \in d$ and $\tau_1,\ldots,\tau_{n+m} \in Tm_d$, $p \in P_d$.
Then

(I) $f^D : {}^nD \longrightarrow D$ and $P^D \subseteq {}^mD$ are the usual interpretations of the symbols f and P in the model $\underline{\underline{D}}$.

(II) We define the value $\tau[k,q,t]_M$ of any $\tau \in Tm_d$ in M at $\langle k,q,t\rangle$ by postulating (i)-(ii) below.

(i) $x_i[k,q,t]_M \stackrel{d}{=} k_i$ and $y_i[k,q,t]_M \stackrel{d}{=} q_i(t)$.

(ii) $f(\tau_0,\ldots,\tau_{n-1})[k,q,t]_M \stackrel{d}{=} f^D\langle \tau_i[k,q,t]_M : i<n\rangle$.

(III) The relation $Mt \Vdash \varphi[k,q]$ is defined by (1)-(7) below, by recursion.

(1) $Mt \Vdash P(\tau_0,\ldots,\tau_{m-1})[k,q]$ iff $\langle \tau_i[k,q,t]_M : i<m\rangle \in P^D$.

(2) $Mt \Vdash \text{Alw}\ \varphi[k,q]$ iff $(\forall b \in T)\ Mb \Vdash \varphi[k,q]$.

(3) $Mt \Vdash \text{First}\ \varphi[k,q]$ iff $M\ \text{first} \Vdash \varphi[k,q]$.

(4) $Mt \Vdash \text{Next}\ \varphi[k,q]$ iff $M\ \text{next}(t) \Vdash \varphi[k,q]$.

(5) $Mt \Vdash \exists y_i \varphi[k,q]$ iff $(\exists g \in {}^{\omega}I)(Mt \Vdash \varphi[k,g]$ and $(\omega\sim\{i\})\upharpoonright g \subseteq q)$.

(6) The definitions for the classical connectives $\exists x_i\varphi$, $\neg\varphi$, $(\varphi\wedge\psi)$ are the usual, see (5) above.

(7) $Mt \Vdash \Box(p,\varphi)[k,q]$ iff (for every output $h \in {}^{\omega}D$ of p with input k in M , we have $Mt \Vdash \varphi[h,q]$).

(IV) $M \models \varphi[k,q]$ iff $M\ \text{first} \Vdash \text{Alw}\ \varphi[k,q]$.

(V) $M \models \varphi$ iff $(\forall k \in {}^{\omega}D)(\forall q \in {}^{\omega}I)\ M \models \varphi[k,q]$. END(Def.6)

We define the first order dynamic language DL_d of type d to be

$$DL_d \stackrel{d}{=} \langle DF_d\ ,\ DM_d\ ,\ \models\rangle$$ where

the relation $\models\ \subseteq (DM_d \times DF_d)$ of validity is defined in Def.6(V).

Let $Th \cup \{\varphi\} \subseteq DF_d$. We define $Th \models \varphi$ to hold iff $(\forall M \in DM_d)(\ M \models Th \Rightarrow M \models \varphi\ )$.

THEOREM 1

(i) There is a strongly complete inference system for DL_d . Namely, there is a decidable set Prf of finite strings of symbols such that for any $Th \cup \{\varphi\} \subseteq DF_d$ we have

$$Th \models \varphi \quad \text{iff} \quad [\langle H,\pi,\varphi\rangle \in Prf \text{ for some finite } H\subseteq Th \text{ and some } \pi].$$

(ii) Let $Th \subseteq DF_d$ be recursively enumerable. Then $\{ \varphi \in DF_d : Th \models \varphi \}$ is recursively enumerable too.

(iii) DL_d is compact, i.e.
$$(\forall Th \subseteq DF_d)(\forall \varphi \in DF_d)[Th \models \varphi \quad \text{iff} \quad (\exists \text{ finite } H\subseteq Th)\ H \models \varphi] .$$

DEFINITION 8

Let $\varphi \in DF_d$. Then φ^* is defined to be the formula
$((\, First\,\varphi \wedge Alw(\ \varphi \rightarrow Next\,\varphi\ )\ ) \rightarrow Alw\,\varphi\ )$.
$IA \overset{d}{=} \{ \varphi^* : \varphi \in DF_d \}$ and $IA^{cl} \overset{d}{=} \{\varphi^* : \varphi \in Fm_d\}$.

THEOREM 2

(i) Let $Th \subseteq F_d$, $p \in P_d$ and $\psi \in F_d$. Assume $Th \models \exists x_1(x_1 \neq x_2)$. Then $\Box(p,\psi)$ is Floyd-Hoare provable from Th iff $IA^{cl} \cup Th \models \Box(p,\psi)$.

(ii) There are finite d , $Th \cup \{\psi\} \subseteq F_d$ and $p \in P_d$ such that $\Box(p,\psi)$ is not Floyd-Hoare provable from Th but $IA \cup Th \models \Box(p,\psi)$ and all models of Th are finite.

For any $Ax \subseteq DF_d$ we let $Mod(Ax) \overset{d}{=} \{ M \in DM_d : M \models Ax \}$.

Dynamic logic is defined to consist of the language $\langle DF_d , Mod(IA) , \models \rangle$, logical axioms IA , and proof concept Prf introduced in Thm.1(i).

By Thm.1(i) and Thm.2(ii), dynamic logic is strictly stronger than Floyd-Hoare logic for proving Floyd-Hoare statements $\Box(p,\psi)$.
By Thm.1(i) we have an explicite characterization of the information implicitely contained in the Floyd-method. By the standard abbrevi-

ation " Stm φ " $\stackrel{d}{=}$ " $\neg$Alw$\neg\varphi$ " , Thm.2 gives an answer to the question "is sometimes sometime better than always" raised in SRI. The answer is no unless we allow arbitrarily complex mixing of these two modalities like " Stm Alw Stm Alw φ " $\in DF_d$.

REMARK

To include " if χ then p_1 else p_2 " into the formation rules of P_d would be superfluous. Let p = " if χ then p_1 " . Let $Th \subseteq F_d$ be arbitrary but such that $Th \models \tau \neq \sigma$ for some $\tau, \sigma \in Term_d$ with free variables not occurring in p . Assume that $z \in X$ does not occur in p, τ, σ . Then

((z $\leftarrow \tau$) ; while ($\chi \wedge z=\tau$) do [p_1;(z $\leftarrow \sigma$)]) is equivalent with p under Th . Similarly

((z $\leftarrow \tau$) ; if χ then (p_1;(z $\leftarrow \sigma$)) ; if ($z=\tau \wedge \neg\chi$) then p_2) is equivalent with " if χ then p_1 else p_2 " . Thus this " if...then...else " rule is derivable from our primitive rule " while...do... " .

Let " $Next^0(\varphi)$ " $\stackrel{d}{=}$ "φ" and $(\forall n \in \omega)$ " $Next^{n+1}(\varphi)$ " $\stackrel{d}{=}$ " Next $Next^n(\varphi)$".

$$Pe \stackrel{d}{=} \{ Alw(\exists y \exists x(Next(x=y) \wedge First(x\neq y))) \} \cup$$
$$\cup\{Alw\ \forall y_1 \forall x_1 \exists y_2 \exists x_2 (Next(x_2=y_2) \wedge (x_1=y_1 \rightarrow Alw(Next(x_2=y_2) \rightarrow x_1=y_1)))\} \cup$$
$$\cup\{Alw\ (\exists y \exists x(Next^n(x=y) \wedge x\neq y)) \ :\ n \in \omega\ ,\ n>0 \}\ .$$

These axioms hold if $\underline{\underline{T}}$ is elementarily equivalent with $\langle \omega, 0, succ \rangle$ and $(\forall t \in T)(\exists g \in I)(\forall b \in T)[g(t)=g(b) \Rightarrow t=b]$. To this we need only $|D| \geq 2$.

PROBLEM 1

Are there d and $Th \subseteq F_d$ such that

$$IA \cup Pe \cup Th \models \Box(p,\psi) \quad \text{and} \quad IA \cup Th \not\models \Box(p,\psi)$$

for some $p \in P_d$ and $\psi \in F_d$ and $Th \models \exists x_1(x_0 \neq x_1)$?

$$Ex \stackrel{d}{=} \{ (\exists x\ \varphi(x) \wedge Alw\ \forall x_0 \exists x_1\ \psi(x_0,x_1)) \rightarrow$$
$$\rightarrow \exists y \exists x_0 (First(x_0=y) \wedge \varphi(x_0) \wedge Alw\ \exists x_0 \exists x_1 (x_0=y \wedge Next(x_1=y) \wedge \psi(x_0,x_1))) \}\ .$$

PROBLEM 2

Is $Ex \cup IA \cup Th \models \Box(p,\psi)$ and $IA \cup Th \not\models \Box(p,\psi)$ true for some d, $Th \subseteq F_d$, $p \in P_d$ and $\psi \in F_d$?

Let $d(f) = \langle 1,F\rangle$ and $d(c) = \langle 0,F\rangle$. Then we define

$$Dia(f,c) \overset{d}{=} \{\forall\bar{x}[(\varphi(c,\bar{x}) \wedge \forall x_0[\varphi(x_0,\bar{x}) \to \varphi(f(x_0),\bar{x})]) \to \forall x_0\varphi(x_0,\bar{x})] : \varphi(x_0,\bar{x}) \in DF_d \text{ and } n \in \omega,\ \bar{x} = \langle x_1,\dots,x_n\rangle\} .$$

Clearly $IA \cup Ex \cup Dia(f,c) \cup Th \models \Diamond(p,TRUE)$ and

$IA \cup Dia(f,c) \cup Th \not\models \Diamond(p,TRUE)$

for some simple Th and p (and $f,c \in Dom\ d$).

PROBLEM 3

Are there $Th \subseteq F_d$, $p \in P_d$, $\psi \in F_d$ and f,c such that

$IA \cup Ex \cup Dia(f,c) \cup Th \models \Box(p,\psi)$ and $IA \cup Dia(f,c) \cup Th \not\models \Box(p,\psi)$?

For a much stronger version of DL_d having all the positive properties of the present one (e.g. strong completeness w.r.t. a decidable set of proofs) see [5], [2], [3], [6]. See also pseudoalgorithmic logic of [7]. Interesting results and problems related to the subject of the present paper can be found in §3 of [7].

R E F E R E N C E S

1. Andréka,H.-Dahn,B.I.-Németi,I., On a proof of Shelah. Bulletin de l'Academie Polonais des Sciences (Series Math.) 27, 1976, pp.1-7.
2. Andréka,H.-Németi,I.-Sain,I., Henkin-type semantics for program schemes to turn negative results to positive. Fundamentals of Computation Theory, FCT'79 (Proc.Conf. Berlin 1979). Ed.: L.Budach, Akademie Verlag Berlin 1979. Band 2. pp.18-24.
3. Andréka,H.-Németi,I.-Sain,I., Completeness problems in verification of programs and program schemes. Mathematical Foundations of Computer Science MFCS'79 (Proc.Conf. Olomouc, Czechoslovakia 1979), Lecture Notes in Computer Science 74, Springer Verlag 1979. pp.208-218.
4. Bowen,K.A., Model theory for modal logic. Kripke models for modal predicate calculi. D.Reidel Publ.Co. 1979. pp.1-127.
5. Németi,I., A complete first order dynamic logic. Preprint, Math. Inst.Hung.Acad.Sci., 1980.
6. Paris,J.B.-Csirmaz,L., A property of 2-sorted Peano models and program verification. Preprint, Math.Inst.Hung.Acad.Sci., 1981.
7. Salwicki,A., Axioms of algorithmic logic univocally determine semantics of programs. Mathematical Foundations of Computer Science MFCS'80 (Proc.Conf. Rydzina 1980), Lecture Notes in Computer Science Springer Verlag 1980.

Elimination of Second-Order Quantifiers for Well-Founded Trees in Stationary Logic and Finitely Determinate Structures

by

Detlef Seese

Institut für Mathematik
AdW der DDR
Mohrenstr. 39
1080 Berlin
GDR

1 Stationary Logic and finitely determinate Structures

Stationary logic is an extension of elementary logic by a generalized second order quantifier aa which was introduced by Shelah in [19]. Shelah wrote in [19] p. 356 that "The regular second-order quantifier s too strong from the point of view of model theory, and so there are no nice model theoretic theorems about it. But there could be generalized second-order quantifiers which are weak enough for their model theory to be nice, for example satisfying Löwenheim-Skolem, compactness or completeness theorems."

A deep investigation of stationary logic can be found in [1], where it is proved that stationary logic has a nice model theory (in the above sense) and a high expressive power. For instance the quantifier Q_1 is definable in stationary logic, where $Q_1 x\, \varphi(x)$ means "there are uncountably many x with $\varphi(x)$".

Some other definitions are necessary before we can define stationary logic.

Let M be a nonempty set. $P_\omega(M)$ denotes the set of all countable subsets of M. $\underline{\underline{C}} \subseteq P_\omega(M)$ is said to be closed if for each sequence $C_0 \subseteq C_1 \subseteq \dots \subseteq C_n \subseteq \dots (n \in \omega)$ of elements of $\underline{\underline{C}}$, $(\bigcup_{n\in\omega} C_n)$ $\underline{\underline{C}}$ holds. $\underline{\underline{C}} \subseteq P_\omega(M)$ is said to be unbounded if for each $C \in P_\omega(M)$ there is an $D \in \underline{\underline{C}}$ with $C \subseteq D$. Kueker [11] defined a filter cub(M) by

---

Some of the results of this article were obtained when the author was a guest of CSAV in Prague.

$\underline{C} \in cub(M)$ *iff* $\underline{C} \subseteq P_\omega(M)$ and there is a $\underline{D} \subseteq \underline{C}$ which is closed ana unbounded.

Let L be a countable first-order language. We expand L by countably many variables $X_1, \ldots,$ the $\in$-symbol and a new quantifier aa. The formulas of the new language L(aa) (tne expansion of L to stationary logic) are formed as usual with the new formation rule: if φ is a formula of L(aa) so is $(aaX)\varphi$ for each set variable X. Note that we allow $\forall x$ and $\exists x$ in the formation rules for individual variables x, but we do not allow $\forall X$ or $\exists X$ for set variables.

Let $\underline{A}$ be a structure for L and let φ be a formula of L(aa). Then $\underline{A} \models \varphi$ is defined in the obvious way. The cruical clause of the induction is:

$\underline{A} \models (aaX)\varphi(X)$ *iff* $\{C\colon C \in P_\omega(A)$ ana $\underline{A} \models \varphi[C]\} \in cub(A)$.

Here A denotes the domain of $\underline{A}$. We write $Th_{aa}(\underline{A})$ for $\{\varphi\colon \varphi$ is a sentence of L(aa) and $\underline{A} \models \varphi\}$, and we denote $\bigcap_{\underline{A} \in K} Th_{aa}(\underline{A})$ by $Th_{aa}(K)$ for a class K of L-structures.

In the model theory of stationary logic the class of finitely determinate structures plays an important role.

A structure $\underline{A}$ for L is finitely determinate (see [9]) if for all formulas $\varphi(\bar{x},\bar{X},Y)$ of L(aa)

$\underline{A} \models (aa\bar{X})(\forall \bar{x})[(aaY)\varphi(\bar{x},\bar{X},Y) \vee (aaY)\neg\varphi(\bar{x},\bar{X},Y)]$

holds.

A deep investigation of finitely determinate structures can be found in [9] and [7].

In [7] it is proved that for finitely determinate structures each formula of L(aa) is equivalent to one in prenex form. Moreover L(aa)-elementary equivalence is preserved by performance of disjoint sums and direct products (see [9],[7]) and each finitely determinate structure has an L(aa)-elementary substructure of cardinality $\leq \aleph_1$ (see [7]). All these properties are not valid for arbitrary structures (see [9] and [7]).

In [7] and in [2] it is proved that all abelian groups are finitely determinate and that the class of all abelian groups has a decidable theory in stationary logic. The same holds for the class of all well-orderings, as it was proved in [16].
Contrary to these results in [18] it was proved that the class of all linear orderings or Boolean algebras has an undecidable theory in stationary logic. This is interesting since the corresponding theories in $L(Q_1)$ (the logic with the quantifier "there exist uncountably many") are decidable (see [3]).

§2 Well-founded trees

A well-founded tree is a partial ordering $\underline{A} = (A, <)$, where for each $a \in A$ the set $\hat{a} := \{b : b \in A \text{ and } b < a\}$ is well-ordered by $< \upharpoonright \hat{a}$.
Let $\underline{A}$ be a well-founded tree and let a be an element of A. The height of a, denoted by $h_{\underline{A}}(a)$ is defined inductively by

$$h_{\underline{A}}(a) := \{h_{\underline{A}}(b) : b < a \text{ and } b \in A\}.$$

The height $h(\underline{A})$ of the well-founded tree $\underline{A}$ is defined by

$$h(\underline{A}) := \bigcup_{a \in A} (h_{\underline{A}}(a) \cup \{h_{\underline{A}}(a)\}).$$

For each ordinal α let WT_α be the class of all well-founded trees of height $\leq \alpha$. Let WT denote the class of all well-founded trees. Well-founded trees of height $\leq \omega$ will be denoted as ω-trees.
In [17] (see also [14] and [3]) it was proved that the theory of WT_ω in the language with the quantifier Q_1 is decidable. The main point of the proof is the construction of a simple class of ω-trees which is dense for the corresponding theory.
It is an open problem whether the class of all well-founded trees has a decidable theory in the language with the quantifier Q_1. This question has a negative answer if we use stationary logic L(aa) instead of $L(Q_1)$.

Theorem 1. $Th_{aa}(WT)$ and $Th_{aa}(WT_\alpha)$ are undecidable for each $\alpha \geq \omega_1$.

This theorem follows by interpretability from the fact that the theory in stationary logic of the class $\{(\omega_1\cdot\omega, <, P): P \subseteq \omega_1\cdot\omega\}$ is undecidable, what was proved in [16]. An example of Shelah from [9] shows that there are well-founded trees of height ω_1 which are not finitely determinate. Mekler found in [13] a tree of height ω_1 without a path of length ω_1 which is not finitely determinate. It seems to be open whether each tree of height $< \omega_1$ is finitely determinate.

<u>Theorem 2.</u> Each ω-tree is finitely determinate.

<u>Theorem 3.</u> $Th_{aa}(WT_\omega)$ is decidable.

Both theorems were proved in [15] (see also [4]). They generalize corresponding results of Baudisch and Tuschnik [5] for graph theoretic trees, i.e. structures with one binary relation which contains no circles. Moreover Baudisch and Tuschnik proved that the quantifier aa is eliminable in the theory of the class of all graph theoretic trees in the language with the quantifier Q_1. It was left open in [15] and [4] whether this result holds also for ω-trees.

<u>Definition 4.</u> Let S be a theory in L(aa). S admits strongly elimination of aa in $L(Q_1)$ iff for each formula $\varphi(\bar{x},\bar{X})$ of L(aa) there is a formula $\psi(\bar{x},\bar{X})$ of $L(Q_1)$ such that

$$S \vdash (aa\bar{X})(\forall \bar{x})\ (\varphi(\bar{x},\bar{X}) \longleftrightarrow \psi(\bar{x},\bar{X})).$$

<u>Lemma 5.</u> Let S be a theory in L(aa). S admits strongly elimination of aa in $L(Q_1)$ if the following holds:

for all models $\underline{A},\underline{B}$ of S and all expansions $\underline{A}',\underline{B}'$ of $\underline{A}$ and $\underline{B}$ respectively by a finite number of individual constants and countable sets $\underline{A}' \equiv \underline{B}'(Q_1)$ implies $\underline{A}' \equiv \underline{B}'$ (aa).

Definition 4, Lemma 5 and its proof were stimulated by corresponding results from [7] (see also [12]).

<u>Theorem 6.</u> There exists a recursive function g: $\omega \longrightarrow \omega$ such that for all models $\underline{A}$, $\underline{B}$ of $Th_{aa}(WT_\omega)$ (eventually extended by a finite number of individual and set constants)

$$\underline{A} \equiv_{g(n)} \underline{B}\ (Q_1) \quad \text{implies} \quad \underline{A} \equiv_n \underline{B}\ (aa).$$

Here $\equiv_n$ means the restriction of $\equiv$ to formulas of quantifier rank $\leq n$ (see e.g. [4] for exact definition).

Theorem 6 is a strict generalization of a result from [4] wnere it was proved for trees $\underline{A},\underline{B}$ from WT_ω instead of models from $Th_{aa}(WT_\omega)$. The proof is similar to the corresponding proof from [4]. We need some definitions before it is possible to present the main steps of the proof.

Let $\underline{A}$ be a model of $Th_{aa}(WT_\omega)$ (eventually extended by a finite number of individual constants and constants for countable sets) and let a be an element of A.

Define $A_a := \{b\colon b \in A \text{ and } a \leq b\}$. $\underline{A}_a$ denotes the restriction of $\underline{A}$ to A_a. Note that the signature of $\underline{A}$ and of $\underline{A}_a$ is not always the same since in $\underline{A}_a$ all individual constants are omited which represent elements $< a$. Let k be a natural number and let a,b,c be elements of A. Then define:

$$a \sim_k b\ (\underline{A},c) \quad \underline{\text{iff}} \quad c \leq a \text{ and } c \leq b \text{ and } (\underline{A}_c,a) \equiv_k (\underline{A}_c,b)(Q_1).$$

$\sim_k (\underline{A},c)$ defines on A_c an equivalence relation with finitely many equivalence classes. Let $C_1(\underline{A},c,k),\dots,C_{l(\underline{A},c,k)}(\underline{A},c,k)$ be an enumeration of the equivalence classes of $\sim_k(\underline{A},c)$.

<u>Lemma 7.</u> Let L be a language for well-founded trees. For each model $\underline{A}$ of $Th_{aa}(WT_\omega)$ (eventually extended by a finite number of individual constants and countable sets) and each natural number k there exists

a closed and unbounded subset $\underline{\underline{E}}(\underline{A},k) \subseteq P_{\omega}(A)$, such that each D from $\underline{\underline{E}}(\underline{A},k)$ has the following properties:

(i) $(\forall a \in D)(\forall b \in A)$ $b < a$ implies $b \in D$;

(ii) $(\forall a \in D)(\forall i\colon 1 \leq i \leq l(\underline{A},a,k))$

$[\mathrm{card}(C_i(\underline{A},a,k)) \leq \aleph_0$ implies $C_i(\underline{A},a,k) \subseteq D]$ and

$[\mathrm{card}(C_i(\underline{A},a,k)) > \aleph_0$ implies $\mathrm{card}(C_i(\underline{A},a,k) \cap D) = \aleph_0]$;

(iii) $(\forall a \in (A \setminus D))(\exists b \in D)$ $b < a$ and

$(\forall c)(b < c < a$ implies $b \notin D)$;

(iv) for each formula $\varphi(x,\bar{y})$ of L(aa) of quantifier rank $\leq k$:

$$(\underline{A},D) \models (\forall \bar{y})\ [(Q_1 x)\varphi(x,\bar{y}) \longrightarrow (\exists z \in D)(Q_1 x)\varphi(x,\bar{y}) \wedge \forall u\ z < u \leq x \longrightarrow u \notin D] \quad .$$

The proof is not difficult and uses a corresponding result for ω-trees. Each model of $\mathrm{Th}_{aa}(\mathrm{WT}_{\omega})$ is finitely determinate, by theorem 2. Hence we can assume that $\underline{\underline{E}}(\underline{A},k)$ is a cub set witnessing the finite k-determinateness of $\underline{A}$.

<u>Lemma 8.</u> There is an $m \in \omega$ such that for all $n \in \omega$ the following holds. If $\underline{A}$, $\underline{B}$ are models of $\mathrm{Th}_{aa}(\mathrm{WT}_{\omega})$ (eventually extended by a finite number of individual constants and countable sets) , $C \in \underline{\underline{E}}(\underline{A},2n+m)$, $D \in \underline{\underline{E}}(\underline{B},2n+m)$ and $\underline{A} \equiv_{2n+m} \underline{B}(Q_1)$, then $(\underline{A},C) \equiv_n (\underline{B},D)(Q_1)$.

This result is the key result in the proof of theorem 6. The lengthy proof usas the game theoretic equivalent of the relation $\equiv_n (Q_1)$ (see [10], [20]). Theorem 6 follows then from lemma 8 using L(aa)-back-and-forth systems from [9] in the same way as in [4]. From theorem 6 follows immedeately the following corollary.

<u>Corollary 9.</u> For all models $\underline{A}$,$\underline{B}$ of $\mathrm{Th}_{aa}(\mathrm{WT}_{\omega})$ (eventually extended by a finite number of individual constants and countable sets) the

following holds: $\underline{A} \equiv \underline{B}$ (Q_1) implies $\underline{A} \equiv \underline{B}$ (aa).

But this together with Lemma 5 gives:

Theorem 10. $Th_{aa}(WT_{\omega})$ admits strongly elimination of aa in $L(Q_1)$.

Remark. Using the weaker version of Theorem 6 from [4] Herre found in [8] another proof of Theorem 10.

It is not possible to generalize this result for trees of greater height. The following structures are used to see this.
Let us assume that $\underline{C}_{\nu}$ (for $\nu < \omega_1$) are linear orderings of order type $\omega + 1$ with disjoint domains.
Define $\underline{A}_0$ to be the linear ordering which we get from $(\underline{C}_{\nu})_{\nu<\omega_1}$ by identification of the minimal elements of all the $\underline{C}_{\nu}$ $(\nu < \omega_1)$.
Let n be a natural number and assume that $\underline{A}_0, \ldots, \underline{A}_{n-1}$ are defined already. Then define $\underline{A}_n$ as follows.
Take an isomorphic copy of $\underline{A}_0$. For each point $a \in A_0$ which is different from the minimal element of $\underline{A}_0$ and which is not a limit element and for each natural number $i < n$ take an isomorphic copy of $\underline{A}_i$, denoted as $\underline{A}_i(a)$. Then identify for each such a the minimal elements of all these copies with a. The resulting tree is denoted as $\underline{A}_n$.
Now we set $\underline{B}_0 := \underline{A}_0$. Assume that $\underline{B}_n$ is already defined for a natural number n. Let a be an element of B_n which has the following property:
(+) $\underline{B}_n \models (\exists b)(\exists c)\, b < a < c \wedge \neg(\exists x)(b < x < a \vee a < x < c)$.
For each $a \in B_n$ with the property (+) take $\underline{A}_0(a)$ and identify the minimal element of $\underline{A}_0(a)$ with a. The resulting tree is $\underline{B}_{n+1}$. Moreover we get a canonical embedding of $\underline{B}_n$ into $\underline{B}_{n+1}$ (take identity).
Let $\underline{B}_{\omega}$ be $\bigcup_{n<\omega} \underline{B}_n$. Then let $\underline{A}'_{\omega}$ be the following extension of $\underline{B}_{\omega}$.
For each path p of $\underline{B}_{\omega}$ which has order type ω take a new element

a(p) and set it over all elements of p. For each $a \in \underline{A}'_{\omega}$, which is not a limit element and not the minimal element of $\underline{A}'_{\omega}$, take $\underline{A}_n(a)$ for all $n \in \omega$ and identify the minimal elements of all $\underline{A}_n(a)$ (for all $n \in \omega$) with a. The resulting tree is denoted as $\underline{A}_{\omega}$.

The following technical result can be proved by induction using the game theoretic equivalent of $\equiv_n$ (Q_1).

Lemma 11. For all $n \in \omega$ $\underline{A}_{\omega} \equiv_n \underline{A}_{n+1}$ (Q_1) and for all $j \geq n$ $\underline{A}_j \equiv_n \underline{A}_n$ (Q_1).

Lemma 12. There is a sentence φ of L(aa) with $\underline{A}_{\omega} \models \neg \varphi$ and $\underline{A}_n \models \varphi$ for all $n \in \omega$.

The lemma can be proved if we choose as φ a sentence expressing the following:

$(aa\ X)(\forall x)[$"x is a limit point" and $(\forall y)(y < x \rightarrow y \in X)] \rightarrow x \in X$.

Using both results it is not difficult to prove:

Lemma 13. $\underline{A}_{\omega} \dot{\cup} (\dot{\bigcup}_{n \in \omega} \underline{A}_n) \equiv (\dot{\bigcup}_{n \in \omega} \underline{A}_n)$ (Q_1), but $\underline{A}_{\omega} \dot{\cup} (\dot{\bigcup}_{n \in \omega} \underline{A}_n) \not\equiv (\dot{\bigcup}_{n \in \omega} \underline{A}_n)$ (aa), where $\dot{\cup}$ denotes the disjoint union of the corresponding structures.

Hence we get:

Theorem 14. There are well-founded trees $\underline{A}$ and $\underline{B}$ with $h(\underline{A})=h(\underline{B})=\omega+1$ such that $\underline{A} \equiv \underline{B}$ (Q_1) and $\underline{A} \not\equiv \underline{B}$ (aa) hold.

The following question remains open.

Problem 15. Is $Th_{aa}(\bigcup_{\alpha < \omega_1} WT_{\alpha})$ decidable and is each model of this theory finitely determinate?

References

[1] J.BARWISE, M. KAUFMANN, M. MAKKAI, Stationary Logic, Annals of Mathematical Logic, vol. 13, Nu. 2, March 1978, 171-224.

[2] A. BAUDISCH, The elementary theory of abelian groups with m-chains of pure subgroups, preprint 1978.

[3] A. BAUDISCH, D. SEESE, H.P. TUSCHIK, M. WEESE, Decidability and Generalized Quantifiers, Akademie-Verlag Berlin 1980.

[4] A. BAUDISCH, D. SEESE, P. TUSCHIK, ω-Trees in Stationary Logic, to appear in Fund. Math.

[5] A. BAUDISCH, P. TUSCHIK, $L(aa) = L(Q_1)$ for trees, unpublished manuscript, 1979.

[6] J.R. BÜCHI, Using Determinacy of Games to Eliminate Quantifiers, Lecture Notes in Computer Science 56, 367-378.

[7] P.C. EKLOF, A.H. MEKLER, Stationary Logic of Finitely Determinate Structures, Annals of Mathematical Logic, vol. 17, Nu. 3, 1979, 227-269.

[8] H.HERRE, Remarks and Problems Concerning the Elimination of Generalized Quantifiers, unpublished manuscript 1980.

[9] M. KAUFMANN, Some results in stationary logic, Ph. D. Dissertation, University of Wisconsin, 1978.

[10] L.D. LIPNER, Some aspects of generalized quantifiers, Doctoral Dissertation, Univ. of California, Berkeley, 1970.

[11] D.W. KEKER, Countable approximations and Löwenheim-Skolem Theorems, Ann. of Math. Logic, 11 (1977) 57-104.

[12] A.H. MEKLER, The Stationary Logic of Ordinals, preprint 1979.

[13] A.H. MEKLER, Another linear order which is not finitely determinate, notes, 1979.

[14] D.SEESE, Decidability of ω-Trees with Bounded Sets - A Surcey, Lecture Notes in Computer Science 56, Berlin 1977, 511-515.

[15] D.G. SEESE, Stationäre Logik - Beschränkte Mengen - Entscheidbarkeit, Dissertation B, Berlin 1980.

[16] D. SEESE, Stationary Logic and Ordinals, Transactions of the AMS, vol. 263, Nu. 1, January 1981, 111-124.

[17] D. SEESE, P. TUSCHIK, Construction of nice trees, Lecture Notes in Mathematics 619, Berlin 1977, 257-271.

[18] D. SEESE, P. TUSCHIK, M. WEESE, Undecidable theories in stationary logic, to appear in Proceedings of the AMS.

[19] S. SHELAH, Generalized quantifiers and compact logic, Transactions of the AMS, vol. 204, 1975, 342-364.

[20] S. VINNER, A generalization of Ehrenfeucht's game and some applications, Israel J. Math. 12, 1972, 279-298.

PROCESSES IN PETRI NETS

Peter H. Starke
Sektion Mathematik der Humboldt-Universität zu Berlin
DDR-1086 Berlin, PSF 1297

1. Introduction

The aim of the present paper is to develop a theory of processes realized by generalized Petri nets. Whereas in the theory of sequential machines and automata the description of processes by finite or infinite sequences of letters is fully adequate this is not the case when concurrency is involved like in Petri nets and related system models. Several attempts have been made to create a mathematical notion which is able to describe concurrent processes. We propose the notion of a semiword (a generalization of the notion of an ordinary word) as a suitable notion for the description of finite processes in unlabelled Petri nets.

We consider a process to consist of occurrences (actions) which are partially ordered by a relation "before". Every occurrence (action) is labelled with the name of the event (transition) of which it is an occurrence (action). We adopt the axiom that two occurrences of the same event can not be concurrent. Two processes will be considered to be identical if they can be mapped onto one another isomorphically. Consequently, a semiword is defined as an isomorphy class of a finite labelled partial ordering within which the labelling of every antichain is 1:1. Hence, words (isomorphy classes of finite labelled orderings) are special semiwords.

Our first result here is that we are able to construct for every semiword a unique canonic representative which allows us to handle semiwords without "isomorphy reasoning". The structure of the set SW(X) of all semiwords labelled with letters from X is investigated then. Besides concatenation a second operation called composition is introduced which constructs concurrent processes from less concurrent ones. We show that every semiword can be built up from letters by concatenation and composition.

We relate semiwords over T to Petri nets with T as set of transitions by the notion of fireability. For a Petri net with the initial marking m_o (and a finite set of designated terminal markings) we call the set of all semiwords which are fireable at m_o (and which result in a terminal marking) the free nonterminal (resp. terminal) semilanguage of that net. We characterize the families of all nonterminal (resp. terminal) Petri net semilanguages by their closure properties.

2. Semiwords

labelled partial ordering over X is a tripel $\sigma=(A,R,\beta)$ where A is a set, R is an irreflexive and transitive binary relation in A (i.e. (A,R) is an irreflexive partial ordering) and $\beta:A\rightarrow X$ is a mapping from A into X. Labelled partial orderings $(A,R,\beta),(A',R',\beta')$ are called isomorphic iff there exists a 1:1 mapping α from A onto A' such that for $a,b\in A$ it holds: $aRB \longrightarrow \alpha(a)R'\alpha(b)$ and $\beta(a) = \beta'(\alpha(a))$.

or labelled partial orderings $\sigma=(A,R,\beta)$ we denote the class of all abelled partial orderings σ' isomorphic with σ by $[\sigma]=[\![A,R,\beta]\!]$.

ollowing (Gra79), PW(X), the set of all partial words over X, is

$$PW(X):=\{[\![A,R,\beta]\!]\mid (A,R) \text{ is a finite partial ordering, } \beta:A\rightarrow X\}.$$

partial word $p=[\![A,R,\beta]\!]$ over X is said to be a semiword iff for ll $a,b\in A$ it holds: $\beta(a)=\beta(b)\rightarrow a=b \vee aRb \vee bRa$.

he set of all semiwords over X is denoted by SW(X). Obviously, SW(X) s a proper subset of PW(X) and, the empty word $e:=[\![\emptyset,\emptyset,\emptyset]\!]$ as well s the word $\underline{x}:=[\![\{1\},\emptyset,\{(1,x)\}]\!]$ (for $x\in X$) are semiwords over X.

e are going to use semiwords to describe the behavior of Petri nets. f $p=[\![A,R,\beta]\!]$ describes a process of the net N then the elements of correspond to the elementary changes of N (i.e. to the firings of ransitions of N) and the mapping specifies that correspondence. hereby the relation R is interpreted as the "earlier-later" relation etween elementary actions of N. Therefore, two firings of the same ransition have to be on one line of the process, i.e. nonconcurrent. bviously, under this interpretation only semiwords can appear as escriptions of processes in Petri nets.

o make the calculus of semiwords lucid we introduce canonic repreentatives for semiwords which allow us to treat semiwords nearly in he same way we are used with words. Words are mappings defined on inite ordered sets. The canonic representative of a word has as its upport a finite initial segment of the set $\mathbb{N}^+$ of all positive inteers in their natural ordering. Therefore we can speak of the first, econd,... letter of a word. Clearly, within a semiword this is not ossible but we can speak of the first, second,... occurrence of a ertain fixed letter $x\in X$. This observation gives rise to the definiion of the canonic representative of a semiword.

et be $\sigma=(A,R,\beta)$ an arbitrary representative of $q=[\sigma]\in SW(X)$, $a\in A$:

$$\nu_\sigma(a) := \max_{\leq}\{\operatorname{card}(B\cap\beta^{-1}(\beta(a)))\mid B\subseteq A,\ B \text{ chain in } (A,R),\ a=\max_R B\}.$$

One shows without difficulties that for $a,b \in A$ with $\beta(a) = \beta(b)$ $\nu_\sigma(a) = \nu_\sigma(b)$ implies $a = b$, and aRb implies $\nu_\sigma(a) < \nu_\sigma(b)$. Moreover, if α is an isomorphism from σ onto σ' then $\nu_\sigma(a) = \nu_{\sigma'}(\alpha(a))$ $(a \in A)$.

For $\sigma = (A,R,\beta)$ with $[\sigma] \in SW(X)$, $x \in X$, $i \in \mathbb{N}$ we put

$$\sigma(x,i) := \begin{cases} \text{that } a \in A \text{ with } \beta(a)=x \text{ and } \nu_\sigma(a)=i, \text{ if } 1 \leq i \leq \text{card}(\beta^{-1}(x)), \\ \text{not defined, else.} \end{cases}$$

The definition of $\sigma(x,i)$ is justified since $\beta^{-1}(x)$ is always a chain in (A,R). If α is an isomorphism from σ onto $\sigma' = (A',R',\beta')$ then for all $x \in X$ and all i with $1 \leq i \leq \text{card}(\beta^{-1}(x)) = \text{card}(\beta'^{-1}(x))$ it holds $\alpha(\sigma(x,i)) = \sigma'(x,i)$.

Now we define the canonic representative of the semiword $[\sigma] \in SW(X)$, $\sigma = (A,R,\beta)$ to be the triple $(\hat{A},\hat{R},\hat{\beta})$ where

$$\hat{A} := \bigcup_{x \in X} \{x\} \times \{1,2,\ldots,\text{card}(\beta^{-1}(x))\},$$

$$\hat{R} \subseteq \hat{A} \times \hat{A} \quad \text{with} \quad (x,i)\hat{R}(y,j) :\leftrightarrow \sigma(x,i) R \sigma(y,j),$$

$$\hat{B}(x,i) := x \qquad\qquad (x,i),(y,j) \in A.$$

One shows that the mapping $\alpha: \hat{A} \to A$ with $\alpha(x,i) = \sigma(x,i)$ is an isomorphism from $(\hat{A},\hat{R},\beta)$ onto σ, and, if (A_1,R_1,β_1) and (A_2,R_2,β_2) are isomorphic (i.e. representatives of the same semiword) then it holds $\hat{A}_1 = \hat{A}_2$, $\hat{R}_1 = \hat{R}_2$ and $\hat{\beta}_1 = \hat{\beta}_2$.

Within a canonic representative of a semiword the mapping $\hat{\beta}$ is always the projection to the first coordinate. Therefore we can omit it and ask for conditions under which a partial ordering in a finite subset of $X \times \mathbb{N}^+$ is the canonic representative of a certain semiword. One proves that this is the case iff for all $x \in X$, $i,j \in \mathbb{N}^+$ it holds:

(SW1) $\quad (x,i) \in A \wedge 1 \leq j \leq i \to (x,j) \in A$

(SW2) $\quad (x,i),(x,j) \in A \to ((x,i)R(x,j) \leftrightarrow i < j)$.

In the sequel we often shall identify semiwords with their canonic representatives and, therefore, consider partial orderings in finite subsets of $X \times \mathbb{N}^+$ with (SW1),(SW2) as semiwords.

3. The structure of SW(X)

In this section we present an algebraic characterization of SW(X) in terms of letters $\underline{x}$ and certain operations, i.e. we define operations for semiwords which allow us to construct each semiword from letters.

First we generalize the PARIKH-mapping ψ to semiwords. For $q \in SW(X)$, $\psi(q): X \to \mathbb{N}$ adjoins with each $x \in X$ the number of actions in q labelled with x, i.e. for $q = (A,R) \in SW(X)$ $\psi(q)(x) := \text{card}(A \cap \beta^{-1}(x))$.

For partial words concatenation has been defined in (Gra79). For semiwords $q_1 = (A_1,R_1)$, $q_2 = (A_2,R_2)$ we obtain $q_1q_2 := (A,R)$ where

$A := A_1 \cup \{(x,\psi(q_1)(x)+i) \mid (x,i) \in A_2\}$, and, for $(x,i),(y,j) \in A$

$$(x,i)R(y,j) :\leftrightarrow (i \leq \psi(q_1)(x) \wedge j \leq \psi(q_1)(y) \wedge (x,i)R_1(y,j)) \vee \vee (i \leq \psi(q_1)(x) \wedge j > \psi(q_1)(y)) \vee (i > \psi(q_1)(x) \wedge \wedge j > \psi(q_1)(y) \wedge (x,i-\psi(q_1)(x))R_2(y,j-\psi(q_1)(y))).$$

Obviously, SW(X) with respect to concatenation forms a semigroup with identity e containing the set W(X) of all words (generated by the set X of all letters) as a proper subsemigroup.

The parallel product $p_1 \times p_2$ of partial words $p_i = [\![A_i,R_i,\beta_i]\!]$ $(i=1,2)$ where without loss of generality $A_1 \cap A_2 = \emptyset$ is defined as

$p_1 \times p_2 := [\![A_1 \cup A_2, R_1 \cup R_2, \beta_1 \cup \beta_2]\!]$.

The set SW(X) is not closed under parallel product but one can see that for semiwords p_1,p_2 the partial word $p_1 \times p_2$ is a semiword again iff p_1,p_2 are X-disjoint, i.e. $\beta_1(A_1) \cap \beta_2(A_2) = \emptyset$. This is the case iff the canonic representatives of p_1,p_2 are disjoint. Therefore we put for $q,r \in SW(X)$, $q = (A_1,R_1)$, $r = (A_2,R_2)$

$$q \times r := \begin{cases} (A_1 \cup A_2, R_1 \cup R_2), & \text{if } A_1 \cap A_2 = \emptyset, \\ \text{not defined, else.} \end{cases}$$

Within its domain of definition parallel product is commutative, associative and has e as an identity.

From a result in (Gra79) it follows that it is not possible to construct every semiword from letters using only concatenation and parallel product. Therefore we introduce the composition of semiwords as an operation which constructs concurrent processes from less concurrent ones. The idea is that the composition of two semiwords should contain the sequential subprocesses of both but not more, hence, it should be as sequential as possible.

For every binary relation R, by $\vec{R}$ we denote the transitive closure of R, i.e. the smallest transitive relation containing R. If A is a finite subset of $X \times \mathbb{N}^+$ with (SW1) then by $\langle A \rangle$ we denote the smallest relation R in A such that (A,R) is a semiword, i.e.

$\langle A \rangle := \{((x,i),(x,j)) \mid (x,i),(x,j) \in A \wedge i < j\}$.

Now, for $q = (A_1,R_1)$, $r = (A_2,R_2)$, $q,r \in SW(X)$ the composition $q \sqcup r$ of q and r is defined by

$$q \sqcup r := (A_1 \cup A_2, \overrightarrow{R_1 \cup \langle A_1 \cup A_2 \rangle} \cap \overrightarrow{R_2 \cup \langle A_1 \cup A_2 \rangle}).$$

<u>Theorem 1</u> (Sta81)

1. SW(X) is closed under composition
2. SW(X) forms a commutative semigroup with respect to composition

3. For $q \in SW(X)$ it holds $q \sqcup q = q$ and

$$q \sqcup e = \bigsqcup_{x \in X} (x^{\Psi(q)(x)}) = q \sqcup \bigsqcup_{x \in X} (x^{\Psi(q)(x)}).$$

Let $q = (A_1, R_1)$, $r = (A_2, R_2)$ be semiwords. We say that q is smoother than r iff $A_1 = A_2$ and $R_1 \supseteq R_2$. The smoothing Sm(r) of the semiword r is the set of all <u>words</u> $q \in W(X)$ which are smoother than r (Gra79). For every semiword r, Sm(r) is a finite language. One can show:

<u>Lemma 2:</u> $q \in SW(X) \rightarrow q = \bigsqcup Sm(q)$

From this it follows that $Sm(q \sqcup r) = Sm(q) \cup Sm(r)$ and

<u>Theorem 3:</u> SW(X) is the least set of partial words over X containing the empty word e and all the letters x from X which is closed under concatenation and composition.

4. Fireability

A tuple $N = (P,T,F,K,V,m_o)$ is said to be a Petri net iff P and T are finite disjoint sets (of places and transitions resp.), $\emptyset \neq F \subseteq$ $\subseteq (P \times T) \cup (T \times P)$ is the flow-relation with $dom(F) \cup cod(F) = P \cup T$, $K: P \rightarrow \mathbb{N}^+ \cup \{\infty\}$ is a mapping which adjoins with each place p its token capacity K(p) which can be infinite, $V: F \rightarrow \mathbb{N}^+$ is a mapping which adjoins with each arc its multiplicity and $m_o: P \rightarrow \mathbb{N}$ is the initial marking of the net whereby $m_o \leq K$ (placewise). With every $t \in T$ we adjoin two markings $t^-, t^+ \in \mathbb{N}^P$ as follows. For $p \in P$

$$t^-(p) := \begin{cases} V((p,t)), & \text{if } (p,t) \in F, \\ 0, & \text{else} \end{cases} \qquad t^+(p) := \begin{cases} V((t,p)), & \text{if } (t,p) \in F, \\ 0, & \text{else.} \end{cases}$$

Moreover we put $\Delta t := t^+ - t^-$ $(\in \mathbb{Z}^P)$.

Let $U \subseteq T$ and let m be a marking with $m \leq K$. We say that U has concession under m in N iff

(1) $U^- := \sum_{t \in U} t^- \leq m$ and (2) $\forall U' (U' \subseteq U \rightarrow m + \Delta U' \leq K)$

where $\Delta U' := \sum_{t \in U'} \Delta t$.

The condition (1) ensures that all the transitions $t \in U$ are concurrently enabled, i.e. each place contains enough tokens to supply all the transitions concurrently, while condition (2) ensures that the capacities are not exceeded even in the case that some transitions fire simultaneously.

The reachability set $R_N(m_o)$ of N, the nonterminal language $L_N(m_o)$ of N and the transition function $\delta_N : \mathbb{N}^P \times W(T) \rightarrowtail \mathbb{N}^P$ are defined in the usual way (cf. (Sta80)). Moreover we put

$$\delta_N(m,U) := \begin{cases} m + \Delta U, & \text{if } U \text{ has concession under } m \text{ in } N, \\ \text{not defined, else.} \end{cases}$$

Let $q = (A,R)$ be a semiword over T. An antichain in q is a subset B of A whose elements are pairwise incomparable by R, i.e. the actions of B are causally independent from oneanother. Then the set

$BR := \{a \mid a \in A \wedge \exists b(b \in B \wedge bRa)\}$

is the set of those actions which depend on actions in B in the sense that at least one action of B must have occurred before, while

$C(B) := A - (B \cup BR)$

contains all the actions which are independent from B although elements of B may depend on the occurrence of elements of C(B) since

$RB := \{a \mid a \in A \wedge \exists b(b \in B \wedge aRb)\} \subseteq C(B)$

and RB obviously contains all the actions which must have occurred before some element of B.

A subset C of A is called leftmonotonic in q iff $a \in C$ and bRa always implies $b \in C$.

One can see that for every antichain B in q, RB and C(B) are leftmonotonic.

Let be $N = (P,T,F,K,V,m_o)$ a Petri net and $q = (A,R)$ a semiword over T. If q is a process in N then to every "state" of q there has to correspond a state (i.e. a marking) of N in such a way that to every change in the "state" of q there is a corresponding change in the marking of N caused by the firing of certain transitions whereby the initial "state" of q corresponds to m_o. A "state" of q obviously is sufficiently described by a cut in q, namely a bipartition of A into A_1, A_2 ($A_1 \cup A_2 = A$, $A_1 \cap A_2 = \emptyset$) where A_1 is leftmonotonic in q. In our interpretation, A_1 contains those actions which in the described state already have occurred while the elements of A_2 are going to occur and the minimal elements of A_2 form an antichain B in q the elements of which are next to occur. From the theory of partial orderings it is known that the correspondence between cuts (A_1,A_2) and antichains $B = \min_R A_2$ is 1:1. Therefore we can use antichains to describe the states of q in our definition of firebility:

The semiword $q = (A,R) \in SW(T)$ is called fireable under m in N iff $m \leq K$ and for every antichain B in q the set $\hat{\beta}(B)$ has concession under $m + \sum_{c \in C(B)} \Delta\hat{\beta}(c)$ in N. The set of all semiwords $q \in SW(T)$ which are firable under m in N is denoted by $SL_N(m)$. Thus, $SL_N(m_o)$ is the nonterminal semilanguage of N.

Obviously, if $q \in W(T)$ is a word then $q \in SL_N(m)$ iff $q \in L_N(m)$.

In our interpretation, $C(B) - RB$ is the set of all actions in q which are concurrent with every element of B. This is reflected by

<u>Lemma 4:</u> Let be $q = (A,R) \in SL_N(m_o)$, B an antichain in q and C a left-monotonic subset of A with $RB \subseteq C \subseteq C(B)$. Then $\hat{\beta}(B)$ has concession under $m_o + \sum_{c \in C} \triangle\hat{\beta}(c)$ in N.

Now we can extend the transition function δ_N to semiwords $q = (A,R)$:

$$\delta_N(m,q) := \begin{cases} m + \sum_{a \in A} \triangle\hat{\beta}(a), & \text{if } q \in SL_N(m) \\ \text{not defined,} & \text{else} \end{cases}$$

<u>Theorem 5:</u> If $q \in SW(T)$, $r \in SL_N(m_o)$ and $q \sqcup r = r$ then $q \in SL_N(m_o)$.

From Theorem 5 it follows that for $r \in SL_N(m_o)$ every word q which is smoother than r is an element of $SL_N(m_o)$, especially $Sm(r) \subseteq L_N(m_o)$. In this sense, the nonterminal semilanguages of Petri nets are closed under smoothing but note that $Sm(r) \subseteq L_N(m_o)$ does not imply $r \in SL_N(m_o)$.

A semiword q is called to be an initial segment of $r \in SW(T)$ iff there exists a semiword u such that $qu = r$. One can show that the nonterminal semilanguages of Petri nets are closed under initial segmentation and

<u>Lemma 6:</u> If $q \in SL_N(m_o)$ then for $u \in SW(T)$ it holds
$$u \in SL_N(\delta_N(m_o,q)) \longleftrightarrow qu \in SL_N(m_o).$$

5. Semilanguages
Every set of semiwords is called a semilanguage while a language always is a set of words. For a Petri net $N = (P,T,F,K,V,m_o)$ and every finite set M of markings of N we define
$$SL_N(m_o,M) := \{q \mid q \in SL_N(m_o) \wedge \delta_N(m_o,q) \in M\}$$
as the semilanguage represented by N with the terminal set M. As usual we put
$$L_N(m_o,M) := \{q \mid q \in L_N(m_o) \wedge \delta_N(m_o,q) \in M\}.$$
Let be FNL resp. FNS the families of all <u>f</u>ree <u>n</u>onterminal Petri net <u>l</u>anguages resp. <u>s</u>emilanguages and FTL resp. FTS the families of all free <u>t</u>erminal Petri net <u>l</u>anguages resp. <u>s</u>emilanguages.

Obviously, $L_N(m_o) \subseteq SL_N(m_o)$ and $L_N(m_o,M) \subseteq SL_N(m_o,M)$. By introducing into N a so-called run-place which is initially marked and which self-loops with every transition of N we obtain from N a net N' such that $L_N(m_o) = SL_{N'}(m_o')$ and $L_N(m_o,M) = SL_{N'}(m_o',M')$ for all $M \subseteq \mathbb{N}^P$. This implies that $FNL \subseteq FNS$ and $FTL \subseteq FTS$, and, therefore the families FNS and

FTS are not closed under union, intersection with regular languages, catenation, catenation closure and (e-free) homomorphisms.

For semilanguages $S \subseteq SW(T)$ let $Sm(S) := \bigcup_{q \in S} Sm(q) \quad (\subseteq W(T))$.

From $Sm(SL_N(m_o)) = L_N(m_o)$ and $Sm(SL_N(m_o,M)) = L_N(m_o,M)$ we obtain

$S \in FNS \rightarrow Sm(S) \in FNL$ and $S \in FTS \rightarrow Sm(S) \in FTL$.

The converses of these implications in general do not hold, even if we assume additionally that S is weak (cf. (Gra79)), i.e. that S contains every semiword which is smoother than some semiword in S. This shows that it is not possible to characterize the representability of semilanguages by Petri nets in terms of Petri net languages. However, one can prove

Theorem 7: Let be $L \subseteq W(T)$ and $S(L) := \{r \mid r \in SW(T) \wedge Sm(r) \subseteq L\}$. Then

$L \in FNL \leftrightarrow S(L) \in FNS$ and $L \in FTL \leftrightarrow S(L) \in FTS$.

5. Closedness

In this section we list some results on the closedness of our families FNS and FTS under certain operations which are applied below to give an algebraic characterization of these families.

The left-derivative $\partial_r(S)$ of $S \subseteq SW(T)$ with respect to $r \in SW(T)$ is

$\partial_r(S) := \{q \mid q \in SW(T) \wedge rq \in S\}$.

The right quotient S/S' of semilanguages S,S' is

$S/S' := \{r \mid \exists s' (s' \in S' \wedge rs' \in S\}$.

Theorem 8: FTS is closed under left-derivatives and under right quotients by finite semilanguages.

FNS is not closed under these operations since a left-derivative as well as a right quotient by a finite semilanguage can be empty while every element of FNS contains the empty word.

For $S \subseteq SW(T)$ let be $T(S) := \bigcup \{\hat{\beta}(A) \mid (A,R) \in S\}$ the set of all letters which appear within some semiword of S. Semilanguages S,S' are said to be letterdisjoint iff $T(S) \cap T(S') = \emptyset$. For letterdisjoint semilanguages S,S' we define the parallel product $S \times S'$ by

$S \times S' := \{s \times s' \mid s \in S \wedge s' \in S'\}$.

The restriction $S_1 \circledR S_2$ of semilanguages S_1,S_2 is defined by

$S_1 \circledR S_2 := (S_1 \times SW(T(S_2)-T(S_1))) \cap (S_2 \times SW(T(S_1)-T(S_2)))$.

Theorem 9: FNS and FTS are closed under intersection, letterdisjoint parallel product and restriction.

Let $\sigma = (A,R,\beta)$ be a labelled partial ordering over X and $h: X \rightarrow W(Y)$ a homomorphism. We define $h(\sigma) := (\tilde{A},\tilde{R},\tilde{\beta})$ where

$\tilde{A} := \{(a,w) \mid a \in A \wedge e \neq w \sqsubseteq h(\beta(a))\}$

$\tilde{\beta}((a,w)) :=$ the last letter of w, $\quad \tilde{R} := \vec{R_o}$

whereby for $(a,w),(b,u) \in \tilde{A}$

$(a,w)R_o(b,u) :\longleftrightarrow (a = b \wedge \exists y(y \in Y \wedge wy = u)) \vee aRb$

and $\sqsubseteq$ denotes the inital segment relation.

Obviously, $h(\sigma)$ is a labelled partial ordering over Y and if σ is isomorphic with σ' then $h(\sigma)$ is isomorphis with $h(\sigma')$. Therefore we can define the image of a partial word $p \in PW(X)$ under the homomorphism h as

$h(p) := [\![h(\sigma)]\!]$ for $p = [\![\sigma]\!]$.

A homomorphism always maps words to words and does this in the same way usual in language theory. The example $h: \{a,b\} \to \{a\}$ shows that the homomorphic image of a semiword is not always a semiword again: $h(a \times b) = a \times a \notin SW(\{a\})$. But one can show

Lemma 10: If $h: X \to W(Y) - \{e\}$ is an e-free homomorphism and $r \in SW(Y)$ then $h^{-1}(r) := \{p \mid p \in PW(X) \wedge h(p) = r\} \subseteq SW(X)$.

Hence, the image of a semilanguage under an e-free inverse homomorphism is a semilanguage. Now one can prove

Theorem 11: FNS and FTS are closed under e-free inverse homomorphisms.

7. Characterization

Let be for $n = 0,1,2,\ldots$

$A_n := \{a,b\} \times \{1,2,\ldots,n\}$, $R_n := \overrightarrow{\{((a,i),(b,i)) \mid i=1,\ldots,n\} \cup \langle A_n \rangle}$.

For every n, (A_n,R_n) is a semiword over a,b . By D we denote the least semilanguage over $\{a,b\}$ which contains all the (A_n,R_n) and which is weak, i.e. which contains with some semiword q all the semiwords smoother than q. Them Sm(D) is the (one-sorted) DYCK-language over $\{a,b\}$, and $D = S(Sm(D))$. Therefore, from $Sm(D) \in FTL$ and Theorem 7 we have $D \in FTS$.

Theorem 12: FTS is the least family of semilanguages containing the semilanguage D and closed under left-derivatives, right quotients by finite languages, restriction and e-free inverse homomorphisms.

Let be E the initial extension of D, i.e.

$E := \{q \mid \exists r(r \in SW(\{a,b\}) \wedge qr \in D)\}$.

One can see easily that $E \in FNS$. For every $i \in \mathbb{N}$ the leftderivative $E_i := \partial_{a^i}(E)$ is an element of FNS too. We can show

Theorem 13: FNS is the least family of semilanguages which contains all the semilanguages E_i $(i \in \mathbb{N})$ and which is closed under restriction and e-free inverse homomorphisms.

Let us finally consider Petri nets N = (P,T,F,K,V,m_o) where the capacity K(p) of every place p is 1 (safeness), the multiplicity of every arc is 1 (ordinary net) and which contain no selfloops. Petri nets of this kind we can interpret as condition-event-systems and vice versa (cf. (NTA80)). Processes in CE-systems have been described by means of occurrence nets. We here confine ourselves to proper processes where every occurrence is a forward occurrence of an event. Obviously, such forward processes can be sufficiently described by semiwords since in this case the conditions (and, therefore, the cases) can be reconstructed from the occurrences of the events, i.e. the elements of the semiword. Therefore the following theorem might be of interest in the theory of condition-event-systems:

Theorem 14: A semilanguage S is the set of all processes leading a certain finite condition-event-system from its initial case into a certain finite set of terminal cases iff S is contained in the least family of semilanguages which contains the languages $(ab)^*$, $(ab)^*a$, $b(ab)^*$ and $b(ab)^*a$ and which is closed under restriction and inverse renaming.

Thereby a renaming is understood as a homomorphism having only letters as values.

The proofs of the assertions given here can be found in (Sta81).

5. References

(Gra79) J. Grabowski: On partial languages. Preprint 40/79, Sekt. Math., Humboldt-Univ., 1979. To appear in Fundamenta Informaticae (Warszawa).

(Maz77) A. Mazurkiewicz: Concurrent program schemes and their verification. DAIMI PB 78, Aarhus 1977.

(NTA80) Net Theory and Application. Lecture Notes in Computer Sci. 84 (1980).

(Rei80) W. Reisig: Processes of marked arc-weighted nets. Petri Nets and Related System Models Newsletter 5 (1980) 13-15.

(STa80) P.H. Starke: Petri-Netze. DVW Berlin, 1980.

(Sta81) P.H. Starke: Processes in Petri nets. Elektron. Informationsverarbeitung u. Kybernetik 17 (1981) No. 4-6.

(Win79) J. Winkowski: An algebraic approach to concurrency. Lecture Notes in Computer Sci. 74 (1979) 523-532.

SOME ALGEBRAIC ASPECTS OF RECOGNIZABILITY AND RATIONALITY

Magnus Steinby
Department of Mathematics
University of Turku
20500 Turku 50
Finland

Introduction

Regular languages arise in so many ways that there can be no doubt about their importance. When I now propose to reexamine the concept from a general algebraic perspective, I certainly do not want to replace it by anything else or even to improve on it. However, the fact that a notion is arrived at from several different directions is always a noteworth phenomenon and it should be of some interest to know what assumptions account for the various properties of the concept. Such an analysis is also needed when one wants to generalize the idea or to transplant it into another environment; the fact that the different definitions lead to the same result may well depend on some special circumstances which do not prevail generally. That this is the case for regular languages was realized already in the 1960's when it was observed that Kleene's theorem cannot be generalized to arbitrary monoids. This led S. Eilenberg to separate between two classes of subsets in a monoid, the recognizable subsets saturated by finite congruences, and the rational subsets which can be obtained from finite subsets by the three rational operations. Although further classes are possible, I shall concentrate on these two. A generalization or transfer of a concept also presupposes a certain interpretation of the original set-up. The two usual interpretations to be considered here will be called the monoid interpretation and the unary interpretation. In the monoid interpretation languages are viewed as subsets of a free monoid. The generalizations of recognizability and rationality to arbitrary monoids follow quite naturally from this interpretation, and the further generalization to general algebras is equally natural.

In much of the algebraic theory of automata the monoid interpretation is taken for granted. However, the theory of tree automata evolved from the unary interpretation in which words are regarded as unary terms. The regular forest operetions introduced by THATCHER and WRIGHT [12] are products of the unary interpretation. Although these yield a valid generalization of the Kleene theorem for term algebras, a furher generalization of the rationality concept to arbitrary algebras does not come so naturally as it did in the monoid interpretation. Moreover, these regular forest operations are not the same one will get by applying to term algebras the general concepts obtained by adopting the monoid interpretation. On the other hand, both interpretations lead to the same definition of recognizable subsets of an arbitrary algebra.

Among the many characterizations of regular languages the theorems of Nerode, Myhill and Kleene are the most fundamental ones and they will be singled out as some kind of touchstones against which the interpretations and generalizations are tested.

A completely self-contained presentation would have made the text too long. Therefore, no formal proofs are presented and some well-known concepts and results will be used without proper definitions or explanations. Some of the proofs are outlined, and I hope that this will enable the interested reader to construct them. Rational forests I plan to treat more extensively elsewhere. Some of the following references may be useful.

The automaton theory reviewed in Section 1 may be found in almost any text-book on the subject. The classic paper by RABIN and SCOTT [9] is a good authentic source. For the universal algebra the reader is referred to COHN [2], GRÄTZER [6] or WERNER [15]. However, the unkowing reader must be warned against the notorious confusion around the concepts 'term' and 'algebraic function'. All results concerning recognizable and rational subsets of general monoids mentioned in Section 1 can be found very conveniently in Sections III.1-2 of BERSTEL [1]. Of course, they appear also in EILENBERG [3]. Recognizable subsets of arbitrary algebras were first considered by MEZEI and WRIGHT [8]. Other papers on the subject include [4], [7], [11], [13] and [14]. The particular formalism for the theory of tree automata used here is fully developed in [5] (cf. also [11]).

1. Preliminaries

The set X^* of all words in a given alphabet X is a monoid freely generated by X with the concatenation of words as the binary operation. The unit element is the empty word e. Subsets of X^* are called (X-) languages. A language is said to be recognizable if it is recognized by a finite-state (Rabin-Scott) recognizer. The set of all recognizable X-languages is denoted by Rec X .

Let θ be an equivalence on a set S. The θ-class of an element a of S is denoted by $a\theta$, and θ is said to be finite if the quotient set $S/\theta = \{a\theta \mid a \in S\}$ is finite. A subset H of S is saturated by θ if H is the union of some θ-classes.

A right congruence of a semigroup S is an equivalence θ on S such that

$$a \equiv b\ (\theta) \text{ implies } ac \equiv bc\ (\theta),$$

for all $a,b,c \in S$. Let L be an X-language. The greatest right congruence of the monoid X^* which saturates L is called the Nerode congruence of L, and we denote it by ρ_L.

Nerode's theorem (A. Nerode 1957). For any X-language L the following three conditions are equivalent:

(a) $L \in \text{Rec } X$.

(b) L is saturated by a finite right congruence of X^*.

(c) The Nerode congruence ρ_L is finite.

The equivalence of conditions (a) and (b) is so immediate that it is justified to adopt (b) as an algebraic definition of recognizability. Let us recall that

$$u \equiv v\ (\rho_L) \quad \text{iff} \quad u^{-1}L = v^{-1}L \qquad (u,v \in X^*).$$

Therefore, Nerode's theorem also says that L is recognizable iff the number of the left quotients of L is finite.

The greatest conguence of X^* saturating a given X-language L will be denoted by θ_L. It is usually called the syntactic congruence of L, but here I prefer to call it the Myhill congruence of L. Also the Myhill congruence can be defined by means of quotients:

$$w \equiv w'\ (\theta_L) \quad \text{iff } (\forall u,v \in X^*)\ w \in u^{-1}Lv^{-1} \Leftrightarrow w' \in u^{-1}Lv^{-1} .$$

Myhill's theorem (J. R. Myhill 1957). For any X-language L the following three conditions are equivalent:

(a) $L \in \mathrm{Rec}\, X$.

(b) L is saturated by a finite congruence of X^*.

(c) The Myhill congruence θ_L is finite.

The name 'regular language' comes from the third characterization of recognizability, which is Kleene's theorem. However, we shall speak about 'rational languages' and reserve the term 'regular language' for a more general use which does not refer to any particular definition. The family Rat X of rational languages over X is the smallest set C which satisfies the following conditions:

1° Every finite X-language is in C.

2° If $U,V \in C$, then $U \cup V \in C$.

3° If $U,V \in C$, then $UV \in C$.

4° If $U \in C$, then $U^* \in C$.

Here UV and U^* are, respectively, the usual product and iteration of the languages.

Kleene's theorem (S. C. Kleene 1956). Rat X = Rec X .

Now we shall recall a few algebraic concepts and introduce some connected notation.

Let Σ be a set of operational symbols. For each $m \geq 0$, Σ_m is the set of all m-ary symbols in Σ. In a Σ-algebra $\mathcal{A} = (A,\Sigma)$, A is the set of elements and every $\sigma \in \Sigma_m$ ($m \geq 0$) is realized as an m-ary operation $\sigma^{\mathcal{A}}: A^m \to A$. In what follows, Σ is always a set of operational symbols, and $\mathcal{A} = (A,\Sigma)$ and $\mathcal{B} = (B,\Sigma)$ are Σ-algebras.

The congruences of $\mathcal{A}$ form with respect to the inclusion relation a complete lattice which will be denoted by $C(\mathcal{A})$. The set of all endomorphisms of $\mathcal{A}$ is denoted by $\mathrm{End}(\mathcal{A})$. Recall that a congruence $\theta \in C(\mathcal{A})$ is said to be fully invariant if

$$a \equiv a' \ (\theta) \quad \text{implies} \quad a\psi \equiv a'\psi \ (\theta) \quad \text{for all } a,a' \in A \text{ and } \psi \in \mathrm{End}(\mathcal{A}).$$

The fully invariant congruences of $\mathcal{A}$ form a complete sublattice of the lattice $C(\mathcal{A})$.

The subalgebra generated by a subset H of $\mathcal{A}$ is denoted by <H>. The algebra $\mathcal{A}$ is said to be finitely generated if $\mathcal{A}$ = <H> for some finite subset H of A.

Let V be a set. The set $T_\Sigma(V)$ of ΣV-terms is the smallest set such that

(1) $V \subseteq T_\Sigma(V)$, and

(2) $\sigma(t_1,\ldots,t_m) \in T_\Sigma(V)$ whenever $m \geq 0$, $\sigma\in\Sigma_m$ and $t_1,\ldots,t_m \in T_\Sigma(V)$.

The ΣV-<u>term algebra</u> $\mathcal{T}_\Sigma(V) = (T_\Sigma(V),\Sigma)$ is defined so that

$$\sigma^{\mathcal{T}_\Sigma(V)}(t_1,\ldots,t_m) = \sigma(t_1,\ldots,t_m) \qquad (m\geq 0,\ \sigma\in\Sigma_m,\ t_1,\ldots,t_m\in T_\Sigma(V)).$$

The algebra $\mathcal{T}_\Sigma(V)$ is freely generated by V over the class of all Σ-algebras. This means that $\langle V\rangle = T_\Sigma(V)$, and that for any Σ-algebra $\mathcal{A}$, every mapping $\alpha: V \to A$ has a unique extension to a homomorphism $\hat{\alpha}: \mathcal{T}_\Sigma(V) \to \mathcal{A}$.

The set of <u>unary algebraic functions</u> $\mathrm{Alg}_1(\mathcal{A})$ of $\mathcal{A}$ is the smallest set F such that

1° F contains the identity mapping $1_A: A \to A$,

2° F contains every constant mapping $c_a: A\to A$, $x\mapsto a$ $(a\in A)$, and

3° $\sigma^{\mathcal{A}}(f_1,\ldots,f_m)\in F$ whenever $m\geq 0$, $\sigma\in\Sigma_m$, and $f_1,\ldots,f_m\in F$.

Note that terms are called 'polynomial symbols' in [6] and 'words' in [2]. Algebraic functions are often called polynomial functions.

2. The monoid interpretation

The monoid interpretation means simply that X^* is viewed as the free monoid generated by X, and that all language concepts are interpreted accordingly. Myhill's and Nerode's theorems do not require any particular interpretations as they are already expressed in terms of semigroup concepts. Myhill's theorem suggests a natural definition of recognizability which does not call for any concept of an automaton. Moreover, it applies unchanged to arbitrary monoids, too: a subset of a monoid M is said to be <u>recognizable</u> if it is saturated by some finite congruence of M. The Nerode theorem remains a true theorem in this generalization. Many of the basic closure theorems for regular languages are easily extended to arbitrary monoids. For example, the Boolean combinations and the inverse homomorphic images of recognizable subsets of monoids are recognizable. Notable exceptions are the homomorphic images, complex products, and submonoids generated by recognizable subsets which may be nonrecognizable.

If the monoid interpretation is adopted, then it is natural to define rational subsets of an arbitrary monoid thus: a subset of a monoid is <u>rational</u> if it can be constructed from finite subsets by forming unions, complex products and submonoids (generated by subsets already known to be rational).

Kleene's theorem does not hold in all monoids, but the following fact can be noted.

McKnights theorem (J. D. McKnight 1964). In a finitely generated monoid every recognizable subset is rational.

The operations that preserve rationality in a monoid are quite different from those preserving recognizability. Of course, the three rational opererations preserve rationality, and homomorphic images of rational subsets are also rational. On the other hand, intersections, complements, or inverse homomorphic images of rational subsets may be nonrational.

Let us now define recognizable and rational subsets of arbitrary algebras as suggested by the monoid interpretation and the above generalization to arbitrary monoids.

Definition 2.1 (MEZEI and WRIGHT [8]). A subset of an algebra is said to be recognizable if it is saturated by a finite congruence of the algebra. We denote the set of all recognizable subsets of an algebra A by Rec A .

The following facts are well-known.

Proposition 2.2. Let A and B be Σ-algebras. Then

(a) $\emptyset, A \in \mathrm{Rec}\, A$;

(b) if $P, Q \in \mathrm{Rec}\, A$, then $P \cap Q, P \cup Q, P - Q \in \mathrm{Rec}\, A$; and

(c) if $\varphi : A \to B$ is a homomorphism and $Q \in \mathrm{Rec}\, B$, then $Q\varphi^{-1} \in \mathrm{Rec}\, A$.

When Definition 2.1 is viewed as a result of the monoid interpretation, then the greatest congruence saturating a given subset P of an algebra A is to be regarded as the Myhill congruence of P. The quotients of languages are largely replaced by the inverses of unary algebraic functions. This can be seen from the following proposition (cf. [11]).

Proposition 2.3. Let P be a subset of an algebra A.

(1) If $P \in \mathrm{Rec}\, A$, then $f^{-1}(P) \in \mathrm{Rec}\, A$ for every $f \in \mathrm{Alg}_1(A)$.

(2) $P \in \mathrm{Rec}\, A$ iff the set $\{f^{-1}(P) \mid f \in \mathrm{Alg}_1(A)\}$ is finite.

(3) For any $a,b \in A$, $a \equiv b\ (\theta_P)$ iff $(\forall f \in Alg_1(A))\ a \in f^{-1}(P) \Leftrightarrow b \in f^{-1}(P)$.

From the case of general monoids we already know that subalgebras generated by recognizable subsets or homomorphic images of recognizable subsets are not always recognizable. Similarly, one cannot expect subsets of the type

$$\sigma^A(P_1,\ldots,P_m) = \{\sigma^A(a_1,\ldots,a_m) \mid a_1 \in P_1,\ldots,a_m \in P_m\}$$

to be in Rec A whenever $m \geq 0$, $\sigma \in \Sigma_m$, and $P_1,\ldots,P_m \in$ Rec A.

The following general notion of rationality has obviously not been studied before although it appears to be a very logical result of the monoid interpretation.

Definition 2.4. The set Rat A of all rational subsets of a Σ-algebra A is defined as the smallest set C satisfying the following conditions:

1° Every finite subset of A is in C.

2° If $R,S \in C$, then $R \cup S \in C$.

3° If $m > 0$, $\sigma \in \Sigma_m$ and $R_1,\ldots,R_m \in C$, then $\sigma^A(R_1,\ldots,R_m) \in C$.

4° If $R \in C$, then $\langle R \rangle \in C$.

The form of Definition 2.4 suggests the useful principle of rational induction by means of which one may prove facts about rational subsets, or define concepts relating to them.

For monoids the recognizable and the rational subsets are obviously the same as those considered above. Hence it is clear that Kleene's theorem cannot be valid for all algebras. As a matter of fact, in Section 4 we shall see that not even McKnight's theorem can be generalized to hold for arbitrary algebras. However, the following weaker generalization is possible and expresses the essence of the theorem.

Proposition 2.5. If Rec A = Rat A, then Rec $A/\theta \subseteq$ Rat A/θ for all congruences $\theta \in C(A)$.

From the case of monoids we also know that most of the basic operations, except for the rational ones, cannot be expected to always preserve rationality in algebras. However, the following fact can easily be established by rational induction.

Proposition 2.6. Let $\varphi: \mathcal{A} \to \mathcal{B}$ be a homomorphism. If $R \in \text{Rat}\, \mathcal{A}$, then $R\varphi \in \text{Rat}\, \mathcal{B}$. If φ is surjective, then there exists for every $S \in \text{Rat}\, \mathcal{B}$ an R in $\text{Rat}\, \mathcal{A}$ such that $S = R\varphi$.

3. The unary interpretation

Let us now consider the unary interpretation. The input alphabet X is now regarded as a set of unary operational symbols. Finite-state recognizers are then finite X-algebras equipped with an initial state and a set of final states. In X^* an X-algebra structure is defined as follows: for each $x \in X$ and $w \in X^*$, put $x^X(w) = wx$. The resulting algebra $\mathcal{X} = (X^*, X)$ is freely generated by e over the class of all X-algebras. The congruences of $\mathcal{X}$ are exactly the right congruences of the monoid X^*. In particular, the Nerode congruence ρ_L of an X-language L is the greatest congruence of $\mathcal{X}$ which saturates L. If we now define recognizable subsets of algebras using Nerode's theorem as the starting point, we get again Definition 2.1.

But what does Myhill's theorem say in the unary interpretation? In order to answer this question we note that the congruences of the monoid X^* are precisely the fully invariant congruences of the X-algebra $\mathcal{X}$. Indeed, for any $u, w \in X^*$, $uw = w\varphi$, where φ is the unique endomorphism of $\mathcal{X}$ such that $e\varphi = u$. This implies that a relation in X^* is left invariant iff it is invariant with respect to the endomorphisms of $\mathcal{X}$. The Myhill congruence θ_L of a language L is now the greatest fully invariant congruence of $\mathcal{X}$ which saturates L. Consequently, the Myhill congruence θ_P of a subset P of any algebra $\mathcal{A}$ is defined as the greatest fully invariant congruence of $\mathcal{A}$ which saturates P. Myhill's theorem has the following generalization:

Proposition 3.1. Let $\mathcal{A}$ be a finitely generated algebra. For any subset P of A, the following three conditions are equivalent:

(1) $P \in \text{Rec}\, \mathcal{A}$.

(2) P is saturated by a finite fully invariant congruence of $\mathcal{A}$.

(3) The Myhill congruence θ_P is finite.

[Proof. The only part of the proposition which is not obvious is the implication (1) $\Rightarrow$ (2). If $P \in \text{Rec}\, \mathcal{A}$, then P is saturated by a finite congruence ρ of $\mathcal{A}$. If we define the relation θ in A so that for all $a, b \in A$, $a \equiv b\ (\theta)$ iff $(\forall \varphi \in \text{End}(\mathcal{A}))\ a\varphi \equiv b\varphi\ (\rho)$, then θ is a finite fully invariant congruence which saturates P.]

The unary algebraic functions of the algebra X are the constant mappings and the mappings f such that $f(w) = wv$ $(w \in X^*)$ for some arbitrary fixed $v \in X^*$. For any X-language L, ρ_L and θ_L satisfy the following conditions:

$$w_1 \equiv w_2 \ (\rho_L) \quad \text{iff} \quad (\forall v \in X^*)\ w_1 v \in L \Leftrightarrow w_2 v \in L,$$

and

$$w_1 \equiv w_2 \ (\theta_L) \quad \text{iff} \quad (\forall u,v \in X^*)\ uw_1 v \in L \Leftrightarrow uw_2 v \in L,$$

for all $w_1, w_2 \in X^*$.

These conditions may be expressed in terms of unary algebraic functions and endomorphisms of X, and are then seen to express special cases of the following general theorem.

Proposition 3.2. The Nerode congruence and the Myhill congruence of a subset P of an arbitrary algebra A are determined by the following conditions: for any $a,b \in A$,

$$a \equiv b \ (\rho_L) \quad \text{iff} \quad (\forall f \in \mathrm{Alg}_1(A))\ f(a) \in P \Leftrightarrow f(b) \in P$$

and

$$a \equiv b \ (\theta_L) \quad \text{iff} \quad (\forall f \in \mathrm{Alg}_1(A))(\forall \varphi \in \mathrm{End}(A))\ f(a\varphi) \in P \Leftrightarrow f(b\varphi) \in P.$$

The rational language operations are more problematic in the unary interpretation. Viewing words as unary trees THATCHER and WRIGHT [12] generalized the product and the iteration to forests (tree languages) and obtained a generalization of Klèene's theorem. However, the regular forests of Thatcher and Wright are not the rational subsets one would get by applying Definition 2.4 to term algebras. Moreover, there does not seem to be any obvious way of extending the idea to arbitrary algebras. TRNKOVÁ and ADÁMEK [13,14] have suggested the following method of defining the product and iteration operations in a quotient algebra of a term algebra. Let φ be the canonical homomorhism. The x-product (as in the case of forests there will be several products) of two subsets P and Q of the quotient algebra is defined to be the set $(P\varphi^{-1} \cdot_x Q\varphi^{-1})\varphi$, where the x-product is the usual forest operation. Iterations are defined similarly. If we apply these constructions to free monoids, regarding them as quotient algebras of suitable term algebras, we will not recapture the usual language operations. Hence it is not surprising that Kleene's theorem does not hold in all quotient algebras of term algebras if rationality is defined using the above described product and iteration operations. However, a form of McKnight's theorem can be found in [13]. Of course, any algebra is isomorphic to a quotient algebra of a term algebra, but such a representation is not unique. One should also note the fact that the construction of a regular forest may require more variables than those which generate the term algebra.

4. Recognizable and rational forests

The notions discussed above will now be applied to forests, that is to say, we shall specialize them to finitely generated term algebras of finite type. First some concepts from the theory of tree automata must be introduced.

Let Σ be a ranked alphabet i.e. a finite set of operational symbols, and let X be an ordinary alphabet, the frontier alphabet. Now ΣX-terms are also called ΣX-trees, and subsets of $T_\Sigma(X)$ are called ΣX-forests (or just trees and forests, respectively). A ΣX-recognizer $\underline{A}$ consists of a finite Σ-algebra A, an initial assignment $\alpha: X \to A$, and a set $A_F \subseteq A$ of final states. We write $\underline{A} = (A, \alpha, A_F)$. The forest recognized by $\underline{A}$ is the ΣX-forest

$$T(\underline{A}) = \{t \in T_\Sigma(X) \mid t\hat{\alpha} \in A_F\},$$

where $\hat{\alpha}: T_\Sigma(X) \to A$ is the unique homomorphic extension of α. A ΣX-forest T is said to be recognizable if $T = T(\underline{A})$ for some ΣX-recognizer $\underline{A}$. Let Rec(Σ,X) be the set of all recognizable ΣX-forests. It is easy to see that the recognizable ΣX-forests are exactly the recognizable subsets of the term algebra $T_\Sigma(X)$ (in the sense of Definition 2.1).

Definition 2.4 yields the following notion.

Definition 4.1. The set Rat(Σ,X) of all rational ΣX-forests is the smallest set C satisfying the following conditions:

1° Every finite ΣX-forest belongs to C.

2° If $R, S \in C$, then $R \cup S \in C$.

3° If $m \geq 0$, $\sigma \in \Sigma_m$, and $R_1, \ldots, R_m \in C$, then $\sigma(R_1, \ldots, R_m) \in C$.

4° If $R \in C$, then $\langle R \rangle \in C$.

The rational operations appearing in 3° and 4° should be explained in terms of trees. The forest $\sigma(R_1, \ldots, R_m)$ consists of all trees of the form $\sigma(r_1, \ldots, r_m)$, where $r_1 \in R_1, \ldots, r_m \in R_m$. For any ΣX-forest R, $\langle R \rangle = R_0 \cup R_1 \cup R_2 \cup \ldots$, where (i) $R_0 = R$, and for all $k \geq 0$,

(ii) $R_{k+1} = R_k \cup \{\sigma(r_1, \ldots, r_m) \mid m \geq 0, \sigma \in \Sigma_m, r_1, \ldots, r_m \in R_k\}$.

We may also write

$$\langle R\rangle = T_\Sigma(\{\xi\})(\xi \leftarrow R),$$

where the right-hand side consists of all trees which can be obtained from some tree $t \in T_\Sigma(\{\xi\})$ by substituting trees from R for every occurrence of ξ. It is not hard to see that these operations preserve recognizability. Since finite forests are recognizable and unions of recognizable forests are recognizable, we get

Proposition 4.2. $\mathrm{Rat}(\Sigma,X) \subseteq \mathrm{Rec}(\Sigma,X)$.

The *yield* mapping $yd: T_\Sigma(X) \to X^*$ is defined as follows:

1° $yd(x) = x$ $(x \in X)$, and

2° $yd(\sigma(t_1,\ldots,t_m)) = yd(t_1)\ldots yd(t_m)$ for all $m \geq 0$, $\sigma \in \Sigma_m$, and $t_1,\ldots,t_m \in T_\Sigma(X)$.

The *yield* of a ΣX-forest T is the X-language $yd(T) = \{yd(t) \mid t \in T\}$. It is a well-known fact that the yield of a recognizable forest is a context-free language and that every context-free language is the yield of a recognizable forest (cf. [8] or [5]). Let us now determine the yields of the rational forests.

First we turn X^* into a Σ-algebra $\Sigma(X) = (X^*,\Sigma)$ by putting

$$\sigma^{\Sigma(X)}(w_1,\ldots,w_m) = w_1 w_2 \ldots w_m \qquad (m \geq 0,\ \sigma \in \Sigma_m,\ w_1,\ldots,w_m \in X^*).$$

In particular, $\sigma^{\Sigma(X)} = e$ for each $\sigma \in \Sigma_0$.

Let us call Σ a *full* ranked alphabet if $\Sigma_0 \neq \emptyset$ and $\Sigma \neq \Sigma_0 \cup \Sigma_1$. It is easy to see that the yield mapping is surjective iff Σ is full. Moreover, it can be shown that $\mathrm{Rec}\,\Sigma(X) = \mathrm{Rec}\,X$ and $\mathrm{Rat}\,\Sigma(X) = \mathrm{Rat}\,X$ when Σ is full. The latter fact combined with Proposition 2.6 gives

Proposition 4.3. Let Σ be a full ranked alphabet. For any rational ΣX-forest R, yd(R) is a regular language. For every language $L \in \mathrm{Rec}X$ there exists a rational ΣX-forest R such that $L = yd(R)$.

The first statement of Proposition 4.3 would be true even without the assumption of fullness. If Σ is full and X contains more than one letter, then there exists always a recognizable ΣX-forest T such that yd(T) is nonregular. Hence, we get from Propositions 4.2 and 4.3 the following

Corollary 4.4. If Σ is full and $|X| \geq 2$, then Rat(Σ,X) is a proper subclass of Rec(Σ,X).

Hence our interpretation-generalization-specialization-process has given a new class of recognizable forests. It is also interesting to compare Corollary 4.4 with McKnight's theorem which gives the opposite inclusion for finitely generated monoids. The classes Rat(Σ,X) enjoy some closure properties not possessed by the classes of rational subsets of arbitrary monoids. For example, intersections and inverse homomorphic images of rational forests are rational.

5. Monadically rational languages

As a conclusion, let us return to our starting point, this time respecializing the general concepts of Section 2 to languages. If we use the monoid interpretation, Definitions 2.1 and 2.4 will both give just the usual regular languages. Let us assume the unary interpretation. Definition 2.1 was the result of this interpretation, too, and the recognizable subsets of the algebra $\mathcal{X} = (X^*,X)$ are exactly the regular X-languages. Since words can be viewed as unary trees we have by Proposition 4.4 Rat $\mathcal{X} \subseteq$ Rec $\mathcal{X}$. Although Corollary 4.4 is not applicable here, one may safely conjecture that the inclusion is usually proper. Let us call the rational subsets of $\mathcal{X}$ monadically rational languages. In order to find out what they are we should determine the rational operations of $\mathcal{X}$.

For any $x \in X$ and $L \subseteq X^*$, $x^{\mathcal{X}}(L) = Lx$. Obviously we get Lw from L by rational operations for any given word $w \in X^*$. The subalgebra generated by a language L is the language LX^*. It is now clear that all languages of the form

$$(*) \qquad L = Q \cup \bigcup_{i=1}^{n} R_{i1}X^*R_{i2}X^*\dots X^*R_{ik_i} ,$$

where $n \geq 0$, $k_i \geq 2$ $(i=1,\dots,n)$, and the languages Q and R_{ij} are finite, are monadically rational. By verifying that the languages of the form (*) include all finite X-languages (put n=0), and that their class is closed under the rational operations of $\mathcal{X}$, we get

Proposition 5.1. The monadically rational X-languages are exactly the languages of the form (*).

It so happens that I considered these languages quite a few years ago (in [10]) for entirely different reasons as generalizations of definite languages.

References

[1] BERSTEL, J., Transductions and context-free languages. - B. G. Teubner, Stuttart (1979).

[2] COHN, P.M., Universal algebra. - Harper&Row, New York (1965).

[3] EILENBERG, S., Automata, languages, and machines. Vol. A. - Academic Press, New York (1974).

[4] EILENBERG, S. and J.B. WRIGHT, Automata in general algebras. - Information and Control 11 (1967), 452-470.

[5] GÉCSEG, F. and M. STEINBY, The theory of tree automata. - Submitted for publication (1981).

[6] GRÄTZER, G., Universal algebra, 2. ed.. - Springer Verlag, New York (1979).

[7] LESCANNE, P., Équivalence entre la famille des ensembles algébriques. - RAIRO Inform. Théor. Sér. Rouge 10 (1976), no.8, 57-81.

[8] MEZEI, J. and J.B. WRIGHT, Algebraic automata and context-free sets. - Information and Control 11 (1967), 3-29.

[9] RABIN, M.O. and D. SCOTT, Finite automata and their decision problems. - IBM J. Res. Develop. 3 (1959), 114-125.

[10] STEINBY, M., On definite automata and related systems. - Ann. Acad. Sc. Fenn., Ser. A I, 444 (1969).

[11] STEINBY, M., Syntactic algebras and varieties of recognizable sets. - Les Arbres en Algebre et en Programmation, Proc. 4éme Coll. de Lille, Lille (1979), 226-240.

[12] THATCHER, J.W. and J.B. WRIGHT, Generalized finite automata theory with an application to a decision problem of second order logic. - Math. Systems Theory 2 (1968), 57-81.

[13] TRNKOVÁ, V. and J. ADÁMEK, On languages, accepted by machines in the category of sets. - Math. Found. Comput. Sci., Proc. 6th Symp., Tatranská Lomnica 1977, Lect. Notes Comput. Sci 53 (1977), 523-531.

[14] TRNKOVÁ, V. and J. ADÁMEK, Tree-group automata. - Fundamentals of computation theory '79, Proc. Conf., Berlin/Wendisch-Rietz 1979, (1979),462-468.

[15] WERNER, H., Einführung in die allgemeine Algebra. - Bibliographisches Institut, Mannheim (1978).

PEBBLING AND BANDWIDTH

Ivan Hal Sudborough†

Department of Electrical Engineering and Computer Science

Northwestern University

Evanston, Illinois 60201

U.S.A.

ABSTRACT

It is shown that a graph with n vertices and bandwidth k requires at most min($2k^2$+k+1,2klogn) pebbles. Furthermore, the pebble problem restricted to and/or graphs of bandwidth f(n) is in NSPACE(f(n)×$\log^2$n) and is log space hard for the class NTISP(poly,f(n)). (NTISP(poly,f(n)) denotes the class of sets accepted by nondeterministic Turing machines in polynomial time and simultaneous f(n) space.)

INTRODUCTION

We consider two pebble games played on directed acyclic graphs (DAGs) and on directed acyclic and/or graphs. The *black pebble game* has been used to model register allocation problems [14], to study flowcharts and recursive schemata [13], and to analyze the relative power of time and space as Turing machine resources [2,6]. The *black-white pebble game* has been used to obtain the lower bound $\Omega(n^{1/4})$ on the space complexity of the solvable path system problem, which is log space complete for deterministic polynomial time [2], on a special computational model [3].

The black-white pebble game is played on a DAG G, with the objective to place a black pebble on a special designated vertex of G, called the *goal*, and then to remove all of the pebbles from G, by the following rules:

RULE 1: One may place a white pebble at any time on any vertex (which does not contain a pebble).

RULE 2: One may remove a black pebble at any time from any vertex.

RULE 3: If all predecessors of a vertex x contain a black or white pebble and x contains no pebble, then one may place a black pebble on x.

RULE 4: If all predecessors of a vertex x contain a black or white pebble and x contains a white pebble, then one may remove the white pebble from x.

RULE 3a: (*sliding rule*) If all predecessors of a vertex x contain a black or white pebble and x contains no pebble, then one may move a black pebble from a predecessor of x to x.

RULE 4a: (*sliding rule*) If all predecessors of a vertex x except one, say vertex y, contain black or white pebbles and x contains a white pebble, then one may move the white

† Work partially supported by NSF grant #MCS-79-08919

pebble from x to y.

In papers by van Emde Boas and van Leeuwen and by Gilbert, Lengauer, and Tarjan [4 ,5], it is shown that the sliding rules 3a and 4a decrease the number of pebbles needed by at most one. The *black pebble game* is the special case of the black-white pebble game in which only black pebbles are used. In both games the objective is to place a black pebble on the goal node and then remove all the pebbles by a sequence of steps that minimizes the number of pebbles on the graph at any one time. The minimum number of pebbles needed in the black-white pebble game for a DAG G and goal node t is denoted by $demand_{b/w}(G,t)$. The minimum number of pebbles needed in the black pebble game is denoted by $demand_b(G,t)$. The best result known concerning the relationship between pebble demand in the black-white pebble game and pebble demand in the black pebble game is given by the following [10]:

Theorem (Meyer auf der Heide): Let G be a DAG and t be a designated vertex. Let $k=demand_{b/w}(G,t)$, then $demand_b(G,t)\leq\frac{1}{2}(k^2+k)+1$.

Let S_m be the pyramid with m levels. (The pyramid S_5 is shown in Figure 1.) Let r be the *apex* of the pyramid. It is known that $demand_b(S_m,r)=m+1$ [2] and $(\frac{1}{2}m)^{\frac{1}{2}}-1 \leq demand_{b/w}(S_m,r) \leq [\frac{1}{2}m]+2$ [3 ,10]. It follows that black-white pebbling is known to save one half the number of pebbles needed for pebbling pyramids using the black pebble game. For the complete k-ary tree with depth n and root r, denoted by T_n^k, it is known that $demand_b(T_n^k,r)=(n-1)(k-1)+k+1$ and $demand_{b/w}(T_n^k,r)=[\frac{1}{2}(k-1)n+k+1]+1$ [9]. For arbitrary trees T with root r, it is known that $demand_{b/w}(T,r)\geq\frac{1}{2}demand_b(T,r)$ [7].

The pebble game may also be played on *and/or graphs*. Let G=(V,E) be a DAG and *label*:V $\longrightarrow$ {*and*,*or*}. (And/or graphs have well-known applications in game theory and artificial intelligence [16].) To play the pebble game on an and/or graph we use rules 3,4,3a, and 4a in the case that a vertex x is such that *label*(x)=*and*. On the other hand, if the vertex x is such that *label*(x)=*or*, then the clause " If all predecessors of a vertex x ... " in the rules 3,4,3a, and 4a is replaced by the clause " If at least one predecessor of a vertex x ... ". That is, if a vertex x has two predecessors y and z and *label*(x)=*or*, then a black pebble may be placed on x if a pebble has already been placed on either vertex y or vertex z.

The *pebble problem* is to determine, for a given DAG G, designated vertex t, and positive integer k, if $demand_b(G,t)\leq k$. The *pebble problem for and/or graphs* is defined analagously. It is known that the pebble problem and the pebble problem for and/or graphs are log space complete for polynomial space [5 , 8].

We consider the relationship between the number of pebbles needed to pebble a graph and the bandwidth of the graph. It is shown that, if G is a DAG with n vertices and bandwidth k and t is a designated vertex, then $demand_{b/w}(G,t)\leq 2k$ and $demand_b(G,t)\leq 2k\log n$. It follows from the theorem of Meyer auf der Heide that, for graphs G of n vertices and bandwidth k, $demand_b(G,t)\leq\min(2k^2+k+1,2k\log n)$. Thus, for example, a graph of 2^{10} vertices and bandwidth 3 requires at most 22 black pebbles and a graph of 2^{10} vertices and bandwidth 12 requires at most 240 black pebbles. It is also shown that the pebble problem restricted to graphs and and/or graphs of bandwidth f(n) is in NSPACE($f(n) \times \log^2 n$) and that the pebble problem for and/or graphs of bandwidth f(n) is log space hard for the class NTISP(poly,f(n)). It should, perhaps, be noted that bandwidth restricted versions of problems complete for NSPACE(log n), P, and NP have appeared in the literature and that, in general, placing a bandwidth restriction on a computational problem has been shown in previous work to correspond to placing a space restriction of the same size on a Turing machine [1 ,11,12,15].

PEBBLE DEMAND AND BANDWIDTH

It can be shown that 3k-1 black pebbles are necessary for pebbling DAGs of bandwidth k. Graphs of bandwidth 2 and 3 requiring 5 black pebbles and 8 black pebbles, respectively, are described in Figures 2 and 3 (in the Appendix). Thus, a lower bound of 3k-1 can be established for black pebble demand in graphs of bandwidth k. For an upper bound on the pebble demand in the black pebble game on graphs of bandwidth k, we consider the following algorithms for pebbling.

Let G be a DAG of bandwidth k under a layout ℓ:vertices(G)$\xrightarrow{1-1}\{1,\ldots,|G|\}$. The following algorithm places black and white pebbles on the vertices of G:

Step 1. Place white pebbles on all of the vertices $\ell^{-1}(1)$, ... ,$\ell^{-1}(2k)$.

Step 2. Remove the white pebbles one at a time from the vertices $\ell^{-1}(1)$, ... ,$\ell^{-1}(k)$ and place black pebbles on these vertices instead.

Step 3. For i=1 step 1 until n-2k do
begin
if $\ell^{-1}(i+2k)$ is a predecessor of $\ell^{-1}(i+k)$ then move the white pebble from $\ell^{-1}(i+k)$ to $\ell^{-1}(i+2k)$ else begin remove the white pebble from $\ell^{-1}(i+k)$ and place a white pebble on $\ell^{-1}(i+2k)$ end
if $\ell^{-1}(i)$ is a predecessor of $\ell^{-1}(i+k)$ then move the black pebble from $\ell^{-1}(i)$ to $\ell^{-1}(i+k)$ else begin remove the black pebble from $\ell^{-1}(i)$ and place a black pebble on $\ell^{-1}(i+k)$ end
end

Step 4. Remove the white pebbles one at a time from the vertices $\ell^{-1}(n-k+1)$, ... , $\ell^{-1}(n)$ and place black pebbles on these vertices instead. Then, remove all of the black pebbles from G.

It is straightforward to verify that the algorithm describes a correct strategy for placing black and white pebbles on G, if G has bandwidth k. Since the algorithm never places more than 2k pebbles on the graph, it follows that G has pebble demand at most 2k in the black-white pebble game. It follows that G has pebble demand $2k^2+k+1$ in the black pebble game, by the theorem of Meyer auf der Heide. An alternative divide-and-conquer algorithm for placing pebbles on the graph G in the black pebble game is now described.

Let B(j,k) denote the *block of k vertices starting with the vertex* $\ell^{-1}(j)$ under the layout ℓ, *i.e.* $B(j,k)=\{y\in \text{vertices}(G) \mid j\le\ell(y)\le j+k-1\}$. The algorithm works by assigning itself the initial goal of pebbling all of the vertices in a given block

B(j,k). Such a block B(j,k) divides the vertices of G into two sets: (1) the vertices assigned integers less than j by the layout ℓ, which we call the set of vertices to the left of B(j,k), and (2) the vertices assigned integers greater than j+k-1, which we call the set of vertices to the right of B(j,k). The division of the vertices is worth noting, since the graph G has bandwidth k under the layout ℓ and, therefore, no edge in G connects a vertex on the left of B(j,k) with a vertex on the right of B(j,k). That is, the graph G has bandwidth k and B(j,k) is a block of k consecutive vertices under ℓ, so no edge can connect vertices on opposite sides.

Let x be a vertex in B(j,k) that can, given a particular configuration of pebbles sitting on the vertices of G, receive a pebble by a sequence $s_1,s_2,\ldots,s_t$ of pebble game moves in which no other vertex in B(j,k) receives a pebble. Clearly, such a sequence of steps must exist, since all vertices in G can be pebbled. Let y be a vertex on the right (left) of B(j,k) that receives a pebble during this sequence of pebble game moves, say at step s_u. Then, the subsequence obtained by deleting all moves from $s_1,s_2,\ldots,s_u$ that add a pebble to a vertex on the left (right) of B(j,k) forms a valid sequence of black pebble game moves and results in a pebble being placed on y by a sequence of moves that add pebbles only to vertices to the right (left) of B(j,k). That this subsequence forms a valid sequence of pebble game moves follows from the observation that adding pebbles to the left (right) of B(j,k) cannot help add pebbles to the right (left) of B(j,k) except by first helping to add a pebble to the block B(j,k), since the graph G has bandwidth k. In this way, we can divide the task of placing a pebble on a particular vertex x into two half size tasks of placing a pebble on a vertex to the left of B(j,k) and a pebble on a vertex to the right of B(j,k).

Let PEBBLE(i,m,j,λ) be a call to a recursive procedure called PEBBLE which, given a particular arrangement of pebbles on the vertices of G, will pebble all of the vertices in B(j,k) that are possible to pebble by a sequence of pebble game steps that never adds pebbles to vertices z such that $\ell(z)<i$ or $\ell(z)>m+k-1$. (It should be noted that, in general, there will be vertices in B(j,k) that cannot be pebbled by <u>any</u> sequence of steps that does not add pebbles to vertices z such that $\ell(z)<i$ or $\ell(z)>m+k-1$.) A pebble placed on G by this procedure PEBBLE will be called a λ-pebble, if it is placed on G during a call to the procedure PEBBLE with the last argument having the value λ, but excluding the times during which there is a recursive call to the procedure with the last argument having a value greater than λ.

<u>Procedure</u> PEBBLE(i,m,j,λ)
<u>if</u> $|i-m|\le 2k$ <u>then</u> place λ-pebbles on all of the vertices in B(j,k) that are possible to pebble by a sequence of steps in the black pebble game which adds no new pebbles to vertices z such that $\ell(z)<i$ or $\ell(z)>m+k-1$;
<u>else</u> <u>begin</u>
place λ-pebbles on all vertices in B(j,k) that have no predecessors;
L ← (i+j)/2;
R ← (j+k+m)/2;

```
FLAG ← 1;
while FLAG=1 do
      begin
      FLAG ←  0;
      PEBBLE(i,j,L,λ+1);
      PEBBLE(j,m,R,λ+1);
      PEBBLE(L,R,j,λ);
      if a λ-pebble has been added to B(j,k) in the last step
            then FLAG ← 1;
      remove all (λ+1)-pebbles from G
      end
end
```

The procedure PEBBLE works in the following way with arguments i,m,j, and λ. We assume that j is midway between i and m. When PEBBLE(i,m,j,λ) is called some of the nodes in G may already be pebbled. In particular, some of the vertices in B(i,k) and B(j,k) may have been previously pebbled. The purpose of PEBBLE(i,m,j,λ) is to add pebbles to B(j,k) by steps in the pebble game that do not add pebbles to the left of B(i,k) or to the right of B(m,k). Let us assume that $|i-m|\geq 2k$. Then PEBBLE(i,m,j,λ) works as follows. It initially adds pebbles to all of the vertices in B(j,k) that have no predecessors (unless these vertices already have pebbles). The values L and R are chosen to be midway between i and j and midway between j and m, respectively. The algorithm then (1) pebbles all of the vertices in B(L,k) that are possible to pebble by steps in the pebble game that do not add new pebbles to vertices to the left of B(i,k) or to the right of B(j,k) , and (2) pebbles all of the vertices in B(R,k) that are possible to pebble by steps in the pebble game that do not add new pebbles to vertices to the left of B(j,k) or to the right of B(m,k). We observe that the number of vertices between B(i,k) and B(j,k) and the number of vertices between B(j,k) and B(m,k) are at most ½ the number of vertices between B(i,k) and B(m,k). Once pebbles have been placed on B(L,k) and B(R,k) by the calls PEBBLE(i,j,L,λ+1) and PEBBLE(j,m,R,λ+1), the algorithm returns to the task of pebbling the vertices in B(j,k). That is, it adds pebbles to all of the vertices in B(j,k) that can be pebbled by a sequence of moves in the pebble game that do not add new pebbles to vertices to the left of B(L,k) or to the right of B(R,k). This is done by the recursive call to the procedure with PEBBLE(L,R,j,λ). Again, we may observe that the number of vertices between B(L,k) and B(R,k) is one half of the number of vertices between B(i,k) and B(m,k). Thus, each of the recursive calls to the procedure PEBBLE are for a portion of the graph that is one half the size of the portion between B(i,k) and B(m,k).

The procedure PEBBLE can be used to pebble all of the vertices of G by initially using it to pebble the vertices in the block of k vertices in the middle of G under the layout ℓ. That is, initially we call the procedure by PEBBLE(1,n-k,[n/2],1). The result is that all of the vertices in the block B([n/2],k) are pebbled. This block of k vertices, B(j,k), divides the graph into two equal size pieces. The vertices to the left and right of B(j,k) can be pebbled by calling the procedure recursively. Furthermore, it is straightforward to verify that the algorithm places at most 2klog n pebbles on G at any one time. That is, at most 2k pebbles are added at any level of the recursive process. Since each time the size of the graph considered is decreased by one half, the depth of recursion is log n, where n is the number of vertices in G. Thus, altogether 2klog n pebbles are used by this procedure to place pebbles on G. Combining the results of the two algorithms, we obtain:

Theorem. Let G be a DAG with n vertices and bandwidth k and let t be an arbitrary vertex of G. Then, $\mathit{demand}_b(G,t) \leq \min(2k^2+k+1, 2k\log n)$.

THE PEBBLE PROBLEM FOR AND/OR GRAPHS OF BANDWIDTH f(n)

We consider the pebble problem for and/or graphs restricted to and/or graphs which have bandwidth f(n) under the layout implicit in their representation. This problem is represented by AND/OR PEBBLE(f(n))

AND/OR PEBBLE(f(n)):

Input: An encoding of a DAG G=(V,E), a labeling function *label*:V → {*and*,*or*}, a designated goal node t, and a positive integer k.

Property: (1) The graph G has bandwidth f(n), where n is the number of vertices in G, under the layout implicit in the encoding of G, and (2) A pebble can be placed on t by playing the pebble game on the and/or graph G by a sequence of steps in which at most k pebbles are ever placed on the vertices of G.

The algorithm presented in the last section, which shows that 2klog n pebbles are sufficient to pebble a DAG with bandwidth k and n vertices, is easily seen to provide the same upper bound on the pebble demand for and/or graphs of bandwidth k. That is, at most O(f(n)×log n) pebbles are required to pebble an and/or graph of bandwidth f(n). It follows that one can build a nondeterministic algorithm to play the pebble game on an and/or graph which uses at most $f(n)\times\log^2 n$ space. The algorithm simply guesses at each step which move to perform. Since f(n)×log n is an upper bound on the number of pebbles and log n space is sufficient to represent the index of the node each pebble is placed on, the algorithm needs at most $f(n)\times\log^2 n$ space to represent the position of all of the pebbles. Thus, we have:

Theorem. Let f be some function on the natural numbers which is space constructible. Then, AND/OR PEBBLE(f(n))$\in$ $\mathbb{N}$SPACE$(f(n)\times\log^2 n)$.

Thus, AND/OR PEBBLE(log n) is in $\mathbb{N}$SPACE$(\log^3 n)$, for example. Is it true that AND/OR PEBBLE(log n) is in $\mathbb{D}$SPACE(log n)? In fact, we will show that AND/OR PEBBLE(f(n)) is log space hard for the class $\mathbb{N}$TISP(poly,f(n)). It follows that AND/OR PEBBLE(log n) is log space hard for $\mathbb{N}$SPACE(log n) and, therefore, is in $\mathbb{D}$SPACE(log n) only if $\mathbb{N}$SPACE(log n)=$\mathbb{D}$SPACE(log n). In order to show that AND/OR PEBBLE(f(n)) is log space hard for $\mathbb{N}$TISP(poly,f(n)) we may reduce any known $\mathbb{N}$TISP(poly,f(n))-hard problem to this problem. We choose to reduce the 3-SATISFIABILITY problem restricted to well-formed formula of bandwidth f(n), denoted by 3SAT(f(n)) [12]:

3SAT(f(n)):

Input: A well-formed formula $w=C_1C_2\ldots C_m$ in 3CNF.

Property: (1) if a positive of negative instance of a variable occurs in clauses C_i and C_j, then $|i-j|\leq f(m)$, *i.e.* w has *bandwidth* at most f(m), and (2) there is a truth assignment to the variables of w such that every clause is true, *i.e.* w is satisfiable.

Since 3SAT(f(n)) is known to be log space complete for NTISP(poly,f(n)) [12], it follows that, by showing 3SAT(f(n)) $\leq_{\log}$ AND/OR PEBBLE(f(n)), we will show that the latter problem is log space hard for NTISP(poly,f(n)). In fact, we show that, for every well-formed formula (wff) w, we can construct an and/or graph G_w with a goal vertex t and a positive integer k (as a bound on the number of pebbles) such that: w is satisfiable if and only if a pebble can be placed on the vertex t by playing the pebble game on the and/or graph G using at most k pebbles. Furthermore, the and/or graph G_w constructed is shown to have bandwidth at most some fixed constant c times the bandwidth of w. That is, there is a *bandwidth preserving reduction* from 3SAT to AND/OR PEBBLE. The reduction described is derived from an earlier reduction from 3QBF to the pebble problem described by Gilbert, Lengauer, and Tarjan [5]. It is, of course, the case that 3SAT is just the restriction of QBF to the situation where all quantifiers are existential. Thus, our contribution is to modify this earlier reduction in such a way that it becomes bandwidth preserving.

An important building block in the construction is the pyramid S_m. (Again, S_5 is shown in Figure 1 of the Appendix. We shall abbreviate the pyramid S_m with a triangle with the number m written inside, as shown in the figure.) Cook [2] has shown that the apex of the pyramid S_m requires m+1 pebbles to pebble. The following observation is from Gilbert, Lengauer, and Tarjan [5]:

> ... *A pyramid can be used to lock a pebble on a given vertex for a given time interval. This is done by making the vertex the apex of a pyramid which is so large that in order to repebble the vertex, so many pebbles have to be taken off the graph for use on the pyramid that the results achieved after the vertex was first pebbled are lost... Note also that if any source of a k-pyramid contains a pebble that cannot be moved, then the apex can be pebbled with k-1 additional pebbles ...*

Let $w=C_1C_2\ldots C_m$ be a wff in 3CNF with bandwidth k. Let x be a variable that occurs in w. Let C_i be the first clause of w containing either a positive or negative occurrence of x and let C_j be the last clause of w containing either a positive or negative occurrence of x. Let [i,j] denote the set of positive integers {i,...,j}. Then, the domain of the variable x, denoted by *domain*(x), is the set [i,j]. A variable x is *active* at clause C_ℓ, if $\ell\epsilon$*domain*(x). There can be at most 3k active variables at any clause, since the wff w has bandwidth k and is in 3CNF. Let $x_1,x_2,\ldots,x_n$ be an enumeration of all the variables in w such that, if i<j, then either (a) *domain*$(x_i)=[s_i,t_i]$ and *domain*$(x_j)=[s_j,t_j]$ with $s_i<s_j$, or (b) *domain*$(x_i)=[s_i,t_i]$ and *domain*$(x_j)=[s_j,t_j]$ with $s_i=s_j$ and the positive or negative occurrence of the variable x_i occurs in clause C_{s_i} before the positive or negative occurrence of the variable x_j. For any

$i \geq 1$, define the successor of a variable x_i in the wff w, which we shall denote by SUCCESSOR(x_i), by: SUCCESSOR(x_i)=x_j, if, for all k ($i<k<j$), either (a) *domain*(x_i) $\cap$ *domain*(x_k) $\neq \phi$, or (b) $\exists \ell < i$ such that SUCCESSOR(x_ℓ)=x_k.

The and/or graph G_w will be constructed with p components, where p is the maximum number of active variable at any clause of w, corresponding to the variables of the wff w. G_w will also have m components corresponding to the clauses of the wff w. To represent a truth assignment inside the component corresponding to variables of the wff, there will be four vertices for each variable of w. Two pebbles placed on the subgraph induced by these four vertices represent a truth value as indicated in Figures 4(a)-(d). Figure 5 illustrates the manner in which subgraphs corresponding to separate variables x, y, and z, such that y=SUCCESSOR(x) and z=SUCCESSOR(y), are connected. The way that this structure works is as follows. Two pebbles are placed on the subgraph corresponding to the variable x to represent a truth assignment. These two pebbles remain there during the time period that pebbles are being placed on vertices in the components corresponding to clauses at which x is active. When x is no longer active, the two pebbles are moved upward to the portion of the graph corresponding to the variable y. Since y=SUCCESSOR(x) and, therefore, y becomes active only after x stops being active, the truth assignment for x is no longer needed in the wff when the two pebbles are shifted upwards. The purpose of the vertex D_x, shown in Figure 5, is as follows. Its purpose is to force at least one of the pebbles used to represent the truth value of the variable x to remain on the vertices x' and $\bar{x}$' during the time that x is active. That is, if both pebbles are moved to the vertices x and $\bar{x}$, then neither can be moved upwards in the structure, as there is no pebble on the vertex D_x. On the other hand, if one of the two pebbles remains on x' or $\bar{x}$', then a new pebble can be moved to D_x and, then, both pebbles can be moved upward in the structure.

Figure 6 illustrates the block of vertices corresponding to a clause $C_i=(a_{i1}+a_{i2}+a_{i3})$. After k-3 pebbles are used on the components corresponding to the variables, the remaining three pebbles are available to pebble these clause components. For each literal a_{ij}, $1 \leq j \leq 3$, there is a fixed pebble on vertex a_{ij}, if the literal a_{ij} is true, or on a'_{ij}, if the literal is false. Thus, if $w=C_1C_2 \ldots C_m$ is satisfied, then the clause pyramids can be pebbled in the order given by the wff w with three pebbles; however, if some clause is false, then the apex of the corresponding 4-pyramid cannot be pebbled with fewer than four pebbles, since the 4-pyramid is empty of pebbles.

Figure 7 illustrates the entire construction. It is sufficiently similar to the construction provided by Gilbert, Lengauer, and Tarjan to require very little additional explanation. However, we need to indicate how the graph is to be layed out so that the bandwidth of G_w is at most a fixed constant c times the bandwidth of w.

First, we observe that a k-pyramid can be layed out to have bandwidth k. Such a layout is illustrated in Figure 1. If we chain together a sequence of vertices as shown in Figure 8, each of the vertices requires the same number of pebbles as the first. In this way, a k-pyramid, a (k-1)-pyramid, a (k-2)-pyramid, ... can be layed out in such a way that consecutive vertices are connected by disjoint paths to the apexes of the successive pyramids and the whole structure has bandwidth k. This is illustrated for a 5-pyramid, 4-pyramid, and a 3-pyramid in Figure 8. As previously indicated, there are at most 3k active variables for any clause in w, since w has bandwidth k and is in 3CNF. Thus, there are at most 6k pyramids needed for the components corresponding to variables, shown for example in Figure 7. The largest pyramid is of size 6k+3. As we have seen, all of the pyramids of size 6k+3, 6k+2, 6k+1, ..., 4 can be layed out with bandwidth O(k) in such a way that consecutive vertices are connected to the apexes of the various pyramids. Furthermore, the components corresponding to the clauses can be layed out in the same order as they appear in the wff w. Thus, all of the vertices corresponding to clause C_i receive smaller integers than the vertices corresponding to clause C_{i+1}. Assign the vertices corresponding to a variable x to integers between those assigned to vertices in the components corresponding to C_i and C_{i+1}, if *domain*(x)=[i,j], for some j≥i. The only edges connecting the component corresponding to a variable x with the component corresponding to a clause C occur when the variable x occurs in clause C. Thus, under this layout the graph G_w has bandwidth at most a fixed constant c times the bandwidth of w.

It follows that 3SAT is reducible to AND/OR PEBBLE by a bandwidth preserving reduction. Thus,

Theorem. AND/OR PEBBLE(f(n)) is log space hard for NTISP(poly,f(n)).

It would be interesting to improve this result by showing that the pebble problem itself, and not for and/or graphs, is log space hard for NTISP(poly,f(n)) under the f(n) bandwidth constraint. It would also be interesting to improve on the upper or lower bounds given for the number of pebbles required for graphs of bandwidth k.

REFERENCES:

1. M-J Chung, W.M. Evangelist, and I.H. Sudborough, Some additional examples of bandwidth constrained NP-complete problems, *Proc. 15th Conf. on Info. Sciences and Systems* (1981), The Johns Hopkins University, Baltimore, Md., U.S.A., to appear.
2. S.A. Cook, An observation on time-storage trade-off, *J. Comput. System Sci.* (1974), pp. 308-316.
3. S.A. Cook and R. Sethi, Storage requirements for determinstic polynomial time recognizable languages, *J. Comput. System Sci.* 13 (1976), pp. 25-37.
4. P. van Emde Boas and J. van Leeuwen, Move rules and trade-offs in the pebble game, Proc. 4th GI Conf., Springer *Lecture Notes in Computer Sci.* 67 (1979), pp. 101-112.
5. J.R. Gilbert, T. Lengauer, and R.E. Tarjan, The pebbling problem is complete in polynomial space, *SIAM J. Comput.* 9,3 (1980), pp. 513-524.
6. J.E. Hopcroft, W. Paul, and L. Valiant, On time versus space, *J. Assoc. Comput. Mach.* 24 (1977), pp. 332-337.
7. T. Lengauer and R.E. Tarjan, The space complexity of pebble games on trees, Preprint, Stanford University (1979).
8. A. Lingas, A PSPACE complete problem related to a pebble game, Proc. 1978 ICALP, Springer *Lecture Notes in Computer Sci.* 62, pp. 300-321.
9. M.C. Loui, The space complexity of two pebble games on trees, TM-133, Laboratory for Computer Sci., Massachusetts Inst. of Technology, Cambridge, Mass., U.S.A. (1979)

10. F. Meyer auf der Heide, A comparison of two variations of a pebble game on graphs, *Theoretical Computer Sci.* 13 (1981), pp. 315-322.

11. B. Monien and I.H. Sudborough, On eliminating nondeterminism from Turing machines which use less than logarithm worktape space, Proc. 1979 ICALP, Springer *Lecture Notes in Computer Sci.* 72, pp. 431-445.

12. B. Monien and I.H. Sudborough, Bandwidth constrained NP-complete problems, *Proc. 13th Annual ACM Theory of Computing Symp.* (1981), to appear.

13. M.S. Paterson and C.E. Hewitt, Comparative schematology, *Record of Project MAC Conf. on Concurrent Systems and Parallel Computation*, Cambridge, MA, U.S.A., pp. 119-128.

14. R. Sethi, Complete register allocation problems, *SIAM J. Comput.* 4 (1975), pp. 226-

15. I.H. Sudborough, Efficient algorithms for path system problems and applications to alternating and time-space complexity classes, *Proc. 21st Annual Symp. on Foundations of Computer Sci.* (1980), IEEE Computer Society, Long Beach, Calif., pp. 62-7

16. P.H. Winston, *Artificial Intelligence*, Addison-Wesley Publishers (1977).

APPENDIX

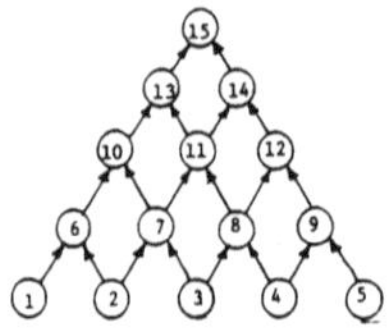

Figure 1. A 5-Pyramid

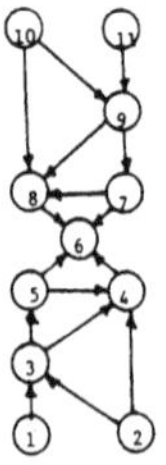

Figure 2. A graph with bandwidth 2 that requires 5 pebbles.

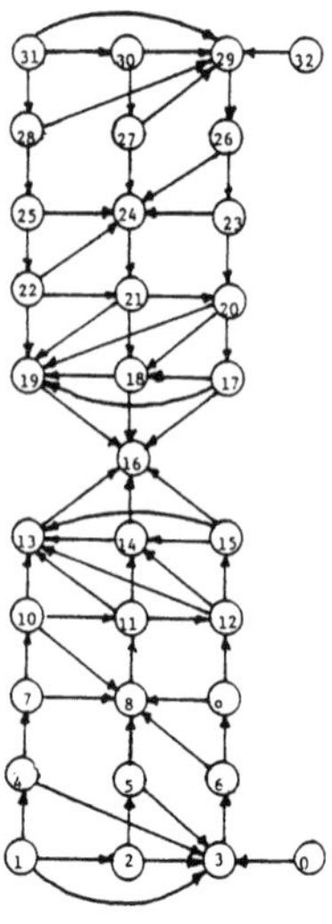

Figure 3. A graph with bandwidth three that requires 8 pebbles.

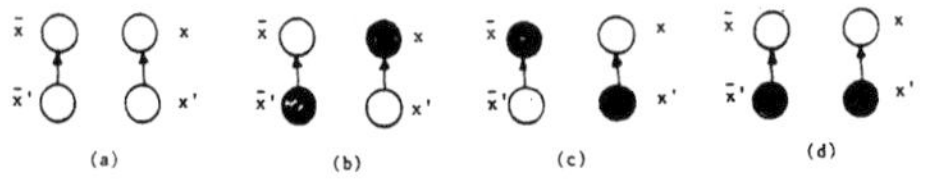

Figure 4 (a) vertices representing a variable; (b) true configuration; (c) false configuration; (d) double false configuration

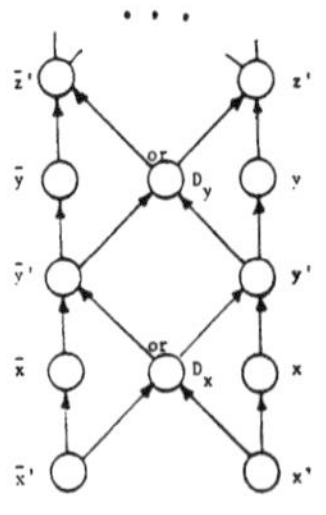

Figure 5. Portion of component corresponding to variables x, y, and z, where y=SUCCESSOR(x) and z=SUCCESSOR(y).

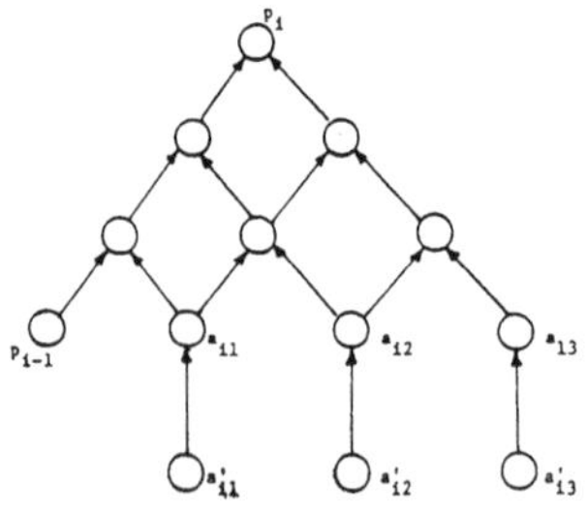

Figure 6. Block of vertices for clause $a_{i1}+a_{i2}+a_{i3}$. Note that the vertices a_{ik} and a'_{ik} occur among the variable blocks. Vertex p_{i-1} is part of the i-1-st clause block; p_0 is the vertex b, which has no predecessors.

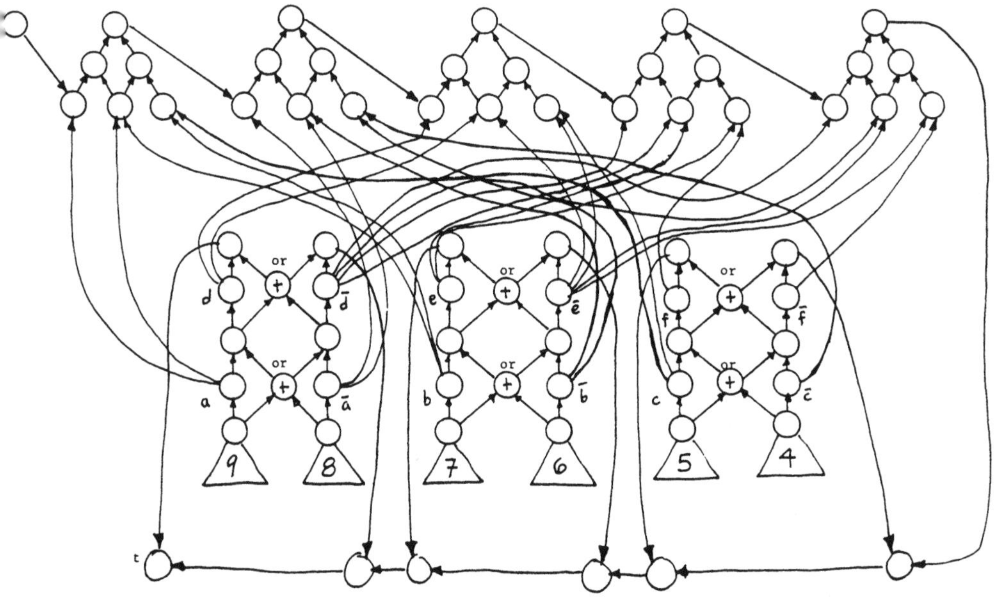

Figure 7. The and/or graph G_w constructed from the wff $w=(a+b+c)(\bar{a}+\bar{b}+\bar{c})(d+\bar{e}+c)(\bar{d}+e+f)(\bar{d}+\bar{e}+\bar{f})$. One is allowed 9 pebbles to place a pebble on the goal node t.

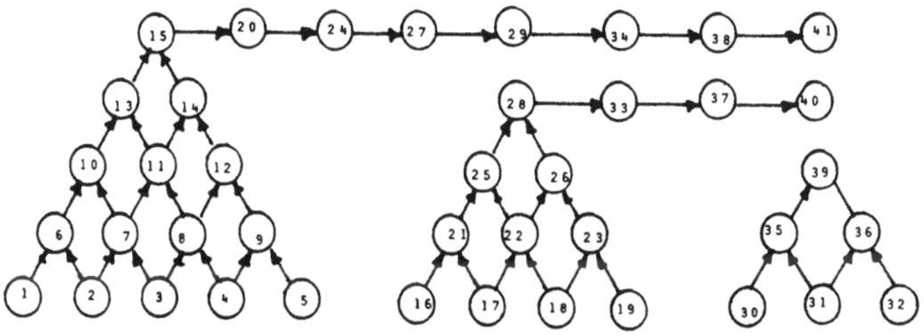

Figure 8. A layout of a 5-pyramid, a 4-pyramid, and a 3-pyramid with bandwidth five so that vertices connected to the apexes of the pyramids are consecutive vertices.

ON CELLULAR GRAPH-AUTOMATA AND SECOND-ORDER DEFINABLE GRAPH-PROPERTIES

Gy. Turán
Research Group on the Theory of Automata
Somogyi u. 7.
H-6720, Szeged

Introduction

Cellular graph-automata model local or "myopic" algorithms on graphs. A cellular graph-automaton consists of a graph and identical copies of a finite automaton placed on each vertex, with different automata communicating only through the edges of the graph. Properties of graphs recognizable by these automata were studied e.g. by Rosenstiehl-Fiksel-Holliger [4] and Wu-Rosenfeld [6].

In this paper we try to find general classes of recognizable properties. A broad definability class is considered, the class of subgraph-definable properties i.e. properties second-order definable with second-order quantifiers restricted to subgraphs of the given graph. This class contains many "natural" graph-properties. It is shown that the class of subgraph-definable properties is properly included in the class of properties recognizable by cellular graph-automata. (In order to prove this we demonstrate that the existence of an automorphism in a graph is not subgraph-definable - this is an example of second-order undefinability.)

Another problem considered is the problem of identification of a center in a graph. It is shown that there exists a finite automaton that can identify a center in every graph where this is at all possible. Some related problems are mentioned in Angluin [1]. (As automata considered here have "centered initial configuration" (see below) this problem is different from identifying the center from "uniform initial configuration" [1].)

The basic cellular graph-automaton model has several interpretations. Here graphs of bounded degree are considered with identical, deterministic and synchronized automata and with exactly one automaton being in a distinguished initial state. Besides determinism it is mainly this last assumption (mentioned above as "centered initial configuration") that is of importance from the point of view of the computing power of these automata (concerning some other possibilities see

Angluin [1]).

In Section 1. the necessary definitions are given. Section 2. contains some constructions needed later and an example of cellular graph-automata recognizing graph-properties, an efficient automaton for planarity (mentioned as a problem in Wu-Rosenfeld [6]). In Section 3. we compare the classes referred to above and state some other facts about recognizable properties and definiability classes. Finally in Section 4. we deal with the identification problem.

Acknowledgement. I am grateful to Prof. L. Lovász and to Prof. F. Gécseg for their support.

1. Definitions

In this section we outline the necessary definitions concerning cellular graph-automata, second-order formulas and the Fraissé-Ehrenfeucht game.

Before giving the exact definition of a cellular graph-automaton we give a short informal description. A cellular graph-automaton consists of

a) a graph G (undirected, connected, without loops and multiple edges and with all degrees $\leq d$) and for every vertex v an ordering of the edges incident with v (a "compass");

b) a finite automaton A (with disjoint sets of states for the unique distinguished vertex and for "ordinary" vertices, with disjoint sets of states for automata on vertices of degree i for each $1 \leq i \leq d$, and with vectors of dimension $\leq d$ of states as input alphabet, plus the transition function satisfying some compatibility and invariance conditions);

c) an assignment of starting states to every vertex of G consistent with the degrees of vertices s.t. exactly one vertex gets a distingushed starting state.

Configurations, computation sequences, accepting and rejecting computations are defined as usual.

Definition 1. A graph G=(V,E) is meant to be (unless explicitly stated otherwise) undirected, connected, without loops and multiple edges. A d-graph is a graph with all degrees $\leq$ d. A compass for G is a set $=\{\varrho(v): v \in V\}$ where $\varrho(v)$ is an ordering $(e_{i_1},\dots,e_{i_{d(v)}})$ of the edges of G incident with v. A basic graph $G^*=(G,\varrho)$ is a 2-tuple where G is a d-graph and ρ is a compass for G.

Definition 2. A basic automaton A is a 6-tuple $A=(Q,\Sigma,S,A,R,\delta)$, where Q is the set of states, Σ is the input alphabet, S,A,R are the sets of starting, accepting and rejecting states respectively and δ is the transition function s.t. the following holds:

(1) $Q=\bigcup_{i=1}^{2}\bigcup_{j=1}^{d} Q_{ij}$, $S=\bigcup_{i=1}^{2}\bigcup_{j=1}^{d} S_{ij}$, $A=\bigcup_{i=1}^{2}\bigcup_{j=1}^{d} A_{ij}$, $R=\bigcup_{i=1}^{2}\bigcup_{j=1}^{d} R_{ij}$

(where union means disjoint union);

$S_{ij}\cup A_{ij}\cup R_{ij} \subseteq Q_{ij}$ and S_{ij}, A_{ij}, R_{ij} are pairwise disjoint for $1\leq i\leq 2$, $1\leq j\leq d$;

(2) $\Sigma=\bigcup_{j=1}^{d} Q^j$

(3) $\delta=\{\delta_{ij}: 1\leq i\leq 2,\ 1\leq j\leq d\}$, $\delta_{ij}: Q_{ij}\times Q^j \to Q_{ij}$

(4) $s_0\in S_{2,j}$, $s_1,\dots,s_j\in S \Rightarrow \delta(s_0,(s_1,\dots,s_j))=s_0$ for $1\leq j\leq d$;

(5) $q_0\in A_{i,j}\cup R_{i,j}$, $q_1,\dots,q_j\in Q \Rightarrow \delta(q_0,(q_1,\dots,q_j))=q_0$

for $1\leq i\leq 2$, $1\leq j\leq d$.

We remark that (4), (5) are usual stability conditions for cellular automata.

Definition 3. A cellular graph-automaton C is a 3-tuple $C=(G^*,A,\mu)$ where $G^*=(G(V,E),\rho)$ is a basic graph, $A=(Q,\Sigma,S,A,R,\delta)$ is a basic automaton and $\mu: V\to S$ is a mapping s.t. for each $v\in V$ if the degree of v is i then $\mu(v)\in S_{1,i}\cup S_{2,i}$ and there is exactly one vertex v_0 s.t. $\mu(v_0)\in \bigcup_{j=1}^{d} S_{1,j}$.

Definition 4. Configurations and computation sequences are defined as usual (if the degree of v is i, then at time t+1 the state $s_{t+1}(v)$ of the automaton on v is $\delta(s_t(v),(s_t(v_1),\dots,s_t(v_i)))$, where $v_1,\dots,v_i$ are the neighbours of v). C is a said to accept if it arrives to a configuration where every automaton is in an accepting state, reject if it arrives to a configuration where every automaton is in a rejecting state.

It can be noted that requiring only the distinguished vertex to accept or reject makes no change in recognizability.

Definition 5. Let P be a property (i.e. a class) of d-graphs. A basic automaton A recognizes P if for every $G \in P$, every $G^* = (G, \rho)$ and every μ the cellular graph-automaton $C = (G^*, A, \mu)$ accepts, and for every $G \notin P$, every $G^* = (G, \rho)$ and every μ the cellular graph-automaton $C = (G^*, A, \mu)$ rejects. P is recognizable if there exists a basic automaton A recognizing P. The class of recognizable properties is denoted by $\mathcal{C}$.

(The parameter d is assumed to be a fixed integer ≥ 3.)

Now we turn to some concepts of logic. The language of graphs contains only one binary relation R(x,y) (and equality). The notion of G being a model of a first-order sentence φ is assumed to be known. A sentence is second-order if it's variables are relations or first-order variables. A second-order sentence is monadic if it's second-order variables are relations of one variable (i.e. sets); dyadic if it's second order variables are relations of at most two variables.

Definition 6. A dyadic sentence φ is weakly dyadic if it's dyadic quantifiers are of the form

$$\exists R_1 (\forall x,y (R_1(x,y) \rightarrow R(x,y)) \wedge \ldots)$$
$$\forall R_1 (\forall x,y (R_1(x,y) \rightarrow R(x,y)) \rightarrow \ldots)$$

(where R is meant to be the adjacency-relation), or, alternatively, if it's dyadic variables when interpreted are restricted to (possibly directed) subgraphs of the given graph G. A property P of d-graphs is subgraph-definable if there exists a weakly dyadic sentence φ s.t. $P = \{G : G \text{ is a d-graph and } G \models \varphi\}$. The class of subgraph-definable d-graph properties is denoted by $\mathcal{S}$.

Thus subgraph-definability is a restricted version of dyadic second-order definability. Properties definable by sentences of the form "there exists a subgraph G_1 such that for every subset V_1 of the vertices..." belong to $\mathcal{S}$. In Section 3. we are going to give some examples of such properties.

The version of the Fraissé-Ehrenfeucht game used here is the following: there are given two graphs G_1, G_2, two players, I and II, and m, the number of moves. In one move, player I selects one of the graphs, and chooses either a vertex (a "vertex move") or a set of vertices (a "set move") or a set of edges (a "subgraph move") of the chosen graph. Player II makes a corresponding choice in the other graph. After m moves consider the vertices chosen during "vertex moves" (their number in the two graphs is equal). They determine a structure containing the subgraph spanned by them, and the restriction of the sets and subgraphs chosen during "set moves" and "subgraph moves". There is a natural correspondence between the two structures on the "chosen vertices". Player II wins if this correspondence is an isomorphism, otherwise player I wins. (For a formal description of a similar Fraissé-Ehrenfeucht game see Ladner [3]). Graphs G_1 and G_2 are m-equivalent ($G_1 \sim_m G_2$) if player II has a winning strategy in the m move game.

2. Some constructions

Consider a cellular graph-automaton $C=(G^*,A,\mu)$. A subgraph G_1 of G is said to be coded in a configuration of C if the automaton of each vertex v stores the adjacencies of v in G_1 (a bounded number of bits).

With this terminology there exists a basic automaton A_0 s.t. for an arbitrary $C=(G^*,A_0,\mu)$ C accepts and in it's final state there is a breadth-first spanning tree of G coded (see Rosenstiehl-Fiksel-Holliger [4] or Wu-Rosenfeld [6]). From this spanning tree and the compass ϱ one gets an order on the vertices of G (e.g. the preorder defined by the tree and ϱ). This gives an order (lexicographically) on the subsets of vertices and on the subsets of edges.

One can construct automata A_1,A_2,A_3 s.t. $C(G^*,A_1,\mu)$ (resp. $C(G^*,A_2,\mu)$, $C(G^*,A_3,\mu)$) always accepts and if vertex v (resp. a subset of vertices V_1, a subset of edges E_1) is coded in it's starting configuration, then in the final configuration of C it is the successor of v (resp. V_1, E_1) that is coded (w.r.t. the corresponding order). We say that these automata enumerate vertices, subsets of vertices and subsets of edges. In the same sense one has

Proposition 1. There exist automata enumerating spanning trees, compasses and connected partitions of vertices (i.e. partitions the subsets of which span connected subgraphs).

The construction of these automata is not complicated and is not detailed here. On the other hand all partitions and permutations cannot be enumerated.

When building cellular graph-automata from simple ones it is useful to be able to connect automata according to a certain program. This is easier if the automata satisfy the synchrony condition, i.e. all vertices reach their final state simultaneously. Every automaton can be modified to satisfy the synchrony condition using solutions of the firing-squad problem [4]. The composition of cellular graph-auto-nata is not formalized here.

An example of an efficient automaton is contained in the following.

Theorem 1. Planarity can be recognized in time $O(n^3)$ by cellular graph-automata. (Here n is the number of vertices.)

Outline of proof: one can adapt the algorithm of Demoucron-Malgrange-Pertuiset [2] that decides planarity of 2-connected graphs. (Finding 2-connected components is solved in Wu-Rosenfeld [6]). Given a subgraph G_1 of a graph G the bridges of G relative to G_1 are the equivalence classes of edges connected by paths not containing vertices of G_1 as internal vertices. Bridges can be constructed in O(n) steps. Deciding whether a bridge can be inscribed into a given face requires also O(n) steps. Counting the number of augmentations necessary gives the bound $O(n^3)$. □

3. On the class $\mathfrak{C}$

Theorem 2.: $\mathfrak{S} \subsetneq \mathfrak{C}$.

Proof:

a) $\mathfrak{S} \subseteq \mathfrak{C}$.

Consider a weak dyadic sentence φ, and

$$G_\varphi = \{G : G \text{ is a d-graph, } G \models \varphi\}.$$

We can suppose φ is in prenex form i.e.

$$\varphi \Leftrightarrow Q_1R_1 \dots Q_kR_k\,\psi$$

where Q_i is $\exists$ or $\forall$, R_i is first-order, monadic or weak dyadic second-order variable and ψ is quantifier-free. The automaton A_φ recognizing G_φ simulates quantifiers by cycling through all possible values of $R_1,\dots,R_k$ (having the R_i coded as described in Section 2.). (Details are omitted.)

b) AUT $\in \mathfrak{C}$.

We define AUT as

$$\text{AUT} = \{G : G \text{ is a d-graph, } G \text{ has a non-trivial automorphism}\}.$$

Consider a graph G=(V,E). A linear order $T \subseteq V \times V$ of the vertices is a tree-order on V if there exists a rooted spanning tree F of G and a compass $\wp$ for G s.t. T is the order defined by the preorder traversal of $(F,\wp)$. In this case T is denoted by $T_{F,\wp}$.

Now let $T_{F,\varrho}$ be a tree-order on V and σ be a non-trivial automorphism of G. Let the edges of F be $\{(v_{i_k}, v_{j_k}): k=1,\dots,n-1\}$ and the root of F be v_0. Then the set of edges

$$\{(\sigma(v_{i_k}), \sigma(v_{j_k})): k=1,\dots,n-1\}$$

forms a spanning tree of G, with root $\sigma(v_0)$ this determines a rooted spanning tree F'. We define the compass ϱ' as follows: for $v \in V$, $e_1, e_2 \in E$ incident with v

$$e_1 <_{\varrho'(v)} e_2 \Leftrightarrow \sigma^{-1}(e_1) <_{\varrho(\sigma^{-1}(v))} \sigma^{-1}(e_2).$$

Now we have a tree-order $T_{F',\varrho'}$ on V. The automorphism σ is the unique order-preserving map from $(V, T_{F,\varrho})$ to $(V, T_{F',\varrho'})$. This shows that G has a non-trivial automorphism defined as the order-preserving map between two tree-orders. Hence to prove b) it suffices to construct an automaton A checking this last condition. Given two tree-orders $T_{F,\varrho}$ and $T_{F',\varrho'}$ the condition can be verified by simulating two pairs of automata "climbing" through the trees and checking simultaneous existence or non-existence of edges. Composing this with an automaton enumerating tree-orders (that exists as remarked in Section 2.) we get b).

c) AUT$\notin \mathfrak{S}$.

To prove this we need some preliminaries, so part c) (completing the proof of Theorem 2.) is stated separately as Lemma 1.. □

The following facts are needed about the Fraissé-Ehrenfeucht game.

<u>Proposition 2.:</u> a) m-equivalence is an equivalence-relation;

b) the number F(m) of m-equivalence classes is at most

$$2^{2^{\cdot^{\cdot^{\cdot^{2^{\sum_{i=0}^{m}\sum_{j=1}^{i} 2^{\binom{j}{2}} (2^{j^2})^{m-i} j!S(i,j)}}}}}} \Big\} m$$

(where S(i,j) is the Stirling-number of the second kind);

c) if $G_1 \underset{m}{\sim} G_2$ and φ is a weak dyadic sentence in prenex form with $\leq m$ quantifiers then

$$G_1 \models \varphi \Leftrightarrow G_2 \models \varphi.$$

The proofs are omitted, proofs of analogous statements can be found in Ladner [3].

(We remark

Proposition 3.: F(m) is not recursive.
This remains true if we consider the simpler first-order game where players are allowed two choose vertices only.)
Some examples of subgraph-definable properties are the following:

Example 1.: HAM={G:G is Hamiltonian}
is subgraph-definable. This example is "sharp" in the sense that (when graphs are considered without the degree-constraint) HAM is not definable neither by a monadic nor by an existential weak dyadic formula.

MATCH={G:G contains a perfect matching}
is subgraph-definable (but not monadic).

An "arithmetic" example:

$\mathrm{MOD}_d^{k,l}=\{G: G$ is (connected, undirected) d-graph, $|V| \equiv k \bmod l\}$
is subgraph-definable. [5]

Lemma 1.: AUT is not subgraph-definable.

Proof: Suppose φ is a weak dyadic sentence s.t.

$$G \in \mathrm{AUT} \Leftrightarrow G \models \varphi$$

where φ is in prenex form and the number of it's quantifiers is n.
Let P_i be a path of length i. Consider the set of graphs

$$G=\{P_i: i=1,\dots,F(n+1)+1\}.$$

Then by Proposition 2. there are two graphs P_i and P_j s.t. $P_i \sim_{n+1} P_j$ (i<j). Let G_1, G_2 be the graphs of Figure 1., composed of two copies of P_i and an edge, resp. a copy of P_i, a copy of P_j and an edge. Then $G_1 \in \mathrm{Aut}$ while $G_2 \notin \mathrm{Aut}$. We claim

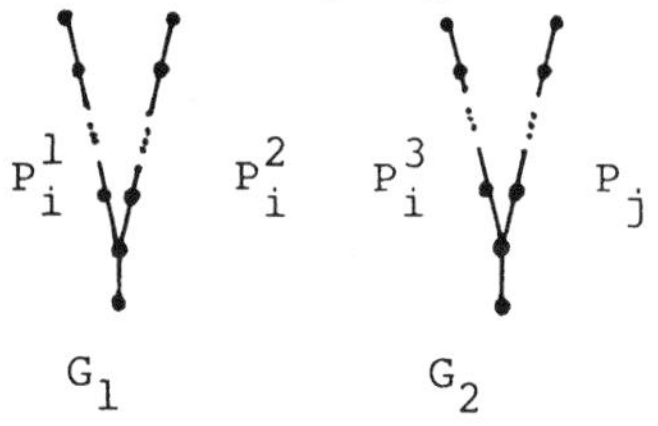

Fig. 1.

$$G_1 \sim_n G_2.$$

The winning strategy of player II is to "match" P_i^1 with P_i^3 and P_i^2 with P_j and to play according to the trivial strategy if I chooses from the first pair and according to the (P_i,P_j) winning strategy if I chooses from the second pair supposing that the endpoints corresponding to the vertices of degree 3 have been selected in a fictive move 0 (as in this case the answer to an endpoint must be an endpoint, this is possible). Thus the two strategies are "compatible". From Proposition 2., $G_1 \models \varphi \Leftrightarrow G_2 \models \varphi$, contradicting the definition of φ. □

To close this section we remark some further facts about class $\mathcal{C}$.

Proposition 4.: $\mathcal{C}$ contains properties complete in PSPACE.

Proposition 5.: $\mathcal{C}$=DSPACE$^{*}(\frac{n}{\log n})$, where $*$ means we restrict ourselves to languages that are d-graph propertes.

Proposition 6.: considering automata with a memory of logn bits (instead of finite automata) the power of cellular graph automata strictly increases.

(By diagonalization from Proposition 5.)

4. Identification of the center

Consider a d-graph G and a basic automaton A. A identifies a center in G if there is a vertex v_0 of G s.t. if C=(G^{*},A,μ) is an arbitrary cellular graph-automaton (i.e. ρ and μ are arbitrary), then C accepts and v_0 is coded in it's final state. I.e. C "finds" v_0 started from any distinguished vertex and compass of G. If P is a class of d-graphs, then A identifies a center in P if A identifies a center in every $G \in P$. A class P is identifiable if there is a basic automaton A that identifies a center in P.

Let FIX be the following class of d-graphs:

FIX=$\{$G: there is a vertex v of G fixed by every automorphism of G$\}$.

Lemma 2.: if $G \notin$ FIX, then $\{G\}$ is not identifiable.

Proof: suppose there is a basic automaton A that identifies center v_0 in G. Let σ be an automorphism of G s.t. $\sigma(v_0)=v_1 \neq v_0$. Then taking an arbitrary $C=(G^{*},A,\mu)$, $G^{*}=(G,\rho)$ define $C'=(G^{*\prime},A,\mu')$, $G'=(G,\rho')$ by

$$e_1 <_{\rho'(v)} e_2 \Leftrightarrow \sigma^{-1}(e_1) <_{\rho(\sigma^{-1}(v))} \sigma^{-1}(e_2);\ \mu'(v)=\mu(\sigma^{-1}(v)).$$

Then C' will have v_1 coded in it's final state. □

Theorem 3.: FIX is identifiable.

Proof: consider a graph $G \in$ FIX and the set

$$\mathcal{T} = \{T_{(F_i,\rho_i)}\ i=1,\dots,r\}$$

of tree-orders on V. To each $T_{F,\rho}$ there belongs an adjacency matrix of G labeled according to $T_{F,\rho}$. Take the set $\mathcal{T}' \subseteq \mathcal{T}$ of tree-orders determining the lexicographically minimal adjacency matrix.

Let

$$V' = \{v : v \in V,\ v \text{ is fixed by every automorphism of } G\}.$$

(As $G \in FIX$, $V' \neq \emptyset$.) Then for every $v \in V'$

(*) there is an index $i(v)$ s.t. v occupies the i-th position in every $T_{F,\wp} \in \mathcal{T}'$

holds. Otherwise the order-preserving map between two orders with different positions for v would give an automorphism of G not fixing v. On the other hand if (*) holds for $v \in V$ then $v \in V'$. Indeed, if $v \notin V'$, take an automorphism σ s.t. $\sigma(v) = v' \neq v$ and $T_{F,\wp} \in \mathcal{T}'$ arbitrary. Then $T'(v_1, v_2) \Leftrightarrow T_{F,\wp}(\sigma^{-1}(v_1), \sigma^{-1}(v_2))$ is a tree-order from $\mathcal{T}'$, with v in different positions in $T_{F,\wp}$ and T'.

Thus $\mathcal{T}'$ defines an order on V'. The center to be identified is the minimal element of V' in this order. The proof of the theorem is completed by constructing automata similar to those used in Theorem 2..□

References

[1] D. Angluin: Local and global properties in networks of processors Proc. 12. ACM Symp. on Th. of Comp., Los Angeles, 1980. pp. 82-93.

[2] Demoucron-Malgrange-Pertuiset: Graphes plein-airs, reconnaissance et construction des representations planaires topologiques, R. Fr. R. Op., 1964, pp. 33-47.

[3] R.E. Ladner: Application of model theoretic games to discrete linear orders and finite automata, Inf. and Cont., 1977, pp. 281-303.

[4] Rosentiehl-Fiksel-Holliger: Intelligent graphs, Graph Theory and Computing, Ac. Press, 1972. pp. 219-265.

[5] Turán Gy.: Gráf-tulajdonságok lokális felismerhetőségéről és definiálhatóságáról, Thesis. Szeged, 1980.

[6] A. Wu-A. Rosenfeld: Cellular graph-automata I-II. Inf. and Cont., 1979, pp. 305-353.

EXTENSIONS OF SYMMETRIC HOM-FUNCTORS TO THE KLEISLI CATEGORY

Jiří Vinárek
Faculty of Mathematics and Physics
Charles University Prague
Sokolovská 83, 186 00 Praha 8
Czechoslovakia

If one wants to study non-deterministic automata using categorical methods, it is necessary to express the situation that the transition relation assigns to a given state and an imput a set of states. In this case a set Q of states of an automaton is canonically imbedded into its power-set PQ. There is also a natural transformation $P^2 \longrightarrow P$ assigning to every set $S \in P^2Q$ a set $\bigcup S \in PQ$.

The above situation was generalized by the definition of a monad and a Kleisli category of a monad. The concept of the Kleisli category is also closely connected with fuzzy-theories.

Given a category $\underline{C}$, a monad $\underline{T}$ and a functor $F : \underline{C} \longrightarrow \underline{C}$ there is a canonical imbedding $K : \underline{C} \longrightarrow \underline{C}_{\underline{T}}$ (where $\underline{C}_{\underline{T}}$ is the Kleisli category of the monad $\underline{T}$). A natural problem arises whether there exists a functor $F_{\underline{T}} : \underline{C}_{\underline{T}} \longrightarrow \underline{C}_{\underline{T}}$ such that the following diagram commutes :

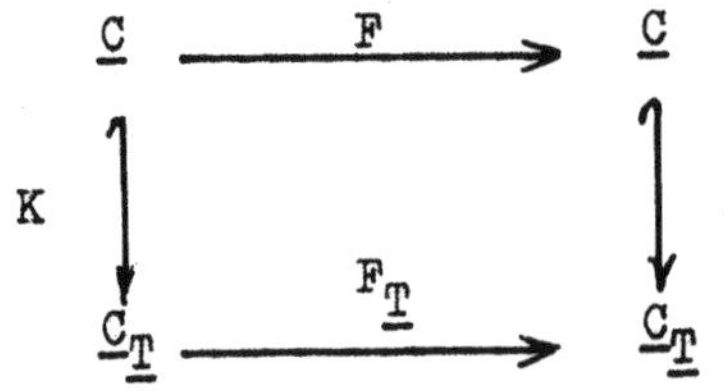

This problem was investigated by M.A.Arbib and E.G.Manes (see[1]). They found a sufficient and a necessary condition for existence of such an extension, viz. commuting of diagrams (called distributive laws) analogous to the Beck distributive laws between monads (see[2]). M.A.Arbib and E.G.Manes proved in [1] that set functors $-\times\Sigma$ satisfy distributive laws with respect to any monad over the category SET of sets and mappings (and therefore they can be extended on a Kleisli category of any monad). On the other hand (see[5]), hom-functors Hom(n,-) with $n \geq 2$ do not satisfy distributive laws with respect to a monad corresponding to the variety of semigroups with units ("a monad of words").

In the present note, there are studied questions of existence of an extension of symmetric hom-functors to the Kleisli category of the monad of words.

0. At first, recall some definitions and notations :

0.1. Let $\underline{C}$ be a category, $T : \underline{C} \longrightarrow \underline{C}$ a functor, $Id : \underline{C} \longrightarrow \underline{C}$ an identity functor, $\eta : Id \longrightarrow T$, $\mu : T^2 \longrightarrow T$ natural transformations. $\underline{T} = (T,\eta,\mu)$ is called a monad iff the following diagrams commute:

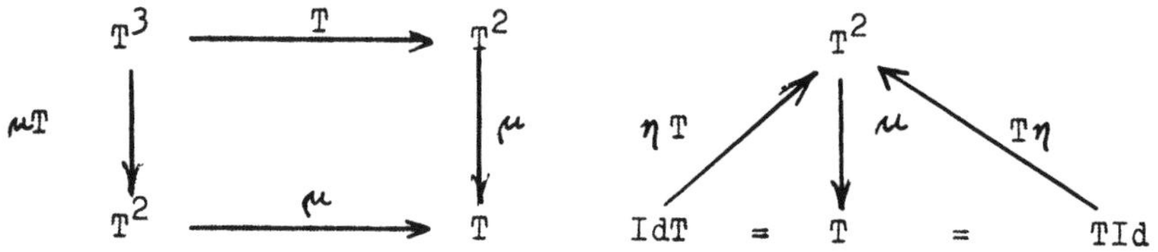

The Kleisli category $\underline{C}_{\underline{T}}$ is defined as follows : the class of objects obj $\underline{C}_{\underline{T}} = \{X_{\underline{T}} ; X \in \text{obj } \underline{C}\}$; $f_{\underline{T}} : X_{\underline{T}} \longrightarrow Y_{\underline{T}}$ is a morphism in $\underline{C}_{\underline{T}}$ iff $f : X \longrightarrow TY$ is a morphism in $\underline{C}$. Given $f_{\underline{T}} : X_{\underline{T}} \longrightarrow Y_{\underline{T}}$ and $g_{\underline{T}} : Y_{\underline{T}} \longrightarrow Z_{\underline{T}}$, the composition $g_{\underline{T}} * f_{\underline{T}} = (\mu_Z \circ Tg \circ f)_{\underline{T}}$.

0.2. Denote (similarly as in [5]) Mon = (M,e,m) a monad which assigns to each set A a monoid of words created by A, i.e. a free monoid over A. (MA = $\{a_1 \dots a_n ; n \in \{1,2,\dots\}, a_i \in A \text{ for } i = 1,\dots,n\} \cup \{\Lambda\}$ where Λ is the empty word, $e_A(a) = a$, $m_A((a_{11} \dots a_{1k_1}) \dots (a_{n1} \dots a_{nk_n})) = a_{11} \dots a_{nk_n}$). The corresponding category of monadic algebras is a variety of all the monoids, the corresponding Kleisli category is its subcategory of free monoids.

0.3. Recall the following definition (Arbib - Manes) :
Let $\underline{C}$ be a category, $F : \underline{C} \longrightarrow \underline{C}$ a functor, (T,η,μ) a monad. F is said to satisfy distributive laws over (T,η,μ) if there exists an assignment to each object A of $\underline{C}$ a morphism λ_A : FTA $\longrightarrow$ TFA such that the following two diagrams commute for every A and $\alpha : A \longrightarrow TB$

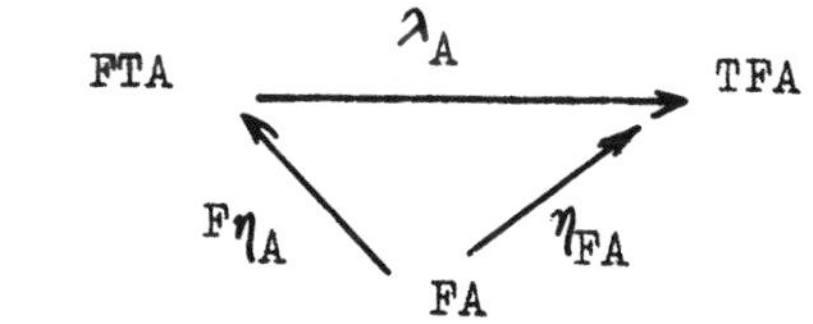

(the first distributive law)

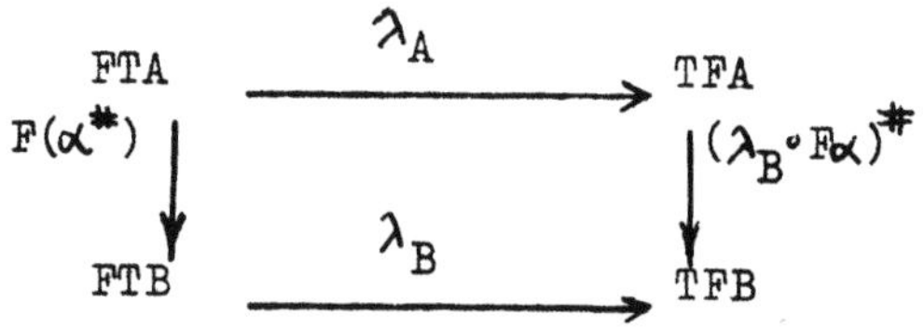

(the second distributive law)

where $\alpha^{\#} = \mu_B \circ T\alpha$.

0.4. Remarks :

a) A functor F can be extended on a Kleisli category over (T,η,μ) iff it satisfies the distributive laws (see [1]).

b) Let $I \neq \emptyset$ be a set, $\underline{N}$ be a set of all the positive integer numbers, $\varphi : I \longrightarrow \underline{N}$ be a bounded mapping, $\max\{\varphi(i)\ ;\ i \in I\} \geq 2$. Then a disjoint sum $F = \coprod_{i \in I} \mathrm{Hom}(\varphi(i),-)$ does not satisfy distributive laws over Mon (see [5]).

c) $F = \coprod_{n=1}^{+\infty} \mathrm{Hom}(n,-)$ satisfies distributive laws over Mon(see [5]).

Symmetric hom-functors

1.1. Definition. For any set $A \neq \emptyset$ define a symmetric hom-functor Sym(A,-) as a factorfunctor Hom(A,-)/$\sim$ where $f \sim g$ whenever there exists a bijection $b : A \longrightarrow A$ such that $f \circ b = g$.

1.2. Remarks.

a) A symmetric hom-functor is a special case of tree-group varietors, investigated by V. Trnková and J. Adámek, which are the only superfinitary varietors for which Kleene theorem holds (see[4]).

b) For the case A = n be finite we shall denote the equivalence class corresponding to an element $(x_0,\dots,x_{n-1}) \in \mathrm{Hom}(n,X)$ by $\langle x_0,\dots,x_{n-1}\rangle$.

Now, we are going to study distributive laws over Mon with respect to sums of finite symmetric hom-functors. While existence of distributive laws for $\coprod_{n\in M} \mathrm{Hom}(n,-)$ where $M \subsetneq \underline{N}$ is unbounded is still open, for the case of symmetric hom-functors a negative solution is presented.

1.3. Theorem. Let $I \neq \emptyset$ be a set, $\varphi : I \longrightarrow \underline{N}$ a mapping. Then $F = \coprod_{i\in I} \mathrm{Sym}(\varphi(i),-)$ satisfies distributive laws over Mon iff either φ is onto, or $\varphi(i) = 1$ for every $i \in I$.

Before proving Theorem we are going to prove some lemmas :

1.4. Lemma. Let A be a set, $n,k_1,\dots,k_n \in \underline{N}$, $x_{11},\dots,x_{1k_1},\dots,x_{n1},\dots,x_{nk_n} \in A$ such that $x_{ri} \neq x_{sj}$ whenever $(r,i) \neq (s,j)$, and $w = \langle x_{11}\dots x_{1k_1},\dots,x_{n1}\dots x_{nk_n}\rangle$. Suppose $\lambda_A(w) = \langle y_{11},\dots,y_{1l_1}\rangle\dots\langle y_{m1},\dots,y_{ml_m}\rangle$. If $k_i = k_j$ then $\mathrm{card}\{q\ ;\ y_{pq} = x_{ir}\} = \mathrm{card}\{q;\ y_{pq} = x_{jr}\}$ for every $p = 1,\dots,m$, $r = 1,\dots,k_i$.

Proof. Define $f : A \longrightarrow MA$ by $f(x_{ir}) = x_{jr}$ for $r = 1,\dots,k_i$, $f(x_{jr}) = x_{ir}$ for $r = 1,\dots,k_i$, $f(x) = x$ otherwise. Then $\lambda_A F(f^{\#})(w) = \langle y_{11},\dots,y_{1l_1}\rangle\dots\langle y_{m1},\dots,y_{ml_m}\rangle$. Hence, $\langle f(y_{p1}),\dots,f(y_{pl_p})\rangle = \langle y_{p1},\dots,y_{pl_p}\rangle$ for any $p = 1,\dots,m$ and $\mathrm{card}\{q ; y_{pq} = x_{ir}\} = \mathrm{card}\{q ; f(y_{pq}) = x_{ir}\} = \mathrm{card}\{q ; y_{pq} = f(x_{ir})\} = \mathrm{card}\{q ; y_{pq} = x_{jr}\}$. Q.E.D.

1.5. Lemma. Let $w = \langle x_{11}\dots x_{1k_1},\dots,x_{n1}\dots x_{nk_n}\rangle$, $\lambda_A(w) = \langle y_{11},\dots,y_{1l_1}\rangle\dots\langle y_{m1},\dots,y_{ml_m}\rangle$. Then $\{y_{p1},\dots,y_{pl_p}\} \subseteq \{x_{11},\dots,x_{1k_1},\dots,x_{n1},\dots,x_{nk_n}\}$ for any $p = 1,\dots,m$. (In particular, if $k_1 = k_2 = \dots = k_n = 0$, then $\lambda_A(\Lambda) = \Lambda$.)

Proof. Let $f : A \longrightarrow A$ ($\subseteq MA$) be an arbitrary mapping such that $f(x) = x$ for any $x \in \{x_{11},\dots,x_{1k_1},\dots,x_{n1},\dots,x_{nk_n}\}$. Then $F(f^{\#})(w) = w$. Hence, $\langle f(y_{p1}),\dots,f(y_{pl_p})\rangle = \langle y_{p1},\dots,y_{pl_p}\rangle$ for $p = 1,\dots,m$. Since f was an arbitrary mapping identical on $\{x_{11},\dots,x_{1k_1},\dots,x_{n1},\dots,x_{nk_n}\}$ there is $\{y_{p1},\dots,y_{pl_p}\} \subseteq \{x_{11},\dots,x_{1k_1},\dots,x_{n1},\dots,x_{nk_n}\}$.

Q.E.D.

1.6. Lemma. Suppose $n > 0$, $k_i > 0$ for $i = 1,\dots,n$, let $x_{11},\dots,x_{1k_1},\dots,x_{n1},\dots,x_{nk_n}$ be distinct elements of A, $w = \langle x_{11}\dots x_{1k_1},\dots,x_{n1}\dots x_{nk_n}\rangle$, $\lambda_A(w) = \langle y_{11},\dots,y_{1l_1}\rangle\dots\langle y_{m1},\dots,y_{ml_m}\rangle$. Then for any $j_1,\dots,j_n$ such that $1 \le j_1 \le k_1$, ..., $1 \le j_n \le k_n$ there exists $p \le m$ such that $\{x_{1j_1},\dots,x_{nj_n}\} \subseteq \{y_{p1},\dots,y_{pl_p}\}$.

Proof. Define $f : A \longrightarrow MA$ by $f(x_{ij_i}) = x_{ij_i}$ for $i = 1,\dots,n$, $f(x) = \Lambda$ otherwise; then $F(f^{*})(w) = \langle x_{1j_1},\dots,x_{nj_n}\rangle$. 1-st distributive law implies that $\lambda_A\langle x_{1j_1},\dots,x_{nj_n}\rangle = \langle x_{1j_1},\dots,x_{nj_n}\rangle \in MFA$ (a word of a length 1). 2-nd distributive law implies that there is exactly one $p \in \{1,\dots,m\}$ such that $\lambda_A\langle f(y_{p1}),\dots,f(y_{pl_p})\rangle = \langle x_{1j_1},\dots,x_{nj_n}\rangle$. By Lemma 1.5, $\{x_{1j_1},\dots,x_{nj_n}\} \subseteq \{f(y_{p1}),\dots,f(y_{pl_p})\} \setminus \{\Lambda\} \subseteq \{y_{p1},\dots,y_{pl_p}\}$. Q.E.D.

1.7. Lemma. Suppose $n > 0$, $k_i > 0$ for $i = 1,\dots,n$, $w = \langle x_{11}\dots x_{1k_1},\dots,x_{n1}\dots x_{nk_n}\rangle$, $\lambda_A(w) = \langle y_{11},\dots,y_{1l_1}\rangle\dots\langle y_{m1},\dots,y_{ml_m}\rangle$. Then for every $x \in \{x_{11},\dots,x_{1k_1},\dots,x_{n1},\dots,x_{nk_n}\}$ there exists $p \in \{1,\dots,m\}$ such that $\mathrm{card}\{i ; (\exists j \le k_i),\ x_{ij} = x\} \le \mathrm{card}\{q ; y_{pq} = x\}$.

Proof. Let $B = \{z_{11},\dots,z_{1k_1},\dots,z_{n1},\dots,z_{nk_n}\}$ be a set with $\sum_{i=1}^{n} k_i$

elements, $v = \langle z_{11}\dots z_{1k_1},\dots,z_{n1}\dots z_{nk_n}\rangle$. Define $f : B \longrightarrow MA$ by $f(z_{ij}) = x_{ij}$. Distributive laws imply that $\lambda_B(v) = \langle u_{11},\dots,u_{1l_1}\rangle\dots\langle u_{m1},\dots,u_{ml_m}\rangle$ where $f(u_{ij}) = y_{ij}$. Suppose that $x_{i_1j_1} = x_{i_2j_2} = \dots = x_{i_tj_t} = x$ ($i_k \neq i_{k'}$ for $k \neq k'$). By Lemma 1.6, there exists $p \leq m$ such that $\{z_{i_1j_1},\dots,z_{i_tj_t}\} \subseteq \{u_{p1},\dots,u_{pl_p}\}$. Hence, $\mathrm{card}\{q \; ; \; y_{pq} = x\} \geq t$. Q.E.D.

1.8. Now, we can prove Theorem 1.3.

1. Let φ be bounded, $n = \max\{\varphi(i) \; ; \; i \in I\} \geq 2$. Suppose existence of a collection $\{\lambda_A : FMA \longrightarrow MFA \; ; \; A \in SET\}$ satisfying distributive laws. Choose an $(n+2)$-point set $A = \{a_{11},a_{12},a_{21},a_{22},a_3,\dots,a_n\}$ and denote $w = \langle a_{11}a_{12},a_{21}a_{22},a_3,\dots,a_n\rangle$. Suppose $\lambda_A(w) = \langle y_{11},\dots,y_{1l_1}\rangle\dots\langle y_{m1},\dots,y_{ml_m}\rangle$. Lemma 1.6 implies that there exists $p \leq m$ such that $\{a_{11},a_{22},a_3,\dots,a_n\} \subseteq \{y_{p1},\dots,y_{pl_p}\}$. On the other hand, the assumption $\max\{\varphi(i) \; ; \; i \in I\} = n$ implies that $l_p = n$ and $\langle y_{p1},\dots,y_{pl_p}\rangle = \langle a_{11},a_{22},a_3,\dots,a_n\rangle$. Hence, $\mathrm{card}\{q \; ; \; y_{pq} = a_{12}\} = 0 \neq 1 = \mathrm{card}\{q \; ; \; y_{pq} = a_{22}\}$ which contradicts Lemma 1.4.

2. Suppose that φ is not onto and $\sup\{\varphi(i) \; ; \; i \in I\} = +\infty$. Then there exists $n \in \underline{N}$ such that $n-1 \notin \varphi(I)$ and $n \in \varphi(I)$. Choose an $(n+2)$-point set $A = \{a_{11},a_{12},a_{21},a_{22},a_3,\dots,a_n\}$ and denote $w = \langle a_{11}a_{12},a_{21}a_{22},a_3,\dots,a_n\rangle$. Suppose existence of distributive laws and suppose that $\lambda_A(w) = \langle y_{11},\dots,y_{1l_1}\rangle\dots\langle y_{m1},\dots,y_{ml_m}\rangle$.

a) Define $f,g : A \longrightarrow MA$ by $f(a_{12}) = f(a_{21}) = \Lambda$, $f(x) = x$ otherwise, $g(a_{21}) = g(a_{22}) = \Lambda$, $g(a_{12}) = a_{22}$, $g(x) = x$ otherwise. Then $F(f^{\#})(w) = \langle a_{11},a_{22},a_3,\dots,a_n\rangle$, $F(g^{\#})(w) = \langle a_{11}a_{22},\Lambda,a_3,\dots,a_n\rangle$. Lemma 1.4 implies that $\mathrm{card}\{q \; ; \; y_{pq} = a_{12}\} = \mathrm{card}\{q \; ; \; y_{pq} = a_{22}\}$ for $p = 1,\dots,m$. Hence, $\langle f(y_{p1}),\dots,f(y_{pl_p})\rangle = \langle g(y_{p1}),\dots,g(y_{pl_p})\rangle$ and $(\lambda_A \circ Ff)^{\#}(\lambda_A(w)) = (\lambda_A \circ Fg)^{\#}(\lambda_A(w))$. According to 2-nd distributive law, $\lambda_A F(g^{\#})(w) = \lambda_A F(f^{\#})(w) = \langle a_{11},a_{22},a_3,\dots,a_n\rangle$.

b) Denote $v = \langle a_{11},\Lambda,a_3,\dots,a_n\rangle$ and define $h : A \longrightarrow MA$ by $h(a_{11}) = a_{11}a_{22}$, $h(x) = x$ otherwise. Suppose that $\lambda_A(v) = \langle b_{11},\dots,b_{1t_1}\rangle\dots\langle b_{d1},\dots,b_{dt_d}\rangle$. Since $(\lambda_A \circ Fh)^{\#}(\lambda_A(v)) = \lambda_A(F(h^{\#})(v)) = \lambda_A\langle a_{11}a_{22},\Lambda,a_3,\dots,a_n\rangle$, there is $d = 1$ and $t_1 \geq 1$. Lemmas 1.4, 1.5 and the assumption $n-1 \notin \varphi(I)$ imply that $\mathrm{card}\{q \; ; \; b_{1q} = a_{11}\} = \mathrm{card}\{q \; ; \; b_{1q} = a_3\} = \dots = \mathrm{card}\{q \; ; \; b_{1q} = a_n\} > 1$. (§)

Consider $\langle h(b_{11}),\dots,h(b_{1t_1})\rangle \in FMA$. Then $\lambda_A\langle h(b_{11}),\dots,h(b_{1t_1})\rangle = (\lambda_A \circ Fh)^{\#}(\lambda_A(v))$. Lemma 1.7 and (§) imply that $1 = \mathrm{card}\{x \in \{a_{11},$

$a_{22}, a_3, \ldots, a_n\}; x = a_{11}\} \geq card\{q ; b_{1q} = a_{11}\} > 1$ which is a contradiction.

3. If $\varphi(i) = 1$ for every $i \in I$ then $\coprod_{i \in I} Sym(\varphi(i),-) = \coprod_{i \in I} Hom(\varphi(i),-) = - \times I$ and distributive laws are satisfied due to [1].

4. If φ is onto then one can define $\lambda_A(\Lambda) = \Lambda, \lambda_A \langle x_{11} \ldots x_{1k_1}, \ldots, x_{n1} \ldots x_{nk_n} \rangle = \langle x_{11}, \ldots, x_{1k_1}, \ldots, x_{n1}, \ldots, x_{nk_n} \rangle$. Distributive laws are clearly satisfied (the proof is similar as in [5]). Q.E.D.

2.1. At the end, recall an open problem concerning sums of hom-functors : Let I be a set, $\varphi : I \longrightarrow \underline{N}$ a mapping. Denote $F = \coprod_{i \in I} Hom(\varphi(i),-)$; characterize when a functor F can be extended on a Kleisli category over Mon.

References

[1] M. A. ARBIB - E. G. MANES, Fuzzy machines in a category (preprint) Univ. of Massachusetts at Amherst

[2] J. BECK, Distributive laws, Lecture Notes in Mathematics, Springer-Verlag, 80(1969), 119-140

[3] S. MACLANE, Categories for the working mathematician, Springer-Verlag, 1972

[4] V. TRNKOVÁ - J. ADÁMEK, Tree-group automata, FCT '79, 462-467

[5] J. VINÁREK, On extensions of functors to the Kleisli category, Comment. Math. Univ. Carolinae, 18(1977), 319-327

A NEW OPERATION BETWEEN LANGUAGES

J. Beauquier
Université de Picardie
L.I.T.P. 248
& U.E.R. de Mathématiques
33, rue St Leu - 80039 Amiens Cedex
FRANCE

ABSTRACT :

Here is introduced a new operation between languages noted ∇ in order to study their structural complexity. Two languages L and M being given, we say that L is of structural complexity greater than M iff the full semi-AFL $\mathcal{F}(M)$ generated by M is included in the full semi-AFL $\mathcal{F}(L)$ generated by L . Then ∇ satisfies the property that, when applied to some structurally comparable languages, it produces a language of intermediary structural complexity.

We use the ∇ operation to build a context-free language which has structural complexity greater than any finite-turn language but is not context-free generator.

Beside the classical notion of complexity of recognization for a language, which is a dynamic notion, it has been developed during the recent years a notion of static or structural complexity (cf. [6] , [3] , [5]). The study of structural complexity is strongly related to AFL and rational cones theory, since a language L is structurally more complex than a language M if and only if the rational cone generated by L contains this one generated by M.

The Transfer Theorems for iterative pairs in the family of context-free languages ([6] , [3] , [5]) or in other families [4] show how the degree of structural complexity of a language depends on local properties of its words. Several open problems in languages theory can be attacked with the tool of the structural complexity and, for instance, the following one : Is the rational cone of languages, that are non generator of the whole rational cone of context-free languages, principal or not ? This question can be reformulated in :
Does there exist a context-free non-generator which is of structural complexity greater than any context-free non-generator ? But this kind of question can also

e asked for other familes : Does there exist a linear non-generator of structural complexity greater than any bounded linear language ? No answer has been given up to day to this kind of problems.

The study of the structural complexity of languages has sometimes be done in relation with some operations between languages : union, product, star operation, substitution, insertion, homomorphic replication, ...

All these operations have the property, when applied to two languages L_1 and L_2 to produce a language L_3 which is structurally more complex than L_1 and L_2. In order to attack the two previous problems, we introduce here a new operation which has the property, when applied in some conditions, to weaken the structural complexity of the languages under consideration.

In particular, when applied to some languages which are structurally comparable, the new operation produces a language of intermediary structural complexity.

The use of this new operation allows us to give some precisions about the first previous open problem and to solve positively the second.

But it also has a theoretical interest in itself and raises a lot of open questions. In particular, we precise here its relation with the non-determinism. From this point of view, the new operation appears as being very akin to the operation introduced by Greibach [16] which associates to each context-free language its "non-deterministic" version.

At the end, we use the ∇ operation in order to define another operation, noted $\perp$, whose basic properties are studied.

- DEFINITIONS :

We note X^* the free monoïd generated by the finite alphabet X and ε its unity, the empty word. $|f|$ notes the length of f . Definitions and classical properties of regular and context-free languages can be found in [23] or [17]. We say that an homomorphism, $\phi : X^* \to Y^*$ is alphabetic iff $\forall x \in X \ |\phi(x)| \leq 1$ strictly alphabetic iff $\forall x \in X \ |\phi(x)| = 1$. A rational cone (= full trio in [8]) is a family of languages closed under homomorphism, inverse homomorphism and intersection with regular sets. We handle rational transductions by using their characterization of [20] : the transduction $T : X^* \to Y^*$ is rational iff there exist an alphabet Z , a regular set K in Z^* and two homomorphisms $\phi : Z^* \to X^*$ and $\Psi : Z^* \to Y^*$ such that the graph of T is : $\hat{T} = \{(\phi f , \Psi f) \mid f \in K\}$. Then we note $T = \langle \phi, \Psi, K \rangle$.

A rational transduction T is length preserving iff $\forall f, \forall g \in T(f)$, $|g| = |f|$. It is known that rational cones are families of languages closed under rational transduction. L being a language, the rational cone generated by L, $T(L)$ is the smallest rational cone containing L. A rational cone C is principal of generator L, iff $C = T(L)$.

$L_1 \subset X_1^*$ and $L_2 \subset X_2^*$ being two languages, the marked union of L_1 and L_2 is the language $L_1 \cup \# L_2$, where $\#$ is a new symbol ($\# \notin X_1 \cup X_2$).

We call full semi AFL [9] a rational cone closed under union. From [2], [3], or [5], it comes that the structural complexity of two languages can be compared by using the relation: "to be rationally greater than". We note $L_1 \geqslant L_2$ and we say that L_1 is rationally greater than L_2 iff $T(L_2) \subseteq T(L_1)$, that is there exists a rational transduction T such that $T(L_1) = L_2$. $\geqslant$ is a preorder relation and can be associated with an equivalence relation, noted $\approx$ and called rational equivalence. Specific definitions for insertion and substitution are in [14]. The IRS (Intersection with Regular Sets) condition appears in [15]: we say that a language L is LRS iff it does not contain any infinite regular set. We shall use some specific languages, like :

- PAL, generated by the context-free grammar
  $S \to z_1 S \bar{z}_1 \mid z_2 S \bar{z}_2 \mid \varepsilon$
- COPY $= \{w w \mid w \in X^*\}$
- the one-sided Dyck set $D_2'^*$, context-free generator, generated by the context-free grammars :
  $S \to z_1 S \bar{z}_1 S \mid z_2 S \bar{z}_2 S \mid \varepsilon$
- the language E, context-free generator, generated by :
  $S \to a S b S c \mid d$
- the Goldstine's language [12] G,

$G = \{a^{n_1} b a^{n_2} b \,..\, a^{n_k} b \mid k \in N, n_i \in N \text{ for } i = 1, \ldots, k, i, i \neq n_i\}$

(G is a linear language).

We note Lin the family of linear languages, QR the family of quasi-rational languages, closure of Lin by substitution (quasi-rational = non-expansive = derivation bounded of [11]).

We note FT the family of finite-turn languages, that is the family of context-free languages recognized by a finite-turn pushdown automaton [10].

It is known that $FT \subsetneq QR$ and that FT is the insertion closure of Lin.

Now, we give some more specific definitions.

Definition 1 : Let X_1 and X_2 be two alphabets (not necessarily disjoint) and let L_1 and L_2 be two languages in X_1^* and X_2^*. The derivative of L_1 and L_2, noted $\nabla(L_1, L_2)$ is the language in $(X_1 \cup X_2)^*$ defined by :

$$\nabla(L_1, L_2) = \{x_1\ y_1\ x_2\ y_2 \ldots x_k\ y_k \mid k \in N,\ x_i \in X_1,\ y_i \in X_2 \quad \text{for } i = 1, \ldots, k \quad x_1\ x_2 \ldots x_k \in L_1 \ \underline{\text{or}}\ y_1\ y_2 \ldots y_k \in L_2\}$$

This definition can be extended to n-languages.

Definition 2 : Let $X_1, X_2, \ldots, X_n$ be n alphabets (not necessarily disjoint) and let $L_1, L_2, \ldots, L_n$ be n languages in $X_1^*, X_2^*, \ldots X_n^*$. The derivative of $L_1, L_2, \ldots, L_n$, noted $\nabla(L_1, L_2, \ldots, L_n)$ is the language in $(X_1 \cup X_2 \cup \ldots \cup X_n)^*$ defined by :

$$\nabla(L_1, L_2, \ldots, L_n) = \{x_{1,1}\ x_{1,2} \ldots x_{1,n}\ x_{2,2} \ldots x_{2,n} \ldots x_{k,1}\ x_{k,2} \ldots x_{k,n} \mid k \in N,\ x_{i,j} \in X_j$$

$$\text{for } i = 1, 2, \ldots, k\ ;\ j = 1, 2, \ldots, n,\ x_{1,1}\ x_{2,1} \ldots x_{k,1} \in L_1$$

$$\underline{\text{or}}\quad x_{1,2}\ x_{2,2} \ldots x_{k,2} \in L_2\ \underline{\text{or}} \ldots \underline{\text{or}}\ x_{1,n}\ x_{2,n} \ldots x_{k,n} \in L_n\}$$

Remark 1 : The derivative of L_1 and L_2 is not the shuffle [14] of L_1 and L_2. In the definition of $\nabla(L_1, L_2)$ appears a logical or ; for the shuffle, it is a logical and. Nevertheless $\nabla(L_1, L_2)$ can be considered as the union of the litteral shuffle of L_1 and X_2^* and the litteral shuffle of X_1^* and L_2.

Remark 2 : The ∇ operation is not associative, since, L_1, L_2, L_3 being three languages, one has generally :

$$\nabla(\nabla(L_1, L_2), L_3) \neq \nabla(L_1, (L_2, L_3))$$

and

$$\nabla(\nabla(L_1, L_2), L_3) \neq \nabla(L_1, L_2, L_3) .$$

Example : Let $L_1 = L_2 = L_3 = PAL$ et let w be the word :

$$w = a\ \bar{a}\ \bar{a}\ \bar{a}\ b\ \bar{b}\ \bar{b}\ \bar{b}\ \bar{b}\ a\ b\ a\ \bar{a}\ a\ b\ b$$

$w \in \nabla(\ (PAL, PAL), PAL)$ since :

$$w = \underline{a}\ \bar{a}\ \underline{\bar{a}}\ \bar{a}\ \underline{b}\ \bar{b}\ \underline{\bar{b}}\ \bar{b}\ \underline{\bar{b}}\ a\ \underline{b}\ a\ \underline{\bar{a}}\ a\ \underline{b}\ b$$

and $a\ \bar{a}\ b\ \bar{b}\ \bar{b}\ b\ \bar{a}\ b \in \nabla(PAL, PAL)$

But $w \notin \nabla(PAL, \nabla(PAL, PAL))$ since

$a\ \bar{a}\ b\ \bar{b}\ \bar{b}\ b\ \bar{a}\ b \notin PAL$

and $\bar{a}\ \bar{a}\ \bar{b}\ \bar{b}\ a\ a\ a\ b \in \nabla(PAL, PAL)$.

Moreover $w \notin \nabla(PAL, PAL, PAL)$ since $|w| \neq 0$ (modulo 3).

Question 1 : How characterize the languages L_1, L_2, L_3 such that :
$\nabla(\nabla(L_1, L_2), L_3) = \nabla(L_1, \nabla(L_2, L_3))$?

II - ELEMENTARY PROPERTIES OF THE ∇ OPERATION.

a) Derivative and union :

The union appears in the definition of the ∇ operation. Informally the question what we are interested in, is : is there a loss of information in $\nabla(L_1, L_2)$ with respect to $L_1 \cup \# L_2$?

That is : Is $\nabla(L_1, L_2)$ of structural complexity smaller than $L_1 \cup \# L_2$?

That leads us to consider the respective positions of rational cones generated by $\nabla(L_1, L_2)$ and $L_1 \cup \# L_2$.

Proposition 1 : Let $L_1 \subset X_1^*$ and $L_2 \subset X_2^*$ be two languages and # a new symbol. Then $\nabla(L_1, L_2)$ belongs to the rational cone generated by $L_1 \cup \# L_2$.

Corollary 1 : Let F be a full semi-AFL. Then, for any languages L_1 and L_2 in F , $\nabla(L_1, L_2) \in F$.

Example : If L_1 and L_2 are context-free (resp. regular) , then $\nabla(L_1, L_2)$ is context-free (resp. regular).

b) Derivative and structural complexity :

The following proposition will allow us to precise more accurately the relation between structural complexity and ∇ operation. This proposition states that, if L_1 is structurally more complex than L_2, then $\nabla(L_1, L_2)$ is of intermediary structural complexity.

Proposition 2 : Let $L_1 \subseteq X_1^*$ and $L_2 \subseteq X_2^*$ be two languages such that $L_1 \geqslant L_2$ and there exists a length preserving rational transduction T such that $T(L_1) = L_2$. Then : $L_1 \geqslant \nabla(L_1, L_2) \geqslant L_2$.

Corollary 2 : For any language L , $\nabla(L, L)$ is rationally equivalent to L.

Proposition 3 : Let $L_1 \subseteq X_1^*$ and $L_2 \subseteq X_2^*$ be two languages such that $L_1 \geqslant L_2$ and let # be a marker $(\# \notin X_1 \cup X_2)$, then $L_1 \geqslant \nabla(L_1, L_2) \cup \# L_2 \geqslant L_2$.

Examples :

1 - $\nabla(D'^*_2, PAL)$ is a context free generator, from corollary 4.

2 - For any language $L_1 \subset X_1^*$, $\nabla(L_1, X_2^*) = (X_1 X_2)^* \approx X_2^*$.

3 - $\nabla(COPY, PAL) \approx COPY \cup \# PAL$, from corollary 4.

4 - $\nabla(D'^*_2, G)$ is not a context free generator (cf. Part. IV).

Question 2 : Is it true that :

$\nabla(D'^*_2, G) \ngeqslant G$ involving $D'^*_2 \ngeqslant \nabla(D'^*_2, G) \ngeqslant G$?

This question can be generalized.

Question 3 : How obtain conditions on L_1 and L_2 such that $L_1 \not\geq L_2$ implies $L_1 \not\geq \nabla(L_1, L_2) \not\geq L_2$.

That question is strongly related to one of the most important open problem about context-free languages :

Question 4 [1] : L_1 and L_2 being two context-free languages such that $L_1 \not\geq L_2$, does there exist a language L_3 such that $L_1 \not\geq L_3 \not\geq L_2$.

The importance of question 4 comes from the fact that a positive answer to it would allow to solve an important open problem about context-free languages.

Conjecture 1 [13], [1] : The rational cone of context-free non generators is not principal.

A positive answer to question 4 would involve that conjecture 1 is true by :

Let us suppose that the rational cone Nge of context-free non generators is principal of generator L . Then :

$$D'^{*}_{2} \not\geq L$$

If Question 4 has a positive answer, there exists M such that :

$$D'^{*}_{2} \not\geq M \not\geq L$$

Then, M is a non-generator since $D'^{*}_{2} \not\geq M$ and $M \notin T(L)$.

III - DERIVATIVE AND NON-DETERMINISM.

In this part, the languages under consideration are supposed to be context free. One could think that a pushdown automaton, trying to analyze a word w in $\nabla(L_1, L_2)$ must necessarily choose whether it decides to analyse the even or the odd occurences of w and then must be nondeterministic. This idea is false in the general case :

Fact : There exist context-free languages L_1 and L_2 such that $\nabla(L_1, L_2)$ is deterministic.

Example : Let $L = \{a^n b^n \mid n \in N\}$ and $L_1 = L_2 = \{a, b\}^* \setminus L$.
Then $\{a, b\}^* \setminus \nabla(L_1, L_2) = \{a^{2n} b^{2n} \mid n \in N\}$ is a deterministic context free language. Since deterministic languages are known to be closed under complementation, $\nabla(L_1, L_2)$ is a deterministic context-free language.

The positive result that we can obtain is the following.

For any language L in X^* let be $\#$ a new symbol, $\# \notin X$. We consider the morphism $\phi_{\#} : (X \cup \{\#\})^* \to X^*$ defined by :

$$\phi_{\#}(x) = x \qquad \text{for } x \in X$$
$$\phi_{\#}(\#) = \varepsilon$$

We set $L_{\#} = \phi_{\#}^{-1}(L)$.

Proposition 4 : Let L_1 and L_2 be two context-free languages in X_1^* and X_2^* and let $\#_1 \notin X_1$, $\#_2 \notin X_2$. Then the language $\nabla(L_{\# 1}, L_{\# 2})$ is a non-deterministic context-free language.

(Note : the proof uses a result by Stearns [22] about iterative pairs in a deterministic language).

IV - FINITE TURN LANGUAGES, BOUNDED LINEAR LANGUAGES AND ∇ OPERATION :

In this part, we use the ∇ operation in order to construct :

1°) A language whose the generated rational cone :
- strictly contains the family of finite turn languages,
- is strictly included in the family of context-free languages.

2°) A language whose the generated rational cone :
- strictly contains the family of bounded linear languages,
- is strictly included in the family of linear languages.

No such languages have been exhibited up to day.

1°) Let $L \subseteq \{x, y, t\}^*$ be the language defined by :

$$L = \{x^{n_1} y^{m_1} x^{n_2} y^{m_2} \dots x^{n_k} y^{m_k} t^p \mid k \in N, n_i, m_i \in N, p \neq k\}$$

L is context-free, generated by

$S_o \to S_1 \qquad S_1 \to T S_1 t \mid S_1 t \mid t \qquad S_2 \to T S_2 t \mid S_2 T \mid T$

$T \to x T \mid x U \qquad U \to y U \mid y$

We consider the language $D'^*_2 \; \#^*$ obtained by adding some $\#$'s to words in D'^*_2.

From corollary 1 $\nabla(D'^*_2 \; \#^*, L)$ is context free.

Let K be the regular set :

$$K = \{z_1 x, z_2 x, \bar{z}_1 y, \bar{z}_2 y\}^* \quad (\# \; t)^*$$

We set $C = (\nabla(D'^*_2 \; \#^*, L)) \cap K$.

Lemma 1 : C is rationally greater than any language obtained by iterative insertions of PAL into PAL.

Lemma 2 : The rational cone generated by C contains the smallest insertion closed rational cone containing PAL.

Lemma 3 [7] : The family of finite turn languages is the smallest insertion closed rational cone containing PAL .

Lemmas 2 and 3 involve :

Proposition 5 : The language C is rationally greater than any finite turn language.

On the other hand, we can prove :

Proposition 6 : C is not a context-free generator.

2°) We will now give an analogous property for linear languages. Let A_k the language defined by :

$A_1 = \{a_1^n b_1^n \mid n \in N\}$

$A_2 = \{a_2^n w b_2^n \mid n \in N, w \in A_1\}$

$\vdots$

$A_k = \{a_k^n w b_k^n \mid n \in N, w \in A_{k-1}\}$

Let K be the regular set :

$K = \{z_1 x, z_2 y\}^* \{\bar{z}_1 y, \bar{z}_2 x\}^* (\# t)^*$

and let :

$F = (\nabla(PAL \#^*, L)) \cap K$

L is linear and, form corollary 1, F is linear.

We prove like above :

Lemma 4 : F is rationally greater than any language A_k . Hence :

Proposition 7 : F is rationally greater than any bounded linear language.

Then, by using the theorem of characterization of linear generators [2] , tating that L is a linear generator iff there exist an homomorphism ϕ , a regular set K such that :

$\phi^{-1}(L) \cap K = \#_1 \, PAL \, \#_2$, where $\#_1$ and $\#_2$ are markers, one proves :

Proposition 8 : L is not a linear generator.

V - THE $\perp$ OPERATION :

Definition 3 : For any integer K and any language $L \subseteq X^*$, we define :

$\nabla_k(L) = \{x_{1,1} x_{1,2} .. x_{1,k} x_{2,1} x_{2,2} .. x_{2,k} .. x_{r,1} x_{r,2} .. x_{r,k} \mid r \in N$

$j \in [1, k] \; x_{1,j} x_{2,j} .. x_{r,j} \in L\}$

Definition 4 : For any language $L \subseteq X^*$, we define :

$\perp(L) = \bigcup_{k=1} \nabla_i (L)$

Definition 5 : For any language $L \subseteq X^*$ and any rational language $R \subseteq X^*$, we define $\perp(L, R) = \perp(L) \cap R$.

Remark : If K is a rational language, $\perp(K)$ is generally not rational. For instance let $K = X^* a\,a\,a\,a\, X^*$. Then :

$\perp(K) \cap ab^+ ab^+ ab^+ ab^+ = \{ab^{r_1} ab^{r_2} ab^{r_3} ab^{r_4} \mid r_1 = r_2 = r_3,\ r_4 \equiv (-1) \bmod(r_1 + 1)\}$

We study the basic properties of the $\perp$ operation.

Proposition 9 : If L is a context-free language then $\perp(L)$ is recognized by an automaton with one pushdown and one counter.

Proposition 10 : If K is a rational language over $X = \{a\}$, then $\perp(K, a^*)$ is a rational language.

Definition 6 : A language $L \subseteq X^*$ is said to be hard if it satisfies :
For any $w \in L$ and any factorization $w = w_1\, x\, w_2$, $x \in X$ then there exists $y \in X$ such that $w_1\, y\, w_2 \notin L$.

Proposition 11 : If $L_1 \subseteq X^*$ and $L_2 \subseteq X^*$ are two hard languages then :

$$\perp(L_1) = \perp(L_2) \Leftrightarrow L_1 = L_2$$

The $\perp$ operation can be naturally extended to families of languages.

Notation : Let F be a family of languages. We note :

$$\perp(F) = \{\perp(L, K) \mid L \in F\}$$

F being supposed to satisfy some closure properties, we are interested in the closure properties of $\perp(F)$.

Proposition 12 : Let Ψ be an alphabetic homomorphism. Then :

$$\Psi^{-1}(\perp(L, R) = \perp(\Psi^{-1}(L), \Psi^{-1}(R)).$$

Definition 6 : A family of languages C is said to be an alphabetic cylinder iff it is closed under inverse alphabetic homomorphism an intersection with rational languages.

Proposition 13 : If C is an alphabetic cylinder, then $\perp(C)$ is an alphabetic cylinder.

CONCLUSION :

The ∇ operation seems to be of a great interest in the study of structural complexity of languages because it allows to "weaken" the structure of a language without destroying the entire information it contains. In fact this weakening ability concern a whole family of operations of same nature. Since we could define, for example, an operation $\nabla_{1,2}$ by :

$$\nabla_{1,2}(L_1, L_2) = \{x_1\, y_1\, z_1\, x_2\, y_2\, z_2 \dots x_k\, y_k\, z_k \mid k \in N,\ x_i \in X_1 \quad y_i, z_i \in X_2 \quad x_1\, x_2 \dots x_k \in L_1 \ \underline{\text{or}}\ \ y_1\, z_1\, y_2\, z_2 \dots y_k\, z_k \in L_2\}$$

More generally an operation $\nabla_{i,j}$ can be defined.

The study of such operations sets a number of questions, which seem not to be simple. For exemple, how characterize a language L such that :

$$\nabla_{1,2}(L, L) \approx L ?$$

BIBLIOGRAPHY :

[1] AUTEBERT J.M., BEAUQUIER J., BOASSON L., NIVAT M. - Quelques problèmes ouverts en théorie des langages algébriques, R.A.I.R.O. Informatique théorique 13 (1979), p. 363-379.

[2] BEAUQUIER J. - Contribution à l'étude de la complexité structurelle des langages algébriques, Thèse d'Etat, Université Paris VII (1977).

[3] BEAUQUIER J. - Générateurs algébriques et systèmes de paires itérantes, Theoretical Computer Science 8, (1979), p. 293-323.

[4] BEAUQUIER J. - Deux familles de langages incomparables, Information and Control 43, n° 2, (1979), p. 101-122.

[5] BERSTEL J. - Transductions and context-free languages, Teubner Verlag (1979).

[6] BOASSON L. - Langages algébriques, paires itérantes et transductions rationnelles, Theoretical Computer Science 2 (1976), p. 209-223.

[7] CRESTIN J.P. - Langages quasi-rationnels, ultralinéaires et bornés, in Langages algébriques, ENSTA Paris (1973).

[8] GINSBURG S. - Algebraic and Automata-Theoretic Properties of Formal Languages Amsterdam, New York, Oxford (1975).

[9] GINSBURG S., GREIBACH S.A., HOPCROFT J.E. - Pre AFL, in : Studies in abstract families of languages, Mem. of the Amer. Math Soc. 87 (1969), p. 41-51.

10] GINSBURG S., SPANIER E.H. - Finite-turn pushdown automata, SIAM J. Control 4 (1966), p. 429-453.

11] GINSBURG S., SPANIER E.H. - Derivation bounded languages, J. Computer System Sciences 2 (1968), p. 228-250.

12] GOLDSTINE J. - Substitution and bounded languages, Journ. of Computer and System Sciences 6 (1972), p. 9-29.

13] GREIBACH S.A. - Chains of full AFL's, Math. Syst. Theory 4 (1970) p. 231-242.

14] GREIBACH S.A. - Syntactic operators on full semi-AFL's, J. Computer System Sciences 6 (1972), p. 30-76.

15] GREIBACH S.A. - One counter languages and the IRS condition, J. of Comp. and Syst. Sciences 10, (1975), p. 237-247.

16] GREIBACH S.A. - The hardest context-free language, SIAM J. of Computing 2, (1973), p. 304-310.

17] HARRISON M.A. - Introduction to formal languages theory, London, Amsterdam, Sydney.

18] NIVAT M. - Transductions de langages de Chomsky, Ann. de l'Institut Fourier 18, (1968), p. 359-456.

19] OGDEN W. - A helpful result for proving inherent ambiguity, Math. System. Theory 2, (1968), p. 191-194.

20] STEARNS R.E. - A regularity test for pushdown machines, Information and Control 11 (1961), p. 323-340.

21] SALOMAA A. - Formal languages, New York, San Francisco, London.

LOGICAL DESCRIPTION OF COMPUTATION PROCESSES

E. Börger
Lehrstuhl Informatik II
Universität Dortmund
Postfach 500 500
D-4600 Dortmund 50

Introduction

When talking about logical description of computation processes we mean above all descriptions within the framework of first order predicate logic. By Gödel's completeness theorem for first order logic one knows that all computation processes are expressible within predicate calculus. Turing [1937] for the first time gave an *explicit description of Turing machine computations by logical formulae* of first order thereby reducing effectively a particular recursively unsolvable problem for arbitrary Turing machines M to the problem whether the describing logical formula α_M is deducible within predicate logic. Turing thus was able to conclude that Hilbert's Entscheidungsproblem cannot be solved algorithmically.

Since at that time a large number of special classes of first order formulae was known already to have a solvable Entscheidungsproblem, Turing found it interesting to observe in op.cit. that for his reduction he really needed only a small portion of predicate logic, namely (rephrased in terms of satisfiability instead of deducibility, as I shall do always in the following) the class of closed prenex formulae with at most binary predicate symbols and a prefix of the form $\vee\wedge\vee\wedge^{6}$-$\wedge^{n}$ denotes a sequence of length n of universal quantifiers; for $n=\infty$ that means of arbitrary finite length. After the appearance of Turing's paper, many efforts were spent to improve this result, on the one hand by producing "smaller" such "*reduction classes*" - therefore with undecidable Entscheidungsproblem, see Suranyi [1959]-on the other hand by exhibiting decision procedures for larger and larger subclasses, see Ackermann [1954]. Many new reductions were found, but not even the decision problem of all formally specified classes of formulae like the prefix classes Π-determined by the form Π of the prefixesof their elements which are closed and in prenex normal form -

ould be settled. Strangeley enough Turing's idea to look for smooth
and direct logical descriptions of machine computations or similar
processes was not really pursued up to 1962.

Büchi [1962] took up again Turing's approach and combined it with
kilful use of *two theorems* **due** *to Skolem*. These theorems tell that a
renex formula of restricted predicate logic (i.e. without function
ymbols and without identity sign) is satisfiable iff its Skolem nor-
al form is, and that in models for such Skolem normal forms one can
estrict attention to the domain of terms built up from the individual
onstants and function symbols occuring in the formula and to interpre-
ation of the terms by themselves. Büchi's observation constituted a
reak through: the unsettled prefix class $\vee\wedge\vee\wedge$ was shown to be a re-
uction class, and above all Büchi's reduction procedure made the re-
ation between the Turing machine computations and the formulae ex-
ressing them very clear and easy to grasp. Indeed following this line
f attack within 4 years the decision problem of all prefix-similarity
lasses could be settled (see Kahr/Moore/Wang [1962], Kahr [1962],
ostyrko [1964], Gurevich [1966]), where a prefix-similarity class
$(m_1, m_2, m_3, \ldots)$ is defined as class of all closed prenex formulae of
estricted predicate logic having a prefix of form Π and containing at
ost m_i predicate symbols of rank i.

Such at first sight surprisingly small classes like $\wedge\vee\wedge(\infty,1)$,
$\vee^{\infty}\wedge(0,1)$ and $\wedge\vee\wedge\vee^{\infty}(0,1)$ turned out to be reduction classes. Following
he spirit of Büchi's reduction technique the logical description of
omputation processes could be simplified above all by the fact that
ne had not to care any more about the formal representation of the
bjects of computation - like numbers, words, sequences, domino posi-
ions and like; these data are represented just as individual terms
ppearing in formulae in Skolem normal form. As a consequence the re-
uction procedures revealed very close connections between structural
roperties of the considered computationprocesses and the logical
tructure (the expressive means) of the formulae describing these pro-
esses. It became clear that in order to show a class to be a reduc-
ion class one had to look above all for an appropriate type of com-
inatorial system - like Turing machines, Thue systems, Post corres-
ondence problems, domino games - whose unsolvable decision problem
ould be encoded smoothly into that class.

At this point Krom [1970], Aanderaa [1971] and myself in Börger
1971] independently came up with a third idea which pursued Büchi's
pproach to its last consequences. Whereas Büchi and his followers
n the 60-ies described computation processes with explicit reference

to the time component, we realized that this is not necessary if one aims at a description of properties of computations where the time needed to reach this property is irrelevant. Such a property is for ex. that a computation just halts, without worring about how many steps this may take. Therefore we tried and succeeded in *describing computation processes without referencing time*. This method allowed enormous simplifications of reduction formulae describing machine and like problems and resulted in almost trivializing many proofs and in much sharper reduction classes as before, defined by imposing restrictions not only on prefix and similarity, but also on the truth-functional structure of reduction formulae, on the structure of atomic subformulae in them and on the number of occurences of atomic subformulae.

In this survey we try to illuminate the intimate connections between computation formalisms and their logical descriptions from the viewpoint of predicate logic, of recursion theory and of complexity theory. We want to make clear above all the structural correlations between combinatorial decision problems and purely logical decision problems: how smooth reduction procedures carry over not only decidability or undecidability, but also all kinds of recursion theoretic properties and the degree of complexity of corresponding decision problems. We have no abstract general theory for that, but we hope to be able to transmit the feeling of these intimate connections through suitably chosen examples.

The *plan* of our paper is as follows. The first paragraph gives examples of how automata can be encoded by logical formulae, followed by a discussion of some logical decision problems thus obtained from corresponding machine problems. We then show examples of automata theoretic problems which can be settled by use of corresponding logical decision problem. In the third paragraph we give results from complexity theory derived from appropriate reductions of combinatorial decision problems to logical ones. Due to the essay character of this paper we do not aim at completeness at all. For most proofs we will refer to the literature. Only a few things are new and will be proved here. Terminology and prerequisites will be explained and mentioned at the places where needed for the first time.

Logical Decision Problems

We begin with a description of how machines M are encoded in formulae α_M. Through an appropriate choice of M and α_M we wish to achieve two things: to get syntactically simple formulae α_M whose logical structure reflects the syntactical structure of M, and - based on such a relation - to make the proof of equivalence of the M-decision problem to the α_M-deducibility question a trivial one. As explained in the introduction, the data of M are represented by the logical terms constituting the universe of the intended model of α_M; the states of the finite control of M are encoded by predicate symbols occuring in α_M. The quantifier free matrix of α_M will be a conjunction of implications where to each possible transition step ("instruction", "rule") I_i defined by the program M corresponds a conjunct ρ_i of α_M assuring that if an instance of the premisses in ρ_i represents a configuration C of M, then the corresponding instance of the conclusions in ρ_i represents the immediately succeding configuration according to I_i.

Consider a *register machine program* M working over 2 registers. Since their invention by Minsky [1961] and Shepherdson/Sturgis [1963] these machines have become widely known, in particular the fact that they are universal for the computation of all partial recursive functions. For convenience of exposition we assume without loss of generality that M consists of instructions $I_i=(i,o_i,p_i,q_i)$ with instruction numbers $1 \leq i \leq r$, operation symbols $o_i \in \{a_1,a_2,s_1,s_2,\text{stop}\}$ and numbers $1 \leq p_i,q_i \leq r$ of the next instruction to be executed. Execute I_i means: test, if the register considered in o_i equals zero or not, execute o_i - i.e. +1 in the j-th register in case $o_i=a_j$, -1 in the j-th register in case $o_i=s_j$ - and then go over to the next instruction with number p_i resp. q_i if the register tested was equal zero resp. not. Assume also that $I_r=(r,\text{stop},r,r)$, that the start instruction is I_1 and that tests are executed only in subtraction instructions (i.e. $o_i \in \{a_1,a_2\}$ implies $p_i=q_i$).

To define α_M we represent natural numbers n by means of logical terms $\underline{n} \equiv 0\underbrace{'\cdots'}_{n\text{-times}}$ built up from a symbol 0 for an individual constant and a symbol ' of a monadic function. Any state i of M is represented by a binary predicate symbol, call it again I_i. Therefore I_ixy represents the M-configuration (i,x,y) with internal state i and register contents x,y. Define

$$\sigma_M :\equiv \bigwedge_x \bigwedge_y (\rho_1 \wedge \ldots \wedge \rho_{r-1})$$

where

$$\rho_i :\equiv \begin{cases} I_i xy \to I_{p_i} x'y & \text{if } o_i = a_1 \\ \rho_{i,o} \wedge \rho_{i,1} & \text{if } o_i = s_1 \end{cases}$$

with $\rho_{i,o} :\equiv (I_i Oy \to I_{p_i} Oy) \wedge (I_i x'y \to I_{q_i} xy)$, and symmetrically for $o_i \in \{a_2, s_2\}$ with interchanged positions of x and y.

σ_M logically describes the effect of possible M-instructions on configurations, and indeed it is easy to show:

Theorem 1. $\forall 1 \leq i \leq r : \forall m,n,p,q$:
(1) $(i,m,n) \vdash_M (r,p,q)$ iff $\sigma_M \vdash_{PL} (I_i \underline{mn} \to I_r \underline{pq})$

where $C \vdash_M C'$ means that M, started in C, after a finite number of steps reaches C', and $\vdash_{PL}$ denotes deducibility in predicate logic.

Proof: Assume $(i,m,n) \vdash_M (r,p,q)$. Then there exists a terminating computation $C_o, C_1, \ldots, C_t$ of M starting with $C_o = (i,m,n)$ und with stop configuration $C_t = (r,p,q)$. We have to show that every model for $\sigma_M \wedge I_i \underline{m}\ \underline{n}$ is also a model for $I_r \underline{p}\ \underline{q}$. By Skolem's theorem (see for ex. Kreisel/Krivine [1967: 18-20]) it is sufficient to consider models $A = \langle \omega; \mathcal{I}_1, \ldots, \mathcal{I}_r \rangle$ with domain $\omega = \{O, O', O'', \ldots\}$. By an easy induction on $s \leq t$ one shows the simulation property: if $C_s = (i_s, m_s, n_s)$, then $\mathcal{I}_{i_s}(\underline{m_s}, \underline{n_s})$ is true in A - note that for C_o this is assured by the conjunct $I_i \underline{m}\ \underline{n}$. Therefore $\mathcal{I}_r(\underline{p}, \underline{q})$ holds in A since by assumption $C_t = (r,p,q)$.

Assume now $(i,m,n) \not\vdash_M (r,p,q)$. Then the following algebra $A = \langle \omega; \mathcal{I}_1, \ldots, \mathcal{I}_r \rangle$ is obviously a model for $\sigma_M \wedge I_i \underline{m}\ \underline{n} \wedge \neg I_r \underline{p}\ \underline{q}$:

$$\mathcal{I}_j(\underline{p}, \underline{q}) : \text{iff} (i,m,n) \vdash_M (j,p,q)$$

Corollary 1 (Church [1936], Turing [1937]) The Entscheidungsproblem for (restricted) first order predicate logic is unsolvable.

Corollary 2 (Aanderaa [1971], Börger [1971]). The Büchi prefix class $\forall\wedge\forall\wedge$ in Krom and Horn formulae is a reduction class, where a *Krom formula* (Kr) is a formula in prenex conjunctive normal form with binary disjunctions, a *Horn formula* (Hf) a formula in prenex conjunctive normal form whose disjunctions contain not more than one non-negated disjunct.

Proof. By theorem 1 the unsolvable problem $(1,m,n) \vdash_M (r,o,o)$ is many-one reduced to the PL-deducibility problem of $\sigma_M \wedge I_i \underline{m}\ \underline{n} \longrightarrow I_r OO$ which is the Skolem normal form of an equivalent formula α_M of restricted predicate calculus. By the universality of 2-register machines there is an M such that for all (Gödel numbers of) first order formulae α : α is satisfable iff M, started in configuration $(1,2^{\neg\alpha},0)$, does not reach the halting configuration (r,0,0) which by theorem 1 is equivalent to the satisfiability of the formula $\sigma_M \wedge I_1 \underline{2^{\neg\alpha}}\, 0 \wedge \neg I_r 00$; but the latter can be brought into Skolem prenex normal form of an equivalent $\alpha_M \in \vee\wedge\vee\wedge \cap \mathrm{Kr} \cap \mathrm{Hf}$. Therefore this class is a reduction class, i.e. the Entscheidungsproblem of PL can be reduced to the Entscheidungsproblem of this class.

With respect to prefix structure the class $\vee\wedge\vee\wedge \cap \mathrm{Kr}$ is already a border case since deletion of any quantifier in it results in a solvable Krom prefix class (i.e. with solvable Entscheidungsproblem), see Aanderaa/Lewis [1973]. Variants and much more sophisticated versions of the fundamental method of encoding machines into formulae shown in theorem 1 have been produced to get smaller and smaller reduction classes up to the point where solvable classes appear. We cite only a few examples: Goldfarb [1974] shows the prefix class $\wedge\vee\wedge^{\infty}$ to be a reduction class even when restricted to formulae with *only 4 occurences of atomic subformulae* (the case with 3 is open); the trick is to find a way of encoding the conjunction of all ρ_i in our σ_M as arguments of a single atomic formula, in such a way that these arguments function as program table and the logical formula encoding the effect of M-instructions I_i provides for a machinery to look up ρ_i in the table. For the case of predicate logic with functions and identity Wirsing [1977] develops a very sophisticated encoding machinery to show the class of all *formulae* $\bigwedge_{x_1}\ldots\bigwedge_{x_6}(s=t \wedge u \neq v)$ *with only one equation and one inequality* to be a reduction class, where the terms s,t,u,v are built up from one binary function symbol and $x_1,\ldots,x_6$; roughly speaking the idea here is to simulate M-transformations by equating terms representing a configuration and its immediate successor, the unequality stating that identiy of initial and final configurations does not hold.

Another direction pursued to produce reduction classes at the boundary to the decidable has been to look not for formulae with few atomic subformulae, but to prefix classes where only very *few kinds of possible atomic subformulae* are used. The most interesting case in this respect is the case of $\bigwedge_x\bigvee_u\bigwedge_y$-subclasses of predicate logic with only one binary and arbitrary monadic predicate symbols, specified by which of the 12 possible forms Rxx,Ryy,Ruu,Rxy,Ryx,Ryu,Ruy,Rxu,Rux,Px, Py,Pu of atomic formulae may occur. The history of this case illuminates very well our thesis about the interplay between combinatorial decision problems and purely logical decision problems. Indeed to settle completely the remained most difficult of the 2^{12} possible particular subcases, Aanderaa and Lewis [1974] especially developed an interesting new combinatorial problem, the so called linear sampling problem, and the hardest part of their work consisted in showing that some special cases of this linear sampling problem - which could be reduced conveniently to the classes of formulae under consideration - are indeed unsolvable. As a result all those $\bigwedge_x\bigvee_u\bigwedge_y$-subclasses are unsolvable where at least three of the twelve forms including the pair Rxy,Ryu or the pair Ryx,Ruy may occure as atomic subformulae, see Lewis [1980]; all other subclasses are solvable (see Dreben/Goldfarb [1980]).

To give at least one more sophisticated version of a reduction in the spirit of theorem 1, but which can still be treated within the scope of this paper, let us present a theorem due to Rödding and myself (for an abstract see Rödding/Börger [1974]). Kostyrko [1964] showed $\wedge\vee^{\infty}\wedge(0,1)$ to be a reduction class, Krom [1970] improved this to $\wedge\vee^{\infty}\wedge(0,\infty)\cap Kr$. By an appropriate modification of 2-register machines we could show that 4 binary predicate symbols are sufficient for this prefix class in Krom formulae. The decision problem of the classes $\wedge\vee^{\infty}\wedge(0,k)\cap Kr$ with $1\leq k\leq 3$ is still open. It is known however that Kr restricted to formulae with a single predicate letter, a dyadic one, is a reduction class; Lewis [1976] shows this by reduction of the Post correspondence problem - in the spirit of the method outlined in theorem 1 - in combination with ideas developed by Shannon for his construction of a universal Turing machine with two internal states. Now to our

Theorem 2. (Rödding/Börger) $\wedge\vee^{\infty}\wedge(0,4)\cap Kr\cap Hf$ is a reduction class.

Proof: As discussed before it is sufficient to show that for an arbitrary 2-register machine M we can construct formulae σ_M, $START_1$, $STOP_r$ such that analogously to theorem 1

(1) $(1,0,0)\vdash_M(r,a,b)$ for some a,b iff $\sigma_M\vdash_{PL} START_1\rightarrow STOP_r$

holds where $\sigma_M\wedge START_1\wedge\neg STOP_r$ can be prenexed in such a way that it becomes Skolem normal form of a formula $\alpha_M\in\wedge\vee^{\infty}\wedge(0,4)\cap Kr\cap Hf$.

We have to find an encoding of configurations (i,p,q) by atomic formulae built up from terms over monadic function symbols; let us choose $Ka_i|^p*x\,|^q*x$ with binary predicate symbol K, variable x and function symbols $a_i,*,|$. For intermediate calculations needed to be able to give a logical description in Skolem normal form we use further monadic function symbols a,b_i,c_i,d_i,e_i for $1\leq i\leq r$ and 3 auxiliary symbols K',K^*,K^c_o. Assume w.l.o.g. that M operates on the first register only when the second register is not empty.

To describe the initial configuration (1,0,0) define

$$START_1 :\equiv \bigwedge_{xy}\bigwedge (Kaxx \wedge (K^*yx \leftrightarrow Ky{*}x) \wedge (K^*axy \rightarrow K^*a_1xy))$$

assuring that $Ka_1{*}u{*}u$ holds for any u in any model for $START_1$. The stop condition is expresses by

$$STOP_r :\equiv \bigvee_{xy}\bigvee Ka_rxy.$$

σ_M is the conjunction of all ρ_i, σ_i defined below, prefixed by $\bigwedge_{xy}\bigwedge$. Operations on the second register can be expressed directly by use of the two auxiliary predicate symbols K',K* encoding just one occurence of | resp. * in front of the second argument of K:

$$\rho_i :\equiv (Ka_ixy \rightarrow K'a_{p_i}xy) \wedge (K'yx \leftrightarrow Ky|x) \qquad \text{for } O_i=a_2$$

$$\rho_i :\equiv (K'a_ixy \rightarrow Ka_{q_i}xy) \wedge (K^*a_ixy \rightarrow K^*a_{p_i}xy) \qquad \text{for } O_i=s_2$$

To describe operations on the first register we are not allowed to write $Ka_i|xy$ or to use new auxiliary predicates K^{a_i} similar to K' and K*. Therefore we have to bring the state information a_i in Ka_ixy in front of the second argument - we encode it as string of i consecutive occurences of * using the fact that at this very moment the second argument begins with an occurence of | -, then to operate on the first register content as asked by I_i and finally to transform the state information $*^i$ in front of the second argument into the correct new state symbol a_{p_i} resp. a_{q_i} in front of the first argument. These intermediate "stack" operations are achieved by means of new function symbols b_i, c_i, d_i, e_i as follows:

$$\rho_i :\equiv (Ka_ixy \rightarrow Kb_ixy) \wedge (K'c_ixy \rightarrow K'a_{p_i}xy) \qquad \text{for } O_i=a_i$$

$$\sigma_i :\equiv (Kb_{i+1}xy \rightarrow K^*b_ixy) \wedge (Kb_oxy \rightarrow K^{c_o}|xy) \wedge (K^{c_o}xy \leftrightarrow Kc_oxy) \wedge$$
$$\wedge (K^*c_ixy \rightarrow Kc_{i+1}xy) \qquad \text{for } 0 \leqq i < r$$

$$\rho_i :\equiv (Ka_ixy \rightarrow Kd_ixy) \wedge (K'c_ixy \rightarrow K'a_{p_i}xy) \wedge (K'e_ixy \rightarrow K'a_{q_i}xy) \quad \text{for } O_i=s_i$$

$$\sigma_i :\equiv (Kd_{i+1}xy \rightarrow K^*d_ixy) \wedge (Kd_oxy \rightarrow Kxy) \wedge$$
$$\wedge (K{*}xy \rightarrow K^{c_o}{*}xy) \wedge (K|xy \rightarrow Ke_oxy) \wedge$$
$$\wedge (K^*e_ixy \rightarrow Ke_{i+1}xy) \qquad \text{for } 0 \leqq i < r$$

By the explanations preceding the definitions of ρ_i and σ_i it should be clear that in any model for $\sigma_M \wedge START_1$ the following simulation property holds: if $(1,0,0) \vdash_M (i,m,n)$, then $Ka_i|^m{*}u|^n{*}u$ is fullfilled in the model for any element u. This yields the implication from left to right of (1) as in theorem 1.

Assuming $(1,0,0) \not\vdash_M (r,a,b)$ for any a,b it is easy to verify that one gets a term model for $\sigma_M \wedge START_1 \wedge \neg STOP_r$ by defining:

$K(v,w)$: iff there are a term u and numbers $t,p,q,1\leq i\leq r,\ j\leq i$ s.t.

$(v\equiv au$ and $w\equiv u)$ or

$C_t=(i,p,q)$ and $\{(v\equiv a_i|^p*u, w\equiv|^q*u)$

or $(O_i=a_1,\ w\equiv *^j|^q*u,\ (v\equiv b_{i-j}|^p*u$ or $v\equiv c_{i-j}|^{p+1}*u))$

or $(O_i=s_1),\ w\equiv *^j|^q*u,\ [v\equiv d_{i-j}|^p*u$ or $(v\equiv c_{i-j}*u,\ p=0)$

or $(v\equiv e_{i-j}|^{p-1}*u, p>0)])$

or $(O_i=s_1,\ w\equiv *^i|^q*u, v\equiv|^p*u)\}$

We want to conclude this section with the observation that by theorem 1 we have as by-product also *Trachtenbrot's* [1953] *theorem on the recursive unseparability of the sets of all contradictory resp. finitely satisfiable formulae*. Indeed consider all 2-register machines with two stop states $I_o=(0,stop,0,0)$ and $I_r=(r,stop,r,r)$. Since there exist recursively unseparable sets, the classes N and F of all such machines which started in $C_o=(1,0,0)$ stop in state r resp. 0 are recursively unseparable. By theorem 1, first half we know already

(i) $M\in N \succ \sigma_M \wedge I_1 00 \wedge \bigwedge\bigwedge_{xy} \neg Krxy$ is contradictory.

Looking again at the second half of the proof for theorem 1 we see that if $M\in F$, then the model constructed there can be cut down to its finite part $\{1,\dots,l\}$ where l is the successor of the maximal register content occuring during the terminating computation started with $C_o=(1,0,0)$ and where k' is defined as k+1 for k<l and l':=l. Therefore we know also

(ii) $M\in F \succ \sigma_M \wedge I_1 00 \wedge \bigwedge\bigwedge_{xy} \neg K_r xy$ is finitely satisfiable.

(i) and (ii) yield Trachtenbrot's theorem, even restricted for ex. to $\vee\wedge\vee\wedge$-Krom formulae.

Modifying slightly the proof just given one obtains the following generalization of Trachtenbrot's theorem:

Theorem 3. The sets of all contradictory resp. finitely satisfiable resp. those formulae which admit only infinite models are recursively unseparable.

Proof. By the unsolvability of predicate logic it is obvious that the classes of all contradictory resp. only infinitely satisfiable formulae are not recursively separable. Therefore we have still to show that the finitely satisfiable formulae and the infinity axioms are not recursively separable. Since F and the set I (of all 2-register machines whose configuration sequence defined by $C_o=(1,0,0)$ does not become periodic) are known to be recursively unseparable, it is sufficient to show that one can define a formula σ'_M fulfilling (ii) and (iii):

(ii) $M\in F \left.\right\}$ $\sigma_M \wedge \sigma'_M \wedge I_1 00 \wedge \bigwedge\bigwedge_{xy} \neg K_r xy$ has a finite model

(iii) $M\in F \left.\right\}$ " " " " is an infinity axiom

σ'_M has to assure that if the M-computation does not become periodic, then any model of our formula has arbitrarily "large" elements corresponding to the arbitrarily big numbers appearing in some register. Define therefore with a new binary predicate symbol G:

$$\sigma'_M :\equiv \bigwedge\bigwedge_{xy}(\neg Gxx \wedge (Gxy \to Gx'y) \wedge \bigwedge_{i\leq r}((I_i xy \to Gx'x) \wedge (I_i yx \to Gx'x)))$$

Thus σ'_M requires for every register content x a "greater" element x' and (ii) and (iii) follow.

By similar methods all known reduction classes can be shown to preserve also finite satisfiability of formulae; for a more difficult case see Aanderaa/Börger/Lewis [1981].

Automata theoretic decision problems

In the previous section we have used reductions of machines to formulae to establish unsolvability of logical decision problems from known unsolvability results for machines. By contraposition one can establish decidability results for machines based on decidable corresponding logical decision problems.

Theorem 4 (Lewis [1972]). The emptiness problem for context-free grammars, finite (tree) automata and pushdown automata is reducible to the decision problem of monadic predicate logic and therefore recursive.

Proof. Take any context-free grammar M with variables $I_1,\ldots,I_r$, terminal letters $a_1,\ldots,a_n$, axiom I_1 and rules $I_j \to I_k I_l$ for $(j,k,l)\in R_1$ and $I_j \to a_k$ for $(j,k)\in R_2$ in Chomsky normal form. (Without loss of generality we assume that no e-rules occur.) To describe M-deductions ending in a terminal word we use I_i as monadic predicate symbol, a_j as individual constants and ( ) as a binary function symbol. Analogously to theorem 1 define:

$$\rho_i :\equiv \begin{cases} I_k x \wedge I_l y \to I_j(xy) & \text{for } i=(j,k,l)\in R_1 \\ I_j a_k & \text{for } i=(j,k)\in R_2 \end{cases}$$

$$\sigma_M :\equiv \bigwedge_{xy} \bigwedge_{i\in R_1 \cup R_2} \rho_i$$

Following the lines of the proof exemplified for theorem 1 it is now routine to show:

(i) $L(M) \neq \emptyset'$ iff $\sigma_M \vdash_{PL} \bigvee_x I_1 x$

(Hint: Think of $I_i t$ as stating $I_i \vdash_M \bar{t}$ where the word $\bar{t}$ over $a_1, \ldots, a_n$ is obtained from the term t by omitting all occurences of parantheses.)

It should be ovious now how the construction of theorem 4 applies to regular grammars. For the *emptiness problem of tree automata* M of arbitrary order k, states i of M are represented by a monadic predicate symbol I_i $(1 \leq i \leq r)$, the letters a_j of M by k-ary function symbols a_j $(1 \leq j \leq n)$, take c to be an individual constant. Define

$$\rho_i :\equiv I_{i(1)} x_1 \wedge \ldots \wedge I_{i(k)} x_k \rightarrow I_1 a_j x_1 \ldots x_k$$

for any instruction $I_i = ((i(1), \ldots, i(k), j) \rightarrow 1)$ of M and

$$\sigma_M :\equiv \bigwedge_{x_1} \ldots \bigwedge_{x_k} (\rho_1 \wedge \ldots \wedge \rho_r)$$

Every term built up from C and the function symbols a_j $(1 \leq j \leq n)$ corresponds to a unique tree $\bar{t}$ of order k over the alphabet $\{a_1, \ldots, a_n\}$, therefore think of $I_i t$ as telling that M, started in its initial state 1 over the tree $\bar{t}$, reaches the state i at the root. Thus it becomes routine to show (assume r to be the stop state of M):

$$L(M) \neq \emptyset \quad \text{iff} \quad \sigma_M \vdash_{PL} (I_1 c \rightarrow \bigvee_x I_r x)$$

For push-down automata the construction is similar, see Lewis [1972].

Following this line Lewis gives also a refinement of Manna's [1968] description of abstract programs by logical formulae yielding for ex. a proof for the solvability of the totality and total-equivalence *problem for* full *program schemata*. Lewis [1972] specifies furthermore an interesting class of formulae with solvable decision problem and such that leaving out any of its specific restrictions yields an unsolvable class (Post corresp. problems can be encoded via automata with an oracle for comparing the contents of their stack to the input read so far, or the emptiness problem for two-pushdown automata or for the intersection of two context free languages).

Another case where a combinatorial decision problem corresponds *exactly* to the decision problem of a class of logical formulae has been found by Börger/Kleine Büning [1980]. A broad class of formulae of extended purely multiplicative arithmetic is specified there by syntactical restrictions on prefix, similarity, truth functional form and form of the terms and it is shown that:

a) the *reachability problem for Petri nets* is reducible to the decision problem of that class,

b) every formula in it expresses a particular reachability question,

c) perturbation of any of the syntactic specifications yields an unsolvable class. Unfortunately we have not been able to find a logical description of the reachability problem for Petri nets by formulae of a class whose decision problem is already known to be solvable. This is a challenging task however.

Complexity of decision problems

We show in this section that by theorem 1 machine programs and logical formulae are so closed to each other that not only mere (un-)decidability, but also the complexity of corresponding decision problems are preserved. We give five typical examples to illustrate this:

1) recursive isomorphy type or r.e. many-one degree complexity of machine problems carry over to the corresponding logical decision problems, yielding also exact localizations of various logical meta-decision problems in the arithmetical hierarchy and a solution to a problem of Wang [1962];

2) inseparability properties for machines result in sharp lower bounds for the complexity of models for the describing logical formulae, yielding among others a very simple solution to a problem of Hilbert/Bernays [1939];

3) application of the method of theorem 1 to the description of finite computations within fixed time bounds yields

a) a sharpening of Cook's NP-completeness theorem for propositional logic with a surprising logical characterization of Turing machine complexity for Boolean functions,

b) an interesting approximation of first and higher order spectra and

c) sharp lower bounds for the subrecursive complexity of many classical logical decision problems.

Careful inspection of theorem 1 shows that the many-one degrees of the halting problem $H_M=\{(1,m,n) \mid (1,m,n) \vdash_M (r,.,.),\ 0\leq m,n\}$ and of the $\vdash$-decision problem of the class $A_M=\{(\sigma_M \wedge \bigwedge_{xy}\bigwedge \neg I_r xy) \rightarrow \neg I_1 \underline{mn} \mid 0\leq m,n\}$ are the same, that the deducibility problem of any r.a. set A of formulae is m-equivalent to an H_M and that under a certain natural assumption on A these relations hold even for recursive isomorphy instead of m-equivalence. On the one hand this means that for arbitrarily prescribed r.e. m-degree there are candidates (namely the A_M) for logical decision problems which are as natural as machine problems or any other problems representing the given degree (see Börger [1979]). This answers the question in Wang [1962], pg. 54. On the other hand the correlation between H_M and A_M permits to deduce from well known recursion theoretic facts about the arithmetical complexity of index sets of classes of machines exact locations of logical meta-decision problems in the Kleene-Mostowski hierarchy like: the property of a set of formulae to contain no deducible formula is Π_1-complete, to contain an infinite number of (resp. only) deducible formulae is Π_2-complete, to contain almost only deducible formulae resp. to have a solvable $\vdash$-decision problem resp. to be a reduction class is Σ_3-complete (see Börger/Heidler [1976], Börger [1974]).

Hilbert/Bernays [1939], pg. 191 raised the problem whether every closed satisfiable formula of restricted predicate logic admits a recursive model. Kreisel [1953], Mostowski [1953], Mostowski [1955] and Rabin [1958] gave rather complicated examples - with complicated proofs - of *satisfiable formulae excluding recursive models*. The method of theorem 1 yields simple such formulae, and without too much effort: Define $\tau_M :\equiv \sigma_M \wedge \bar{\sigma}_M$ where $\bar{\sigma}_M$ results from σ_M by inverting all implication signs. Let M have a second stop instruction $I_o=(0,\text{stop},0,0)$. Paraphrasing the proof for theorem 1 one shows that for all $1 \leq r$ and m,n holds:

(1) $(i,m,n) \vdash_M (r,0,0)$ iff $(\tau_M \wedge \neg I_r 00 \wedge I_o 00) \vdash_{PL} \neg I_i \underline{mn}$

(2) " " " $(0,0,0)$ iff " " " " $\vdash_{PL} I_i \underline{mn}$

(Hint: from right to left in (1) interprete $I_j \underline{pq}$ as $(j,p,q) \not\vdash_M (r,0,0)$, in (2) as $(j,p,q) \vdash_M (0,0,0)$; from left to right in (2) use $\bar{\sigma}_M$ instead of σ_M.)

Take τ^o_M as conjunction of all conjuncts $\rho_{i,0}$ in σ_M and $\bar{\sigma}_M$, of $\neg I_r 00$ and of $I_o 00$ after substitution of 0 by u; take τ^1_M as conjunction of all other conjuncts in σ_M and $\bar{\sigma}_M$ after substitution of x' by v. For a new binary predicate variable S (for "successor") define

$$\gamma_M :\equiv \bigvee_u \bigwedge_y \tau^o_M \wedge \bigwedge_x \bigvee_v Nxv \wedge \bigwedge_x \bigwedge_v \bigwedge_y (Nxv \rightarrow \tau^1_M).$$

<u>Corollary 4</u> (Börger [1975], Aanderaa [1971]). γ_M is satisfiable but without recursive models, if M is a 2-register machine M with recursively unseparable sets $S_i=\{n \mid (1,2^n,0) \vdash_M (i,0,0)\}$ for i=0,r).

<u>Proof.</u> $I := \{n \mid \mathcal{A} \text{ verifies } I_1 \underline{2^n} 0\} \notin \Sigma_o$ for any model $\mathcal{A}$ of γ_M since by (1),(2) $S_o \subseteq I$, $I \cap S_1 = \emptyset$.

Note that $\gamma_M \notin Kr$ and that every satisfiable Krom formula has a Σ_o-model

(Aanderaa/Jensen [1973], Ershov [1973]).

Corollary 5. $\tau_M \wedge \neg I_r 0\ldots 0 \wedge I_o 0\ldots 0$ is $E_{n+1}-E_n$ satisfiable for any register machine M with E_n-unseparable E_{n+1}-sets $S_i=\{n \mid (1,n,0,0,\ldots) \vdash_M (i,0,0,\ldots)\}$ for i=0,r, where E_n denotes the n-th set of the Grzegorczyk-hierarchy, $n\geq 3$.

Proof. Use the fact that a set X is in E_n iff it can be decided (i.e. $X=S_r$) by a register machine within E_n-time bound and that there are E_n-inseparable E_{n+1}-sets.

Adapting the reduction method of theorem 1 to terminating computations within fixed time bounds, one can obtain a refined version of Cook's NP-completeness of propositional logic: machine computations of length l over input $q_1,\ldots,q_n$ are expressed by formulae $\alpha_{M,l,n}(x_1|q_1,\ldots,x_n|q_n)$ with variables $x_1,\ldots,x_n$, $y_1,\ldots,y_m$ such that all instances $\alpha_{M,n,l}(x_1|q_1,\ldots,x_n|q_n)$ of $\alpha_{M,n,l}$ are *Horn formulae*. The decision problem for Horn formulae is in P; measuring the minimal length of such pseudo-Horn definitions for arbitrary Boolean functions f yields a complexity measure which has been shown equivalent to network and therefore Turing machine complexity of f, see Aanderaa/Börger [1981].

Another interesting way of encoding finite computations by small formulae in the spirit of Büchi can be found in Rödding/Schwichtenberg [1972]. Long but terminating computations of register machines M are here described by formulae of type theory, where the type structure helps to encode the length of the computations. The uphsot of this nice construction is an interesting approximation of the classes of all spectra at type level n by a natural hierarchy between E_2 and E_3.

Finally let us mention applications of smooth reductions of combinatorial decision problems to logical decision problems in the field of sharp complexity bounds for various decision problems. Most lower bound results in the literature for decidable logical theories or classes of formulae depend heavily on the quality of reduction of machines to formulae, which mostly means efficiency of encoding of "long" computations by "small" formulae. Since space does not allow us any more to be detailed we have to refer just to some surces, for ex. Ferrante/Rackoff [1979], Lewis [1980], Fürer [1980], where also more literature is cited.

References

NB. Due to the lack of space we cite only papers from which all other references mentioned in this survey can be found.

Aanderaa, S.O., Börger, E. [1981]: The equivalence of Horn and network complexity for Boolean functions. Acta Informatica (to appear)

Aanderaa, S.O., Börger, E., Lewis, H.R. [1981]: Conservative reduction classes of Krom formulas. The Journ. of Symb. Logic (to appear)

Börger, E. [1975]: On the construction of simple first-order formulae without recursive models. Proc. Coloquio sobra logica simbolica, Madrid, 9-24

--- [1979]: A new general approach to the theory of the many-one equivalence of decision problems for algorithmic systems. ZMLG 25, 135-162

Börger, E., Kleine Büning, H. [1980]: The reachability problem for Petri nets and decision problems for Skolem arithmetic. Theoretical Computer Science 11, 123-143

Dreben, B., Goldfarb, W.D. [1979]: The decision problem. Reading

Ershov, Yu.L. [1973]: Skolem functions and constructive models. Algebra y Logika 12, 644-654

Ferrante, J., Rackoff, Ch.W. [1979]: The Computational Complexity of Logical Theories. Springer LNM 718

Lewis, H.R. [1979]: Unsolvable classes of quantificational formulas. Reading

--- [1980]: Complexity results for classes of quantificational formulas. JCSS

Rödding, D., Börger, E. [1974]: The undecidability of $\wedge\vee^{\infty}\wedge(0,4)$-formulae with binary disjunctions. Journ. of Symb. Logic 39, 412-413

Rödding, D., Schwichtenberg, H. [1972]: Bemerkungen zum Spektralproblem. ZMLG 18, 1-12

Wirsing, M. [1977]: Das Entscheidungsproblem der Klasse von Formeln, die höchstens zwei Primformeln enthalten. manuscripta math. 22, 13-25

AN ALGORITHM TO IDENTIFY SLICES, WITH APPLICATIONS TO VECTOR REPLACEMENT SYSTEMS

Jan Grabowski

Sektion Mathematik, Humboldt-Universität, DDR 1086 Berlin

This paper is concerned with the theory of reachability relations in finitely generated free commutative monoids. Such monoids are of the form $(\mathbb{N}^K,+)$, where $\mathbb{N}$ denotes the natural numbers, and $K \in \mathbb{N}$.

Definition. A binary relation R on $\mathbb{N}^K$ is called a reachability relation iff

(1) R is reflexive,
(2) R is transitive,
(3) R is a submonoid of the product $\mathbb{N}^K \times \mathbb{N}^K = \mathbb{N}^{2K}$, and
(4) R is the least relation satisfying (1),(2),(3) and containing some finite relation R_o.

The finite defining relation R_o is usually called a vector replacement system, or a generalized Petri net (cf. Hack /8/). Throughout this paper, let some finite relation

$$R_o = \{(u^1,v^1),\ldots,(u^M,v^M)\}$$

be fixed, and R be the reachability relation defined by R_o.

The theory of vector replacement systems investigates the structure of reachability relations, in particular the decision problems connected with them. The most famous problem is the membership problem for R (the Reachability Problem). [1]

Of particular interest are those cases in which R obeys an especially simple structure. The most preferable structures in $\mathbb{N}^K$ are the semilinear sets. A subset of $\mathbb{N}^K$ is said to be semilinear iff it is definable in Presburger Arithmetic (PA), or equivalently, iff it admits a representation

$$S = \bigcup_{i=1}^{n} a_i + \underline{H}(B_i),$$

where $a_i \in \mathbb{N}^K$, and $\underline{H}(B_i)$ denotes the submonoid generated by the finite

---

[1]) Recently, Mayr /12/ presented an algorithm to solve the Reachability Problem. His solution does not cover the considerations to be presented here.

set B_i in $\mathbb{N}^K$.

It is interesting that in many cases where R or the set aR = $\{b \mid (a,b) \in R\}$ for some fixed a is a semilinear set, a representation of this set can be inferred from an enumeration of R. The specific way of utilizing the enumeration depends on the type of vector replacement system under consideration. Often it consists in evaluating some version of the Karp-Miller graph /10/.

We are going to study two such cases, illustrating how such algorithms can be derived from general theorems on inductive inference. These two cases (the symmetric and the persistent one) have been investigated by several authors (/1/,/3/,/9/,/11/,/13/) which, step by step, improved the understanding of the situation. Complete transparency will be achieved by means of our inference principle.

1. Slices and their identification

Jantzen and Valk /9/ were the first to recognize the importance of a special class of semilinear sets relative to the two cases mentioned above.

Definition. $S \in \mathbb{N}^K$ is said to be a slice iff the following implication is satisfied by all $u,v,w \in \mathbb{N}^K$:

If u, u+v, and u+w are in S, then u+v+w is in S.

The following was proved by Eilenberg and Schützenberger /4/.

Fact 1. Every slice is a semilinear set.

This enables us to state our main theorem.

Theorem 1. Let $S \in \mathbb{N}^K$ be a slice. Given an enumeration of S, i.e. an ascending chain

$$F_1 \subsetneqq F_2 \subsetneqq F_3 \subsetneqq \ldots.$$

of finite sets exhausting S, we can compute an ascending chain

$$S_1 \subsetneqq S_2 \subsetneqq S_3 \subsetneqq \ldots$$

of semilinear sets such that $S_j = S$ holds for almost all j. (Computing a semilinear set means computing a representation of it.)

Proof. Let

$$\underline{S}(F) := \bigcup_{a \in F} a + \underline{H}(B_a),$$

where

$$B_a = \{b \mid b \in \mathbb{N}^K \text{ and } a+b \in F\}.$$

If F is finite, then $\underline{S}(F)$ is semilinear. If F is contained in some slice S, so is $\underline{S}(F)$. Setting $S_j := \underline{S}(F_j)$ we obtain a chain with the desired property: $S_j \subseteq S$ follows from $F_j \subseteq S$. On the other hand, since S is semilinear, there exists some finite $F \subseteq S$ such that $S \subseteq \underline{S}(F)$. Let $F \subseteq F_{j_o}$. Then $S \subseteq \underline{S}(F) \subseteq S_{j_o} \subseteq S_j$ for all $j \geq j_o$.

Q.e.d.

Adopting terminology from inductive inference /5/, we could say that the class of all slices is identifiable in the limit by arbitrary enumeration.

As a corollary we observe the following: In order to compute a slice S, it is sufficient to have two things:

(i) an enumeration of S,

(ii) a rule to decide of an arbitrary semilinear subset Q of S if Q equals S.

The method is the following: Apply Theorem 1 to the enumeration given by (i), and terminate once, by (ii), you have $S_j = S$.

Item (ii) could also be given in terms of a first-order formula in the language $\underline{PA}$+Q consisting of Presburger Arithmetic plus a predicate variable Q of appropriate arity. The formula H(Q) should contain Q, and, for any subset Q of S, H(Q) should be true iff Q equals S. If H has this property, we call it a test formula for S.

Example. Assume we are to compute the least slice S containig some finite F. Obviously, an enumeration of S is available. A test formula is obtained by formalizing the phrase

"$F \subseteq Q$, and Q is a slice"

in $\underline{PA}$+Q.

In the next two paragraphs we shall apply our Theorem 1 to the symmetric and persistent cases mentioned above.

3. The symmetric case

Assume an arbitrary vector replacement system R_o to be given, R to be its reachability relation, and $Z = R \cap R^{-1}$ to be the symmetric kernel of R. What can we learn about the structure of Z? - Obviously, Z is a congruence relation on $\mathbb{N}^K$.

First let us assume that the defining relation R_0 be symmetric. The R is symmetric, and Z = R. For this case the membership problem was solved by Malcev /2/, and Biryukov /2/ showed how to compute the congruence classes, which are always semilinear. We will strengthen that result by observing the

Fact 2 /4/. Every congruence relation on $\mathbb{N}^K$ is a slice in $\mathbb{N}^{2K}$.

Fact 2 encourages us to apply the slice identification method. From R_0, an enumeration of R is obviously available. A test formula for R is a formula of PA+Q that says

"Q contains R_0 and is a congruence relation".

This is obviously expressible in PA+Q. Hence we have

Theorem 2. If R_0 is symmetric, R is a computable semilinear set.

Now let us return to the general situation in which R_0 is arbitrary, and assume that we are to compute Z. The congruence relation Z is, in general, not the one defined by R_0. However, it is finitely defined, namely by

$$Z_0 := \mathrm{Min}\left\{(z+u,z+v) \mid (u,v) \in R_0 \text{ and } (z+u,z+v) \in Z\right\}.$$

(Here Min denotes the set of minimal elements, according to componentwise ordering in $\mathbb{N}^K$.)

In /7/ it is shown that, for any finite relation R_0, Z_0 is computable. (The proof requires rather involved techniques and cannot be reproduced here.) Applying the arguments given above to Z_0 instead of R_0, we obtain

Theorem 3. If R_0 is an arbitrary vector replacement system, $Z = R \cap R^{-1}$ is a computable semilinear set. In particular, if R is symmetric, R is a computable semilinear set.

This covers several results on decision problems for Z given by Araki and Kasami /1/.

3. The persistent case

Let us first recall the basic notions connected with persistence. We call the elements of R_0 transitions. If $(u,v) \in R_0$, $y \in \mathbb{N}^K$, we say that the transition (u,v) transforms y+u into y+v. Since R is defined by the finite relation R_0, $(y,z) \in R$ holds iff y can be transformed into z by a finite sequence of transitions.

A conflict of R_o is an element $c \in \mathbb{N}^K$ such that, of two distinct transitions (u^j,v^j) and (u^k,v^k), either can be applied to c individually, while one disables the other. Formally this means that $c \geqq u^j$ and $c \geqq u^k$ and not $c-u^j+v^j \geqq u^k$. - We see that the property of being a conflict can be expressed in PA.

Definition. R_o is called persistent at a iff aR contains no conflict of R_o.

Landweber and Robertson /11/ observed that, when concerned with persistence, it is useful to consider a vector replacement system in extended form. The extended form of R_o is

$$\hat{R}_o := \{((0,u^j),(e^j,v^j)) \mid j = 1,\ldots,M\},$$

where e^j denotes the jth unit vector of $\mathbb{N}^M$. Thus $\hat{R}_o$ is a vector replacement system in $\mathbb{N}^{M+K}$. - The reader easily verifies that no information on R is lost if we consider (instead of R) the reachability relation $\hat{R}$ defined by $\hat{R}_o$. In particular, if a is the image of $\hat{a}$ under projecting $\mathbb{N}^{M+K}$ canonically to $\mathbb{N}^K$, then aR is the image of $\hat{a}\hat{R}$ under the same projection, and R_o is persistent at a iff $\hat{R}_o$ is persistent at $\hat{a}$.

Let us now assume R_o to be given in extended form already. As Jantzen and Valk /9/ observed, persistence of R_o at a implies that aR is a slice. This is the point where Theorem 1 comes to work.

Suppose, first, that R_o is persistent at a. Obviously, from a and R_o an enumeration of aR is available. Further, aR possesses a test formula in PA+Q, namely a formula that says

"$a \in Q$, and if $j \in \{1,\ldots,M\}$ and $y+u^j \in Q$ then $y+v^j \in Q$ ".

Hence aR is a computable semilinear set.

Now let us assume that we do not know if R_o is persistent at a, and we want to decide this question. We proceed as follows. We apply the algorithm of Theorem 1 to the enumeration of aR, as if the assumption of Theorem 1 were fulfilled. It generates a chain of semilinear sets $S_1 \subsetneqq S_2 \subsetneqq S_3 \subsetneqq \ldots$ Now we apply to each of these sets the decision procedure of the test formula given above. As soon as the formula becomes true, we halt.

Assume that this procedure eventually halts. Let the result be the semilinear set S. Then

$$S = \bigcup\{S_j \mid j \in \mathbb{N}\}.$$

We prove that the following conditions are equivalent:

(i) S contains no conflict (of R_0).

(ii) aR contains no conflict.

(iii) S = aR, and aR contains no conflict.

Proof: (i) $\rightarrow$ (ii) is clear since, by construction, aR $\subseteq$ S. (ii) $\rightarrow$ (iii) was shown above. (iii) $\rightarrow$ (i) is trivial.

Thus, if our algorithm terminates, we can decide the persistence of R_0 at a, since S is semilinear. If, however, our algorithm does not terminate, R_0 cannot be persistent at a. To recognize this, we run another algorithm in parallel, which enumerates aR until it detects a conflict in this set. In the case under consideration, this latter algorithm must halt and give the answer. We have proved:

Theorem 4. It is decidable whether aR contains no conflict, and in the positive case aR is a computable semilinear set.

Cf. /6/,/13/. In /7/ this result is strengthened by showing that even the relation $R|_{aR}$ (the restriction of R to aR) is a computable semilinear set. In particular, if R_0 has no conflict at all (is conflict-free), it is shown that R is a computable semilinear set.

4. Conclusions

Our results rely on Theorem 1 which says that the class of all slices is identifiable in the limit. Obviously, the class of all slices shares this property wich the class of all finite sets in $\mathbb{N}^K$, and also with the class of all finitely generated submonoids of $\mathbb{N}^K$. This suggests that the same "inductive" approach can help to solve computational problems for other classes of vector replacement systems, too. For example, if R_0 is context-free (i.e. for every (u,v) $\in R_0$ u is a unit vector) then, by Parikh's Theorem, aR is semilinear. Again, an inductive identification technique can be applied to compute aR.

On the other hand it cannot be expected that arbitrary large classes of semilinear reachability sets can be handled by such an approach. This is due to the fact that the class of all semilinear sets is not identifiable in the limit, for it contains all finite plus some infinite sets (cf. Gold /5/).

Summary: If a reachability relation R, or the set aR for some fixed a, is semilinear, one can often compute R (resp. aR) from an enumeration of it. To do so, we have to show that R (resp. aR) is contained in some identifiable class of semilinear sets. An important example of such a class is the class of slices.

The question arises whether such an approach is suitable for computing non-semilinear reachability sets as well. Our answer is pessimistic. The main difficulty arises from the truncation rule which was expressed by the "test formula" in the language PA+Q. As long as Q is required to be interpreted as a semilinear relation, the test formula is decidable. But even a very limited amount of nonlinearity makes PA+Q an undecidable theory.

References

/1/ Araki,T., and T.Kasami: Decidable properties on the strong connectivity of Petri net reachability sets, Theoretical Computer Science 4 (1977) 99-119.

/2/ Бирюков,А.П.: О некоторых алгорифмических задачах для конечно-определенных коммутативных полугрупп. Сиб. мат. журнал III (1967) 525-534

/3/ Crespi-Reghizzi,S., and D.Mandrioli: A decidability theorem for a class of vector addition systems. Information Processing Letters 3 (1975) 78-80.

/4/ Eilenberg,S., and M.P.Schützenberger: Rational sets in commutative monoids. Journal of Algebra 13 (1969) 173-191.

/5/ Gold,E.M.: Language identification in the limit. Information and Control 10 (1967) 447-474.

/6/ Grabowski,J.: The decidability of persistence for vector addition systems. Information Processing Letters 11 (1980) 20-23.

/7/ Grabowski,J.: Linear methods in the theory of vector addition systems III. Submitted to Elektronische Informationsverarbeitung and Kybernetik.

/8/ Hack,M.: Decision problems for Petri nets and vector addition systems. MAC-TM-59, Project MAC, M.I.T., Cambridge 1975.

/9/ Jantzen,M., and R.Valk: Formal properties of place transition nets. In: Net Theory and Applications. LCS 84, Springer-Verlag 1980

/10/ Karp,R.M., and R.E.Miller: Parallel program schemata. Journal of Computer and System Sciences 3 (1969) 167-195.

/11/ Landweber,L.H., and E.L.Robertson: Properties of conflict-free and persistent Petri nets. Journal of the ACM 25 (1978) 352-364.

/12/ Mayr,E.W.: Ein Algorithmus für das allgemeine Erreichbarkeitsproblem bei Petrinetzen und damit zusammenhängende Probleme.

TUM-I8010, Technische Universität München, September 1980.

/13/ Müller,E.: Decidability of reachability in persistent vector replacement systems. In: Mathematical Foundations of Computer Science 1980. LCS 88, Springer-Verlag 1980, 426-438.

ONE PEBBLE DOES NOT SUFFICE TO SEARCH PLANE LABYRINTHS

Frank Hoffmann , Institut f. Mathematik der Akademie
der Wissenschaften der DDR
1080 Berlin , Mohrenstr. 39, G.D.R.

0. Abstract

In 1975 L. Budach proved in a groundbreaking paper ([7]) that there is no finite automaton which is able to search (to master) all finite (cofinite) plane labyrinths. On the other side we have a result of M. Blum, D. Kozen (1978, [4]) saying that the search can be implemented with just two pebbles. The aim of our paper is to show that one pebble does not suffice, answering a question of Blum,Kozen. Furthermore we present a new construction for universal traps (see H.Antelmann, L.Budach, H.-A.Rollik 1979, [2]).

1. Notations, the Main Theorem and Budach Traps

1.1 Labyrinths and pebble automata (for a formal treatment see [6] [7], [9])

Let $D=\{n,e,s,w\}$ be the set of the four compass directions. If $r\in D$, then by $\bar{r}$ or r^{-1} we denote the opposite direction: $\bar{n}=s$, $\bar{w}=e$, $\bar{\bar{r}}=r$.
A labyrinth L is a graph $L=(Z,E)$,where Z is the set of points and $E\subset Z\times D\times Z$ the set of edges satisfying two conditions:

(i) If $(P,r,Q)\in E$, then $(Q,\bar{r},P)\in E$.

(ii) If $(P,r,Q),(P,r,Q')\in E$, then holds $Q=Q'$.

An edge is also denoted by $Pr=Q$ or $P\xrightarrow{r}Q$. The usual lattice of integer coordinates defines in a natural way a labyrinth. For every subset $M\subset\mathbb{Z}^2$,the so called obstrucles, $\mathbb{Z}^2\setminus M$ generates a plane labyrinth L_M , that is a full sublabyrinth of $\mathbb{Z}^2$. If M is finite and L_M connected, then we call L_M a cofinite maze ; if L_M is finite and connected, then L_M is a finite maze.

A finite automaton $\mathfrak{A}$ walking in labyrinths works as follows: $\mathfrak{A}$ is started in its internal initial state $s_0\in S$ in a point $P_0\in L$. It recognizes the free directions in P_0 ,that means $val_L P_0=\{r\mid\exists Q: (P_0,r,Q)\in E\}$

and depending on its transition function $\mathfrak{A}$ moves in one of the free directions $r_o \in val_L P_o$ to the next point $P_1 = P_o r_o$ and enters a new internal state s_1. We get a sequence (P_k, $val_L P_k$, s_k, r_k) , the so called behaviour of $\mathfrak{A}$ in the pointed labyrinth (L,P_o).
We are concerning in this paper with finite 1-pebble automata $\mathfrak{A}^*$, that means finite deterministic automata as above but equiped with just one pebble. $\mathfrak{A}^*$ can drop its pebble on a point P it is visiting and after returning to P the automaton senses the pebble and if desired $\mathfrak{A}^*$ moves it to the next point. More precisely, $\mathfrak{A}^*$ recognizes in each point P of its path the valency $val_L P$ and the pebble situation $\alpha(P) \in \{0,1,2\}$:

0 = ($\mathfrak{A}^*$ is carrying the pebble)
1 = (the pebble is lying in P)
2 = (the pebble is lying somewhere else)

After that $\mathfrak{A}^*$ determines a new direction $r \in val_L P$ and a pebble action $\beta(P) \in \{0,1,2\}$. Applying an action means:

1 = ($\mathfrak{A}^*$ drops the pebble on P, provided $\alpha(P) \in \{0,1\}$)
0 = ($\mathfrak{A}^*$ carries the pebble, provided $\alpha(P) \in \{0,1\}$)
2 = (no action iff $\alpha(P)=2$)

So we obtain a sequence (P_k, ($val\ P_k$,$\alpha(P_k)$),s_k, (r_k,$\beta(P_k)$)) describing the behaviour of $\mathfrak{A}^*$ in (L,P_o). The sequence (P_k, (r_k, $\beta(P_k)$))) is called the path of $\mathfrak{A}^*$ in (L,P_o), the subsequence containing all points P_k with $\beta(P_k)=0$ is the corresponding pebble path.
We say, an automaton $\mathfrak{A}^*$ with initial state s_o masters a pointed, infinite labyrinth (L,P) if the support of its path starting in P is infinite. $\mathfrak{A}^*$ searches (L,P) if $\mathfrak{A}^*$ reaches every point of L starting in P.

1.2 <u>The Main Theorem</u>:
To an arbitrary finite 1-pebble automaton $\mathfrak{A}^*$ with initial state s_o can be constructed a pointed cofinite (finite) maze (L,O), the so called trap, such that the automaton does not master (search) L starting in O.

Obviously we can assume, that $\mathfrak{A}^*$ starts with pebble situation 1.
Proof: (An extended version is contained in [9]; if not stated explicitly, notations are used as in Budach [7].)
We start with a basic lemma, which allows to make use of some techniques (quasiplanarity, $\mathfrak{A}$-flexibility) developed by Budach.

<u>Lemma</u>: Let $\mathfrak{A}^*$ be a finite 1-pebble automaton with s states and T= T(s)=lcm(2,...,s-1). For every cyclically reduced word $p \in D^+$ and an arbitrary prelabyrinth L containing an edge $P \xrightarrow{p} Q$ holds:
$\mathfrak{A}^*$ cannot distinguish L from L'= L(P, $p^T \vdash p^{kT}$,Q) ,$k \in \mathbb{N}$, starting in any P' $\in$ L and in the corresponding point of L' in any state.

Let i: $D^+ \longrightarrow D^+$ be the orthogonal mapping defined by i(e)=n ,i(n)=w.

We define:

$$wh(r) = (r^T\ (ir^T)\ \bar{r}^T\ (i\bar{r}^T))^T \quad \text{is the r-whirl, } r \in D$$
$$w_1 = wh(e)\ e^T n^T w^T s^T\ \overline{wh(n)}$$
$$w_2 = wh(w)\ w^T s^T e^T n^T\ \overline{wh(s)}$$

Figure 1a shows the test labyrinth used by Budach $\mathcal{L}_o$ and the universal covering $\mathcal{L}_1$ of $\mathcal{L}_o$ is illustrated in figure 1b.

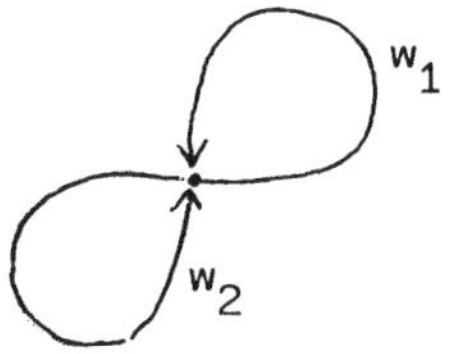

Fig. 1a

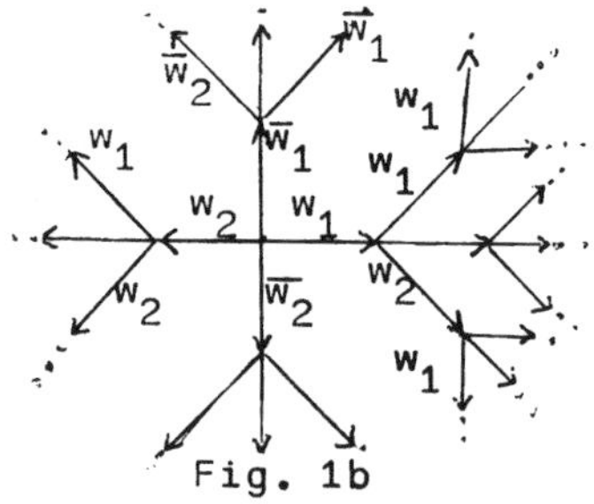

Fig. 1b

Obviously there exists a 1-pebble automaton which is able to check whether or not powers of elements of a finite set of words are contained in the fundamental group $\pi_o(L,P)$ of a finite pointed labyrinth, and we know the simple "go to the left"-algorithm. For that reason we test $\mathfrak{A}^*$ in $(\mathcal{L}_1, 0)$.

Tests: S is the set of states, $\mathfrak{A}^*_s$ denotes $\mathfrak{A}^*$ with initial state s.

I_{*s} =(Path of $\mathfrak{A}^*_s$ starting in 0 with $\alpha(0)=1$)
I^*_s =(Path of $\mathfrak{A}^*_s$ starting in 0 with $\alpha(0)=o$)
I_s =(Path of $\mathfrak{A}^*_s$ starting in 0 without pebble)

For every $s \in S$ we have to distinguish the following cases in each test

(1) The prelabyrinth distance between pebble and automaton is bounded by card(S).

(2) $\mathfrak{A}^*_s$ leaves the pebble in a point P and the following path $P \xrightarrow{pq^\infty} \ldots$ without pebble has a nontrivial prefix p, that means $P \notin \text{supp}(\ldots \xrightarrow{q^\infty} Pp \xrightarrow{q^\infty} \ldots)$

(3) $\mathfrak{A}^*_s$ leaves the pebble and the following path has not a nontrivial prefix.

We start with the easier cases:
Let us consider I_{*s_o}, s_o the initial state of a given $\mathfrak{A}^*$, and assume case (1) or (2) happens. A Budach trap will catch $\mathfrak{A}^*_{s_o}$, in the first case a trap corresponding to the pebble path and in case (2) a trap corresponding to the period without pebble.

.3 Roots of the Budach traps

We present only a list of these labyrinths, for details see [7], [10].
A finite initial automaton is tested in $(\mathcal{L}_1, 0)$:

Cyclically reduced period q	Root R(q) of the trap

1) q is empty

1) $R(q) = \{0\}$

2) $q = w_1^n$, $q = w_2^m$

2) R(q) :

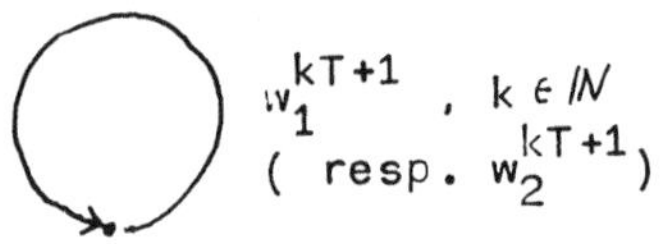

Fig. 2a

3) $q = w_1^{n_1} w_2^{m_1} w_1^{n_2} \ldots w_2^{m_t}$
$n_1 \geqslant 2$

3) R(q) :

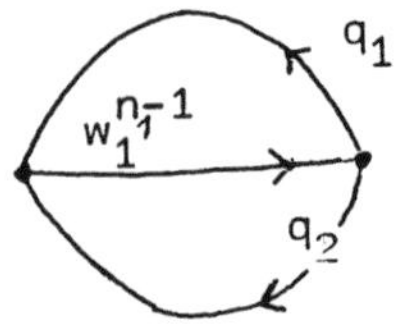

Fig. 2b

where $q_1 = w_1 \; w_2^{m_1} \ldots w_2^{m_t} q^{kT-1} \; w_1$

$q_2 = w_2^{m_1} \ldots w_1^{n_t} w_2^{m_t} q^{k'T-1}$, $k, k' \geqslant 1$

4) $q = (w_1 \; w_2)^n$

4) R(q) :

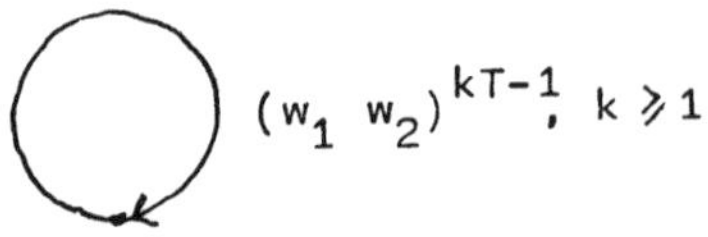

Fig. 2c

5) see Müller [11]
$q = v_1^{\varepsilon_1} v_2^{\varepsilon_2} \ldots v_{2n}^{\varepsilon_{2n}}$; where all $\varepsilon_i \in \{\pm 1\}$, $\varepsilon_1 = +1$ and for all i: $v_i \in \{w_1^{\pm 1} w_2^{\pm 1}$
. $v_i \neq v_{i+1}$, $\varepsilon_1 = \varepsilon_2 = \ldots = \varepsilon_{n_o} = 1$, $\varepsilon_{n_o+1} = \ldots = \varepsilon_{n_o+n_1} = -1$

5.1) $n_1 \leqslant n_o$

5.1) R(q)

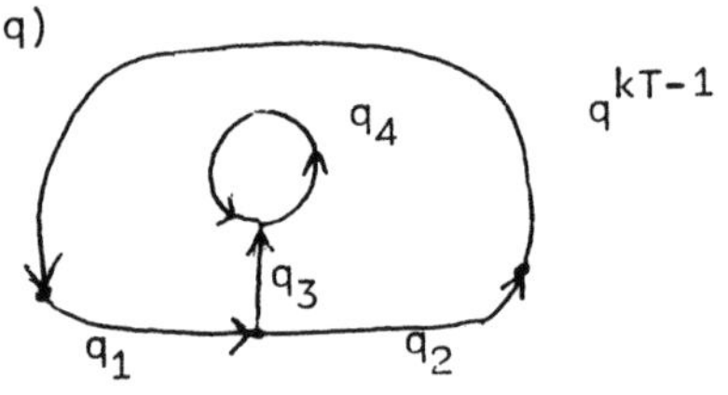

Fig. 2d

where $q_1 = v_1 \ldots v_{n_o - n_1 + 1}$,

$q_3 = v_{n_o+2-n_1} \ldots v_{n_o}$, if $n_o = n_1$, then q_3 is empty

$q_2 = v_{n_o+n_1+1} v_{n_o+n_1+2}^{\varepsilon_{n_o+n_1+2}} \ldots v_{2n}^{\varepsilon_{2n}}$

and $q_4 = v_{n_o+1}\bar{q}_3 q_2 q^{k'T-1} q_1 q_3 v_{n_o+1}$

$k, k' \geqslant 1$

5.2) $n_o \lneq n_1$

5.2) R(q)

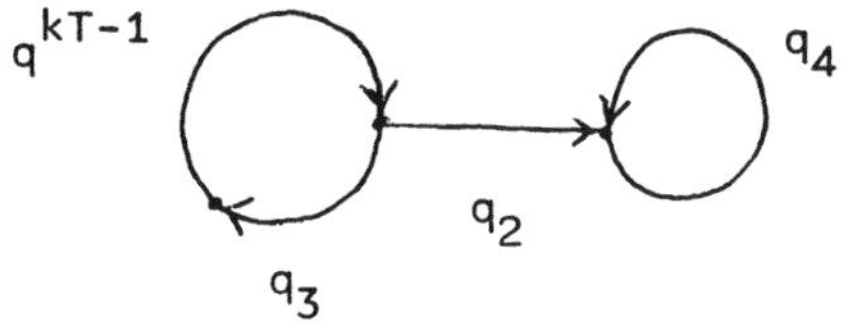

Fig. 2e

where $q_2 = v_1 \ldots v_{n_o-1}$ if $n_o = 1$ then q_2 is empty

$q_3 = v_{2n_o}^{\varepsilon_{2n_o}} \ldots v_{2n}^{\varepsilon_{2n}}$

$q_4 = v_{n_o}\bar{q}_2 q_3 q^{k'T-1} q_2 v_{n_o}$, $k, k' \geqslant 1$

Remarks:

i) A W-labyrinth is a prelabyrinth such that the labels of all edges are words over $W = \{w_1, \bar{w}_1, w_2, \bar{w}_2\}$. A W-labyrinth L is called W-quasiplane if we can multiply whirls in L such that we obtain a quasiplane prelabyrinth (see 8.5.4 in [7]). All roots above are W-quasiplane W-labyrinths for any k,k' .

ii) We can break up roots R(q) in a point P of the periphery such that the arising labyrinth $(\mathring{q})$ starts and ends with label q^T.

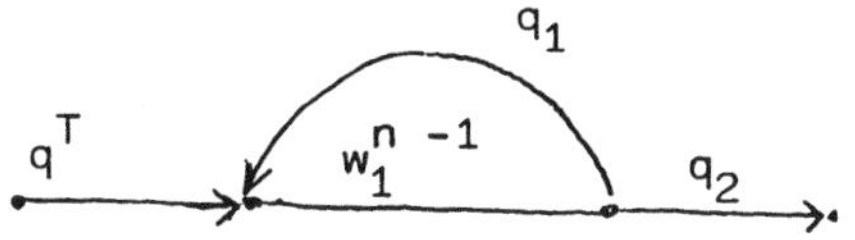

Fig. 3 $(\mathring{q})$ corresponding to fig. 2b

iii) We inflate a root that means: Increase the numbers k,k'.

.4 The case (3)

his is the most complicated case and first of all we construct an abstract trap" , which is not a quasiplane labyrinth. $\mathfrak{A}^*_{s_0}$ leaves the ebble in a point $P \in \mathcal{L}_1$ using a path $P \xrightarrow{p_1^\infty} \ldots$, therefore the automaton ill recover the pebble in the labyrinth $(\mathcal{L}_1, 0)/(P = Pp_1^{kT})$ in a state $_1$.(resp. $(\mathcal{L}_1, 0)/(P = Pp_1^{kT\pm 1})$ if $p_1 \in W^+_{max}$, see 3.1 and 1.3) .
e consider I_{*s_1} and let us assume that we get again case (3). This rocedure is iterated and hence we obtain a periodic pebble path $\xrightarrow{pq^\infty} \ldots$ attached to each of its **points** a finite set of circles, the

so called side paths . We identify the points Op and Opq^{kT} (resp. $Opq^{kT\pm 1}$) and a neighbourhood of the root of this labyrinth (Fig. 4) yields an abstract trap.

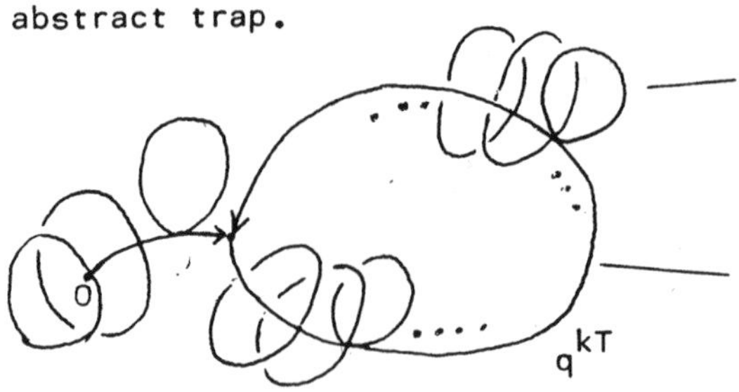

Fig. 4

Using consequently results of section 2,3,4 we finish the proof in section 5 .

2. Quasiplane labyrinths

Two words $p,q \in D^+$ are called similar iff $p= r_1^{n_1}\dots r_k^{n_k}$ and $q= r_1^{m_1}\dots r_k^{m_k}$, where $r_i \in D$ and $n_i, m_i > 0$ for all $1 \leq i \leq k$. A finite labyrinth L is quasiplane iff there is an isotopy $\alpha: L \longrightarrow L'$, L' a plane labyrinth, such that for every edge $P \xrightarrow{p} Q$ the label of $\alpha(P \xrightarrow{p} Q)$ and p are similar. The number T defined in 1.2 yields a relation between automata and quasiplanarity. (see 6.10.3 in [7])

We recall the definition of the rotation index χ:

$$\chi(\ ri(r)\ ) = +1\ ,\ \chi(\ ri^3(r)\ ) = -1\ ,\ \chi(\ rr\ ) = 0 \quad \text{for } r \in D$$

$$\chi(p) = \chi(r_1 r_2) + \chi(r_2 r_3) + \dots + \chi(r_{n-1} r_n) \quad \text{for any reduced } p = r_1 \dots r_n$$

2.1 Characterization

A face of a finite labyrinth is a closed path $P_0 \xrightarrow{r_1} P_1 \xrightarrow{r_2} \dots \xrightarrow{r_n} P_n = P_0$ (modulo initial point) such that for all $0 \leq i \leq n$ there is no $r \in$ val with $\chi(r_i r_{i+1}) - \chi(r_i r) < 0$. $\chi(p' f(p'))$, where p' is the cyclically reduced word corresponding to $p = r_1 r_2 \dots r_n$ and f(p') is the first letter of p' , is called the rotation index of this face.

Theorem: (K. Kriegel, see [10])

A finite labyrinth L is quasiplane $\Longleftrightarrow$ (i) the Euler characteristic of L is 2 and (ii) there is exactly one face of L with rotation index -4 (the so called periphery) and all other faces have rotation index + (the cells).

2.2 Corollary:

Let L be a finite W-quasiplane root of a W-labyrinth, $P_0 \xrightarrow{r_1} P_1 \dots \xrightarrow{r_n} P_n$ a cell (or the periphery) where all $r_i \in W$. Further we consider two points P_i, P_j , $0 \leq i \leq j \leq n-1$, of this face together with $r, r' \in W$

atisfying the condition: $\omega(\bar{r}_i, r, r_{i+1}) = \omega(\bar{r}_j, r', r_{j+1}) = -1$, see 3.1 .
hen for any cyclically reduced word $q \in W^+$ such that its first letter $f(q) = \bar{r}$ and the last letter $l(q) = r'$ holds:
here exists a natural number k such that the W-labyrinth $L\overset{*}{V}(P_j \xrightarrow{(\mathring{q})^k} P_i)$ s W-quasiplane.
roof: Use the fact that $\chi(\bar{r}pr'f(\bar{r})) = 4m$, $m \in \mathbb{Z}$, see 8.2.2 in [7], here p denotes the word $r_{i+1} \ldots r_j$.

3. Crossings

he purpose of this section is to describe the crossings arising between pebble path and side paths in Fig.4 .

.1 Notations

et $r_1, r_2, r_3 \in D$ be pairwise different elements:

$$\omega(r_1, r_2, r_3) = \operatorname{sgn}(\chi(\bar{r}_1 r_3) - \chi(\bar{r}_1 r_2)),$$ see 5.9 in [7] .

e can extend this symbol to arbitrary cyclically reduced words over W. specially , if $p, q \in W^+$ are cyclically reduced, then $\omega(\bar{q}, p, q)$ is defined iff in $(\mathcal{L}_1, 0)$ there is a branching between the paths$(\ldots \xrightarrow{q^\infty} 0 \xrightarrow{q^\infty} \ldots)$ nd $0 \xrightarrow{p^\infty} \ldots$. $\omega(\bar{q}, p, q) = +1$ (=-1) means $0 \xrightarrow{p^\infty} \ldots$ branches off to the right left) side. If $p = r_1 r_2 \ldots r_n \in W^+$, then $p_{[k}$, $1 \leq k \leq n$, denotes the word $_k r_{k+1} \ldots r_n r_1 \ldots r_{k-1}$.

efinition: Let $p, q \in W^+$ be cyclically reduced words.

) (p,q) is a crossing $\Leftrightarrow$ $\omega(\bar{q}, p, q)$ is defined and it holds: $\omega(\bar{q}, p, q) = -\omega(\bar{q}, \bar{p}, q)$.

) Let (p,q) be a crossing. A crossing (p',q') is called an inverse rossing to (p,q) $\Longleftrightarrow$ (i) there are numbers k,l such that $p' = p_{[k}$, $q' = q_{[l}$ and (ii) $\omega(\bar{q}, p, q) = \omega(\bar{q}', p', q')$.

y $C_1^+(q)$ we denote the set of all crossings such that $\omega(\overline{l(q)}, f(p), f(q))$ +1 and $C^+(q) = \bigcup_k C_1^+(q_{[k})$, where $1 \leq k \leq |q|$.
bviously there is no crossing (p,q) such that p or q is an element f $W^+_{max} = \{w_1^n, w_2^n, (w_1 w_2)^n, (w_2 w_1)^n \mid n \in \mathbb{Z} - \{0\}\}$. Therefore we call W^+_{max} he set of words with maximal curvature.

.2 Theorem:

et $p, q \in W^+ - W^+_{max}$ be finite, cyclically reduced words. It holds:

i) There are numbers k,l such that $(p_{[k}, q_{[l})$ is a crossing;

ii) every crossing $(p_{[k}, q_{[l})$ possesses an inverse crossing;

iii) it exists an involution $\mu(q): C^+(q) \longrightarrow C^+(q)$ such that for any $(v, q_{[l})$ the image $\mu(q)(v, q_{[l})$ is an inverse crossing to $(v, q_{[l})$.

Proof (sketch): To prove (i) we have to consider subwords of maximal urvature in p and q . If p=q compare with figures 2b,2d,2e .

(ii),(iii): We construct a finite $|q|$-sheeted covering labyrinth $\mathcal{L}_1(q)$ of $\mathcal{L}_0$, in such a way that $\mathcal{L}_1(q)$ contains the circle $0 \xrightarrow{q} 0$ and any $r \in W$ branching off to the right (left) side of the circle also ends on the right (left), compare with Fig.5 . Consider the images in $\mathcal{L}_1(q)$ of the infinite paths in $\mathcal{L}_1$ representing a crossing.

Example: Figure 5 illustrates this costruction $\mathcal{L}_1(q)$ for $q = w_1 w_2^3 w_1^{-1} w_2^{-1} w_1^2 w_2^{-1}$

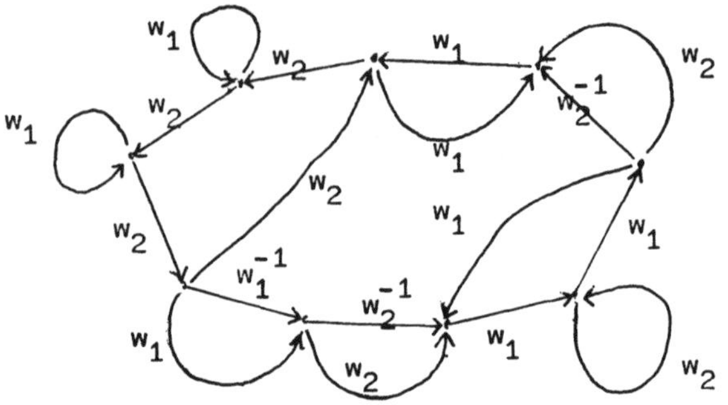

Fig.5

4. Universal traps

Let $\mathcal{A}$ be a finite set of finite automata ("mice"). An universal trap is a labyrinth (L,0), which is a trap for any $\mathfrak{A} \in \mathcal{A}$. In [2] the existence of universal traps is proved using an inductive process. We present a construction testing every $\mathfrak{A} \in \mathcal{A}$ only once in $(\mathcal{L}_0, 0)$.

Theorem: For any finite set $\mathcal{A}$ of mice there exists an universal trap.

Proof (sketch): We consider the tests in $(\mathcal{L}_0, 0)$: $p_{\mathfrak{A}} q_{\mathfrak{A}}^{\infty}$, $\mathfrak{A} \in \mathcal{A}$, (see 8.1 in [7]) assuming all $p_{\mathfrak{A}}, q_{\mathfrak{A}}$ are of minimal length, reduced and $l(p_{\mathfrak{A}}) \neq \overline{f(q_{\mathfrak{A}})}$, $l(q_{\mathfrak{A}}) \neq \overline{f(q_{\mathfrak{A}})}$.
Now we attach to a point 0 all paths $0 \xrightarrow{p_{\mathfrak{A}} q_{\mathfrak{A}}^{T}} P_{\mathfrak{A}}$, $0 \xrightarrow{p_{\mathfrak{A}} q_{\mathfrak{A}}^{-T}} P'_{\mathfrak{A}}$, $\mathfrak{A} \in \mathcal{A}$.
Then we can apply corollary 2.2 and theorem 3.2 in the following way: We connect $P_{\mathfrak{A}}$ and $P'_{\mathfrak{A}}$ using $(\mathring{q}_{\mathfrak{A}})^{n(\mathfrak{A})}$ and it arises a W-quasiplane root of an $\mathfrak{A}$-trap. All remaining paths $0 \to P_{\mathfrak{A}'}$, $0 \to P'_{\mathfrak{A}'}$, $\mathfrak{A}' \in \mathcal{A}$, lying inside this root, such that $q_{\mathfrak{A}'} \notin W^{+}_{max}$, are crossed with the periphery, (Fig.6b) using once more $(\mathring{q}_{\mathfrak{A}})$ and if necessary we inflate the root. Easy to see that we can connect $P_{\mathfrak{A}}$ and $P'_{\mathfrak{A}}$ if $q_{\mathfrak{A}}$ has maximal curvature.
In the end the procedure is iterated and we get the W-quasiplane root of an universal trap for the given set $\mathcal{A}$.

Remark: We have assumed without restriction of generality that all $\mathfrak{A}$ have s states , hence $T=T(s)$, and all $q_{\mathfrak{A}}$ are not trivial.

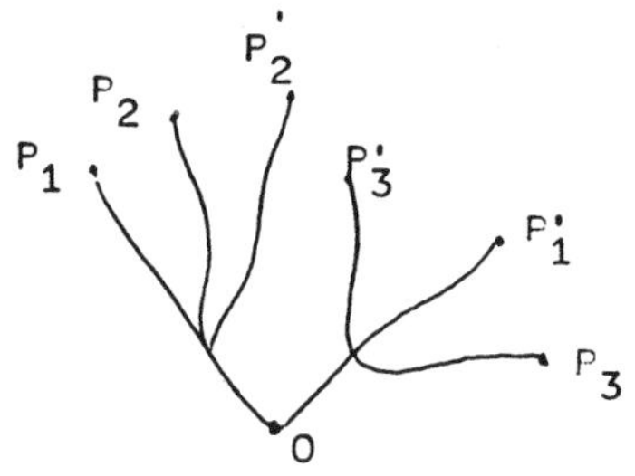

Fig. 6a

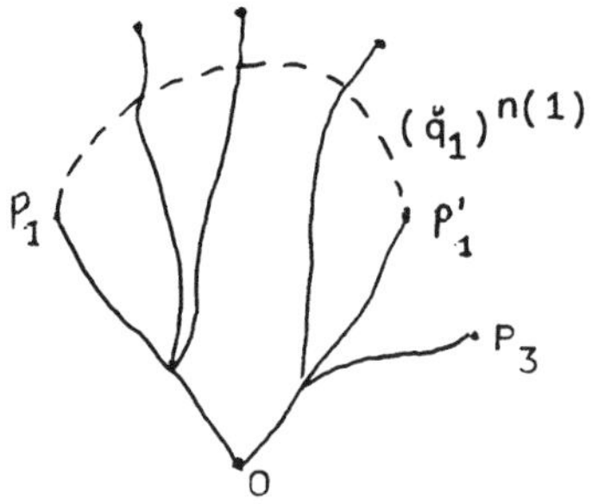

Fig. 6b

5. Proof of the Main Theorem

ow we are prepared to finish the proof. We only discuss the "worst" ituation , for other details see [9] .

et us start with the labyrinth of figure 4. If all side paths do not dhere to the pebble path, then we take a Budach trap L(q) corresponding o the period q of the pebble path. It is possible to choose for all ide paths defining crossings with the pebble path corresponding inverse rossings, such that the prelabyrinth distance (in L(q)) between any rossing and its inverse is greater than 2card(S). Otherwise the auto- aton can find the pebble using L(q), see Fig.7 , contradictory to he abstract trap. We must assume further that in L(q) all side paths o not adhere to the pebble path.

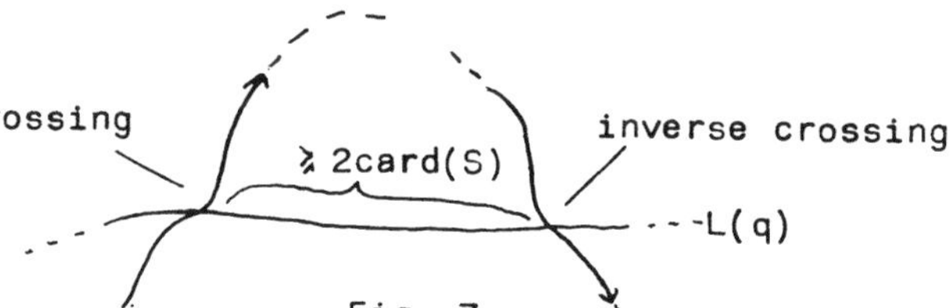

Fig. 7

sing a technique as in section 4 we connect every crossing with its nverse crossing such that we get a W-quasiplane W-labyrinth. By 6.9.2 n [7] and our basic lemma one can define a quasiplane labyrinth and herefore a plane realization corresponding to a finite tree over this -labyrinth. Done.

f a side path adheres to the pebble path $q_1^{k_1 T}$ in a Budach trap $L(q_1)$, hen obviously the automaton can sense the branchings in $R(q_1)$.

emark:The question is still open wether or not there are traps with t most two components of obstrucles for finite automata.(see Müller 11], Asser [3])

Therefore the automaton finds the pebble in $L(q_1)$ in a point P_1 and in an internal state different from the test and from the abstract trap.

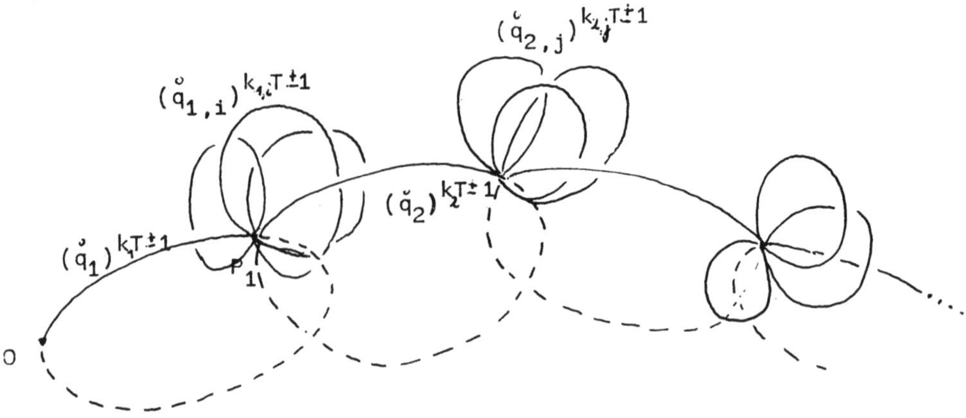

Fig. 8) A schematic representation of the total pebble period

$\mathfrak{A}^*$ is started as before in the W-tree over the root $0 \xrightarrow{(\mathring{q}_1)^{k_1T\pm1}} 0$.
In the point P_1 , exactly: corresponding to P_1 , the automaton changes the behaviour and so we obtain a new periodic pebble path $(q_{1,1}^{k_{1,1}T\pm1})$.
Attaching this closed W-quasiplane pebble path to P_1 we have to consider three cases:

(i) $\mathfrak{A}^*$ senses a branching in $(\mathring{q}_{1,1})^{k_{1,1}T\pm1}$ and changes its behaviour

(ii) $\mathfrak{A}^*$ senses a branching in $(\mathring{q}_1)^{k_1T\pm1}$ near the point P_1 after returning to this point and changes its behaviour

(iii) The pebble path and $(\mathring{q}_{1,1})^{k_{1,1}T\pm1}$ never diverge.

Hence we assume - the remaining easier cases are left to the reader- that $\mathfrak{A}^*$ leaves the neighbourhood of P_1 using a path $(\mathring{q}_{1,k})^n$, $k \geqslant 1$, and then case (i) happens.
So after all we get a total finite pebble period described by tripels

$$(\mathring{q}_i , \{\mathring{q}_{i,1},\dots,\mathring{q}_{i,m_i}\}, \mathring{q}_{i+1})$$

and we suppose that the period is reduced: $q_i \neq \overline{q_{i+1}}$ for all i.
One can extend the W-quasiplane closed path $(\mathring{q}_i)^{k_iT\pm1}$ to a W-quasiplane root $R(\mathfrak{A}^\circ)$ of a W-labyrinth $L(\mathfrak{A}^\circ)$ corresponding to a trap for the whole pebble period.
In the end we want to sketch that it is possible to attach all side paths to $L(\mathfrak{A}^\circ)$ such that we get a W-quasiplane W-labyrinth and the distance between any crossing defined by a side path and the corresponding inverse crossing is greater than 2card(S) , see Fig.7 .

First we observe that any $R(\mathcal{O}^{\circ})$, especially roots of Budach traps, contains a finite set of trees $\{U_i\}$ such that $R(\mathcal{O}^{\circ})-\bigcup_i U_i$ is a disjoint union of edges $q_j^{k_j T}$, see Fig.9a . Making use of inflations we can suppose that for all i,j: $|q_j^{k_j T}| \gg d(U_i)$.

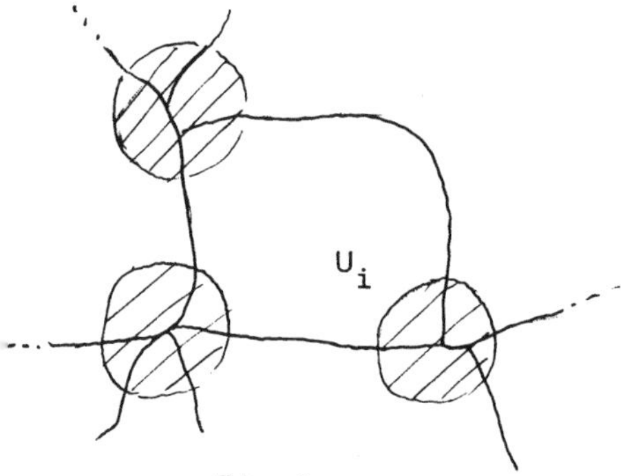

Fig.9a

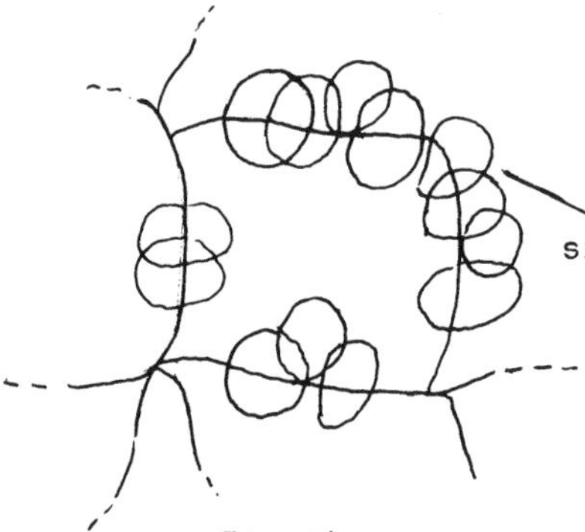

Fig.9b

Then we imprint all side paths restricted to edges $q_j^{k_j T}$. This can be done using theorem 3.2(iii) and section 4 ,see Fig.9b .
If we attach a side path arising from a point in U_i we have to cross some edges $q_1^{k_1 T},\ldots,q_n^{k_n T}$. Obviously there exist crossings but generally we must assume that these crossings are already used in the first step.

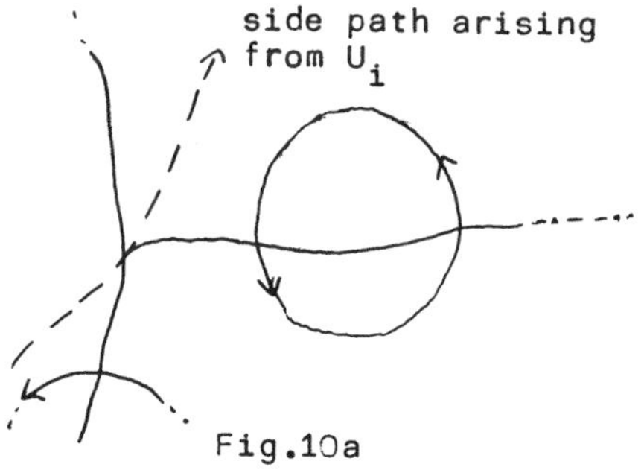

Fig.10a

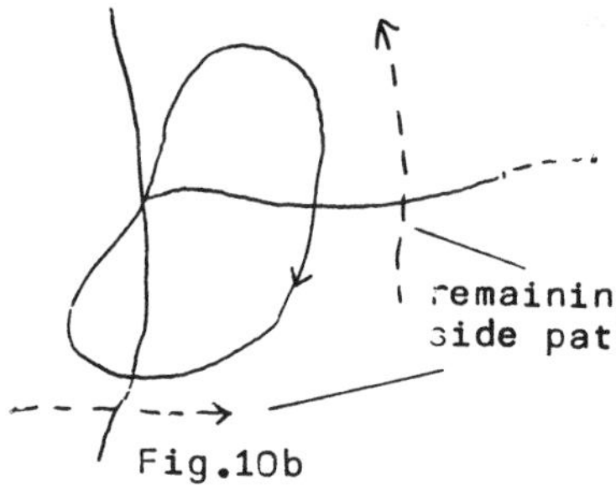

Fig.10b

In this case we break up a corresponding side path attached to the edge and use it to realize the side path arising from U_i , see Fig.10a, 10b . Doing such an exchange for all side paths attached in any U_i we can suppose that the distance between the remaining crossings with edges is pairwise greater than 2card(S). Because $L(\mathcal{O}^{\circ})$ has a finite number of cells an iterated application of 3.2(iii) will finish the construction.It is evident that a W-tree over the arising W-quasiplane labyrinth corresponds to a trap for the given 1-pebble automaton $\mathcal{O}^{*}_{s_0}$.

Acknowledgement

I am indebted to Professor Lothar Budach and Klaus Kriegel, Berlin, for helpful discussions.

References

[1] H. Antelmann, Über den Platzbedarf von Fallen für endliche Automaten, Diss., Berlin 1980

[2] H. Antelmann, L. Budach, H.-A. Rollik, On universal traps, EIK 15/3, 123-131, 1979

[3] G. Asser, Bemerkungen zum Labyrinth-Problem, EIK 13/4-5, 203-216, 1977

[4] M. Blum, D. Kozen, On the Power of the Compass, 19th IEEE FOCS Conf. 1978, 132-142

[5] M. Blum, W. Sakoda, On the capability of finite automata in 2 and 3-dimensional space , 18th IEEE FOCS Conf. 1977

[6] L. Budach, Environments,Labyrinths and Automata, in Lect.Notes in Computer Science 56, 54-64 , Springer 1977

[7] L. Budach, Automata and Labyrinths, Math. Nachrichten 86, 195-282, 1978

[8] L. Budach, Ch. Meinel, Umwelten und Automaten in Umwelten, Seminarberichte Sekt. Mathematik d. Humboldt-Universität zu Berlin, Nr.23, 1980

[9] F. Hoffmann, 1-Pebble automata and labyrinths, to appear 1981

[10] F. Hoffmann, K. Kriegel, Quasiplane labyrinths, to appear 1981

[11] H. Müller, Automata catching labyrinths with at most three components, EIK 15/1-2, 3-9, 1979

ABOUT THE BY CODINGS OF ENVIRONMENTS INDUCED POSETS $[\mathfrak{U}_Z, \leq]$ AND $[\mathcal{L}_Z, \leq]$

Chr. Meinel
Sektion Mathematik
Humboldt-Universität
DDR-1086 Berlin, PSF 1297

If one considers different mathematical models of computation devices one realizes that they essentially consist of two parts: the control part which possesses the structure of a finite automaton and the (processing and) memory part which possesses a structure which we will call an environment. So the independent consideration of control and (processing and) memory part leads to consideration of interactions of finite automata with finite or infinite environments. The effectiveness of such a pair automaton-environment, which will be called a machine, depends mainly of the computational power of the underlaying environment. Therefore it suffice to compare these in order to compare the effectiveness of the whole machines. Such a comparison is possible by the help of a very general kind of morphisms of environments, which will be called codings. These codings guarantee that everything that can be done in a point of the first environment also can be done in the second environment by the help of a fixed finite family of automata. Codability on the set of classes of environments with the same computational power provides to a partial ordering, which will be considered in this paper. The results give interesting informations about the hierarchy of computational devices defined by there effectiveness. They shed new light on classical results and lead to new knowledges by considerations of non standard environments.

0. Environments and codings

An environment $U = (X,Y,Z,d,l)$ consists of a finite set X of "actions", a finite set Y of "local informations", a finite or enumerable set Z of "points", a partial function $d : Z \times X \overset{\supseteq}{\longrightarrow} Z$ and a function $l : Z \longrightarrow Y$. An automaton $\mathfrak{A}$ with input set Y and output set X is able to walk in such an environment. If one put $\mathfrak{A}$ in

a point $P \in U$ then $\mathfrak{A}$ reads the local information $l(P)$ and makes in dependence of its internal state a step in direction x to point $Px := d(P,x)$.
We restrict our consideration to R-environments and R-automata in order to guarantee that $\mathfrak{A}$ chooses in a point P only such actions x that Px is defined. That means: Given a correspondence R from Y to X . An environment U and an automaton $\mathfrak{A}$ will be called R-environment and R-automaton, respectively, if for all points $P \in U$ hold $R \cdot l(P) = \text{val } P := \{x \mid Px \text{ is defined}\}$, and if for all states s of $\mathfrak{A}$ and all $y \in Y$ for the output function λ of $\mathfrak{A}$ holds $\lambda(s,y) \in R(y)$, respectively.
The pair automaton-environment is called a machine. It will be sufficient to compare only the underlying environments in order to compare the computational power of such machines. This will be done by the help of the following morphisms of environments. A <u>coding</u> of an environment $U = (X,Y,Z,d,l)$ into the environment $U' = (X',Y',Z',d',l')$ is a injective mapping

$$\mathfrak{f} : Z \longrightarrow Z'$$

such that there is a finite family of recognizing automata (that are automata with terminal states in which marks from a finite set will be signalized) $\Phi = (\mathfrak{A}, \{\mathfrak{A}_x \mid x \in X\})$ satisfying the following conditions:

(i) $\mathfrak{f}(P)\mathfrak{A}$ = (endpoint of $\mathfrak{A}$ starting in $\mathfrak{f}(P)$) = $\mathfrak{f}(P)$

$\mathfrak{A}(\mathfrak{f}(P))$ = (the mark which $\mathfrak{A}$ signalizes after it had stopped

= $l(P)$ for all $P \in U$ and

(ii) for $x \in \text{val } P$ holds $\mathfrak{f}(Px) = \mathfrak{f}(P)\mathfrak{A}_x$.

That is, if we are given a coding $\mathfrak{f} : U \longrightarrow U'$ everything that can be done in U starting in the point P can also be done in U' starting in the point $\mathfrak{f}(P)$ by the help of a finite set of automata. (Details and motivations can be found in /1/ and /2/.) Environments together with codings form a category, because the product of two codings is a coding too.
The following theorem proves the assertion that it is possible to characterize the computational power of a machine by the help of codings.

<u>SIMULATIONTHEOREM:</u> Let $\mathfrak{f} : U \longrightarrow U'$ be a coding and let $\mathfrak{B}$ be an automaton working in U , then there is an automaton $\Phi(\mathfrak{B})$ working in U' so that the machine $(\mathfrak{B}, U)$ will be simulated by the machine $(\Phi(\mathfrak{B}), U')$.

1. The posets $[\mathfrak{U}_z, \leq]$ and $[\mathfrak{L}_z, \leq]$

.et $\mathfrak{U}$ denote the class of all (R-)environments and let's write J $\leq$ U' (U, U' $\in \mathfrak{U}$) iff U is codable in U'. Environments are :alled to be equivalent U $\equiv$ U' if U $\leq$ U' and U' $\leq$ U . Because ve are only interested in investigations of the computational power of ›nvironments it suffices to regard the equivalence classes $\overline{U} \in \mathfrak{U}/_\equiv$ ›f $\mathfrak{U}$ by $\equiv$. The relation "$\leq$" above defined induces a partial order-.ng on $\mathfrak{U} := \mathfrak{U}/_\equiv$ which will also be denoted by $\leq$. Since recogni-:ing automata are able to work only in the connected components of there starting points we restrict our investigations to the poset $[\mathfrak{U}_z, \leq]$ ›f all elements of $\mathfrak{U}$ which possess a connected representative.

Very high structured environments are of great importance. Examples ıre R-labyrinths, which are defined by the fact, that for every action : $\in$ X there is another action $\overline{x} \in$ X such that the partial function ıefined by x (P $\longmapsto$ Px) is inverse to the partial function defined ›y $\overline{x}$ and that holds im x = rg $\overline{x}$. Labyrinths are R-labyrinths with (P) = val P for all points P . The subposet $[\mathfrak{L}_z, \leq]$ of $[\mathfrak{U}_z, \leq]$:onsisting of all elements which can be represented by R-labyrinths ›lays a very important role because of the following theorem.

'HEOREM: Every connected environment is codable in a plane labyrinth.

'aper /6/ is dedicated the PROOF of this theorem.
he posets $[\mathfrak{U}_z, \leq]$ and $[\mathfrak{L}_z, \leq]$ contain a minimum, namly the equi-alence class of trivial environments. Because of the fact, that every lement has continuum many upper bounds and only a finite or enumerable umber of lower bounds there exist no maximum or maximal elements.

Further the posets $[\mathfrak{U}_z, \leq]$ and $[\mathfrak{L}_z, \leq]$ possess incomparable lements.

ROPOSITION: The elements $\overline{\mathbb{N}}$ and $\overline{\mathbb{Z}}$ defined by the counter environ-ent $\mathbb{N}$ = ({-1,+1}, {0,1}, $\mathbb{N}$, +, sgn) and the environment ' = ({-1,+1}, {*}, $\mathbb{Z}$, +, *) , respectively, are incomparable.

ROOF: The counter environment which possesses points with different ocal informations is not codable in $\mathbb{Z}$ because all points of Z are ndistinguishable by finite automata.
everse, if an environment U is codable in an environment U' f : U $\longrightarrow$ U') then there exists to every U-computable function : U $\longrightarrow$ U an U'-computable function g : U' $\longrightarrow$ U' with

$$g\big|_{\mathrm{im}\,\varphi} = f^{-\varphi} := \varphi \cdot f \cdot \varphi^{-1} .$$

One realizes by a detailed investigation that there is no $\mathbb{N}$-computable function g to the $\mathbb{Z}$-computable function $f(z) = z + 1$ with $g\big|_{\mathrm{im}\,\varphi} = f^{-\varphi}$ for all injctive functions φ .

Another example of incomparable elements are the finite elements $\overline{\mathbb{Z}/(6)}$ and $\overline{[1]}$ which are defined by the factor environment $\mathbb{Z}/(6)$ of $\mathbb{Z}$ and the initial segment $[1]$ of length 1 of counter environment $\mathbb{N}$. A detailed analyse of this example results in the following proposition.

PROPOSITION: The posets $[\mathfrak{U}_z, \leq]$ and $[\mathfrak{L}_z, \leq]$ are not a lattice. They define neither a upper nor a lower semilattice.

2. The sets $E_{\mathfrak{U}}$ and $E_{\mathfrak{L}}$ of finite elements of $[\mathfrak{U}_z, \leq]$ and $[\mathfrak{L}_z, \leq]$, respectively

Because it is possible to store a finite environment U fully in the internal memory of an automaton working over U , one can prove following propositions about the sets $E_{\mathfrak{U}}$ and $E_{\mathfrak{L}}$ of finite elements of $[\mathfrak{U}_z, \leq]$ and $[\mathfrak{L}_z, \leq]$, respectively.

PROPOSITION: The element $\overline{[n]}$ which is defined by the initial segment of length n of the counter environment $\mathbb{N}$ is the maximum of the set of all finite elements of $E_{\mathfrak{L}} \subset E_{\mathfrak{U}}$ with cardinality not greater then $n+1$.

PROPOSITION: The element $\overline{[n]}$ is smaller then all finite elements $\overline{L}$ of $E_{\mathfrak{L}}$ with $\mathrm{red}\, L \geq n+1$.

Further we obtain following information about the fine structure of the elements $\overline{[n]}$.

PROPOSITION: For an R-labyrinth L with $\mathrm{card}\, L = n+1$ holds $L \in \overline{[n]}$ if and only if L is connected and reduced.

Since the identical function obviously provides a coding of $[n]$ in $\mathbb{N}$ the counter element $\overline{\mathbb{N}}$ is an upper bound of the sets $E_{\mathfrak{U}}$ and $E_{\mathfrak{L}}$ Moreover:

PROPOSITION: The counter element $\overline{\mathbb{N}}$ is in $[\mathfrak{L}_z, \leq]$ a least upper bound of $E_{\mathfrak{L}}$.

'ROOF: If the R-labyrinth $L = (X,Y,Z,d,l)$ defines a upper bound of $E_{\mathcal{L}}$ with $L \leq \mathbb{N}$ then almost all points of L possess the same ocal information y. Discussing the structure of the environments $L_x = (\{x\}, Y,Z,d_x,l)$ for $x \in R(y)$ with $d_x = d\,|_{Z \times \{x\}}$ we find n $x_0 \in R(y)$ so that $\mathbb{N}$ is codable in $L_{x_0,\bar{x}_0} \hookrightarrow L$.

ut the counter element is not the supremum of $E_{\mathcal{L}}$ because there exists upper bound of $E_{\mathfrak{A}}$ and $E_{\mathcal{L}}$ which is incomparable with $\mathbb{N}$. The en- 'ironment $Z_{\nu_2} = (\{-1,+1\}, \{0,1\}, \mathbb{Z}, +, l_{\nu_2})$ defines such an element 'here the definition of l_{ν_2} is evident by following picture

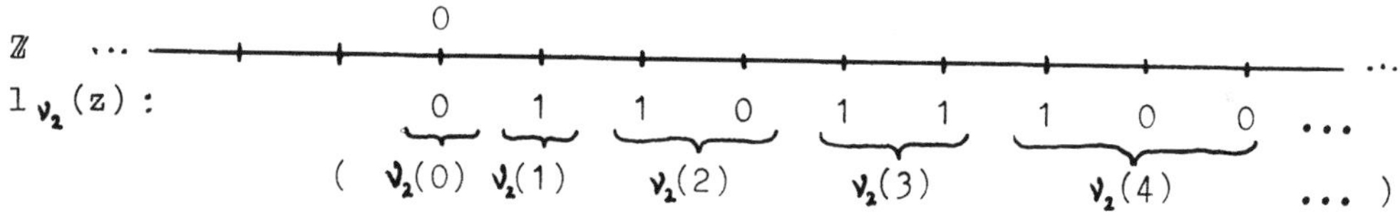

nd the fixation $l_{\nu_2}(z) = l_{\nu_2}(-z)$ ($\mathbb{Z}_{\nu_2}$ is an R-labyrinth about the abyrinth $\mathbb{Z}$). Obviously to any sequence $(z, z+1, \ldots, z+n)$ of points f Z_{ν_2} exist infinitely many other sequences $(z', z'+1, \ldots, z'+n)$ ith the same sequence of marks $l_{\nu_2}(z+i) = l_{\nu_2}(z'+i)$, $0 \leq i \leq n$. urther it is possible to find for a given word $w \in \{0,1\}^*$ any se- uence of points $(z, z+1, \ldots, z+|w|-1)$ with $w = l_{\nu_2}(z)\ldots l_{\nu_2}(z+|w|-1)$.

ROPOSITION: The elements $\overline{\mathbb{Z}_{\nu_2}}$ and $\overline{\mathbb{N}}$ which are defined by the -labyrinth Z_{ν_2} and the counter environment $\mathbb{N}$ are incomparable.

ROOF: Because of $Z \leq Z_{\nu_2}$ and $\mathbb{Z} \not\leq \mathbb{N}$ we obtain immediately $_{\nu_2} \not\leq \mathbb{N}$. In order to prove that $\mathbb{N} \not\leq \mathbb{Z}_{\nu_2}$ we show that any point of $_{\nu_2}$ is the endpoint of at most a finite number of $\mathfrak{A}$-paths in Z_{ν_2} or any automaton $\mathfrak{A}$. This will be done by make the most of the fact hat to every two-way automaton exists a one-way automaton, which ccepts the same language. But for one-way automata it is rather easy o find traps in Z_{ν_2} because of the special structure of this R-la- yrinth. You can find the delicate considerations in /7/. Altogether rom this follows that there exist no Z_{ν_2}-computable function g to he $\mathbb{N}$-computable function $f : n \longmapsto 0$ for all n with $g\,|_{\mathrm{im}\,f} = f^{-1}$ or any injective mapping $f : \mathbb{N} \longrightarrow \mathbb{Z}_{\nu_2}$ and therefore $\mathbb{N} \not\leq Z_{\nu_2}$.

3. The Turing element and Church's universe

The importance of Turing machines as these computational devices on which all partial recursive functions are computable is well known from recursive function theory. In the interacting model automaton-environment a Turing machine with tape alphabet A is a recognizing machine over the (A,b)-store $Tt_A := \text{red } \mathbb{Z}_{(A,b)}$ over the environment $\mathbb{Z}$. The equivalence relations

$$Tt_A{}^{l} \equiv Tt_{A'}{}^{l'} \qquad \text{for } l,l' \geq 1 \text{ and } A,A' \text{ finite alphabets}$$

proved in /2/ reflect the result of recursive function theory, that the number of Turing tapes or the cardinality of tape alphabet doesn't change the computational power of Turing machines. So all Turing tapes define the same element T of $[\mathfrak{U}_Z, \leq]$ which is called the Turing element. The result of the theorem of Minsky will be reflected by the equivalence relations

$$Tt_A \equiv \mathbb{N}^k \qquad \text{for } k \geq 2$$

which show that the Turing element T is also an element of the poset $[\mathcal{L}_Z, \leq]$ (see /2/).

The descending cone

$$\mathfrak{F}_T = \{\bar{U} \mid \bar{U} \leq T\}$$

which will be called Church's universe is of great importance because of the following theorem.

THEOREM: Church's universe $\mathfrak{F}_T$ is closed under all our constructions of environments (see /2/ and /7/).

This leads to the following structural version of Church's Theses:

CHURCH'S THESES (structural version): The Church's universe $\mathfrak{F}_T$ is the smallst descending set in $[\mathfrak{U}_Z, \leq]$ containing $\mathbb{N}$, which is closed under all computable constructions of classes of environments.

(For details see /7/.)

. Embedding of the poset of degrees of unsolvability in the poset $[\mathcal{L}_z, \leq] \subset [\mathcal{U}_z, \leq]$

'rom recursive function theory it is well known that two sets M and I' are called Turing-reducible

$$M \leq_T M'$$

.f M' is computable on a Turing machine with oracle set M . Let's ssume without restriction of generality that the tape alphabet A of uch a Turing machine equals $\{0,1,b\}$. Therefore $M \subseteq \{0,1\}^*$. If ν enotes the lexicographical enumeration of $\{0,1\}^*$ then we are able o prove (by taking advantage of the results of the last paragraph) hat the environment underlying a Turing machine with oracle set M s equivalent to the R-labyrinth

$$\mathbb{N}^2 \times \mathbb{N}_{\nu(M)}$$

ith $\mathbb{N}_{\nu(M)} := (\{-1,+1\}, \{0,1\}^2, \mathbb{N}, +, \mathrm{sgn} \times c_{\nu(M)})$ where c_A de-otes the characteristic function of a set A .
he following propositions give informations about the relationship of uring-reducibility and codability.

ROPOSITION: If set M' is Turing-reducible to set M then the -labyrinth $\mathbb{N}^2 \times \mathbb{N}_{\nu(M')}$ is codable in the R-labyrinth $\mathbb{N}^2 \times \mathbb{N}_{\nu(M)}$.

ROPOSITION: If the R-labyrinth $\mathbb{N}^2 \times \mathbb{N}_{\nu(M')}$ is codable in the -labyrinth $\mathbb{N}^2 \times \mathbb{N}_{\nu(M)}$ then is set M' Turing-reducible to set M .

etailed PROOFs can be found in /2/ and /5/.

y help of these propositions we can lift the function

$$\underline{e} : 2^{\{0,1\}^*} \longrightarrow \mathcal{U}$$
$$M \longmapsto \mathbb{N}^2 \times \mathbb{N}_{\nu(M)}$$

o a function from degrees of unsolvability into $[\mathcal{L}_z, \leq] \subset [\mathcal{U}_z, \leq]$

$$e : \deg M \longmapsto \mathbb{N}^2 \times \mathbb{N}_{\nu(M)} .$$

ecause we can deduce from the first proposition that e is injective nd because the second proposition entails that e is a homomorphism f posets we have obtained a proof of the following theorem.

MBEDDING THEOREM: There is an embedding of the poset of degrees of nsolvability into the poset $[\mathcal{L}_z, \leq] \subset [\mathcal{U}_z, \leq]$.

This theorem provides us with the opportunity to take advantage of results of degree theory for characterizing posets $[\mathcal{L}_z, \leq]$ and $[\mathfrak{U}_z, \leq]$. For instance we get:

PROPOSITION: Every element of Church's universe possesses continuum many, pairwise incomparable upper bounds.

and

PROPOSITION: Every countable poset is embeddable in $[\mathcal{L}_z, \leq] \subset [\mathfrak{U}_z, \leq]$.

References:

/1/ Budach,L., Meinel,Chr.: "Environments and Automata", EIK in print (1981)

/2/ Budach,L., Meinel,Chr.: "Umwelten und Automaten in Umwelten", Seminarberichte Sekt. Math. d. Humboldt-Univ., Nr.23, 1980

/3/ Epstein,R.: "Degrees of Unsolvability: Structure and Theory", LN in Math., Springer-Verl. Berlin Heidelberg New York 1979

/4/ Hopcroft,J., Ullman,J.: Formal Languages and their Relation to Automata", Addison-Wesley Publ.Comp. 1969

/5/ Meinel,Chr.: "Embedding of the poset of degrees of unsolvability in the poset $[\mathfrak{U}, \leq]$ ", EIK in print (1981)

/6/ Meinel,Chr.: "On importance of plane labyrinths", to appear in EI

/7/ Meinel,Chr.: "Über die durch Kodierungen induzierte Hierarchie vo Umwelten und Labyrinthen", Diss., Berlin 1981

/8/ Rogers,H.: "Theory of recursive functions and effective computability", McGraw-Hill, New York 1967

THE COMPLEXITY OF AUTOMATA AND SUBTHEORIES OF MONADIC SECOND ORDER ARITHMETICS

by A.Włodzimierz Mostowski

Institute of Mathematics
University of Gdańsk
POLAND Wita Stwosza 57

1. Introduction.

1.1 Rabin's results in relation to the paper.

Each formula $F \in SkS$ is represented by a set U_F of trees. If $U_F = T(\mathfrak{A})$ for some automaton $\mathfrak{A}$, then we shall say that $\mathfrak{A}$ represents F ; and vice versa, if for some automaton $\mathfrak{A}$, $T(\mathfrak{A})=U_F$ for some formula F of SkS, we shall say that the formula F represents $\mathfrak{A}$.

Rabin's result is as follows. For each formula F of SkS, a Rabin automaton $\mathfrak{A}(F)$ can be effectively constructed, representing the formula F (cf. [12-14] especially [13] for a short survey).

The crucial point in the construction is that of $\mathfrak{A}(\neg F)$ from $\mathfrak{A}(F)$, known as the "complementation lemma". The orginal proof [12] is very complicated. There are also other proofs e.g. [15] or [16] .

Since F is satisfiable iff $U_F = T(\mathfrak{A}_F)$ is non empty, then, according to the other famous Rabin result: $T(\mathfrak{A}) = \emptyset$ is uniformly decidable for each Rabin automaton $\mathfrak{A}$ (known as the "emptiness problem"), the theory SkS is decidable.

Two kinds of problems arise from this result.

1° What is the measure of complexity of $\mathfrak{A}(F)$ for some method of construction. As the measure of complexity we shall mean here a number of states, and a Rabin index.

A good starting point for the second kind of problems are the following further results of Rabin: The formulas of SkS are exactly that which are represented by Rabin automata, [12]. The formulas of weak SkS are exactly those formulas F of SkS, for which both F and $\neg F$ are represented by special automata (i.e. of index 1-empty), [14].

There is the second kind of problems:

2° Connect with a subtheory of SkS a class of automata, such that the set of formulas of the subtheory is exactly the set of formulas which are represented by the class of automata.

The next two parts of the lecture are devoted to the above two groups of problems. The rest of the introductory part is devoted to a brief explanation of notions and notations.

Some topics are not investigated in this lecture, i.e. the history of decidability of monadic arithmetics, comparision between automata on k-ary trees and that on ω-sequences, as well as the comparision with decidability of monadic theory of ordinals c.f. [2].

1. Preliminaries.

1.2 A monadic k successor arithmetics SkS.

It is a theory based on two kinds of variables: individual $x_1, x_2, \ldots$ and set variables $A_1, A_2, \ldots$. If finite set variables are allowed $\alpha_1, \alpha_2, \ldots,$ and no quantification over set variables we speak of weak SkS. The valuation of variables ranges over suitable items of the set $\{0,1,\ldots,k-1\}^*$ i.e. elements, subsets and finite subsets. As operations there are k successor functions $r_i(x) = x_i$ for $i=0,\ldots,k-1$, and as relations $x_i \in A_j$ $(x_i \in \alpha_j)$, $x \leqslant y$ and $x \not\leqslant y$. For details cf. [12 - 14].

Formulas of the form $(\exists x)[(x\,\omega_1\,\eta_1 A_{j_1}) \cap \ldots \quad x(\omega_s\,\eta_s A_{j_s})] = P(A_1,\ldots,A_r) = P(\vec{A})$ where $\omega, \omega_i \subset \{0,\ldots,k-1\}^*$, $\eta_1,\ldots,\eta_s \in \{\in, \notin\}$, $1 \leqslant j_i \leqslant r$, as well as formulas $\omega \in A$ and $\omega \notin A$ are called principial. Each formula $F(\vec{A})$ of SkS can be reduced to normal form:

(1) $\quad F(\vec{A}) = (Q_1\, X_1) \ldots (Q_r\, X_r)\; B(\vec{A}, X_1, \ldots, X_r)$

where B is a boolean combination of principial formulas, $X_1,\ldots,X_r$ are set (finite set) variables and $Q_1,\ldots,Q_r \in \{\exists, \forall\}$ are quantifiers.

A smallest subtheory containing principial formulas, and closed under $\vee$, $\wedge$, $\exists A$, $\forall A$, (note that the negation $\neg$ is excluded), is called positive SkS cf. [7]. A negation of a positive formula is called a negative formula. In [9], [10] it is proved that

positive SkS, negative SkS $\subset$ weak SkS .

The intersection: positive SkS $\cap$ negative SkS is called bounded SkS.

A set $T = \{0,\ldots,k-1\}^*$ is called a k-ary tree. Any $x \in T$ is called a node. A path π is: $\Lambda \in \pi$, and for each node $x \in \pi$, exactly one from the nodes $x0, x1, \ldots, x(k-1)$ belongs to π.

A valuation $v = (v_1,\ldots,v_n)$ is a function $v : T \longrightarrow \{0,1\}^n$.

A set U_F of v valued trees $t = (T,v)$ represents a formula

$F(A_1,\dots,A_n)$ SkS iff for each $(T,v) = t \in U_F$ is equivalent to:
1° $F(A_1,\dots,A_n)$ holds for $\chi_{A_i} = v_i$, for $i=1,\dots,n$, and
2° types of variables are compatible with cardinalities of sets.

Here and in the sequel $\chi_A(x) = 1$ for $x \in A$, $\chi_A(x) = 0$ for $x \notin A$. In $\varphi = \{y : \varphi^{-1}(y)$ is infinite $\}$.
For trees $t_1 = T(v_1)$ and $t_2 = (T,v_2)$ the function $\varrho(t_1,t_2) = [1 + \min\{l(x) : v_1(x) \neq v_2(x)\}]^{-1}$ is a metric. Formulas represented by closed (open, ect.) sets are called closed (open, ect.) formulas. In the sequel we shall deal with sets of trees represented by automata. A formula represented by a set of trees represented by the automaton of some type will be called a formula fo the type of the automaton.

1.3 Automaton over k-ary trees.

There are two kinds of automata, climbing and sinking. A climbing table $\mathcal{T}$ is a triple, $\langle S,M,S_I\rangle$, where S is a set of states, S_I is a distinguished subset of initial states, and M is a function: $M : S \times \{0,1\}^n \longrightarrow 2^{S^k}$. A sinking table is a triple $\langle S,M,S_F\rangle$ where a distinguished set S is a set of final states, and M is a function $M : S^k \times \{0,1\}^n \longrightarrow 2^S$. A run over a tree $t = (T,v)$ is a function $r : T \longrightarrow S$, such that:

$$(1.2)\quad \begin{cases} \langle r(x0),\dots,r(x(k-1))\rangle \in M(r(x),v(x)), \quad r(\Lambda)\in S_I \text{ ; for climbing, and } r(x)\in M(\langle r(x0),\dots,r(x(k-1))\rangle,v(x)) \text{ for sinking} \\ \text{automata, respectively.} \end{cases}$$

Moreover it is naturally requested that $v(x) \in \{0,1\}^n$, where n is the n in the definition of the table.

An automaton is a pair $\langle \mathcal{T},\mathrm{Cond}\rangle$ where $\mathcal{T}$ is table and Cond are conditions for a run. For sinking automata it is requested that conditions contain $r(\Lambda)\in S_F$. A tree $t = (T,v)$ is accepted by an automaton $\mathfrak{A}$ iff there exists a run on t satisfying the conditions. A set of all trees accepted by an automaton $\mathfrak{A}$ will be denoted by $T(\mathfrak{A})$.
If card $M = 1$ for all arguments, then we speak of a deterministic automaton. For deterministic climbing automata card $S_I = 1$ is requested. An automaton such that for some arguments $M = \emptyset$ is called partial.
Let us note that both notions of climbing and sinking automata are not equivalent in general for total automata. But for a climbing (sinking) table a sinking (climbing) may be partial can be constructed, with the same set of states and the same set of runs.

If a condition for a run is given by collection $\Omega = (U_i, L_i)_{i<r}$ where $U_i, L_i \subset S$, $i=0,\dots,r-1$. Then the automaton is called a pair automaton. The number r is called a Rabin index of an automaton. A climbing pair automaton is called a Rabin automaton [12 - 14]. Automata with $r = 1$ and $L_0 = \emptyset$ are called 1-empty automata or Büchi automata. For automata of the index 0 no special conditions are requested.

A tree $t = (T,v)$ is accepted by an automaton $\mathfrak{A} = (\mathcal{T}, \Omega)$ iff there exists a run r on t such that for each path π :

there exist i such that $In(r|\pi) \cap U_i \neq 0$, $In(r|\pi) \cap L_i = \emptyset$.

A tree t is dually accepted iff for each run r over t there exists a path π and an i such that $In(r|\pi) \cap U_i \neq \emptyset$, $In(r|\pi) \cap L_i = \emptyset$. A set of trees dually accepted by $\mathfrak{A}$ is denoted by $D(\mathfrak{A})$.

For the other notions of automata-automata and set automata we refer the reader to [15]. We define only one new notion of a finitary automaton. For a finitary automaton the conditions are given by a distingished subset D of S. A tree t is accepted iff there exists a run r on t such that $In(r) \cap D = \emptyset$, (and $r(\Lambda) \in S_F$ for a sinking automaton).

We shall say that a finite tree $\gamma \subset t = (T,v)$ where $\gamma = (\Delta, v')$ if $\Delta \subset T$ is a finite nonempty set such that $y \leqslant x$ and $x \in \Delta$ follows $y \in \Delta$, and $v' = v|\Delta$. The set $F_t(\Delta)$ of maximal (respective to $\leqslant$) elements of Δ is called the frontier of Δ. The length of a maximal path for Δ is called night of Δ.

We shall say that a sinking automaton $\mathfrak{A} = \langle S, M, S_F, \emptyset \rangle$, of the order 0, with a distinguished initial state $\lambda \in S$ accepts γ iff there is a run r on γ such that $r(x) = \lambda$ for $x \in F_t(\Delta)$ and $r(\Lambda) \in S_F$. For various notions of automata on finite trees cf. e.g. [3 - 4], [14], [17].

An infinite tree t is positively accepted by $\mathfrak{A}$ iff there exists a finite subtree $\gamma \subset t$ accepted by $\mathfrak{A}$. The set of all (infinite) trees positively accepted by $\mathfrak{A}$ will be denoted in the sequel by $T_{pos}(\mathfrak{A})$. The notion of positive acceptance is introduced and investigated in [7].

The equivalence: $t \in T_{da}(\mathfrak{A})$ iff there exists $\Delta_i \subset t$, $i=1,2,\dots$ such that $F_t(\Delta_{i+1})$ is above $F_t(\Delta_i)$ and Δ_i is accepted by $\mathfrak{A}$ for $i=1,2,\dots$, gives descending-automaton-representability defined and investigated in [8 - 11].

2. Complexity problems.

2.1. The estimation of the number of states.

If B is a boolean combination of principial formulas, and B has m symbols, $m > 4$ then a 1-empty Rabin automaton $\mathfrak{A}_B$ representing B can be constructed having at most 2^m states, see [8]. If $\mathfrak{A}_F$ is 1-empty Rabin automaton, then $(\exists \vec{\alpha})F(\vec{\alpha},\vec{A})$ and $(\forall \vec{\alpha})F(\vec{\alpha},\vec{A})$ are represented by effectively constructed 1-empty Rabin automata having at most 3^m and 12^{m+1} states [8]. From this follows, that a weak formula F having m symbols in a normal form (1) can be represented by an effectively construced automaton, having at most $2^{2^{\cdot^{\cdot^{2cm}}}}\}s+1$ states, where s is a number of changes $\exists\forall$ and $\forall\exists$, and c is a constans. For all proofs see [8].

For a boolean combination of principial formulas B, define $B_o = B$, $B_{i+1} = \neg(\exists\vec{A})B_i(\vec{A},\vec{B})$. Let us suppose that a pair of automata $\mathfrak{A}_i$ and $\bar{\mathfrak{A}}_i$ is effectively constructed such that $T(\mathfrak{A}_i)$ and $D(\bar{\mathfrak{A}}_i)$ represents B_i, and the number of states does not exceed m_i, and the Rabin index does not exceed r_i, for both automata. From the above $m_o = 2^{\text{lenghtB}}$, $r_o = 1$. The $\bar{\mathfrak{A}}_{i+1}$ is a projection of the automaton $\mathfrak{A}_i$.

The construction of $\mathfrak{A}_{i+1}$ is given in Rocoff [15] page 22-23. Denote $\bar{\mathfrak{A}}_i = \mathfrak{A}$, $r_i = r$, $m_i = m$, $n = 2^{\frac{r(r+1)}{2}}$ as in the l.c. The construction of $\mathfrak{A}_{i+1}$ gives the following estimations:

$$(3)\quad m_{i+1} \leqslant (m2^{mn})^{mn} \cdot (2^{2+nm^2})^{1+(2+nm^2)2^r+nm^2+2}, \quad r_{i+1} \leqslant 2^{mn^2}\, 2^{r+mn^2} .$$

The details of the estimations are not very easy, but must be ommited here, because of the lack of place in this paper. The first factor of (3) estmates a number of states of the automaton-automaton $\iota^n_{\mathfrak{A}}$ of [15]. The second factor after the dot estimates a number of states of the sequential automaton of the $\mathfrak{M}^n_{\mathfrak{A}}$, according to McNaugton's construction [5] (the other similar to McNaughton's construction is in [18]).

From (3) follows, that for a formula F of SkS having at most m symbols in a normal form (1) the automaton $\mathfrak{A}(F)$ can be effectively constructed, having at most $2^{\cdot^{\cdot^{2^m}}}\}\lceil cs \rceil$ states, where s is a number of changes $\exists\forall$ and $\forall\exists$ in (1), and c is a constant near to the number 2).

In [15] it is shown (page 16) that for an automaton $\mathfrak{A}$ over binary trees, having q states the decision: $T(\mathfrak{A}) = \emptyset$ needs $q2^{(q+1)2^{2q}}$ computational steps at most. All this gives the basic for estimating an upper bound for full tape complexity of a decision procedure for

formulas and weak formulas of SkS. These bounds must be very high. Some lower bounds were estimated by A.R.Meyer [6]. It seems that the differences are in the value of constans.
The upper estimation of full tape complexity for decision procedure for positive formulas i.e. when there are no negations in B from the form (1), is given in [7].

2.2. Determinictic climbing automata.

For each $r=1,2,\ldots$ there are sets of ω-sequences (i.e. sets of unary trees) accepted by a derministic sequencial Rabin automaton of the Rabin index r, and not accepted by any determinictic automaton of the index less than r, cf. e.g. [19]. This gives immediately a very simple.

Theorem 1. For any $k,r = 1,2,\ldots$ there exists a set $U_{k,r}$ of k-ary trees, which is accepted by a deterministic Rabin automaton of the index r, and by no deterministic automaton of a smaller index.

Proof. e.g. $U_{k,r}$ is a set of such trees that the most left path (i.e. the path 0^{ω}) belongs to U . End of the proof.

The main difference between 1-ary case of ω-sequences and k-ary case for $k \geqslant 2$, is that each set of ω-sequences accepted by a Rabin automaton is accepted by a deterministic Rabin automaton, which is not the case for $k \geqslant 2$. So each regular set of ω-sequences has two indices: a nondeterministic Rabin index which is always 1-empty (cf. Büchi theorem [1]) and a deterministic Rabin index, being any number $\geqslant 1$. But for $k \geqslant 2$ there are sets for which only the nondeterministic index can be defined and sets of trees having nondeterministic index different then 1-empty. The examples for the first case are commonly known. But not for the second. To be sincere the author must say that he knows only the example of a set of nondeterministic index 1 , which is not of nondeterministic index 1-empty, given in [14], and does not know whether or not the nondeterministic Rabin index can be greater than 1 for some k .

All the author knows is the following theorem which seems to be an exact analogon of Büchi theorem [1].

Theorem 2. A negation of a deterministic formula is a formula of nondeterministic index 1-empty. A construction of the automaton is effective.

Proof. For a deterministic automaton $\mathfrak{A}$ the acceptability of a tree can be stated as follows: For each run r , and each path π conditions $(r, \pi, \mathfrak{R})$, since for exactly one run: "there exists" is the same as "each". The conditions are: there exists $i = i(\pi)$ and

a place $j = j(\pi)$, that beginning from this place no states from L_i appear in $r|\pi$ and $In(r|\pi) \cap U_i \neq \emptyset$. The negation
$\neg$ conditions $(r, \pi, \Omega) \equiv$ conditions $(r, \pi, \bar{\Omega})$, for a suitably chosen $\bar{\Omega}$.

Hence the negation of acceptability by $\mathfrak{A}$ can be stated: there exists a run r and a path π, and $i = i(\pi)$, $j = j(\pi)$ such that no states from $\bar{L}_i$, beginning from the j-th place of π, appear in $r|\pi$ and $In(r|\pi) \cap \bar{U}_i \neq \emptyset$ for $(\bar{U}_i, \bar{L}_i) \in \bar{\Omega}$. Evidently this can be represented by an 1-empty automaton. This finishes the proof.

As a corrolary from theorem 1, it follows:

Corrolarry. For each $r=1,2,\ldots$ there exists a formula F_r of non-deterministic index 1-empty, such that the formula $\forall \vec{A}\, F_r(\vec{A}, \vec{C})$ is of a deterministic index r.

Proof. The dual acceptance of a tree belonging to $U_{r,k}$ is described by a formula of the form: for each run r there exists a path π : $Cond(r, \pi, \Omega)$, i.e. of the form $\forall \vec{A}\ \exists \vec{B}\ G(\vec{A}, \vec{B}, \vec{C})$, where $\vec{C}$ are the variables of the F representing $U_{r,k}$.
The $F_r(\vec{A}\, \vec{C}) \equiv \exists \vec{B}\, G(\vec{A}, \vec{B}, \vec{C})$.

But note the following:

Theorem 3. If $H(\vec{X}, \vec{Y})$ is a formula of the deterministic index 1-empty, then so is $\forall \vec{X}\, H(\vec{X}, \vec{Y}) = G(\vec{Y})$. The construction of the automaton is effective. The proof is for simplicity of the notations for k=2. Let $\mathfrak{A} = \langle S, M, s_0, F \rangle$ be a deterministic special automaton representing H (have $\Omega = \{(F, \emptyset)\}$). We shall construct the $\mathfrak{B}$ representing G. The set of states $W = \{\langle S_1, S_2 \rangle ;\ S_1, S_2 \subset S\}$. The transition function is $N(\langle S_1, S_2 \rangle, y) = \langle\langle \bar{S}_1, \bar{S}_2 \rangle, \langle \bar{\bar{S}}_1, \bar{\bar{S}}_2 \rangle\rangle$ where for $S_2 \neq \emptyset$:

$s \in \bar{S}_1$ iff (there exists $s' \in S_1$ and x such that $\langle s, \text{something} \rangle = M(s', \langle x, y \rangle)$ or there exists $s'' \in S_2$ and x such that $\langle s, \text{something} \rangle = M(s'', \langle x, y \rangle)$ and $s \in F$)

$s \in \bar{S}_2$ iff (there exists $s'' \in S_2$ and x such that $\langle s, \text{something} \rangle = M(s'', \langle x, y \rangle)$ and $s \notin F$.

For $S_2 = \emptyset$: $N(\langle S_1, \emptyset \rangle, y) \underset{df}{=} N(\langle \emptyset, S_1 \rangle, y)$. The $\bar{\bar{S}}_1$ and $\bar{\bar{S}}_2$ are defined analogously. The initial state is $W_0 = \langle \{s_0\}, \emptyset \rangle$ and the set of frequent states is $\{\langle S_1, \emptyset \rangle : S_1 \subset S\}$.

Lemma 1. The states $\langle S_1, \emptyset \rangle$ are frequent on some path π of a tree valued by $\vec{Y}$, iff for each $\vec{X}$, some states from F are frequent on the same path π of tree valued by $\vec{X}, \vec{Y}$.

Proof. Let us only observe that if on the path π there are only finitely many states $\langle S_1, \emptyset \rangle$ then there exists some valuation on $\vec{X}$, such that on the path π of a tree valued by $\vec{X}, \vec{Y}$, there are only finitely many states from F. So by the determinism of $\mathfrak{A}$, the single run on the tree cannot be the accepting one.

At the end of this section we state the following:

Theorem 4. For a boolean combination B of principial formulas, the following B, $(\exists \vec{X})B$, $(\forall \vec{X})B$, $(\exists \vec{X})(\forall \vec{Y})B$ are 1-empty formulas. The automata under the question can be effectively constructed.

Proof. Especially for $(\forall \vec{X})B$ needs a different construction, and is omitted.

3. Subtheories of SkS and classes of automata.

3.1. Properties of finitary sinking automata and sinking automata of index 0 .

For sinking automata $\mathfrak{A}_1 = \langle S_1, M_1, S_{F1}, D_1 \rangle$ and $\mathfrak{A}_2 = \langle S_2, M_2, S_{F2}, D_2 \rangle$ such that $S_1 \cap S_2 = \emptyset$ let us define:
$\mathfrak{A}_1 \cup \mathfrak{A}_2 = \langle S_1 \cup S_2 \cup \{\text{wrong}\}, M, S_{F1} \cup S_{F2}, D_1 \cup D_2 \rangle$ where $M \mid S_1^k \times V = M_1$, $M \mid S_2^k \times V = M_2$, $M =$ "wrong" for all other cases, where "wrong" $\notin S_1 \cup S_2$. Moreover, let us define
$\mathfrak{A}_1 \times \mathfrak{A}_2 = \langle S_1 \times S_2, M_1 \times M_2, S_{F1} \times S_{F2}, D_1 \times S_2 \cup S_1 \times D_2 \rangle$.
Moreover for $D_1 = \emptyset$ let us define $2^{\mathfrak{A}} = \langle U, N, U_F, \emptyset \rangle$ for $U = 2^S$, $N(\langle u_0, \ldots, u_{k-1} \rangle, v \rangle = u$ where $s \in u$ iff there exists $s_j \in u_j$, $j = 0, \ldots, k-1$ such that $s \in M(\langle s_0, \ldots, s_{k-1} \rangle, v)$; $u \in U_F$ iff $u \cap S_F \neq \emptyset$.
For $\mathfrak{A}_2$ being the automaton over $V_1 \times V_2 = V$ let us define a projection of $\mathfrak{A}_2$: $pr_{V2}(\mathfrak{A}_2) = \langle S_2, M', S_{F2}, D_2 \rangle$ where $s \in M'(\langle s_0, \ldots, s_{k-1} \rangle, v_1)$ iff there exists $v_2 \in V_2$ such that for $v = v_1 \times v_2$, $s \in M(\langle s_0, \ldots, s_{k-1} \rangle, v)$.

Theorem 5. $T(\mathfrak{A}_1 \cup \mathfrak{A}_2) = T(\mathfrak{A}_1) \cup T(\mathfrak{A}_2)$; $T(\mathfrak{A}_1 \times \mathfrak{A}_2) = T(\mathfrak{A}_1) \cap T(\mathfrak{A}_2)$; $t = (T, v_1) \in T(pr_{V2}(\mathfrak{A}_2))$ iff there exists a valuation v_2 such that $t_1 = (T, v_1 \times v_2) \in T(\mathfrak{A}_2)$.

The proof follows exactly from the definitions.

Theorem 6. For a sinking automaton $\mathfrak{A}$ of index 0 , $T(\mathfrak{A}) = T(2^{\mathfrak{A}})$.

Proof. If r' is an accepting run for $2^{\mathfrak{A}}$ then there exists a run r for $\mathfrak{A}$ such that $r(x) \in r'(x)$ for $x \in T$, and $r(\Lambda) \in S_F$.
Let now be r an accepting run for $\mathfrak{A}$. Let us define $r_n(x) = r(x)$ for $l(x) > n$, and $r_n(x) = N(\langle r_n(x0), \ldots, r_n(x(k-1)) \rangle, v(x))$ for $l(x) \leq n$. Then $r'(x) = \bigcup_{n=1}^{\infty} r_n(x)$ is an accepting run for $2^{\mathfrak{A}}$.
Note that $2^{\mathfrak{A}}$ is a deterministic automaton. Note that the theorems

and 6 can be proved for the so called free structure tree automa-a, especially theorem 6 is proved only for automata on finite rees of. [3], [4] but the fact is not explicitely mentioned.

et $\mathcal{A} = \langle S, M, S_F, \emptyset \rangle$ be a deterministic sinking automaton over $_1 \times V_2$ of index 0 . Define $\mathcal{B} = \langle U, N, U_F, \emptyset \rangle$ over V_1 as follows: $= 2^{S \times V_2} \cup \{wrong\}$. For $\{u\} = N(\langle u_o, \ldots, u_{k-1} \rangle, v_1)$, the set of is defined as follows:

or each $x_o, \ldots, x_{k-1}, x \in V_2$ we choose some collection of k-tuples $_o, \ldots, s_{k-1}$ such that $s_i \times x_i \in u_i$ for $i = 0, \ldots, k-1$; for each ollection we set the $s = M(\langle s_o, \ldots, s_{k-1} \rangle, v_1 \times x)$ into the u . Different choices of the collection of k-tuples $s_o, \ldots, s_{k-1}$, ives different u). We request moreover that for each u and for ach i : each $s_i \times x_i \in u_i$ is used for some k-tuple $s_o, \ldots, s_{k-1}$ f the collection.

f this condition cannot be fulfilled or if for some $x_o, \ldots, x_{k-1}, x$ he set of k-tuples $s_o, \ldots, s_{k-1}$ with the desired properties is mpty, we set $N(\langle u_o, \ldots, u_{k-1} \rangle, v_1) = wrong$. The construction is ompleted by defining $U_F = \{S_1 \times v_1 \cup \ldots \cup S_m \times v_m\}$ for all $_1, \ldots, S_m \subset S_F$. (Here $V_2 = \{v_1, \ldots, v_m\}$). For the constructed utomaton the following theorem holds:

heorem 7. $t = (T, v_1) \in T(\mathcal{B})$ iff for all valuations v_2 from V_2 ll $t' = (T, v_1 \times v_2) \in T(\mathcal{A})$.

he proof follows directly from the construction.

.2. Representability of negative SkS by a class of sinking automata of the index 0 .

The power of representability of climbing automata of the index is very poor. But for sinking automata that is not so, accoording o.

heorem 8. Each formula $\neg P$, where P is a principial formula, is epresented by a sinking automaton of the index 0 .

roof. Let

4) $\neg P(A_1, \ldots, A_m) = (\forall x)[(x\, \omega_1\, \eta_1 A_{i_1}) \vee \ldots \vee (x\, \omega_r\, \eta_r\, A_{i_r})]$.

et $\mathcal{C}_j^{(e)} = \langle Q, M, q_j^{(e)}, F_j^{(e)} \rangle$ for $e = 0, 1$, $j = 1, \ldots, m$, be sequen-ial automata over the alphabet $0, \ldots, k-1$ based on the same set Q f states, such that each automaton $\mathcal{C}_j^{(e)}$ accepts, from right to eft, the set of ω such that $x\,\omega \notin A_j$ (for e = 0), and $x\,\omega \in A_j$ for e = 1) appears in (4). Let us define $Q_i = \varphi_i^{-1}(Q)$ for $\varphi_i(x) = M(x, i)$, $i = 0, \ldots, k-1$.

Now we shall construct an automaton $\mathcal{B} = \langle U,N,U_F,\emptyset \rangle$ representing (4). The states $u \in U$ are subsets of Q ; $u \in U_F$ iff $u \cap F_j^{(e)} \neq \emptyset$ for some $j = 1,\dots,m$ and $e = 0,1$.

$$N(\langle u_0,\dots,u_{k-1}\rangle,(e_1,\dots,e_m)) = \begin{cases} \emptyset & \text{if some } u_0,\dots,u_{k-1} \notin U_F \\ \bigcup_{i=0,\dots,k-1} M(Q_i,i) \cup \bigcup_{j=1,\dots,m} \{q_j^{(e_j)}\} \end{cases}$$

Theorem 9. The formulas of the sinking index 0 , are exactly that of negative SkS. There is an effective construction of the automaton.

Proof. Theorems 5, 7 and 8 allow effective construction of the automaton. It remains for a sinking automaton of the order 0 , $\mathfrak{A} = \langle S,M,S_F,\emptyset \rangle$ to construct a $F \equiv F_{\mathfrak{A}}(\vec{X})$ such that $F \in$ negative SkS.

Let us fix some coding of $s \in S$ as $c(s) = (c_0(s),c_1(s),\dots,c_m(s))$ each $c_i(s) \in 0,1$ for $i = 0,1,\dots,m$. (In the sequal for the proof of this theorem the coordinate $c_0(s)$ will not be used and is supposed to be $y_0 = 0$. The y_0 is needed for the proof of theorem 13). Then for a run $r(x)$ the code $c(r(x))$ is a valuation for some sets $(Y_0,\dots,Y_m)$. The condition $RUN_{\mathfrak{A}}(\vec{X},\vec{Y})$ for $\vec{Y}$ being a code for a run of $\mathfrak{A}$ on a tree valued by $\vec{X}$, can be written as a coniunction of negations of principipial formulas. Similarly the condition $Acc(\vec{Y})$ saying that $(y_0(\Lambda),\dots,y_m(\Lambda)) = c(s)$ for some $s \in S_F$ can be so written. But $F_{\mathfrak{A}}(\vec{X}) \equiv (\exists \vec{Y})[RUN_{\mathfrak{A}}(\vec{X},\vec{Y}) \wedge Acc_{\mathfrak{A}}(\vec{Y})] \in$ negative SkS. The end of the proof.

Let now $\mathfrak{A} = \langle S,M,S_F,\emptyset \rangle$ be a deterministic sinking automaton with an initial state $\lambda \in S$ distinguished. Let us define $\mathcal{B} = \langle U,N,U_F,\emptyset \rangle$ a sinking automaton on infinite trees as follows: $U = 2^S$, $U_F = \{X : X \cap S_F = \emptyset\}$; $N(\langle u_0,\dots,u_{k-1}\rangle,v) = u$ where $s \in u$ iff there are $s_0 \in u_0,\dots,s_{k-1} \in u_{k-1}$ such that $s = M(\langle s_0,\dots,s_{k-1}\rangle,v)$, always λ is added to u . Generally for $y_0 = 1$, the λ is notadded to u . (For this theorem $y_0 = 0$).

The following is immediate:

Lemma 2. $t = (T,v) \in T(\mathcal{B})$ iff $t \notin T_{pos}(\mathfrak{A})$.

Theorem 10. A formula $F \in$ Positive SkS iff F is positively represented iff $\neg F$ is a sinking formula of the index 0 .

The theorem says that the two expressions: positive formula, meaning "being of positive SkS" , or "positively represented" are the same.

Proof. Since $F \in$ positive SkS follows F is positively represented (see [7]), the theorem follows from theorem 9 and lemma 2 .

3.3. Bounded SkS and bounded automata.

Let $\mathfrak{A} = \langle S, M, S_F, \emptyset \rangle$ be a deterministic sinking automaton. Definig $\neg\mathfrak{A} = \langle S, M, S\setminus S_F, \emptyset \rangle$, we have $T(\mathfrak{A}) \cup T(\neg\mathfrak{A})$ = all infinite trees, but generally $T(\mathfrak{A}) \cap T(\neg\mathfrak{A}) \neq \emptyset$. If $T(\mathfrak{A}) \cap T(\neg\mathfrak{A}) = \emptyset$ then the automaton is called bounded (b.a). E.g. formulas $\omega \in A$ are represented by b.a. Bounded formulas are closed under $\vee, \wedge, \neg, \exists, \forall$.

Theorem 11. (1) - The smallest theory containing $\omega \in A$ and closed under $\vee$, $\wedge$, $\neg$, $\exists$, $\forall$ is equal to (2) - the set of formulas represented by bounded automata and is equal to (3) - bounded SkS, and is equal to (4) - the set of closed-open formulas.

The theorem states that the notion of bounded formula is free from ambiguity.

Proof. The inclusions $(1)\subset(2)\subset(3)\subset(4)$ are easy consequences of the definitions. Only the middle inclusion needs theorem 9. There remains to prove $(4)\subset(1)$. We shall prove that each U which is closed-open is represented by some formula of (1) (even not containing $\exists$, $\forall$). Since U is open, then there exists a collection D of finite trees Δ, such that $U = \{t : \Delta \subset t$ for some $\Delta \in D\}$. Let us define $h_t = \min_{\Delta \subset D} \{\text{hight } \Delta \subset t\}$, $H = \max_{t \in U} h_t$.

Lemma 3. For a finite alphabet of valuation: $H < \infty$.

Proof. If for a sequence $t_i \in U$, $i = 1,2,\ldots$, $\lim_{i\to\infty} h_{t_i} = \infty$ then by the compactnees there exists t_{i_j}; $j = 1,2,\ldots$ such that $\lim_{j\to\infty} t_{i_j} = t$ exists. Since U is closed $t \in U$ in spite of : for no $\Delta \in D$ is $\Delta \subset t$. This proves the lemma.

Now $D_H = \{\Delta : \Delta \subset D$, hight $\Delta \leqslant H\}$ is a finite set of finite trees, and by lemma 3, $U = \{t : \Delta \subset t$ for some $\Delta \in D_H\}$. But " there exists $\Delta \subset t$ such that $\Delta \in D_H$ " , can be described by a boolean combination of formulas $\omega \in A$ where $l(\omega) \leqslant H$. So $U = U_B$ where $B \in (1)$.

Corrolarry. Each closed-open set of trees (valued by $\{0,1\}^n$) represents some formula (of bounded SkS).

3.4. Finitary automata and finitary formulas.

A class of formulas represented by finitary automata has a very neat characterisation, according to the theorem

Theorem 12. F is represented by a finitary automaton iff $F(\vec{X}) \equiv (\exists\alpha)N(\alpha,\vec{X})$ where $N \in$ negative SkS.

Proof. $F \equiv (\exists\alpha)N(\alpha,\vec{X}) \equiv (\exists Y)N(Y,\vec{X}) \wedge \mathrm{Fin}(Y)$, where $\mathrm{Fin}(Y)$ means "Y is a finite set". Let us define $S = S_F = \{0,1\}$, $D = \{1\}$

$M(<s_0,\dots,s_{k-1}>,v)=v$, $v\in\{0,1\}$. The $\mathfrak{B} = <S,M,S_F,D>$ represents Fin(Y). Hence from theorem 4 and 9, the F is represented by a finitary automaton.

Let now $\mathfrak{A} = <S,M,S_F,D)>$ be a finitary automaton over trees valued by $\vec{X}$. Let $\vec{Y}$ be a code of S. The formula:
$A(\vec{X},\vec{Y}) \equiv RUN_{\mathfrak{A}}(\vec{X},\vec{Y}) \wedge Acc(\vec{Y})$ as well the formula:
$B(Y,\alpha) = (\forall x)[x \notin \alpha \vee (y(x) = c(s)$ for $s\in D)]$ belongs to negative SkS. Then $t\in T(\mathfrak{A}) \equiv F(\vec{X})$ where
$F(\vec{X}) \equiv (\exists\alpha)(\exists\vec{Y})[A(\vec{X},\vec{Y}) \wedge B(\vec{Y},\alpha)]$ is of a desired form.

Finitary formulas form a very wide class of formulas.
E.g. the positive and the negative formulas are finitary formulas. The latter fact is proved in [9], [10]. Moreover finitary formulas are very near to deterministic sinking formulas. $N(\alpha,\vec{X})$ is represented by a deterministic sinking automaton of the index 0, hence the automaton representing $(\exists\alpha)N(\alpha,\vec{X})$ acts in a determined way up to the first α coordinate equal 1. Therefore only a finite number of nondetermined choices of states is allowed. One can easily see that vice versa; the acceptation by nondeterministic sinking automaton of the index 0, in such a manner that only a finite number of nondetermined choices is allowed, leads to the representation by a finitary automaton.

The following theorem also shows how wide the class of finitary formulas is.
Theorem 13. $\neg F$ is descending-automaton-representable then F is a finitary formula.

Proof. Let us define a negative formula $FULL(\alpha,Y_0) \equiv$
$\equiv (\forall x)\,[[\{(x0\notin\alpha)\wedge \dots \wedge(x(k-1)\notin\alpha)\}\vee(x\in\alpha)]\wedge$
$\wedge\,[\{(x\in\alpha)\wedge(x\in Y_0)\}\vee\{(x\notin\alpha)\wedge(x\notin Y_0)\}]]$.

Now for the deterministic automaton $\mathfrak{A}$ representing (in a descending automaton representability manner) the formula $\neg F$, the formula:
$(\exists\alpha)(\exists Y)[FULL(\alpha,Y_0) \wedge RUN_{\neg\mathfrak{A}}(\vec{X},\vec{Y}) \wedge Acc_{\neg\mathfrak{A}}(\vec{Y})] \equiv F(\vec{X})$, is a finitary formula. Here $\vec{Y} = (Y_0,Y_1,\dots,Y_m)$.

References.

[1] Büchi J.R., On a decision method in restricted second ordre arithmetic, Proc. of the Int. Congr. on Logic, Math. and Phil. of Sc. 1960, Stanford Univ. Press, Stanford Calif. 1962.

[2] Büchi J.R.,Siefkes D., The Monadic second Order Theory of all countable ordinals, Lecture Notes in Math. 328, Decidable Theories II, 1973 pp. 89.

[3] Karpiński M., Free structure tree automata. I. Equivalence, Bull de l'Academie des Sciences, vol. XXI, 1973, pp. 441-446.

[4] Karpiński M., Free structure tree automata. II Nondeterministic and deterministic regularity, Vol. XXI, 1973, pp. 447-450.

[5] McNaughton R., Testing and generating infinite sequences by a finite automaton. Inform. and Control 9, 1966, p. 521-530.

[6] Meyer A.R., Weak Monadic Second Order Theory of Successor is Not Elementary Recursive. Proc. Boston University Logic Coll. 1972 - 1973, Lecture Notes of Math. 453, 1975, p. 137-154.

[7] Mostowski A.W., A note concerning the complexity of a decision problem for positive formulas in SkS, Les Arbres en Algebre et en programmation, 4-eme Colloque de Lille, p. 173-180.

[8] Mostowski A.W., Nearly deterministic automata acceptation of infinite trees and a complexity of a weak theory of SkS, Les Arbres en algebre et en programmation, 5-eme Colloque de Lille, 1980, p. 54-62.

[9] Mostowski A.W., Finite automata on infinite trees and subtheories of SkS, Les Arbres en Algebre et en programmation, 5-eme Colloque de Lille, 1980, p. 228-240

[10] Mostowski A.W., Types of finite automata acceptances and subtheories of SkS, 3rd Symposium on Mathematical Foundations of Computer Science, Zaborów, January 21-26, 1980, ICS PAS reports, 411, 1980, Warszawa, p. 53-57.

[11] Mostowski A.W., Positive properties of Infinite trees and nearly deterministic automata, Preprint N° 38, Univ. of Gdańsk, p. 1-11, (ibidem also text of [7 - 9]).

[12] Rabin M.O., Decidability of Second-Order Theories and Automata on Infinite Trees, Trans. of Amer. Math. Soc. 141, 1969, p.1-35.

[13] Rabin M.O., Decidability and Definiability of Second-Order Theories Actes, Congres Intern. Math. 1970, tome I, p. 239-244.

[14] Rabin M.O., Weakly Definiable Relations and Special Automata, Math. Logic Foundations Set Theory, North Holland, 1970, p.1-23.

[15] Raceff C.W., The Emptiness and Complementation Problems for

Automata on Infinite Trees, MIT 1972 Thesis.

[16] Stupp J., The Lattice Model is Recursive in the Orginal Model, 1975 Manuscript.

[17] Thatcher J.W., Wright J.B., Generalised Finite Automata Theory with an Application to a decision Problem of Second Order Logic, Math. System theory, 2, 1968, p. 57-82.

[18] Trachtenbrodt B.A. and Barsdin J.M., Finite Automata, Behavior and Synthesis (in Russian) Moscow 1970.

[19] Wagner K., On ω-regular Sets, Inform. and Control, 43, 1979, p. 123-177.

TAPE COMPLEXITY OF WORD PROBLEMS

S. Waack [1]

1. Introduction

Computational complexity in algebraic structures has become an important field of research. That's why it is natural to consider the tape complexity of word problems in finite group presentations using the model of Turing machine. One aim of this paper is to show that there is a hierarchy similar to that of formal languages.

On the other hand the construction of an "easy" finite group presentation which is complete in a certain sense for the L=?NL-problem would be an important step towards a solution of this problem. Another result is that such finite presentation exist. But it is still very complicated.

Another possibility of working towards a solution of the L=?NL-problem by algebraic methods is to look for a group theoretical characterisation of all finitely generated groups having a word problem solvable in logspace. (Note that the consideration for logspace-groups does not depend on the special recursive presentation.) This problem seems to be very hard. An "easy" finite presentation with a word problem not belonging to LOG-TAPE is still unknown. One has been describing only families of finitely generated (f.g.) groups belonging to LOG-TAPE.

In [1] Lipton and Zalcstein proved that every f.g. subgroup of a full linear group Gl(n,K), where K is a field of characteristic 0, has a word problem solvable in logspace. This result was extended by Simon [4] to f.g. subgroups of Gl(n,K) for an arbitrary field K. Especially all f.g. free groups belong to this class.

The last result the proof of which is sketched here is that there is a finitely presented (f.p.) group having a word problem solvable in logspace which is not linear.

2. Notations and notions

The reduction notion used in this paper is the well-known logspace-reduction. We write $L \leq L'$ or $L \leq_f L'$, if the reduction function f plays a role.

Given a recursive presentation $G=\langle X,R\rangle$, that is the set of generators X is finite and the set of relators R is recursive enumerable (r.e.). Let H be a subgroup of G. Then we denote by

$$WP(\langle X,R\rangle,H) \text{ or by } WP(G,H),$$

if the underlying presentation is known, the following set:

$$\{w\in (X\cup X^{-1})^*;\ w\in H\}.$$

We write WP(G) instead of WP(G,{1}).

In addition to the ordinary notions of hardness and completeness we define that a formal language A is complete for the inclusion $K'\subseteq K$, where K is a class of formal languages, iff

(i) A is hard for K, and

(ii) K'=K iff $A\in K'$.

We refer to [2] as to all group theoretical notions and to [3] as to notions and notations of complexity theory.

3. The main lemmas

There are the following lemmas which are a powerful tool in constructing finite group presentations with a word problem of a given complexity.

Main lemma 1:

Let G=<X,R> be a recursive group presentation, H a subgroup of G generated by a r.e. subset E. We define $\hat{G}$ to be the following HNN-extension of G:

$$\hat{G} := \langle G,t;\ t.h.t^{-1} = h \text{ for all } h\in E\rangle.$$

Furthermore we assume that

$$WP(\langle X,R\rangle)\in DTAPE(\log n), \text{ and}$$

$$WP(G,H)\in DTAPE(s(n)),$$

where s(n) is a nondecreasing tape-constructable function. Then holds:

(i) $WP(g,H)\leq_f WP(\hat{G})$ with $|f(w)|\leq 2|w|+2$.

(ii) $WP(\hat{G})\in DTAPE(s(n))$.

Proof:

First we see that an element $g\in G$ lies in H iff

$$t.g.t^{-1}.g^{-1} = 1 \text{ in } \hat{G}.$$

In order to prove the second claim we use the fact that every element

$$\hat{g}=g_0 t^{e_1} g_1 \ldots t^{e_n} g_n,\quad e_i=+1,-1,\ g_i\in G,$$

of $\hat{G}$ can be uniquely represented as follows:

$$\hat{g}=n.g, \text{ where}$$

$$n=g_0 t^{e_1} g_0^{-1} \ldots g_0 \ldots g_{n-1} t^{e_n} (g_0 \ldots g_{n-1})^{-1}\in\langle\langle t\rangle\rangle,$$

$$g=g_0 g_1 \ldots g_n\in G.$$

It is apparent that the above representation can be obtained by a logspace-transformation. Furthermore we see that the normal subgroup $\langle\langle t\rangle\rangle$ is freely generated by

$\{g.t.g^{-1};\ g\in G/H\}$.

The assumption that the generalized word problem of H in G is s(n)-space-bounded allows us to decide for two elements

$$g_0g_1 \ldots g_i t(g_0g_1 \ldots g_i)^{-1} \text{ for } i=j,k\ j<k,$$

whether or not they are equal in $\hat{G}$ using $s(|g_j..g_k|)\leq s(|g|)$ space only. Now we can apply Lipton's algorithm for the problem "n=1?", because the normal subgroup generated by {a} of the free group <a,b> is freely generated by $\{b^{-r}ab^{r};\ r\in\mathbb{Z}\}$. The problem "g=1?" is clear.

Main lemma 2:

Let $L\subseteq\Sigma^*$ be a recursive formal language. Then there is a finite presentation G=<X,R> having the following properties:

(i) If $L\in$DTAPE(s(n)) for a nondecreasing tape-constructable function s(n), then WP(G)$\in$DTAPE(s(n)) too.

(ii) $L\leq_f$ WP(G) with $|f(w)|\leq 4|w|+10$.

Proof:

We sketch here only the rather easy existance-proof of a recursively presented group. The case of a finite presentation is essentially more complicated and tricky.

We consider the free group

$$F(\Sigma\cup\{\sigma\}),\sigma\notin\Sigma,$$

and the subgroup U freely generated by

$$\{w.\sigma.w^{-1};\ w\in L\}.$$

It is clear that $w\in L$ iff $w\sigma w^{-1}\in U$. We see that the generalized word problem of U in $F(\Sigma\cup\{\sigma\})$ is s(n)-space-bounded. Now we have to apply main lemma 1 only.

. The main theorems

Main theorem 1:

Let s(n), s'(n), S(n) be three nondecreasing tape-constructable functions with

$$s'(n) = s(4n+10), \text{ and}$$

$$\liminf\ (s'(n)/S(n)) = 0.$$

Then there is a finite presentation G=<X,R> such that

$$WP(G)\in DTAPE(S(n))\setminus DTAPE(s(n)).$$

Proof:

This theorem is an easy consequence of the hierarchy-theorem for formal languages and main lemma 2.

Main theorem 2:

There is a finitely presented group G the word problem of which is complete for the inclusion $L \subseteq NL$.

Proof:

Let K be a formal language which is complete for NL. Then there is a finitely presented group G such that by main lemma 2 the following holds:

(i) $K \leq WP(G)$, that is WP(G) is hard for NL,

(ii) $WP(G) \in L$ iff $K \in L$.

Main theorem 3:

The group

$$G = \langle x,\ s_1,\ s_2;\ s_i.x.s_i^{-1} = x^2\ (i=1,2)\rangle$$

has a word problem solvable in logspace and is not a subgroup of Gl(n,R) for any integer domain R.

5. Proof of main theorem 3

Note that

$$G_0 := \langle x,s;\ sxs^{-1} = x^2\rangle$$

is isomorphic to

$$H := \mathbb{Z}[1/2] \rtimes_\phi \langle s\rangle = \{\begin{pmatrix} 2^i & k \\ 0 & 1 \end{pmatrix};\ i \in \mathbb{Z},\ k \in \mathbb{Z}[1/2]\},$$

where $\phi(z) = 2.z$ for all $z \in \mathbb{Z}[1/2]$.

The following lemma can be verified by using basic properties of HNN-extensions and free products with amalgamation.

Lemma 1:

The homomorphism

$$\Phi: \langle x,s_1,s_2;\ s_i x s_i^{-1} = x^2\rangle \longrightarrow \langle G_0,t;\ txt^{-1} = x\rangle$$

defined by

$$\Phi(s_1) = s,\ \Phi(s_2) = tst^{-1},\ \Phi(x) = x$$

is one-to-one.

In order to prove the first part of main theorem 3 it is sufficient to show that the generalized word problem of $\langle x\rangle$ in G_0 is solvable in logspace, because the word problem of G_0 can be solved by Lipton's and Zalcstein's algorithm.

Lemma 2:

The generalized word problem of $\langle x\rangle$ in G_0 is solvable in logspace.

'roof(sketch):

Jsing the fact that G_O is isomorphic to H we can reduce the problem to :he following question:

;iven a number

$$\sum_{i=1}^{n}(-1)^{e_i}2^{a_i},\ e_i=0,1,\ a_i\in\mathbb{Z},$$

n such a way that the number 2^{a_i} for all i is represented in $O(a_i)$-pace.

:an one decide in logspace whether or not this number belongs to $\mathbb{Z}$?

bviously $\sum(-1)^{e_i}2^{a_i}$ can be transformed by a logspace-transformation o

$$(\sum(-1)^{e_i'}.2^{a_i'})/2^{a_{i'}},\ a_{i'},a_i'\geq 0.$$

ow the assertion is apparent, because the numerator of the expression n binary form can be computed in logspace.

he following lemma concludes the proof of our main theorem 3, ecause all f.g. linear groups over an integer domain are residually inite (see [5]).

emma 3:

$x,s_1,s_2;\ s_i x s_i^{-1}=x^2\ (i=1,2)>$ is not residually finite.

ne proof of this lemma is based on the fact that the element

$$g=s_1^{-1}xs_1s_2^{-1}x^{-1}s_2\neq 1$$

ies in all normal subgroups of G with finite index.

eferences:

.] Lipton, Zalcstein, Word Problems solvable in Logspace, Research Report 60, Jan. 1976.

2] Lyndon, Schupp, Combinatorial Group Theory, Springer-Verlag, 1977.

3] Paul, Komplexitätstheorie, Teubner, Stuttgart, 1978.

4] Simon, Word Problems for Groups and contextfree Recognition, FCT'79, Proceedings.

5] Wehrfritz, Infinite linear Groups, Springer-Verlag, 1973.

Institut für Mathematik der Akademie der Wissenschaften der DDR

DDR 1080 Berlin

Mohrenstr. 39

Vol. 77: G. V. Bochmann, Architecture of Distributed Computer Systems. VIII, 238 pages. 1979.

Vol. 78: M. Gordon, R. Milner and C. Wadsworth, Edinburgh LCF. VIII, 159 pages. 1979.

Vol. 79: Language Design and Programming Methodology. Proceedings, 1979. Edited by J. Tobias. IX, 255 pages. 1980.

ol. 80: Pictorial Information Systems. Edited by S. K. Chang and . S. Fu. IX, 445 pages. 1980.

ol. 81: Data Base Techniques for Pictorial Applications. Proceedgs, 1979. Edited by A. Blaser. XI, 599 pages. 1980.

ol. 82: J. G. Sanderson, A Relational Theory of Computing. VI, 47 pages. 1980.

ol. 83: International Symposium Programming. Proceedings, 1980. dited by B. Robinet. VII, 341 pages. 1980.

ol. 84: Net Theory and Applications. Proceedings, 1979. Edited y W. Brauer. XIII, 537 Seiten. 1980.

ol. 85: Automata, Languages and Programming. Proceedings, 1980. dited by J. de Bakker and J. van Leeuwen. VIII, 671 pages. 1980.

ol. 86: Abstract Software Specifications. Proceedings, 1979. Edited y D. Bjørner. XIII, 567 pages. 1980

ol. 87: 5th Conference on Automated Deduction. Proceedings, 80. Edited by W. Bibel and R. Kowalski. VII, 385 pages. 1980.

ol. 88: Mathematical Foundations of Computer Science 1980. oceedings, 1980. Edited by P. Dembiński. VIII, 723 pages. 1980.

ol. 89: Computer Aided Design - Modelling, Systems Engineering, AD-Systems. Proceedings, 1980. Edited by J. Encarnacao. XIV, 31 pages. 1980.

ol. 90: D. M. Sandford, Using Sophisticated Models in Resotion Theorem Proving.
, 239 pages. 1980

ol. 91: D. Wood, Grammar and L Forms: An Introduction. IX, 314 ges. 1980.

l. 92: R. Milner, A Calculus of Communication Systems. VI, 171 ges. 1980.

ol. 93: A. Nijholt, Context-Free Grammars: Covers, Normal Forms, d Parsing. VII, 253 pages. 1980.

l. 94: Semantics-Directed Compiler Generation. Proceedings, 80. Edited by N. D. Jones. V, 489 pages. 1980.

ol. 95: Ch. D. Marlin, Coroutines. XII, 246 pages. 1980.

l. 96: J. L. Peterson, Computer Programs for Spelling Correction: 213 pages. 1980.

l. 97: S. Osaki and T. Nishio, Reliability Evaluation of Some Faultlerant Computer Architectures. VI, 129 pages. 1980.

l. 98: Towards a Formal Description of Ada. Edited by D. Bjørner d O. N. Oest. XIV, 630 pages. 1980.

l. 99: I. Guessarian, Algebraic Semantics. XI, 158 pages. 1981.

l. 100: Graphtheoretic Concepts in Computer Science. Edited by Noltemeier. X, 403 pages. 1981.

l. 101: A. Thayse, Boolean Calculus of Differences. VII, 144 pages. 81.

l. 102: J. H. Davenport, On the Integration of Algebraic Functions. 197 pages. 1981.

l. 103: H. Ledgard, A. Singer, J. Whiteside, Directions in Human ctors of Interactive Systems. VI, 190 pages. 1981.

l. 104: Theoretical Computer Science. Ed. by P. Deussen. VII, 1 pages. 1981.

l. 105: B. W. Lampson, M. Paul, H. J. Siegert, Distributed Systems – chitecture and Implementation. XIII, 510 pages. 1981.

l. 106: The Programming Language Ada. Reference Manual. X, 3 pages. 1981.

Vol. 107: International Colloquium on Formalization of Programming Concepts. Proceedings. Edited by J. Diaz and I. Ramos. VII, 478 pages. 1981.

Vol. 108: Graph Theory and Algorithms. Edited by N. Saito and T. Nishizeki. VI, 216 pages. 1981.

Vol. 109: Digital Image Processing Systems. Edited by L. Bolc and Zenon Kulpa. V, 353 pages. 1981.

Vol. 110: W. Dehning, H. Essig, S. Maass, The Adaptation of Virtual Man-Computer Interfaces to User Requirements in Dialogs. X, 142 pages. 1981.

Vol. 111: CONPAR 81. Edited by W. Händler. XI, 508 pages. 1981.

Vol. 112: CAAP '81. Proceedings. Edited by G. Astesiano and C. Böhm. VI, 364 pages. 1981.

Vol. 113: E.-E. Doberkat, Stochastic Automata: Stability, Nondeterminism, and Prediction. IX, 135 pages. 1981.

Vol. 114: B. Liskov, CLU, Reference Manual. VIII, 190 pages. 1981.

Vol. 115: Automata, Languages and Programming. Edited by S. Even and O. Kariv. VIII, 552 pages. 1981.

Vol. 116: M. A. Casanova, The Concurrency Control Problem for Database Systems. VII, 175 pages. 1981.

Vol. 117: Fundamentals of Computation Theory. Proceedings, 1981. Edited by F. Gécseg. XI, 471 pages. 1981.

This series reports new developments in computer science research and teaching – quickly, informally and at a high level. The type of material considered for publication includes:

1. Preliminary drafts of original papers and monographs
2. Lectures on a new field or presentations of a new angle in a classical field
3. Seminar work-outs
4. Reports of meetings, provided they are
   a) of exceptional interest and
   b) devoted to a single topic.

Texts which are out of print but still in demand may also be considered if they fall within these categories.

The timeliness of a manuscript is more important than its form, which may be unfinished or tentative. Thus, in some instances, proofs may be merely outlined and results presented which have been or will later be published elsewhere. If possible, a subject index should be included. Publication of Lecture Notes is intended as a service to the international computer science community, in that a commercial publisher, Springer-Verlag, can offer a wide distribution of documents which would otherwise have a restricted readership. Once published and copyrighted, they can be documented in the scientific literature.

Manuscripts

Manuscripts should be no less than 100 and preferably no more than 500 pages in length.
They are reproduced by a photographic process and therefore must be typed with extreme care. Symbol not on the typewriter should be inserted by hand in indelible black ink. Corrections to the typescrip should be made by pasting in the new text or painting out errors with white correction fluid. Authors receiv 75 free copies and are free to use the material in other publications. The typescript is reduced slightly i size during reproduction; best results will not be obtained unless the text on any one page is kept withi the overall limit of 18 x 26.5 cm (7 x 10½ inches). On request, the publisher will supply special paper wit the typing area outlined.
Manuscripts should be sent to Prof. G. Goos, Institut für Informatik, Universität Karlsruhe, Zirkel 2, 7500 Karls ruhe/Germany, Prof. J. Hartmanis, Cornell University, Dept. of Computer-Science, Ithaca, NY/USA 1485C or directly to Springer-Verlag Heidelberg.

Springer-Verlag, Heidelberger Platz 3, D-1000 Berlin 33
Springer-Verlag, Neuenheimer Landstraße 28–30, D-6900 Heidelberg 1
Springer-Verlag, 175 Fifth Avenue, New York, NY 10010/USA

ISBN 3-540-10854-8
ISBN 0-387-10854-8

Printed in the United States
By Bookmasters